Questions and Answers in Medical Physiology

THIS BOOK IS DEDICATED TO HERMAN RAHN, HJØRDIS AND BERTEL, MOM AND DAD, KIRSTEN, AND MY DAUGHTERS CARINA, NINA AND PERNILLE.

Questions and Answers in Medical Physiology

Poul-Erik Paulev
MD, DSci
Department of Medical Physiology,
The PANUM Institute, University of Copenhagen

W. B. Saunders Company Ltd.
London • Philadelphia • Toronto • Sydney • Tokyo

W. B. Saunders Company Ltd 24–28 Oval Road
London NW1 7DX

The Curtis Center
Independence Square West
Philadelphia, PA 19106-3399, USA

Harcourt Brace & Company
55 Horner Avenue
Toronto, Ontario, M8Z 4X6, Canada

Harcourt Brace & Company, Australia
30–52 Smidmore Street
Marrickville, NSW 2204, Australia

Harcourt Brace & Company, Japan
Ichibancho Central Building, 22–1 Ichibancho
Chiyoda-ku, Tokyo 102, Japan

This book is printed on acid-free paper

A catalogue record for this book is available from the British Library

ISBN 0-7020-2043-5

Illustrations prepared by Hardlines, Charlbury, Oxfordshire.

Typeset by P&R Typesetters Ltd, Salisbury, Wiltshire.
Printed and bound in Great Britain by WBC, Bridgend, Mid Glamorgan.

Contents

Preface

Medical physiology is the oldest of the biological sciences and is of utmost importance to medical scientists and doctors. Medical physiology integrates classical biochemistry, chemistry, physiology and modern molecular biology into **physiological mechanisms** governing human life. Integration of these basic principles increases our understanding of the mysteries of Nature.

This book describes **quantitative physiology**, which combines **physiology of total man** with **molecular biology.** The Question and Answer system has been developed by teachers and students over the past decade.

I acknowledge the kind assistance of: Peter Bie, Kirsten Berg, Jesper Brahm, Pamela Chow, Kirsten Christiansen, Ivan Divac, Steen Dissing, Jørgen Drejer, Poul Dyhre-Poulsen, Karen Ejrnæs, Sune Frederiksen, Ole Frederiksen, Jørgen Frøkjær, Rolf Gideon, Jørn Giese, Albert Gjedde, Niels Grove Vejlstrup, Jørgen Hedemark Poulsen, Carl Jørgen Hedeskov, Jens J. Holst, Y. Honda, Mette Hornum, Jørn Hounsgaard, Hans Hultborn, Jens Ingeman Jensen, Anne Juel Christensen, J.D.C. Lambert, Henriette Mersebach, Kirsten McCord, Peter Mirakian, Poul Leyssac, Joop Madsen, Ole Munck, Michala Fog Pedersen, M. Pokorski, Peter Robbins, Y. Sakakibara, Jeanet Sheller, Erik Skakkebæk, Ole Skøtt, Peter Thams, Alice Theilgaard, Niels A. Thorn, Jørgen Warberg, Brian Whipp, Eva Wojtal, Thomas Zeuthen and other colleagues and students.

This book is far from being a single person's accomplishment. There are many quotations from Nobel prize laureates and there has been much useful discussion with colleagues. I apologize to the scientists who have contributed without being mentioned here.

To the student

This book is written for students concerned with the essentials of medical physiology. For a basic knowledge of physiology, a general textbook should first be consulted. It is important for students to work **quantitatively and critically** and to use data for logical deductions. This develops their intuition and problem-solving skills.

Physiological reasoning based on facts is, in my experience, a necessity for both medical scientists and doctors. Further intellectual development depends upon individual scholarship.

The book is aimed at both preclinical and clinical students. Preclinical students may wish to delay undertaking the **case histories** until relevant for their clinical courses. During their clinical study, the student will hopefully develop tolerance toward other people, a sense of proportion, and good patient communication. The ultimate goal for the student is to develop **human and intellectual skills**, combined with psychological tolerance. Insight is essential for the benefit of the patients.

Poul-Erik Paulev, MD, DSc
Copenhagen

SECTION I.
Cellular Physiology

Cellular physiology – or **biophysics** – is the discipline covering basic characteristics of most cells. The following seven chapters do not pretend to include all essential topics; however, it is important to clarify basic concepts at the outset. Such concepts (the **Na^+-K^+-pump**, radioactive decay, etc.) will be referred to throughout the book. Frequently used abbreviations are **A** for area, **AP** for action potential, **C** for concentration, **P** for pressure, π for osmotic pressure, **RMP** for resting membrane potential, **T** for absolute temperature in Kelvin or **K**, and **V** for volume in litres (l). An alphabetic list of abbreviations and symbols is present in Chapter 75.

CHAPTER 1. MEMBRANE TRANSPORT

Nutrients and oxygen are transported to the cell interior (the intracellular fluid) from the extracellular fluid through the cell membrane. The **extracellular fluid volume** (ECFV or ECV) consists of the circulating blood plasma and the **interstitial fluid volume** (ISFV or ISF) in the spaces between cells. Membrane transport refers to transfer across both cell membranes and capillary membranes.

What is concentration? What is convective flux of a substance across a barrier?

Concentration (C or brackets around a substance $[Na^+]$) is the mass or moles per unit of fluid volume. The **molar concentration (molarity)** is the number of moles per litre (l) of solution. The **molality** is the number of mol per kg of solution, frequently water. One **equivalent** is the molar mass of all the ions that contain (6.02×10^{23} or Avogadro's number) single charges or valencies when fully dissociated. The **normality** of a solution is the number of equivalents per litre ($\mathrm{Eq\,l^{-1}}$).

Flux (J) is the rate of transport of a substance through an area unit ($1\ \mathrm{m^2}$) of a membrane in moles per second (s). **Convective flux** is the net rate of transport of molecules through an area unit ($\mathrm{mol\,s^{-1}\,m^{-2}}$), caused by **fluid (liquid or air) volume transport.**

What are the driving forces responsible for the passive transport of substances in solution?

Migration of molecules is determined by mechanical, electrical, thermal and gravitational forces. They move the molecules in a direction determined by the vector of the force.

Diffusion is a transport of atoms or molecules caused by their random thermal motion. Diffusion tends to equalize concentration differences (ΔC).

Diffusive flux (J^{dif}) is the movement of molecules by diffusion caused by a concentration gradient (dC/dx) in the direction x. The **diffusion coefficient (D)** is a proportionality constant that relates flux to the concentration gradient (dC/dx). **Einstein** defined D as ($k \cdot T \cdot B$), where T is absolute temperature and B is motility of molecules. Diffusing molecules move from higher to lower concentration, that is **down its concentration gradient**. Accordingly, dC/dx has a negative slope, when molecules diffuse in the direction x. Now, it is easy to calculate the diffusive flux (per $\mathrm{m^2}$):

$$J^{\mathrm{dif}} = -D\, dC/dx.$$

This relationship was first recognized as early as in 1855 by the anatomist and physiologist, Adolf Fick, and it has since been named after him: **Fick's first law of diffusion**. The dimension of D is found by dimension analysis of the equation for J^{dif}:

$$(J^{\mathrm{dif}}\ \mathrm{mol\,s^{-1}\,m^{-2}}) = D(dC/dx\ \mathrm{mol\,m^{-3}\,m^{-1}}).$$

Accordingly, D has the dimension: $m^2 s^{-1}$. D is small, when the molecules are large and when the surrounding medium is viscous.

Einstein's relation states that for average molecules in biological media, the displacement squared, $(dx)^2$, is equal to 2 multiplied by D and by the time (t) elapsed, since the molecule started to diffuse:

$$(dx)^2 = 2Dt$$

For molecules with $D = 10^{-9}\, m^2 s^{-1}$, the time required to diffuse 1 μm is 0.5 millisecond (ms). To diffuse 10 and 100 μm, the time required increases 100-fold each time: 50 ms and 5000 ms, respectively.

What is a permeable, a semipermeable and a selectively permeable plasma membrane?

A **permeable membrane** allows the passage of all dissolved substances and the **solvent** (mainly water).

An **ideal semipermeable membrane** is permeable to water only, but impermeable to all solutes. Most real **semipermeable membranes** are permeable to water and to low molecular substances (crystalloids), but not to macromolecular substances (colloids such as proteins).

A **selectively permeable membrane** is permeable to a particular compound (sucrose, anions only or to cations only).

What is understood by osmosis and osmotic pressure?

Osmosis is transport of solvent molecules (mainly water) through a semipermeable membrane. The water flows from a compartment of high water concentration (or low solute concentration) to one of low water concentration (or high solute concentration). The greater the difference between the solute concentration of the two compartments, the greater is the tendency for water to be unevenly distributed between the two compartments. Water diffuses down its chemical potential gradient into the compartment with higher solute concentration, causing the chemical potential gradient to be reduced until equilibrium is reached.

Osmotic pressure is the hydrostatic pressure that must be applied to the side of a semipermeable membrane with higher solute concentration in order to stop the water flux, so that the net water flux is zero.

What are the colligative properties of water?

The **colligative properties of water** are strictly related to the solvent or water concentration alone. With decreasing water concentration, the water vapour tension, and the freezing point is reduced, whereas the boiling point, and the osmotic pressure of the solution is increased as compared with pure water. The size of the osmotic pressure of a solution depends on the number of dissolved particles per volume unit.

How is the osmotic pressure calculated?

The osmotic pressure (π) depends on the absolute temperature (T Kelvin or K) and on the number of dissolved particles per volume unit (N/V):

$$\pi = TRN/V, \text{or } \pi = TR\Delta C$$

where R is the ideal gas constant ($0.082\,l\cdot atm\cdot Osmol^{-1}\cdot K^{-1}$ or $8.31\,J/(K\ mole)$), and ΔC is the **concentration gradient.**

This relationship was first recognized by **van't Hoff** and applies to ideal solutions only. Real physiological solutions, such as the cytotolic phase and ECF, differ from the ideal solutions which are very dilute.

A correction factor called the **osmotic coefficient** (ϕ) corrects for these differences in osmolality. For physiological electrolytes it is 0.92–0.96, and for carbohydrates it is 1.01.

What is osmolarity and osmolality?

A solution has the ideal osmolarity of **one,** when it contains (6.025×10^{23}) osmotically active particles per litre. Diluted solutions have an **osmolar concentration** or **osmolarity** (**osmol** l^{-1}) numerically equal to the **molarity (mol** l^{-1}**)**. In biological solutions the molarity is different from osmolar concentration. The number of osmol/l is: ($\phi\ N/V$). Now, the corrected van't Hoff law reads:

$$\pi = TR\phi N/V.$$

Osmolality is simply the number of mol per kg of water in the fluid.

Define the colloid osmotic pressure in capillary plasma

The **colloid osmotic pressure** is equal in magnitude to the hydrostatic pressure, which must be applied at the luminal side of the capillary barrier, in order to stop net transport of water caused by uncharged colloids in the blood plasma. **Colloids** are mainly **plasma proteins.**

Calculate the osmotic pressure in biological solutions

The **osmotic pressure** is equal in magnitude to a certain hydrostatic pressure. This pressure column must be applied to the solution to restore the free energy or **chemical potential** of its water to that of pure water. The tendency of water to pass a membrane depends on its **chemical potential** (i.e. vapour pressure). The chemical potential of water decreases with solutes present and with decreasing temperature.

Uniformly distributed substances in diluted solutions behave like gas molecules at atmospheric pressure (atm). The osmotic pressure can be expressed as $P_{osmot}=C\cdot RT$, **which is the eqivalent of the ideal gas equation (**$P\cdot V=nRT$**)**. Here C is the concentration of dissolved solute, and the derivation of the relationship is based on the chemical potential of water. R is the gas constant ($=0.082\,l\cdot atm\cdot Osmol^{-1}\cdot K^{-1}$). At **standard temperature, pressure, dry (STPD)** the volume occupied by 1 mol of any ideal gas is **22.4 l**. STPD is an abbreviation for a volume at standard temperature of 273 K, standard pressure of 101.3 kPa or 760 Torr, and dry air. In an ideal solution, 1 mol of solute will, by analogy, be dissolved in 22.4 l of water, and will exert a P_{osmot} of 1 atmosphere.

In biological solutions at 310 K, such as an ultrafiltrate of plasma (interstitial fluid, ISF), with an osmolality of 0.300 Osmol/kg water, the P_{osmot} must be:

$$P_{osmot} = 0.3\ (Osmol/l) \times 0.082\ (l\ atm\ Osmol^{-1}K^{-1}) \times 310\ (K) = 7.63\ atm\ (= 773\ kPa)$$

or the pressure exerted by a **column of water 76 m high.**

Only **net gradients** across endothelial and plasma membranes are important, and they depend upon **protein concentration gradients.** This is because all the electrolytes (crystalloids) have diffused to equilibrium across the capillary endothelial membrane, whereas proteins (colloids) cannot.

What is the size of the colloid osmotic pressure of plasma?

The average colloid osmotic pressure (π_{coll}) of plasma is approximately **3.6 kPa (27 Torr).** The dissolved proteins have a molality of 1 mmol/kg water, and a net average of **17 negative charges per molecule (1 mmol kg^{-1} = 17 mEq kg^{-1}).** Milliequivalents are abbreviated mEq. The proteins are directly responsible for **2.4 kPa (18 Torr).** The remaining **1.2 kPa (9 Torr)** of the colloid osmotic pressure is due to the unequal distribution of permeable ions, the **Gibbs–Donnan law** or the **Donnan effect.**

What is the Donnan effect?

Let us consider two compartments separated by a membrane that is permeable to water and to small ions. If solutions of differing NaCl concentrations are present, water and ions will diffuse rapidly in both directions across the membrane. Electrical neutrality requires that any net movement of Na^+ is matched by the simultaneous movement of Cl^-. The number of times the two ions collide with one side of the membrane is **proportional to the product of their concentrations: $[Na^+] \cdot [Cl^-]$.** At equilibrium the fluxes of NaCl in each direction are identical, and ultimately the concentrations are the same all over.

Let us now add protein to one compartment (compartment$_p$ modelling streaming plasma), which is separated by a membrane (the capillary endothelial membrane) from the other compartment (compartment$_{ISF}$ modelling interstitial or tissue fluid). The model still contains only diffusible Na^+ and Cl^-, and they still cross the membrane. At equilibrium the product of concentrations of the two ions on either side of the membrane must be equal. On the plasma side, which contains **non-diffusible anions (protein),** the concentration of diffusible anion (Cl^- is the model) must always be less than on the interstitial fluid side. The concentration of diffusible cation (Na^+ is the model) must always be greater than in the ISF. The concentration product of any pair of diffusible ions is identical on either side of the capillary membrane at equilibrium:

$$[Na^+]_{ISF} \cdot [Cl^-]_{ISF} = [Na^+]_P \cdot [Cl^-]_P.$$

The sum of diffusible anion and cation concentrations in plasma is always greater than the sum of the same anion and cation concentrations in ISF:

$$[Na^+]_{ISF} = [Cl^-]_{ISF}\ ;\ [Na^+]_p + [Cl^-]_p > [Na^+]_{ISF} + [Cl^-]_{ISF}$$

This is a simple mathematical argument. It explains why the osmotic pressure in plasma exceeds that of the tissue fluid. This is not due to the plasma proteins alone, but is also due to the **higher concentration of diffusible ions in the plasma.** The **Donnan effect** is the **extra colloid osmotic pressure of proteins caused by the uneven distribution of small, diffusible cations and anions in plasma.** The Donnan effect causes a 5% and 10% concentration difference between the plasma and ultrafiltrate concentrations of monovalent and divalent ions, respectively. In the above equations Na^+ and Cl^- are model ions for all the cations and anions. In our body other anions and cations are present and the Na^+ and Cl^- concentrations are not alike (see calculations in Chapters 54 and 58). The Donnan equilibrium implies an accumulation of electrons on the side with negatively charged proteins. This potential difference across the membrane is termed the **Donnan potential.**

What is facilitated diffusion?

Facilitated diffusion takes place through **transport proteins** not linked directly to metabolic energy processes. Facilitated diffusion shows saturation kinetics, because the number of transport proteins is limited. The **saturation kinetics** is different from the energy limitation in primary active transport. Facilitated diffusion of ions against an electrochemical potential gradient, or of uncharged substrates against a concentration gradient, is negligible. **Glucose, galactose and other monosaccharides** cross the muscle cell membrane by facilitated diffusion.

What is the Na^+-K^+-Pump?

The **Na^+-K^+-pump** is an **integral membrane protein** in the plasma membrane. The pump contains a channel, which consists of two subunits. The catalytic subunit is a Na^+-K^+-activated ATPase of 112 000 Daltons (Da), and the other subunit is a glycoprotein of 35 000 Da.

The pump is a **primary active transporter**, because it uses the energy in the terminal phosphate bond of ATP. The **Na^+-K^+-pump** transports 3 Na^+ out of the cell and 2 K^+ into the cell for each ATP hydrolysed. This is a net movement of positive ions out of the cell, and therefore called **electrogenic transport.** Intestinal and kidney tubule cells transport substrates, such as glucose and amino acids, in **a substrate Na^+ cotransport** in the luminal membrane, linked to the **Na^+-K^+-pump** of the basolateral membrane. This is called **secondary active transport** of substrate. Such transport is powered by an actively established gradient (i.e. the **Na^+ gradient**).

The many **ion-transporting ATPases** form classes or families showing **amino acid sequence homology.**

What is known about glucose transport proteins?

Glucose transport is mediated by a family of homologous transport proteins that are coded by distinct genes. The transporter proteins show a marked tissue specificity which reflects differing transport needs of various tissues.

Five human **glucose transporters** are cloned and identified (GLUT 1–5). The GLUT 1 resides in placenta, brain, perineurial sheaths, red cells, adipose and muscle tissues. GLUT 2 is found in the liver, pancreatic β-cells, proximal renal tubule cells, and the basolateral membranes of small intestinal cells. GLUT 3 is ubiquitously distributed, found predominantly in the brain, and in lower concentrations in fat, kidney, liver and muscle tissues. GLUT 4 is confined to tissues with insulin-responsive glucose uptake (muscle, heart and fat stores). GLUT 5 is found in the luminal membrane of small intestinal cells, and also in brain, muscle and adipose tissues. Some of these transporters also allow fructose and galactose to pass (Fig. 53-1).

GLUT transporters reside on intracellular vesicles. When stimulated by insulin or muscle contraction, the intracellular vesicles, which have a high number of membrane-penetrating glucose transporters, translocate from the intracellular pool to the cell membrane. This can increase the glucose uptake at least 10 times.

What is special about the insulin receptor and the insulin sensitive glucose transporter in the cell membrane?

The **insulin receptor** is a glycoprotein found in the cell membranes. The T-shaped receptor protruding from the cell membrane contains 1370 amino acids forming two α- and two β-subunits. The α-subunits are entirely extracellular, whereas the β-subunits span the membrane. Insulin binding on α-subunits stimulates a **protein kinase** on the intracellular part of the receptor to phosphorylate tyrosine residues on the β-subunit and on endogenous proteins. The exact molecular mechanism linking the receptor kinase activity to changes in cellular enzyme activity and transport processes remains uncertain; but it is shown that the kinase activity is essential for signal transduction.

The **glucose transporters** are five homologous carrier proteins (GLUT 1–5) that penetrate the vesicle membrane. In adipocytes and muscle cells, glucose transport is profoundly influenced by insulin. When insulin binds to its large membrane receptor, many intracellular vesicles translocate to the cell membrane from the cytosol. When these vesicles, rich in glucose transporters, fuse with the cell membrane, the number of glucose transporters increases substantially, thereby increasing glucose uptake.

How are water and ions transferred through the epithelial layer?

The **Na^+-K^+-pump** is located in the **basolateral exit-membrane** of the cell. The primary active ion-transport provides metabolic energy for the secondary water absorption through the brush border membrane (Fig. 1-1).

Hereby, the active pump in the exit-membrane drives the luminal transport across the entry membrane (Fig. 1-1). This transport of NaCl and water is nearly isotonic. The bulk flow can take place against a large osmotic gradient, and increases in diluted solutions. The **entry membrane** is highly permeable to water.

Zeuthen has proposed an interesting hypothesis concerning the ionic mechanism driving water through the exit-membrane. The **Na^+-K^+-pump** builds up a high cellular electrochemical gradient for K^+ and indirectly for Cl^- (Fig. 1-1). Zeuthen couples the water outflux to the outward transport of K^+ and Cl^-. Each ion passes with at least 400 water molecules (Fig. 1-1).

The interstitial fluid receives ions and glucose, causing its osmolarity to increase. The osmotic force causes water to enter the interstitial fluid via the cells and tight junctions. This in turn causes the hydrostatic pressure in the interstitial fluid to rise. The hydrostatic force transfers the **bulk of water, ions and molecules** through the thin-walled, tubular capillaries to the blood. When excess of water (solvent) passes through tight junctions, they lose part of their **tightness** and **water drags many Na^+/Cl^- ions out (solvent drag).**

How is an elimination rate constant defined?

The elimination rate constant (***k***) is the fraction of the total amount of a given substance in the distribution volume of the body eliminated per unit time. An elimination with a constant rate is exponential. The half-life for a substance eliminated exponentially is equal to **$0.693\ k^{-1}$**. This is just a simple mathematical deduction.

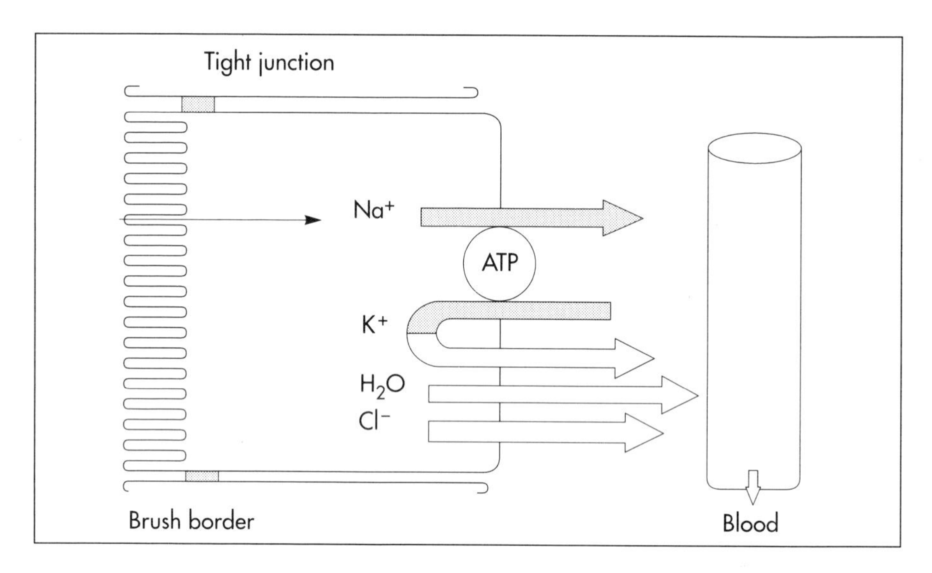

Fig. 1-1: Proximal salt and water transport (modified from Zeuthen, 1992).

What is understood by the law of radioactive decay?

Some nuclei are **unstable** or **radioactive**, because they release certain particles such as helium nuclei or electrons. All radioactive decay processes follow an exponential law. If N_0 is the initial number of unstable nuclei, the number of nuclei remaining after a time $t(N)$ is given by $N = N_0\, e^{-kt}$, where k is a constant characteristic of each nuclide, called the **disintegration constant**. This is the **law of radioactive decay**. Disintegration rates were previously expressed in Curies (Ci), in honour of Marie Curie, who discovered radium. One Ci is defined as the activity of a substance (about 1 g radium) in which $3.7 \cdot 10^{10}$ nuclei disintegrate per second (s). Today the preferred unit is **Becquerel (Bq)**, which is **one disintegration per second**.

References

Carruthers, A. (1990). Facilitated diffusion of glucose. *Physiol. Rev.* 70: 1135.

Christiansen, K., J. Tranum-Jensen, J. Carlsen and J. Vinten (1991). A model for the quarternary structure of human placental insulin receptor deduced from electron microscopy. *Proc. Natl. Acad. Sci. USA,* **88**: 249–52.

Stein, W.H. (1990) *Channels, Carriers, and Pumps: An Introduction to Membrane Transport.* Academic Press, San Diego.

Thomas, H.M., A.M. Brant, C.A. Colville, M.J. Seatter & G.W. Gould (1992). *Biochemical Society Transactions* **20**: 1991–93.

Zeuthen, T. (1992) From contracting vacuole to leaky epithelia. Coupling between salt and water fluxes in biological membranes. *Biochim. Biophys. Acta* **1113**: 229–58.

1. Membrane Transport

A membrane, which is only permeable for water (ideally semipermeable) separates a sucrose solution from a glucose solution in water. The sucrose and glucose concentrations at the beginning are 150 and 50 mol m^{-3}, respectively, and the temperature is 20°C.

1. Calculate the size of the hydrostatic pressure necessary to block any net movement of water.
2. Should the hydrostatic pressure be applied over the sucrose or the glucose solution?

1. Case History

A previously healthy male, 23 years of age, suddenly faints at his job and is brought to hospital deeply unconscious. His body temperature is 310 K, and the vital cardiopulmonary functions are maintained. The plasma [glucose] is 720 mg per 100 ml. In an ultrafiltrate of plasma is found an osmolality of 0.315 Osmol kg^{-1} of water. The patient has lost water and glucose through the urine. The gas constant R is 0.082 (l atmosphere $Osmol^{-1}$ K^{-1}).

1. *Calculate the height of a water column in balance with the osmotic pressure of the plasma ultrafiltrate.*
2. *Calculate the plasma [glucose] in mol m^{-3}. Compare the result to the mean value of 5.3 mmol l^{-1} for a normal population.*
3. *Explain why the patient has lost water.*

Try to solve the problems before looking up the answers in Chapter 74.

CHAPTER 2. RESTING MEMBRANE POTENTIALS

When a microelectrode penetrates a membrane, it records a negative potential with respect to an external reference electrode. This is the **resting membrane potential (RMP)**. The RMP is about **−80 mV** in most myocardial and skeletal muscle cells, whereas in nerve cells the RMP is about **−70 mV**. In smooth muscle cells the RMP is about **−55 mV**.

What is a diffusion potential?

Positive ions (cations) diffuse more rapidly than negative ions (anions) through a membrane. The charge separation creates an **electrical potential gradient** across the membrane. The electrical field enhances the transport of the slow anion and inhibits the transfer of the rapid cation, until the two ions move with the same speed through the membrane. The size of this **diffusion potential** depends on the permeability through the membrane of the two ion types.

Any diffusion that continues long enough to eliminate the **concentration gradient** will also eliminate the diffusion potential!

What is an equilibrium potential across a membrane?

The equilibrium potential for a certain diffusible ion across a membrane, which has a concentration gradient over the membrane, is precisely that **membrane potential difference**, which opposes the flux due to the concentration gradient so that the net transport of the ion concerned is zero.

How is the equilibrium potential calculated for an ion over a membrane?

The **equilibrium potential (Eq)** is simply calculated by balancing the diffusion potential of the ion with the opposing electrical force. The electrical force working on the ion is proportional to the electromotive force of the field. As a consequence, the driving force on the ion and its **diffusion flux is zero**.

This **equilibrium potential** was introduced by Walther Nernst shortly before 1900.

A **membrane potential difference** is conventionally defined as the **intracellular** (φ^i) **minus the extracellular** (φ^o) **electrical potential**. The ion activities (concentrations) inside the cell and outside the cell are called C^i and C^o, respectively.

The Nernst equation for the equilibrium potential of Na^+ across a selective permeable membrane at 310 K is as an example:

$$\varphi^i - \varphi^o = (R\,T/z\,F)\ \ln(C^o{}_{Na^+}/C^i{}_{Na^+})\,\text{Volts (V)}$$

$$Eq_{Na^+} = 61.5\ \log(C^o{}_{Na^+}/C^i{}_{Na^+})\,\text{mV}$$

$$\mathbf{Eq_{Na^+} = 61.5\ \log(140/14) = +61.5\ mV}$$

In the equation above R is the ideal gas constant (0.082 $l \cdot atm \cdot Osmol^{-1} \cdot K^{-1}$ or 8.31 J/(K · mole)), T is the absolute temperature, z is valence of the ion with sign, and F is the Faraday constant (96 500 coulombs/equivalent).

The equilibrium potentials can be calculated from the Nernst equation, when the ion activities (concentrations) inside and outside the cell are known. In a resting skeletal muscle or myocardial cell the equilibrium potential for K^+ is **−94 mV**, and for Cl^- it is **−80 mV**. The equilibrium potentials for Na^+ and Ca^{2+} are +60 and +130 mV, respectively.

What does it mean when the equilibrium potential for an ion over a membrane equals the RMP?

The electrochemical driving force for each ion is the difference between the RMP and its equilibrium potential. When the equilibrium potential equals RMP, the electrochemical driving force is zero. The consequence is that the **net passive transport** of the ion across the cell membrane is **zero**.

In skeletal muscle cells the RMP is −80 mV, and the **equilibrium potential** of Na^+ is +60 mV. Hence, the driving force is: (−80 − (+60)) = −140 mV. Accordingly, there is a net passive influx of Na^+ into these cells down an electrochemical gradient.

How is the RMP of cells maintained?

The Na^+-K^+ pump located in the cell membrane, is responsible for maintaining the high intracellular $[K^+]$ and the low intracellular $[Na^+]$. The energy of the terminal phosphate bond of ATP is used to actively **extrude Na^+ and pump K^+ into the cell.**

The membrane also contains many K^+- and Cl^--channels, through which the two ions **leak out of the cell.**

The RMP is calculated from the **Millman equation**. A convenient version at body temperature is:

$$RMP = (g_{K^+} \cdot Eq_{K^+} + g_{Na^+} \cdot Eq_{Na^+} + g_{Cl^-} \cdot Eq_{Cl^-})/(g_{K^+} + g_{Na^+} + g_{Cl^-})$$

This equation shows that the RMP is determined by the conductance (g) of the membrane to K^+, Na^+ and Cl^-, and by their equilibrium potentials.

The permeability for Na^+ is low (0.2 nm s^{-1}) compared to that of K^+ and Cl^- (5–40 nm s^{-1}). For striated muscle and myocardium, the permeability ratio P_{Na^+}/P_{K^+} is 0.2/5=0.04.

What is the ionic basis of the myocardial resting membrane potential?

In myocardial cells, as in nerve and skeletal muscle cells, K^+ plays a major role in determining the **RMP**. At physiological concentrations of K^+ outside the myocardial cell ($[K^+]^o$ about 4 mM), the RMP is determined by a dynamic balance between the membrane conductance to K^+ and to Na^+. As $[K^+]^o$ is reduced, the membrane depolarizes causing voltage-dependent inactivation of K^+ channels and activation of Na^+ channels, allowing Na^+ to make a proportionally larger contribution to the RMP. Because the equilibrium potential for Na^+ is positive (+60 mV), it tends to depolarize the RMP. Since RMP is −90 mV at normal $[K^+]$, and the equilibrium potential of K^+ is always more negative (−94 mV), there is a small efflux of K^+ from the cell at equilibrium (see also Chapter 23). Eventually, the loss of K^+, and the previously mentioned gain of Na^+ to the cell, would lead to depolarization of the membrane. However, the K^+ and Na^+ gradients are maintained by the **Na^+-K^+-pump** (Chapter 1).

Why don't cells swell and burst?

The ions K^+ and Cl^- both have equilibrium potentials (−94 mV and −80 mV) close to the RMP of most myocardial and skeletal muscle cells (−80 mV). The uneven distribution of small ions across the cell membrane is determined by **negatively charged, non-diffusible macromolecules** such as proteins in the cytosol. According to the Donnan effect, the total osmotic pressure of the cell must draw water into the cell until it bursts (see the Donnan effect in Chapter 1).

The reason that this does not happen is the fact that the cell actively pumps Na^+ out of the cytosol, thereby decreasing its osmotic pressure. The job is performed by the **Na^+-K^+-pump, which extrudes 3 Na^+ and pumps 2 K^+ into the cell per ATP hydrolysed**. This is the **electrogenic action of the pump**. Note that 1 Cl^- tends to move with the 3 Na^+, with consequences for both electrogenicity and osmolarity. Digoxin and low oxygen tensions inhibit the pump and make cells swell.

Further Reading

Aidley, D.J. (1990) *The Physiology of Excitable Cells,* 3rd edn. Cambridge University Press, Cambridge.

Kandel, E.R. and J.H. Schwartz (1966). *Principles of Neural Science*, 3rd edn. McGraw-Hill, New York.

2. Multiple Choice Questions

Each of the following five statements have True/False options:

A. Positive ions (cations) diffuse more rapidly than negative ions (anions) through a membrane.

B. The Na^+-K^+-pump located in the cell membrane, is responsible for maintaining the high intracellular [K^+] and the low intracellular [Na^+].

C. The permeability for Na^+ in cell membranes is high compared to that of K^+ and Cl^-.

D. In skeletal muscle cells the RMP is −80 mV, and the equilibrium potential of Na^+ is +60 mV.

E. A membrane potential difference is conventionally defined as the extracellular minus the intracellular electrical potential.

Try to solve the problems before looking up the answers in Chapter 74.

CHAPTER 3.
THE ACTION POTENTIALS

What is a subthreshold response?

A **subthreshold positive stimulus** depolarizes the membrane and creates a **local response** in the vicinity of the stimulus electrode. The **local, subthreshold response is graded** according to the stimulus. The local response shows **decremental conduction** in that its size decreases with increasing distance from the stimulus electrode (space constant 1–3 mm). The local response cannot in itself initiate an action potential (AP), but if progressively larger depolarizing currents are applied, the **threshold stimulus** is finally reached, and an AP is triggered (Fig. 3-1).

What is the shape of the AP?

When the membrane is depolarized to the threshold potential, an extremely rapid potential change (milliseconds, ms) occurs, resulting typically in a **positive overshoot** at the peak of the AP (Fig. 3-1). This is followed by the **repolarization phase**, when the potential returns toward the resting membrane potential (RMP). The potential may overshoot the RMP, causing a transient hyperpolarization known as the **hyperpolarizing after-potential or after-hyperpolarization** (Fig. 3-1).

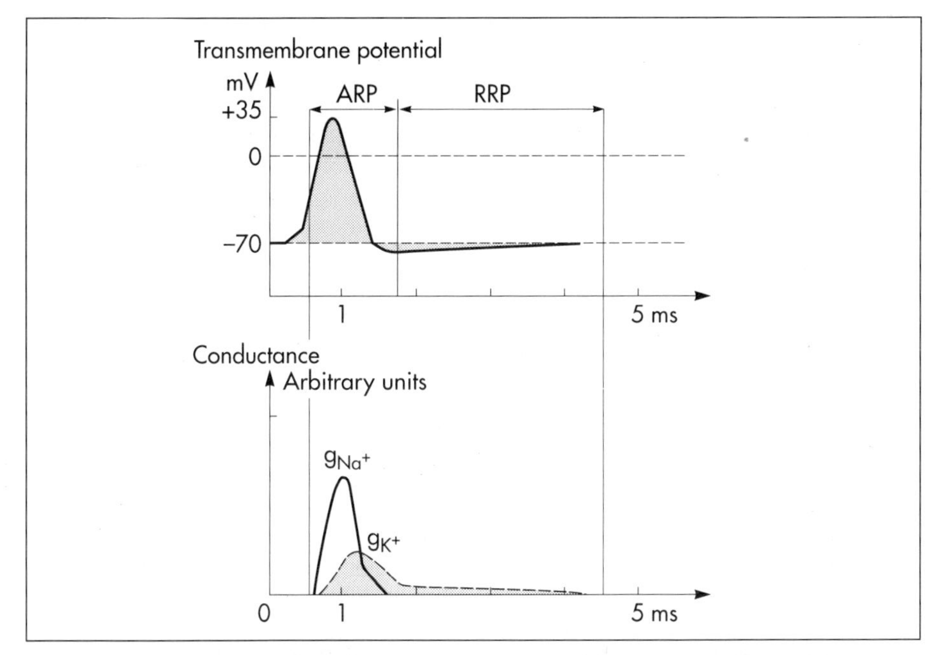

Fig. 3-1: Transmembrane potentials and Na^+-K^+ conductance.

What is an action potential?

The **AP** is an **all-or-none electrical signal**, which appears as a positive wave when recording internally. The **membrane potential difference** is transiently reversed during an AP. The AP is conducted with the same shape and size along the whole length of a muscle cell or a nerve fibre.

What is the ionic basis of the AP?

The **membrane conductance to Na^+** increases very rapidly during the early part of the AP, and reaches a peak almost simultaneously with the peak of the AP ($g_{Na}{}^+$ in Fig. 3-1). The **membrane conductance to K^+ increases slowly**, and reaches a peak in the repolarization phase ($g_K{}^+$ in Fig. 3-1). An increase of the conductance to either of these small ions will cause the membrane potential to move towards the respective equilibrium potential. This is why the rising phase of the AP moves toward +60 mV (the **equilibrium potential for Na^+**). The AP does not, however, reach +60 mV, owing to the opposing K^+-flux, which has started to increase. The Na^+ conductance rapidly returns to normal, whereas the K^+ conductance is still high, pulling the membrane potential closer toward the **equilibrium potential for K^+ (−90 mV)**, and thus creating the **hyperpolarizing after-potential** (Fig. 3-1).

What is the basis of the refractory periods?

During the early part of the AP the cell membrane is completely refractory. A new stimulus, regardless of its size, cannot evoke an AP. Almost all Na^+-channels are inactivated, and will not reopen until the cell membrane is repolarized. This is the **absolute refractory period** covering most of the peak and lasting until well into the repolarizing phase (**ARP** in Fig. 3-1).

During the hyperpolarizing afterpotential, a **suprathreshold stimulus** is able to trigger a new AP, albeit of smaller amplitude than the first AP. This period is called the **relative refractory period** (**RRP** in Fig. 3-1). The cell membrane is relatively refractory, because some Na^+-channels are voltage inactivated and at the same time K^+-conductance is increased.

How does an ion-channel function?

The typical Na^+-channel **opens promptly in response to depolarization**, then after a delay of a few milliseconds (ms) the channel closes even though the cell is still depolarized. The channels then remain inactivated for a period.

Opening of Na^+-channels increases the flux of Na^+ into the subsynaptic neuron. Since this flux depolarizes the membrane (approaches the membrane potential towards the threshold) the effect is excitatory.

Closure of K^+- or Cl^--channels decreases the flux of K^+ out of the subsynaptic neuron or decreases the flux of Cl^- into the cell. These events also depolarize the membrane, and again the effect is excitatory.

Obviously, **closure of Na^+-channels** or **opening of K^+- or Cl^--channels** has an inhibitory effect.

What is the molecular structure of voltage-gated cation channels?

Voltage-gated Na^+, K^+ and Ca^{2+} channels are comprised of subunits with membrane-spanning domains. There is amino acid sequence homology in the transmembrane helices of these channels. The **channel protein** includes a charged group which is sensitive to the electric field across the membrane. During depolarization the gate opens, which changes the whole channel, rendering it much more conductive to specific ions. Each channel continues to open, close and reopen several times during depolarization. The **fast Na^+-channels** close rapidly and are inactivated during depolarization owing to a **channel polypeptide**, located on the cytosolic side.

What is voltage inactivation of APs?

Opening of Na^+-channels requires a rapid change of potential. Partial and slow depolarization inactivates a critical fraction of the Na^+ channels. This is called **voltage inactivation.** Voltage inactivation of Na^+-channels is involved in the accommodation and in the refractory periods.

What is accommodation?

Accommodation is a progressive decrease in firing frequency despite maintained depolarization. Accommodation occurs when a proportion of the **Na^+-channels are voltage inactivated.** This occurs simultaneously with an opening of K^+-channels, which tends to repolarize the cell membrane. This makes the cell still more refractory to stimulation. Accommodation is also related to the conductance of Ca^{2+} and K^+.

How are action potentials conducted?

The lipophilic core of the cell membrane is an electrical insulator, but the salt solutions of the cytoplasm and the extracellular fluid act as conductors of electrical current.

Depolarization spreads along the membrane of excitable cells by local currents flowing to the adjacent segments of the membrane. This is called the **local response** or **electrotonic conduction.** The depolarization decreases monoexponentially from the excitation site. Na^+-channels will be recruited in all areas of membrane, where the threshold potential is exceeded. The Na^+-channels behind the peak of the AP are refractory. This explains why an AP travels in both directions, when it is evoked in the middle of a nerve.

What are the effects of myelination on conduction velocity?

The **myelin sheath** consists of up to 100 layers of membrane formed by Schwann cells wrapping round the axon. The **nodes of Ranvier** are the lateral spaces (1 μm wide) between adjacent Schwann cells, which stretch 1–2 mm. The effects of this arrangement are as follows:

1. Very little current is lost through the electrical insulation of the **myelin sheath.** Thus the **electrotonic conduction** is rapid with only a small decrement in amplitude. The electrotonic conduction is virtually instantaneous. Because of the insulation the depolarization can spread much faster (i.e. the space constant is increased).

2. **Saltatory or leaping conduction** occurs because the AP is generated only at the nodes. The cell membrane below the myelin sheaths has a low density of Na^+-channels and is relatively inexcitable. Saltatory conduction is up to 50 times faster than the conduction through the fastest unmyelinated axons. The AP can also jump over a number of nodes to that furthest away. The Na^+-channels there are activated by the **electrotonic conduction**.
3. Since the ionic currents are restricted to the nodes of Ranvier in the myelinated axons, this minimizes disturbances in the Na^+- and K^+-gradients, which have to be restored by the **Na^+-K^+-pump**. Myelination of the nerve fibre reduces energy cost of maintaining the RMP.

Further Reading

Stuhmer, W. (1991) Structure–function studies of voltage-gated ion channels. *Annual Rev. Biophys Chem.* **20**: 65.

Kandel, E.R. and J.H. Schwartz. (1991) *Principles of Neural Science*, 3rd edn. Elsevier, New York.

Guyton, A.C. (1995) *Textbook of Medical Physiology*, 9th edn. W.B. Saunders, Philadelphia.

3. Multiple Choice Questions

Each of the following five statements have True/False options:

A. The local, subthreshold response is graded according to the stimulus.
B. Accommodation is a progressive decrease in firing frequency despite maintained depolarization.
C. Voltage inactivation of Na^+-channels is involved in the accommodation and in the refractory periods.
D. During the early part of the AP the cell membrane is relatively refractory.
E. Voltage-gated Na^+, K^+ and Ca^{2+} channels are comprised of subunits with membrane-spanning domains.

Try to solve the problems before looking up the answers in Chapter 74.

CHAPTER 4. NEUROMUSCULAR AND SYNAPTIC TRANSMISSION

What is synaptic transmission?

Synaptic transfer refers to the transmission of signals from one neuron to another. The site of contact between the two neurons is called the **synapse.**

Neurotransmission is mediated electrically or chemically.

Describe the structure of neuromuscular junctions

The **neuromuscular endplate** is the **contact zone** between the axons of motor neurons and striated muscle fibres. Axon terminals have vesicles containing acetylcholine (ACh-vesicles in Fig. 4-1). The vesicles dock on the **active zones** or **release sites** of the presynaptic membrane with high affinity. The muscle cell membrane at the endplate is folded in **junctional folds or crypts** (Fig. 4-1). **Receptor proteins for acetylcholine (ACh)** are concentrated at the openings of these junctional crypts. The release sites are located directly over the acetylcholine receptors (Fig. 4-1). The subsynaptic membrane has **acetylcholinesterase** all over its surface.

The **receptor protein for ACh** is an **integral membrane protein** fixed in the postjunctional membrane, whereas cholinesterase is loosely attached to its surface. The protein has five subunits, surrounding a **central ion channel pore** that is opened by the binding of ACh to the protein. Opening of the ion channel increases the conductance for small cations across the postjunctional membrane. These ion channels are **ligand-gated** (gated by ACh) and not voltage-gated (not dependent on changes in membrane potential), like most cation channels in muscle cell membranes.

How is the AP transmitted at the plate?

The ACh-vesicles are probably already stored close to the **release zones**, awaiting the release signal (Fig. 4-1). When the action potential (AP) reaches the axon terminals, the axon membrane is depolarized, and voltage-gated Ca^{2+}-channels are transiently activated. This causes Ca^{2+} to flow down its concentration gradient from the outside into the axon terminal. The influx of Ca^{2+} at the release zones causes the vesicles to fuse with the axon membrane, and empty ACh into the 50-nm wide cleft by **exocytosis** (Fig. 4-1).

After crossing the synaptic cleft by diffusion, ACh binds to its **receptor protein** on the muscle cell membrane. This binding complex opens the ion channel and increases the conductance for small cations across the muscle cell membrane. The influxes of Na^+ depolarize the **endplate** temporarily, the transient depolarization is termed the **endplate potential, EPP**. EPP dies away when acetylcholine is hydrolysed to acetate and choline by the enzyme, **acetylcholinesterase**. The EPP has a large safety margin, as a single AP in the motor axon will produce an EPP that always reach the threshold potential for firing an action potential in the muscle fibre.

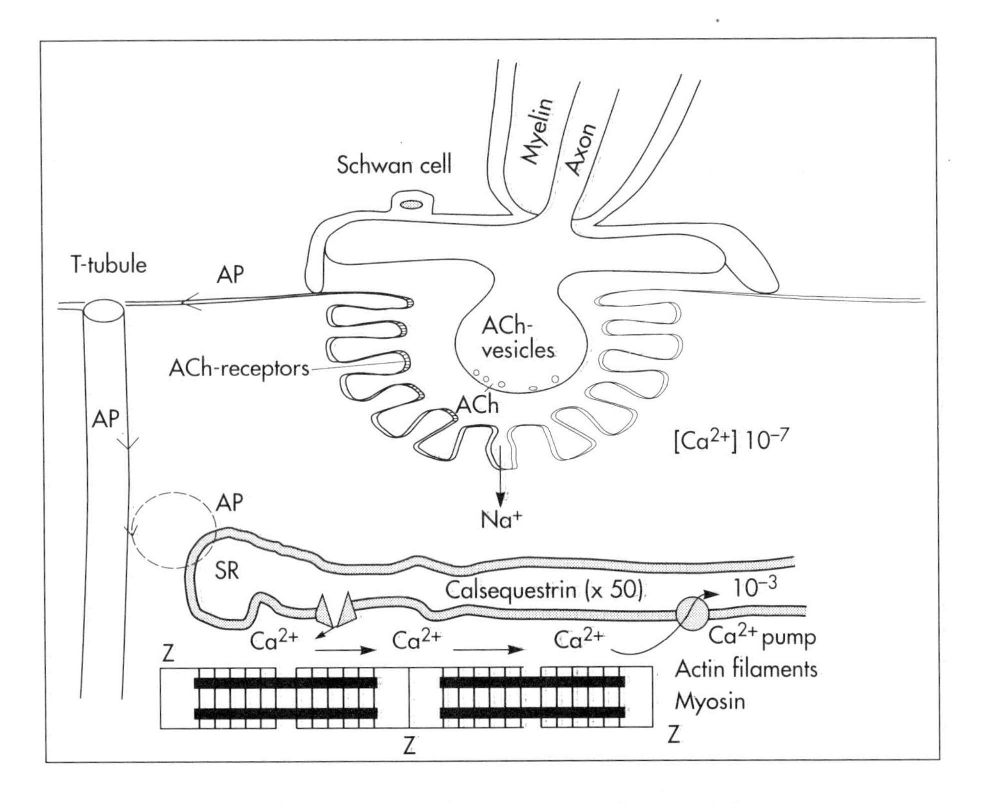

Fig. 4-1: The neuromuscular junction and intracellular events.

Rapid contraction of the muscle fibre is achieved by propagation of the AP along the whole length of the muscle fibre membrane and into the transverse tubules (T-tubules in Fig. 4-1).

Events inside the myofibril are described in Chapter 6.

How is ACh synthesized and released?

Neurons with motoric function have the ability to synthesize ACh, because they contain **choline-acetyltransferase**. This enzyme catalyzes the production of ACh from acetyl-CoA and choline. Almost all cells produce acetyl-CoA and choline. Choline is also actively taken up from the extracellular fluid via a mechanism indirectly powered by the **Na^+-K^+-pump**. There is a 50% reuptake of choline from the synaptic cleft, hence some choline must be synthesized in the motor nerve.

The postjunctional membrane depolarizes spontaneously resulting in so-called **miniature endplate potentials (MEPPs)**. A MEPP is probably caused by the spontaneous release of a single vesicle into the cleft. This is called **quantal release**.

What is the function of cholinesterase?

An EPP is prolonged when **cholinesterase inhibitors** are present in the synaptic cleft. This is because these substances (eserine, edrophonium, malathion, parathion etc.) inhibit the enzyme and thereby protect ACh from being hydrolysed by the enzyme. Under normal conditions, the EPP is terminated by the rapid hydrolysis of ACh by acetylcholinesterase.

Is acetylcholine an important neurotransmitter

Yes. **ACh** is a transmitter in the CNS, in all motor neurons, in all preganglionic neurons of the autonomic nervous system and postganglionic parasympathetic fibres, and in a few postganglionic sympathetic fibres.

What is myasthenia gravis?

Myasthenia gravis is a serious defect in neuromuscular transmission. Many of these patients have an increased blood concentration of antibodies against their own **acetylcholine receptor protein**. The use of radioliganded toxins from poisonous snakes – α-bungarotoxin (which bind irreversibly to the acetylcholine receptor protein) – has shown that there is a **decreased density** of receptor proteins on the **postjunctional** membrane. The symptoms and signs of myasthenia gravis are alleviated by ACh-esterase-inhibitors. Alternative approaches to the treatment of myasthenia gravis include the inhibition of antibody production with steroids (eg prednisolone) or immunosuppressant drugs (eg azalthioprine); or the removal of circulating antibody by plasma exchange.

What is a chemical synapse?

A **chemical synapse** consists of a neuronal presynaptic terminal, a synaptic cleft and subsynaptic or postsynaptic membrane with associated receptor proteins. The chemical synapse is highly developed in the CNS of vertebrates. It conducts the signal one way only, and has a characteristic **synaptic delay**.

The **presynaptic axon terminal** typically broadens to form a **bouton terminaux** (terminal button). When the **AP** reaches the presynaptic membrane Ca^{2+} enters the terminal through voltage-gated Ca^{2+}-channels. Vesicles containing transmitter fuse with the presynaptic membrane and release their contents into the synaptic cleft (Ca^{2+}-induced exocytosis). Transmitter molecules diffuse across the synaptic cleft and bind to specific receptors, which are located on the subsynaptic membrane. This binding elicits a transient opening of **ionophores** which are specifically permeable to small ions. This results in a transient change of the postsynaptic membrane potential. If the presynaptic AP results in a postsynaptic depolarization, the transient is called an **excitatory postsynaptic potential (EPSP)**. If the AP results in a postsynaptic hyperpolarization, the transient is called an **inhibitory postsynaptic potential (IPSP)**. Excitatory synapses often use **glutamate as the transmitter**. The ionophores are mainly permeable to Na^+, which enters the cell and produces an EPSP.

What is a gap junction?

A gap junction is a pathway of low resistance that connects cytoplasm of adjacent cells. A junction allows the membrane potential of the adjacent cells to be **electrically coupled**.

Gap junctions or **electrical synapses** differ from chemical synapses in that transmission is instantaneous and the distance between the two membranes must be less than 3 nm (the synaptic cleft of a chemical synapse is about 30 nm).

Gap junctions contain channels which close in response to increased intracellular [Ca^{2+}] or [H^+] in a cell, thereby increasing their resistances. Gap junctions exchange ions and small molecules up to a molecular weight of 1000 Da.

Gap junctions are found in simple reflex pathways, where rapid transfer of the electrical potential is essential.

Gap junctions are also found between non-neural cells such as myocardial cells, smooth muscle cells and hepatocytes.

Are amino acids used as neurotransmitters?

Yes.

1. **GABA** (gamma-aminobutyric acid) and glycine are **inhibitory neurotransmitters** in the brain and spinal cord (CNS).
2. **Glutamate** and **aspartate** are the most important **excitatory transmitters** in the brain and spinal cord (**excitatory amino acids (EAA) mediators**).

Excitatory neurons possess **EAA receptors**. Glutamate, aspartate and related acidic amino acids are called **EAAs**. In the CNS these **EAA-mediated synapses** predominate.

EAA receptors are a family of receptors with at least four different ion-channels: The ***N*-methyl-D-aspartate-receptor (NMDA)**, and three so-called **non-NMDA receptors**.

What makes a synapse inhibitory?

Inhibitory synapses have **activated transmitter–receptor complexes**, which are permeable to K^+ or to Cl^-. An efflux of K^+ or an influx of Cl^- across the subsynaptic cell membrane, causes **hyperpolarization** which reduces its excitability. The increase in conductance to Cl^- stabilizes the membrane potential and decreases the efficacy of excitatory transmission.

GABA in the brain and glycine in the spinal cord are predominantly inhibitory.

What makes a synapse excitatory?

Excitatory synapses have excitatory receptors on the subsynaptic membrane. Excitation results from **opening of Na^+-channels** which allow an influx of Na^+ across the subsynaptic cell membrane causing the membrane potential to approach the threshold level for excitation. Reduced conductance through Cl^-- and K^+ channels on the subsynaptic cell membrane also causes excitation. Both a reduced Cl^--influx to the neuron and a reduced K^+-outflux move the membrane potential towards the threshold level.

Where on the neuron are voltage-gated channels located?

The axon hillock on the **cell body** has a high density of voltage-gated Na^+- and K^+-channels. The **dendrites** have voltage-gated channels for K^+ and for Ca^{2+}. Recent evidence suggests that dendrites also contain voltage-gated Na^+-channels which are involved in **electrogenesis**.

How are synaptic inputs summated?

Each neuron in the CNS is in contact with up to 10^5 presynaptic axon terminals. Synaptic inputs are integrated by either **spatial** or **temporal summation**.

Spatial summation occurs when inputs from several axons arrive simultaneously at the same postsynaptic cell. Their postsynaptic potentials are additives. EPSPs summate and move the membrane potential closer to the threshold level for firing. Conversely, EPSPs and IPSPs cancel each other out.

Temporal summation occurs when successive APs in a presynaptic neuron follow in rapid succession, so that the postsynaptic responses overlap and summate. Summation is possible because the synaptic potential lasts longer than action potentials by a factor of 10–100 times.

Each individual synapse contains receptors, ion channels, and other key molecules, which are sensitive to the neurotransmitters released at the site. These specific protein molecules are involved in synaptic plasticity and summation.

What characterizes the neuropeptides as transmitters?

Neuropeptides are built by a sequence of amino acids. Neuropeptides are synthesized in the cell bodies of the neurons and transported to the terminal buttons by rapid axonal transport. A large **mother-peptide** is cleaved into several active neuropeptides. Neuropeptides are released from the nerve terminal near the surface of its target cell, and diffuse to the receptors of the target cell. Low concentrations of neuropeptides typically affect the membrane potential by changing the conductance of the target cell to small ions. The action of neuropeptides usually lasts longer than that of enzyme-inactivated transmitters. Following prolonged synaptic transmission, neuropeptides are inactivated by **proteolysis**. The **gut–brain peptides** are described in Chapter 46, and the **hypothalamo-pituitary peptides** in Chapter 64.

Are amine molecules used as neurotransmitters?

Yes.

1. Catecholamines (dopamine, noradrenaline (NA), and adrenaline (Ad)) are neurotransmitters both in the sympathetic system and in the CNS. NA is the transmitter for **most postganglionic sympathetic fibres** (some of these fibres use ACh). In the CNS catecholamines are found in several brain nuclei: **dopaminergic neurons** are found in the substantia nigra, **noradrenergic neurons** in locus coeruleus, and **serotonergic neurons** in the raphe nuclei and in many midbrain structures.

   It is mainly the **loss of dopamine-containing neurons** in the substantia nigra that results in the lack of dopamine in the dopaminergic synapses of the striatum. These structures degenerate in Parkinson's disease causing **muscular rigidity and tremor** (Fig. 12-3).
2. Histamine is a neurotransmitter in the hypothalamus.
3. Serotonin is a transmitter in brainstem nuclei.

Further reading

Aidley, D.J. (1990) *The Physiology of Excitable Cells*, 3rd edn. Cambridge University Press, Cambridge.

Kupfermann, I. (1991) Functional studies of cotransmission. *Physiol. Rev.* **71**: 683.

Steward, O. (1993) Synapse growth as a mechanism for activity-dependent synaptic modification. In: *Memory Concepts 1993. Basic and Clinical Aspects. Novo Nordisk Foundation Symposium* **7**: 281–301, Elsevier, Amsterdam.

4. Multiple Choice Questions

Each of the following five statements have True/False options:

A. Motor neurons synthesize ACh unrelated to their content of choline-acetyltransferase.

B. There is a high density on the subsynaptic membrane of specific ACh receptors.

C. The receptor protein for ACh contains a voltage-gated channel for cations.

D. Binding of ACh elicits a transient opening of ionophores which are specifically permeable to small ions.

E. Parkinson's disease is possibly caused by loss of dopamine containing neurons in the substantia nigra.

Try to solve the problems before looking up the answers in Chapter 74.

CHAPTER 5. SIGNAL TRANSDUCTION

What is signal transduction?

Signal transduction is a cascade of processes from the **receptor–hormone** binding to the final cellular response. Many hormones, such as catecholamines and peptides, raise the concentration of a **second messenger** in the target cell, and act through it. **The cascade** consists of hormone–receptor–second messenger–protein kinase–target enzyme. The receptor–hormone complex activates a **guanosine triphosphate or GTP-binding protein** (so-called G-protein) which controls and amplifies the synthesis of second messenger. Hereby, each hormone molecule can produce many molecules of second messenger such as cAMP. Furthermore, each **protein kinase unit** can phosphorylate many molecules of its substrate, resulting in a great amplification factor.

What is the function of GTP-binding proteins?

G-proteins function as **molecular switches**, regulating many cellular processes which are involved in signal transduction with hormones and neurotransmitters.

G-protein-linked receptors form a family, which has evolved from a common ancestor. Most G-proteins are **membrane-bound heterotrimers** (α β γ) and exist in an **activated state**, where it has high affinity for GTP, and an **inactive state**, where the molecule prefers GDP (Fig. 5-1).

1. Many hormones or transmitters bind to **specific membrane receptor proteins,** which activate a specific G-protein. This causes the exchange of bound GDP with GTP and dissociation of the α subunit (Fig. 5-1).
2. The activated G-protein then binds to the hormone–receptor controlled **adenylcyclase.**
3. The enzyme catalyses formation of cAMP and pyrophosphate (PP_i) from ATP (Fig. 5-1).
4. The increased intracellular [cAMP] activates **protein kinase A** (Fig. 5-1).
5. Protein kinase A has two cAMP-binding subunits and two catalytic subunits. The two catalytic subunits now dissociate and disclose their kinase activity. Protein kinase A catalyses the transfer of phosphate from ATP to its substrate proteins (Fig. 5-1) termed **effector proteins (ion channels, enzymes, etc.).**
6. Following activation of adenylcyclase, the GTP is eventually hydrolysed to GDP, so that the G-protein reverts to its inactive state. This is possible because activated G-protein has GTPase activity.

Examples of G-protein-linked receptors are **β-adrenergic receptors, muscarinic receptors (M_2) and rhodopsin** (Chapter 13). Stimulation of α_2-adrenergic receptors **inhibits** adenylcyclase via an inhibitory G protein.

The target molecules for activated G-proteins are enzymes or ion channels, such as (1) adenylcyclase, (2) guanylate cyclase (stimulated by NO, which is again stimulated by Ca^{2+}-calmodulin), (3) cGMP phosphodiesterase, (4) phospholipase C, (5) phospholipase A_2 and (6) Ca^{2+}-, K^+- and other ion channels.

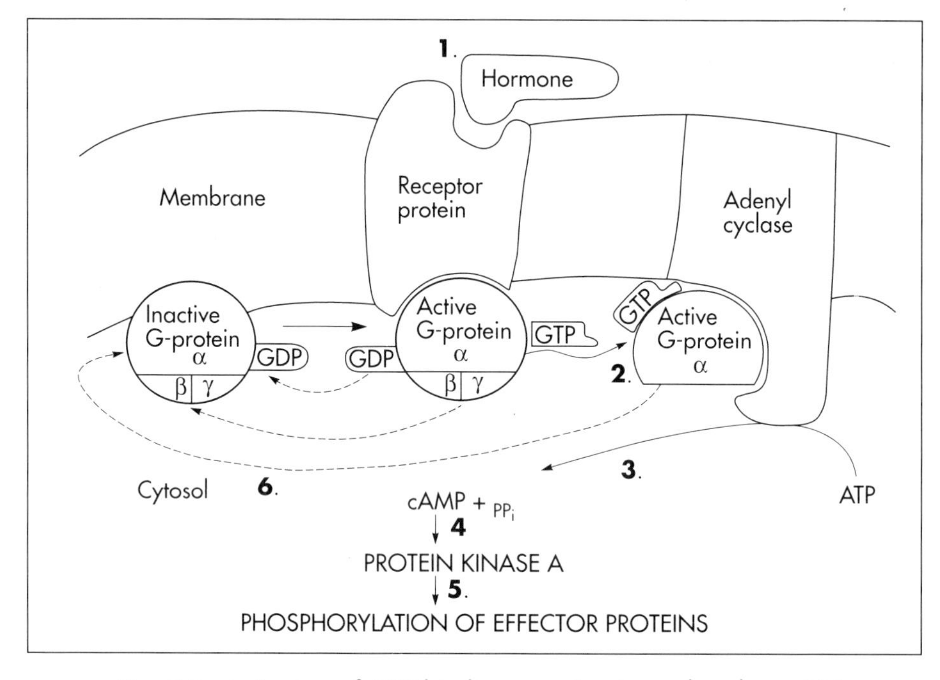

Fig. 5-1: Activation of GTP-binding proteins (pyrophosphate=PP_i).

What is the function of phospholipase C?

Hydrophilic (lipophobic) hormones such as ACh and many peptides bind to membrane receptor proteins, and the hormone–receptor binding activates the enzyme **phospholipase C** via active G-protein (Fig. 5-2).

1. Interaction with the hormone–receptor complex causes the exchange of bound GDP to GTP and dissociation of the α subunit (Fig. 5-2).
2. The activated G-protein binds to the hormone–receptor controlled enzyme **phospholipase C**.
3. Two phosphorylated derivatives of **phosphatidylinositol (PI)** on the inner side of the membrane are essential in signal transduction: **PI-phosphate (PIP) and PI-diphosphate (PIP_2) as is PI itself.**

   Phospholipase C (Fig. 5-2) cleaves (PIP_2) into **inositoltriphosphate (IP_3)** and **diacylglycerol (DAG).**
4. IP_3 is a second messenger that binds to Ca^{2+}-channels in the endoplasmic reticulum (ER), so that Ca^{2+} is released to the cytosol (Fig. 5-2).
5. DAG and Ca^{2+} are second messengers that activate **protein kinase C** (Fig. 5-2), which is involved in the regulation of cellular metabolism, growth and many other processes.
6. Inactive cytosolic **protein kinase C** is activated by Ca^{2+}, and binds to the inner surface of the membrane, where it is activated by DAG (see 5. in Fig. 5-2).

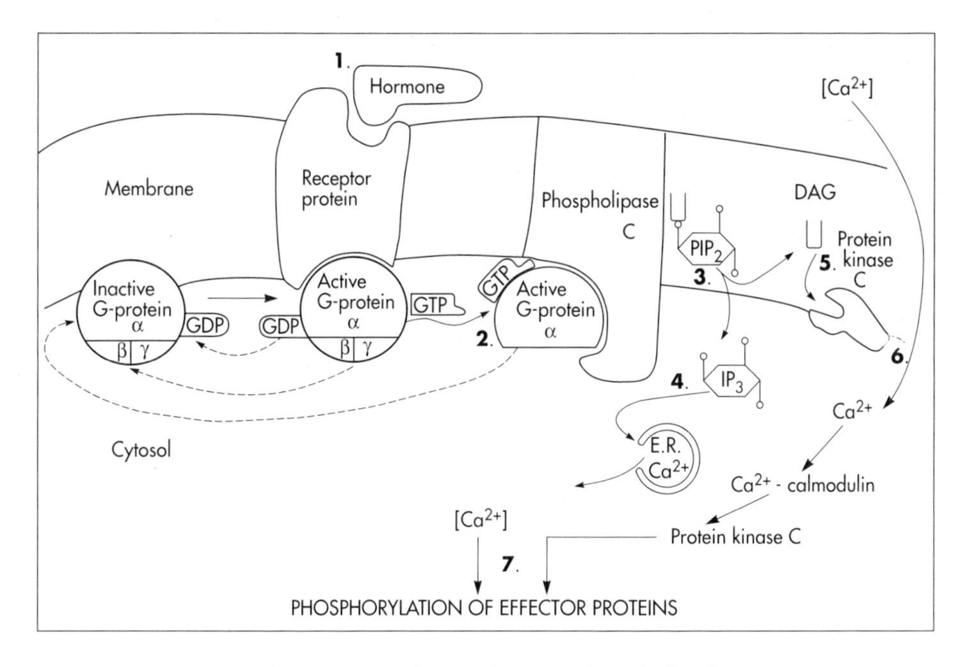

Fig. 5-2: Signal transducting phospholipids.

7. Ca^{2+} and proteinkinase C catalyses the transfer of phosphate from ATP to the effector proteins (Fig. 5-2).

Specific receptor–ligand bindings also activate **phospholipase A_2** via a G-protein. Phospholipase A_2 cleaves membrane phospholipids, and releases **arachidonic acid (AA)** in the cells. AA activates a precursor to **platelet activating factor (PAF)** termed **lyso-PAF**. AA is also the precursor for the synthesis of **endoperoxides, prostacyclin, thromboxanes (mediates platelet aggregation and vasoconstriction) and leucotrienes** (see Chapters 18–20).

Which of the second messengers are influenced by the G-protein cascade?

The cellular concentration of the following second messengers is increased or decreased: (1) cAMP, (2) cGMP, (3) IP_3, (4) DAG and (5) Ca^{2+}. The second messengers change the activity of one or more protein kinases or protein phosphatases.

How does tyrosine kinase work?

Insulin and related growth factor peptides bind to membrane receptors that are glycoproteins protruding from the membrane.

The **insulin receptor**, described in Chapter 1, is typical for this **receptor family**. Peptide binding to the outer receptor subunit stimulates a **protein tyrosine kinase** on the inner receptor subunit. This phosphorylates tyrosine residues, both on the receptor itself and on other proteins. The **tyrosine kinase activity** is essential for **signal transduction**.

Examples of growth factors are: **EGF** (epidermal growth factor), **FGF** (fibroblast growth factor), **IGF-II** (insulin-like growth factor-II), **NGF** (neural growth factor) and **PDGF** (platelet-derived growth factor) – see Chapters 8 and 68.

Protein tyrosine kinase activity is abnormally high in certain types of cancer and cellular modification. This can be caused by growth factors or by a mutation of the tyrosine kinase part of the transmembraneous receptor. Mutations of one gene localized on **chromosome 10** can lead to four different syndromes: multiple endocrine neoplasia, Hirschprung's disease, medullary thyroid carcinoma, and **pheochromocytomas** (see Chapter 71).

What is the final step in signal transduction?

The final step is often phosphorylation or dephosphorylation of a particular **key or effector protein**. Phosphorylation is accomplished by **protein kinases**, and dephosphorylation by **protein phosphatases**. Second messengers (cAMP, cGMP, IP_3, DAG, and Ca^{2+}) control the activities of protein kinases such as **cAMP-dependent protein kinase A, cGMP-dependent protein kinase, calmodulin-dependent protein kinase, and protein kinase C**. Calmodulin binds 4 Ca^{2+}. The phosphorylation level of an enzyme or an ion channel determines and triggers the **physiological response**.

What is the function of protein phosphatases?

Protein phosphatases reverse the effect of protein phosphorylation. The phosphatases dephosphorylate the key proteins, and thus oppose or stop the physiological response.

Is nitric oxide (NO) a neuronal messenger?

The free radical gas NO is a neuronal messenger in both the central and the peripheral nervous system. The NO gas is membrane permeable, and can bypass normal signal transduction in synapses.

Describe the biosynthesis of NO and its transduction mechanism in cell–cell signalling

Two types of **NO synthase (NOS)** have been identified: constitutive Ca^{2+}–calmodulin dependent enzyme, and **inducible** Ca^{2+}-independent enzyme. Both enzymes are flavoproteins containing bound **flavin mononucleotide (FMN)** and **flavin adenine dinucleotide (FAD)**. Both enzymes require the cofactors **NADPH** and tetrahydrobiopterin (BH_4). NOS catalyses the conversion of L-arginine to citrulline and NO in two steps when activated by Ca^{2+}–calmodulin complex or muscarinic agonists, or other activators (Fig. 5-3).

NO diffuses to the target cell, where it activates **guanylcyclase** resulting in the formation of cGMP (Fig. 5-3). NO is labile. Hence, a carrier for NO has been postulated. The biological effect of NO is mediated by an increase in cGMP levels, and the effects on target cells are shown in Fig. 5-3. **Nitric oxide effects** are further developed in Chapters 7, 9, 11, 19, 20, 25, 27, 45, 46, 47, 49, 59 and 68.

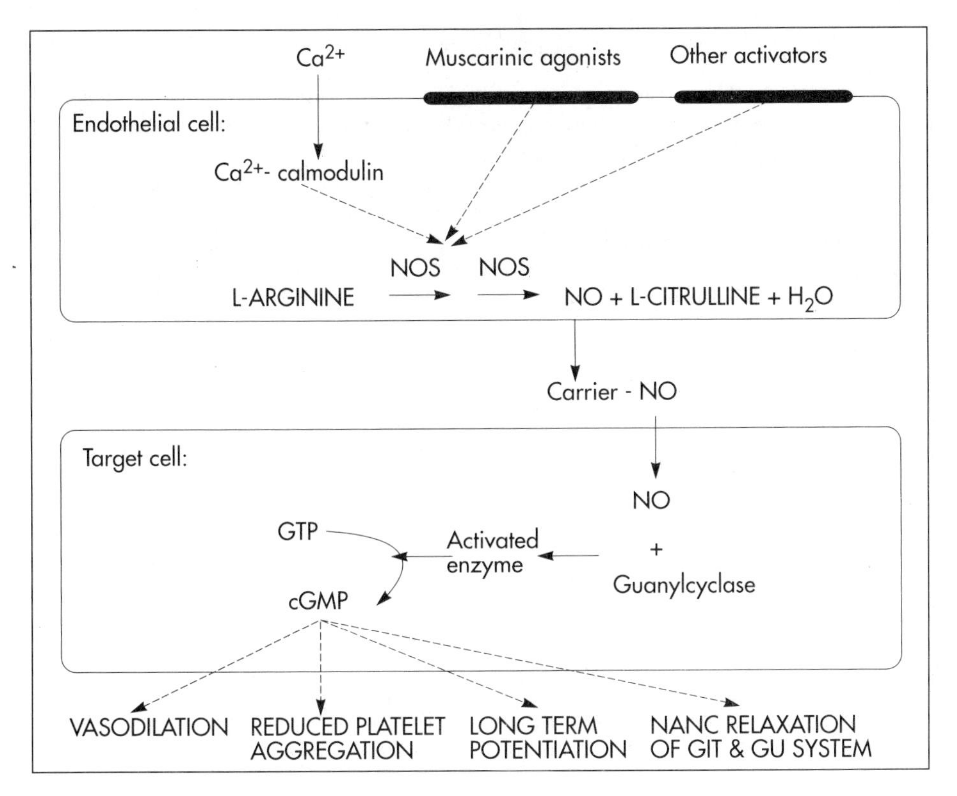

Fig. 5-3: The biosynthesis of NO with cell–cell effects on target cells, such as smooth muscle cells, etc. Non-adrenergic non-cholinergic (NANC) relaxation of the gastrointestinal tract (GIT) and the genito-urinary (GU) system is shown.

Therapeutic intervention on NO biosynthesis

The **NO biosynthetic pathway** can be interfered with at several points.

Nitrovasodilatators, such as **nitroglycerin**, have been used for over a century to treat cardiac cramps or angina pectoris (see Chapter 19). Nitrovasodilatators act by releasing NO and thereby causing coronary vasodilatation. NOS can be inhibited by L-arginine analogues (Fig. 5-3).

One of the effects of NOS inhibitors is an increase in blood pressure. NOS inhibitors are effective treatment for endotoxic shock (Fig. 5-3). This is a condition caused by increased NO synthesis by inducible NO synthases, where the sympathetic vasoconstrictors are often ineffectual. The co-factors, like BH_4, can also be manipulated, e.g. by **anticancer drugs**.

Further reading

Alberts, B. *et al.* (1989) *Molecular Biology of the Cell*, 2nd edn. Garland Publishing, New York.

Apps, D.K., B.B. Cohen and C.M. Steel. (1992) *Biochemistry*, 5th edn. Baillière Tindall, London.
Drejer, J. (1992) Therapeutic opportunities in modulators of excitatory amino-acid mediated neurotransmission. In: *Excitatory Amino Acid Receptors* (Ed. P. Krogsgaard-Larsen and J.J. Hansen). Ellis Horwood, New York.
Hall, A. (1990) The cellular functions of small GTP-binding proteins. *Science* **249**: 635.
Houslay, M.D. and G. Milligan (Eds). (1990) *G Proteins as Mediators of Cellular Signalling Processes*. Wiley, New York.
Van Heyningen, V. (1994) Genetics. One gene, four syndromes. *Nature* **367**: 319–20.

5. Multiple Choice Questions

Each of the following five statements have True/False options:

A. Phosphatidylinositol (PI) is not essential in signal transduction.
B. Activated G-protein has no GTPase activity.
C. Catecholamines and peptide hormones bind to membrane receptors on the cell surface.
D. Second-messenger systems are used for signal transduction by steroids.
E. Lipophilic hormones bind to membrane receptors.

Try to solve the problems before looking up the answers in Chapter 74.

CHAPTER 6. SKELETAL MUSCLES

Describe essential characteristics of striated muscles

Skeletal or striated muscles are attached to a skeleton. Striated muscles are called **striated,** because they have a striking banding pattern. Microscopy with polarized light reveals dark (optically anisotropic) striations or **A bands** alternating with light or optically isotropic striations or **I bands.** Running along the axis of the muscle cell or **muscle fibre** are the **myofibril bundles** of **filaments** which are visible in electron micrographs. The A band contains the **thick filaments** of myosin, and the I band contain **thin filaments** of actin and tropomyosin (Fig. 6-1). The thin filaments are anchored to a transverse structure termed the **Z disc** (Fig. 6-1). Each contractile unit contains the halves of two I bands with the A band in between. This unit is a **sarcomere.** Sarcomes have a length of 2.3–2.5 μm between the two Z discs at rest. The central A band is a relatively isotropic substance – also termed the **H band** – with an **M line** of darkly stained proteins that link the thick filaments into a fixed position.

What is the thin and the thick filament?

The thin filaments are 1 μm long and consist of small globular proteins that form two helic strings. The **double-helix of actin** is supported by a long, thin molecule of **tropomyosin.**

The thick filaments are 1.6 μm long, and consist of large myosin molecules. Myosin is a dimer of almost 500 kDa. Each monomer consists of one heavy chain and two light chains. The heavy chain consists of a helical tail and a global head. The light chains are associated with the head of the heavy chain. Since myosin is a dimer, the **double-helix tail must end in two heads.** The heads contain the ATPase activity and the actin binding site.

Describe the crossbridge cycle

The crossbridge cycle hypothesis states that there are multiple cycles of **myosin-head** attachment and detachment to **actin** during a muscle contraction. When myosin binds to actin, an actomyosin complex is formed – an extremely active ATPase. The interaction between actin and myosin and the hydrolysis of ATP is the basic process that converts chemical energy into mechanical energy.

Each crossbridge consists of two heads. At rest the crossbridge is not attached to actin. Stimulation of a muscle liberates Ca^{2+} in the myoplasma. This alters the stucture of the myofibrils, so the crossbridges are now bound to the thin filaments. The binding accelerates the release of ADP and P from the actin–myosin complex; the head of the crossbridge drags the thick filament 10 nm along towards the Z disc.

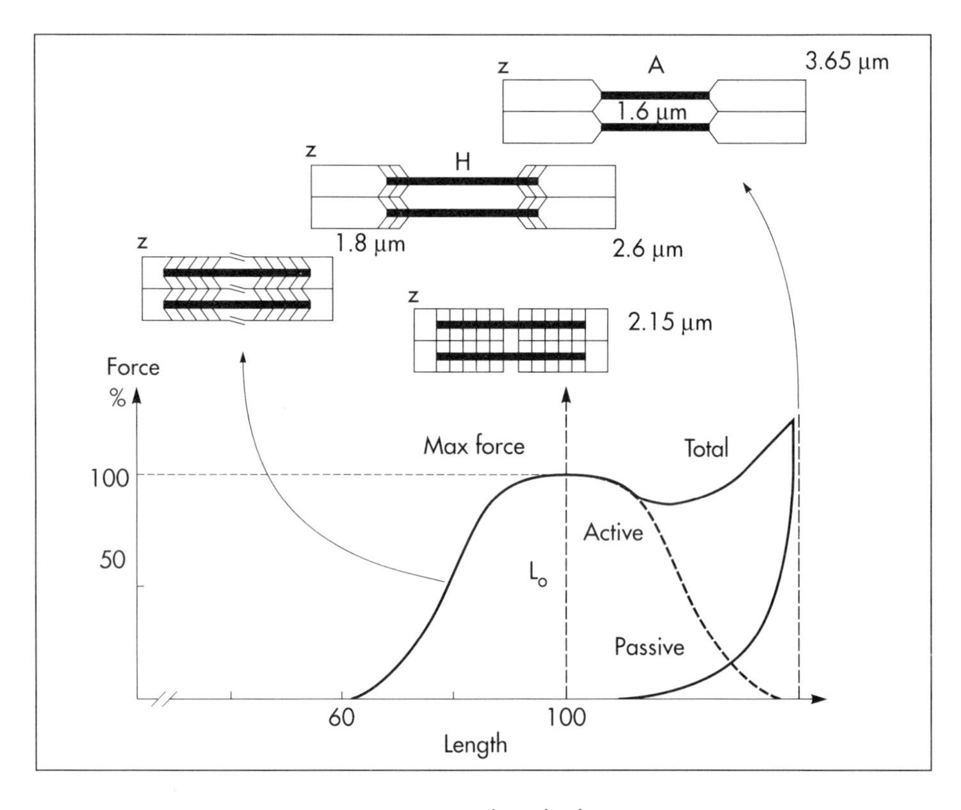

Fig. 6-1: Force–length diagram.

What is the force–length relationship of muscles?

Force is required to stretch a relaxed muscle, because muscle tissue is elastic, and the force increases with increasing muscle length (Fig. 6-1). The passive curve reflects the properties of the elastic, connective tissue, which becomes less compliant or more stiff with lengthening (Fig. 6-1).

A muscle contraction at constant length is termed **isometric**. Force is measured in Newton (N), and **1 N** is the force required to accelerate a mass of 1 kg with an acceleration of 1 m s^{-2}. In muscles the traditional expression is **stress** or **tension** in N per cross-sectional area of the muscle (N m^{-2}), which is actually **pressure (Pascal, Pa)**. Here, the ordinate is force expressed as a percentage of the maximal force (max force in Fig. 6-1).

1. The length at which maximum active contractile force (max Force) is developed is called L_O (Fig. 6-1). L_O is the length of the muscle in the body when at rest. At this length there is a maximum number of active crossbridges (Fig. 6-1). When an isolated muscle in an **isometric stress-meter** is stimulated, the active muscle force decreases with the decrease in overlap between thin and thick filaments (Fig. 6-1). The force is proportional to the number of crossbridges interacting with the thin filament.

2. Force also declines at muscle lengths less than L_o (Fig. 6-1). This is caused by thin filaments overlapping, and thick filaments colliding against Z-discs.
3. When the active muscle length is stretched beyond any overlapping between the thin and the thick filaments the muscle can only develop a force of zero (dashed curve crossing the abscissae in Fig. 6-1).

The lengths of the thick and thin filaments of human striated muscles are similar. They generate maximal tension forces at L_o, namely 300 kN m^{-2} or kPa. Muscle power or work-rate is the product of muscle force and shortening velocity (m s^{-1}). The maximal work rate of human muscles is reached at a contraction velocity of 2.5 m s^{-1}. The maximal work-rate is thus (300 kPa · 2.5 m s^{-1}) = 750 kW per square meter of cross-sectional area.

The sliding of filaments against each other is called the **sliding filament hypothesis.**

What is the force–velocity relationship of muscles?

A preload in kilograms determines the length of an isolated and unstimulated muscle. Consequently, the maximum force is developed at the initial length (Fig. 6-2 right). This is Hill's **force–velocity diagram**, where the shortening velocity decreases rapidly as the afterload is increased following a hyperbola (Fig. 6-2 right). After the muscle is stimulated increasing loads (afterloads of 4 and 9 g) are lifted (Fig. 6-2 left). At the start there is a constant rate of shortening, which suddenly declines to zero as the muscle reaches its final shortening length at maximal afterload (18 g in Fig. 6-2). An unloaded crossbridge can cycle at maximal rate, indicated by maximal shortening velocity (Fig. 6-2 right).

This maximal velocity of shortening is directly proportional to the **myosin ATPase activity**. We increase the velocity of muscle shortening under a given load by the recruitment of additional motor units. The long human arm muscles shorten at a rate of 8

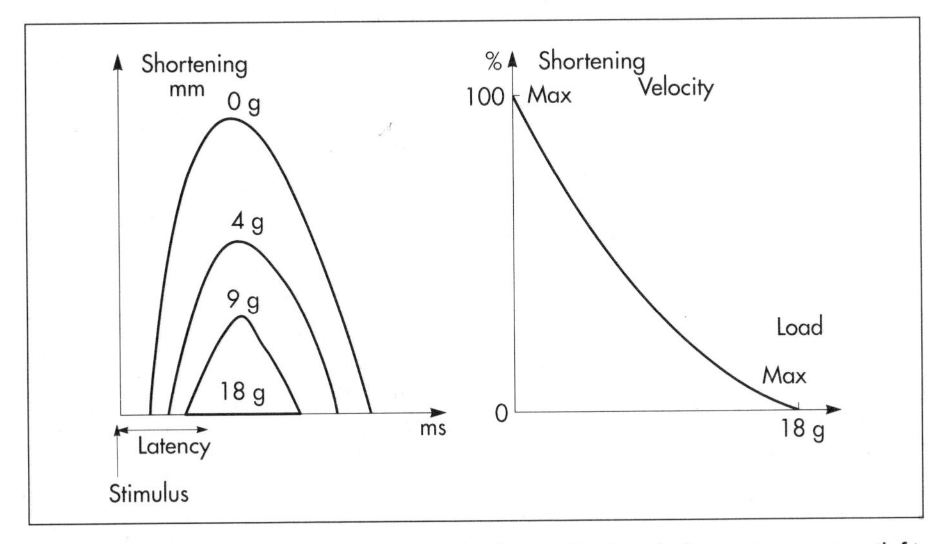

Fig. 6-2: Hill's force–velocity diagram (right) and related shortening curves (left).

m s^{-1}. Muscles can bear a load of 1.6 times the maximal force before the crossbridges are broken, but under such extreme conditions the work-rate (power) of the muscle is minimal (no shortening in Fig. 6-2 left). The **work-rate** of the muscle is equal to the product of force (afterload) times velocity. Maximal work-rate occurs at a load of one-third of the maximal isometric force of the muscle. Here the contractile system has optimal efficiency in converting chemical energy into mechanical energy.

A further rise in filament velocity seems to reduce the potential for actin–myosin interaction. The **crossbridge cycling rate** falls as the load on the crossbridges increases (Fig. 6-2 right).

What triggers the excitation–contraction coupling?

The acetylcholine binding at the motor endplate increases endplate conductance and generates an AP in all directions from the end plate (Fig. 4-1). The electrical excitation of the sarcolemma and the transverse tubules (T-tubules) during the AP triggers the **sarcoplasmic reticulum (SR)** to release Ca^{2+} (Fig. 4-1). The T-tubules penetrate all the way through the muscle fibre. The depolarization transiently opens Ca^{2+}-channels in the SR in the vicinity of the myofibrils (Fig. 4-1). The myoplasmic $[Ca^{2+}]$ increases from 10^{-7} mol l^{-1} to 10^{-5}. This Ca^{2+} diffuses to the adjacent myofibrils, where they bind strongly to troponin C. This enables crossbridges to work as long as the high $[Ca^{2+}]$ remains. A continually active Ca^{2+}-pump returns Ca^{2+} to the SR (Fig. 4-1). Then the thin filament is **off duty**, because Ca^{2+} is withdrawn from its troponin C, and **relaxation** ensues.

How is the force of contraction graded?

In a muscle the force of contraction is graded by **increasing the frequency of APs**, and by **recruiting more muscle cells**. Prolonged crossbridge contraction results in **tetanus**.

Fast twitch muscle fibres are recruited only during maximal efforts such as sprint. The production of ATP by glycolysis matches the high rate of ATP consumption in these fast twitch fibres. During light work the smaller motor units are recruited first, and these highly excitable units activate slow, oxidative cells suited for prolonged activity.

Endurance training increases the oxidative capacity of the activated motor units, whereas strength training increases cellular hypertrophy.

What are the energy sources for contractility?

We have three major metabolic sources of ATP:

1. **Phosphocreatine**, which is an immediate energy source used for intense activity such as sprinting. Lohmann's creatine kinase catalyses the efficient reforming of ATP from ADP by the conversion of a small phosphocreatine pool to creatine. Following exercise the **oxygen debt** is repaid and the phosphocreatine pool is restored.
2. **The glycogen stores** of the muscle produce ATP rapidly but inefficiently by **glycolysis**, with lactate as the end product.
3. **Glucose, free fatty acids, triglycerides and amino acids in plasma** are substrates for oxidative phosphorylation. This is a most efficient pathway and the slowest source of energy due to the many steps in the process.

See also Fig. 42-3 and Chapter 42 for mechanical efficiency, etc.

Further reading

Epstein, H.F. and D.A. Fischman (1991) Molecular analysis of protein assembly in muscle development. *Science* **251**: 1039.

Huang, C.L.H. (1988) Intermembrane charge movements in skeletal muscle. *Physiol. Rev.* **68**: 1197.

Pollack, G.H. (1990) Muscles and molecules: Uncovering the principles of biological motion. Ebner, Seattle, WA.

6. Case History

I. An athlete is exercising maximally with all his muscles for 5 min. His total muscle mass is assumed to represent 40% of the body weight 70 kg. In the muscle tissues the activity hydrolyses 20 mmol ATP kg^{-1} s^{-1}, and this is assumed to be the essential energy involved. The enthalpy of the ATP reaction is 42 kJ mol^{-1}. The specific heat content of the body is 3.47 kJ kg^{-1} $°C^{-1}$.

1. *Calculate the heat energy liberated from the ATP hydrolysis.*
2. *Calculate the theoretical rise in body temperature in 1 min, if heat dissipation does not occur.*

II. Following a marathon, the athlete delivers a muscle biopsy from the left thigh. The biopsy weighs 1234 mg of which 954 mg is water and 35 μmol of Na^+. The extracellular phase of the biopsy is determined to 193 mg, and the content of Na^+ here is 19 μmol. The athlete eliminates 0.2 mol of Na^+ per day, and his body content of exchangeable Na^+ is 3 mol.

3. *Calculate the molality of Na^+ in the muscle cells.*
4. *Calculate the half-life of NaCl in the body of this person.*
5. *What assumptions are implied in calculation 2?*

Try to solve the problems before looking up the answers in Chapter 74.

CHAPTER 7. SMOOTH MUSCLES

Contraction in smooth muscle is essentially caused by the same molecules as in striated muscle, but the intracellular organization and the dynamic characteristics are entirely different.

What characterizes the smooth muscle fibres?

1. Smooth muscles are called so because they **lack the sarcomeric bands of striated muscles**.
2. Smooth muscle cells are spindle-shaped and line the hollow organs and the vascular system; the smooth muscle cells are extremely small (diameter up to 5 μm; length up to 200 μm) compared to striated muscle cells (diameter up to 100 μm; length up to 200 mm), so they do not need T-tubules.
3. Smooth muscle cells contain a few thick myosin-filaments, and many thin actin-filaments attached to **dense bodies** by α-actin (helical sarcomers). The cells are without **regular** sarcomers, Z disc's, myofibrils and T-tubules. Smooth muscle cells lack troponin. Dense bodies are analogous to Z disc's, and some dense areas are attached to the cell membrane.
4. Smooth muscle cells do not contain a typical endoplasmic reticulum that can store and release Ca^{2+}. Instead some fibres possess an analogous **reticular system** located near the **caveoli** of the cell membrane. Caveoli are small invaginations of the membrane, similar to the T-tubules of striated muscles. The more extensive the reticular system is in the smooth muscle fibre, the higher is its shortening velocity.
5. Smooth muscle cells are frequently involved targets in diseases such as hypertension, stroke, asthma, and many gastrointestinal diseases.
6. Smooth muscle cells maintain large forces almost continually at extremely low energy costs. The same tension is maintained for days in smooth muscle organs (intestine, urinary bladder, gall bladder) and can be obtained in striated muscle at high energy cost (up to 300 times the smooth muscle rate of ATP consumption).
7. The resting membrane potential in smooth muscle cells is -50 mV in contrast to that of striated muscle cells (-80 mV).
8. Electrical activity is spread through hundreds of **gap junctions** between neighbour cells belonging to the **single unit smooth muscle tissues**.
9. Thousands of smooth muscle cells belonging to the multi-unit type join in a **functional syncytium**.
10. Smooth muscle cells are **extremely sensitive to extracellular [Ca^{2+}]**.

How do smooth muscle cells communicate?

Smooth muscle cells can be divided into multi-unit smooth muscle and single-unit smooth muscle.

In **multi-unit smooth muscle tissues** each cell operates entirely independent of other cells and the cell does not communicate with other muscle cells through **junctions**. The discrete cells are separated by a thin basement membrane and often innervated by a single neuron, and their main control is through nerve signals. Multi-unit smooth muscle is found in the eye (the ciliary muscle and sphincters such as the iris muscle of the eye), in large arteries, in the vas deferens, and in the piloerector muscles that cause erection of the hairs.

Single-unit smooth muscle tissues are arranged in bundles. These smooth muscle cells communicate through **gap junctions**, separating the cell membranes by only 2 nm. APs generated in one cell can activate adjacent cells by ionic currents spreading rapidly over the whole organ through the many gap junctions. **Visceral smooth muscle**, undergoing peristalsis, generates propagating APs from cell to cell. Other cell-to-cell contacts are **desmosomes** and **intermediary junctions** subserving structural contact. Mechanical force is transferred from one smooth muscle cell to another by these **intermediary junctions** on the plasma membrane, causing the smooth muscle cell to function like a **stretch transducer**. During an AP the inward flux of ions is not Na^{+}, but Ca^{2+} through slow Ca^{2+}-channels. They open mainly in response to a ligand-binding, although they are also voltage-dependent.

Do smooth muscles contain neuromuscular junctions?

No. The nerve fibre branches on a bundle of smooth muscle fibres, and forms a **diffuse junction**. The neurotransmitters are acetylcholine and noradrenaline. Multi-unit smooth muscles have developed a **contact junction** with shorter latency than the diffuse junctions mainly found in the single-unit type.

What is the force–length relation for smooth muscle fibres?

The force–length relation is qualitative similar to that of striated muscles, so the **sliding-filament mechanism** is probably analogous (Fig. 6-1).

Describe the crossbridge regulation

The smooth muscle mechanism is special, because stimulation results in a maintained isometric force with strongly reduced velocities. The **force–length diagram** for smooth and skeletal muscles is similar (Fig. 6-1). Smooth muscle contractions are extremely slow. The number of active crossbridges in smooth muscle is probably regulated slowly and indirectly by Ca^{2+}.

What characterizes the smooth muscle metabolism?

Smooth muscle cells contain some mitochondria, and they show a slow contraction pattern. Smooth muscle contractions typically last for 3 s, in contrast to striated muscle with total contraction periods of 10–100 ms. Since the energy demand in smooth muscle is extremely low, it is balanced by the **oxidative ATP synthesis**. Smooth muscle cells do not have an oxygen debt as striated muscles, although they produce **large amounts of lactate**. This is probably because the ATP-synthesizing glycolytic mechanism is located in the cell membrane and is linked in an energy balance to the ATP-utilizing **Na^{+}-K^{+}-pump**.

Why is smooth muscle contraction so slow?

In a **velocity–stress diagram** as the one for striated muscle (Fig. 6-2 right), the velocities for smooth muscle are all located close to the abscissae for any stress or load.

1. Smooth muscle contains far fewer myosin filaments than striated muscle.
2. The myosin crossbridge heads of smooth muscle contain an isoenzyme with much less ATPase activity than that of striated muscle.
3. Ca^{2+} entry through the cell membrane is much slower than internal release of Ca^{2+}.

How is myoplasmic [Ca^{2+}] regulated?

Stimulation of a smooth muscle fibre to contract, releases Ca^{2+} from two pools. The **large extracellular fluid pool** is essential, and in the fibres which possess a **reticulum** similar to the endoplasmic reticulum of striated muscle there is a fast intracellular pool (Fig. 4-1).

The intracellular release of Ca^{2+} is caused by the second messenger, IP_3. This substance is produced when a stimulus acts on reticular receptors via a G-protein to activate phospholipase C (Fig. 5-2). **Phospholipase C** hydrolyses **phosphatidyl inositol diphosphate (PIP_2)** into IP_3 and DAG (Fig. 5-2). These second messengers elicit a controlled release of Ca^{2+} from the reticulum. With this the myoplasmic [Ca^{2+}] rapidly increases. The crossbridge cycling is regulated by a **Ca^{2+}-dependent myosin kinase.**

The smooth muscle cell membrane contains a **$3Na^+$-$2K^+$-pump, a delayed K^+-channel, a ligand-activated and a voltage-dependent Ca^{2+}-channel, a sarcolemmal Ca^{2+}-pump, and a Na^+-Ca^{2+}-exchanger**. When the high intracellular [Ca^{2+}] during an AP is lowered again towards the resting level, the cell relaxes. This is accomplished by stimulation of the **sarcolemmal Ca^{2+}-pump**, and by blockade of Ca^{2+}-input and Ca^{2+}-release.

What is endothelial-derived relaxing factor (EDRF)?

EDRF is recently shown to be **nitric oxide (NO)**. Activation of endothelial cells produces NO from arginine, and NO diffuses into the smooth muscle cells. NO stimulates directly the enzyme **guanylatecyclase**, and by that intracellular [cGMP] elevates (Fig. 5-3). The action on smooth muscle cells is described in Chapter 20.

How does nitroglycerine cause vasodilatation?

Nitroglycerine, nitroprusside and similar drugs relax smooth muscles by transfer of NO from endothelial cells. NO increases **intracellular [cGMP]** (Fig. 5-3) and is the basis for the beneficial effect of the drugs on cardiac cramps (Chapter 19). These second messengers activate protein kinases that phosphorylate effector proteins such as Ca^{2+}-pumps and K^+-channels. Such vasodilatators **stimulate the sarcolemmal Ca^{2+}-pump,** inhibit Ca^{2+}-influx and stimulate K^+-efflux through the delayed K^+-channel (reduces the excitability). Hereby, the high intracellular [Ca^{2+}] during an AP is lowered towards the resting level (10^{-7} mol l^{-1}), and the cell relaxes producing vasodilatation.

Is it possible to change the calibre of vessels without nerve supply?

Metarterioles and precapillary sphincters without nerve fibres, can still respond to the needs of the tissue by the action of local tissue vasodilatators. The following factors cause smooth muscle relaxation, and therefore vasodilatation: adenosine, NO, lack of oxygen, excess CO_2, increased $[H^+]$, increased $[K^+]$, diminished $[Ca^{2+}]$, and increased [lactate].

Junctional regions are found between the membranes of smooth muscle cells and the endothelial cells of their capillaries. ACh in the blood acts on **endothelial receptors**, and produces smooth muscle relaxation. NO mediates the vasodilatation (Fig. 5-3). Circulating ACh **contracts** the arterial smooth muscles when bound to **cholinergic receptors**.

The **metabolic theory** for the control of local bloodflow is described in Chapter 20.

How does smooth muscle grow?

Smooth muscle cells grow (hypertrophy) as a response to the needs of the body, and they also retain the capacity to divide.

During **hypertension** the lamina media of the arterioles hypertrophies which increases the **total peripheral vascular resistance (TPVP)** in the systemic circulation. These topics are further developed in Chapter 20.

During **pregnancy** the smooth muscles of the myometrium are quiescent and contain few gap junctions under the influence of progesterone. At term the myometrium grows and the number of **gap junctions increases**, due to the high oestrogen concentration. Now the myometrium is well prepared for the co-ordinated contractions during **parturition** (see Chapter 69).

What is meant by stress-relaxation of smooth muscle?

Smooth muscle changes length without marked changes in tension. Initially, there is a high tension developed upon stretching; then the tension falls as the myosin and actin filaments are reorganized by slowly sliding against each other. A sudden expansion of the venous system with blood results in a sharp rise in pressure followed by a fall in pressure over minutes. The smooth muscle fibres in the walls of the venous system are highly compliant, because they have accepted a large blood volume without much rise in pressure (**delayed compliance**).

Further reading

Paul, R.J. (1989) Smooth muscle energetics. *Ann. Rev. Physiol.* **51**: 331.
Motta, P.M. (Ed.) (1990) *Ultrastructure of Smooth Muscle*. Kluwer, Lancaster.
Ruegg, J.C. (1986) *Calcium in Smooth Muscle Activation*. Springer-Verlag, Berlin.
Sperelakis, N. and J.D. Wood (1990) Frontiers in smooth muscle research. *Prog. in Clin. Biol. Res.* **327**. Wiley-Liss, New York.

7. Case History

A 23-year-old female has arterioles with an average length of 4 mm, and an average diameter of 20 μm, while she is relaxing in the sun. The total number of arterioles in her body is 10^8.

1. Calculate the total blood volume located in all arterioles.

At another event the female is stressed, and she contracts the smooth muscle cells of all arterioles to 50% of the original diameter.

2. Calculate the change in the total blood volume located in all arterioles.

3. Is this change consequential to her arterial blood pressure?

Try to solve the problems before looking up the answers in Chapter 74.

SECTION II.
The Nervous System

The **nervous system** is the essential control system for the human body. It consists of a **central nervous system (CNS)** and a **peripheral nervous system**. All information originates in sensory receptors and, enters the CNS via the peripheral nervous system. The CNS controls all the activities of the human body ranging from contractions of striated and smooth muscles to the exocrine and endocrine secretions.

CHAPTER 8. GENERAL NEUROPHYSIOLOGY

What are the main functions of the nervous system?

The **nervous system** is an extremely rapid signal transduction system, and is the most important communication network in our body. The **main integrative functions** allow selection and processing of incoming signals to produce an **appropriate response**.

The **nervous system** includes sensory receptors that detect events in the body as well as in the outer world. Signals or action potentials (APs) from the sense organs travel through peripheral, afferent nerves to the CNS, where they are processed. The CNS controls the various activities of the body by motor mechanisms that generate movements and glandular secretions through efferent nerves. The afferent and efferent nerve fibres distributed all over the body form the **peripheral nervous system**, which is subdivided into the **somatic** and the **autonomic** part.

What characterizes neurons and neuroglia?

Neurons are very specialized cells that are **excitatory** and **neurosecretory**. Neurons **receive and transmit signals (APs)** to other neurons or effectors. Neuronal networks account for information in a **memory**, evaluation of available knowledge, decision making, and transmission of response signals to appropriate effectors. The human nervous system contains about 10^{12} neurons forming at least 10^{15} synapses. The typical neuron consists of a **cell body (soma), dendrites** which have multiple synapses, and an axon. The **axon** of the neuron connects to other neurons or to effector cells. The soma contains a nucleus, a nucleolus, a prominent Golgi-apparatus, mitochondria, microfilaments and microtubules. Stacks of rough endoplasmic reticulum (the **Nissl bodies**) are dispersed in the soma and in the dendrites.

Neuroglia are supportive cells that sheath and protect neurons. Myelinated axons propagate APs up to 50 times faster than unmyelinated with the same diameter. They also eliminate neurotransmitters more rapidly from the synapse. The neuroglia constitute half of the brain volume, and we have about 10^{12} to 10^{13} glial cells. One of the functions of the glial cells is to **buffer $[K^+]$**, that is the neuroglia tries to eliminate any rise in extracellular $[K^+]$ beyond the normal. Secondly, the foot processes of the astrocytes help to close the **blood–brain barrier** of the brain capillaries. The brain capillaries are lined by an endothelial layer, the basement membrane, and the astrocytes.

Is nervous tissue dependent upon the oxygen supply?

Nervous tissues are extremely dependent upon the oxygen supply from the arterial blood as the nerve cells use oxygen continually, and their oxygen stores are minimal. Retina is a typical nervous tissue; following abrupt circulatory arrest its oxygen store is used within 5 s. The rapid oxygen depletion is easily demonstrated by arresting retinal circulation with pressure on the eye ball through the upper lid. This leads to **black out** (temporary blindness) in 5 s. Similar anoxic events take place in the grey matter (cortex), which lead to loss of consciousness – termed **grey out**.

What is a sensory receptor?

Sensory receptors are either neurons in the case of vision, smell and cutaneous senses, or modified epithelial cells in the case of auditory, vestibular and taste senses.

The stimulus elicits a receptor potential (**generator potential**), which is graded according to the stimulus strength. When the stimulus is strong enough to reach the threshold, an AP is fired. In neurons the stimulus intensity is coded by the frequency of APs.

Sensory receptor systems are **biological transducers with a dynamic range of up to** 10^{12} in the most sensitive organ: the ear. The **threshold** is the reciprocal of the sensitivity. The **threshold** is the weakest stimulus to which the receptor will react. The **sensitivity of a sensory receptor** is greater the smaller its threshold stimulus is.

Further information on sensory receptors in the nervous system is given in Chapter 9.

The special sensory receptors for vision are described in Chapter 13, and those for hearing and balance in Chapter 14.

Some sensory receptors have characteristics similar to the well-known **plasma membrane receptors**. Plasma membrane receptors consists of a protein molecule, a channel pore and a specific enzyme (Chapters 3 and 4).

What is understood by exitatory amino acid receptors?

EAA receptors are defined in Chapter 5.

How does a person judge the strength of a stimulus?

Stevens proposed the **power law**, since most physiologic mechanisms are non-linear. The **interpreted stimulus strength (ISS) is equal to a constant (K) multiplied by the actual stimulus strength (SS) raised to the power *n*:**

$$\mathbf{ISS = K\ SS^{n}}.$$

In his original version, only the exponent ***n*** differed for each type of sensation. The equation can be modified by subtracting different constants from SS before raising it to the power n, or by changing the value of K.

Perception of **taste, heat and angular acceleration** are described by **power functions** or **transfer functions with $n>1$**, whereas hearing and smell are described by functions with $n<1$. The only sensation with a particularly large value of n (about 3) is **pain**.

$\mathbf{n = 1}$ describes a linear relation between stimulus and resulting activity in the conducting neuron.

Note, however, that the power law and curve fitting with transfer functions have not improved our understanding of sensory modalities.

Describe the different nerve fibre types and their function

Both a general and a specific classification are frequently used by neurophysiologists. The fibres are divided into types A, B and C in the general classification based on the three main conduction velocities shown in the record of the combined AP from a nerve. Type A fibres are the fast conducting myelinated fibres (subdivided into α, β, γ, and δ), and type C are the small, unmyelinated fibres. Another classification is based on the thickness of the axons (I–IV). The special classification became necessary, when Aα fibres were separated into two subgroups: Ia and Ib. In the following paragraph the general classification is written first, and the specific classification is given in parentheses.

1. The **Aα fibres (motor α fibres and sensory Ia/Ib)** are **motor α-fibres** and proprioceptors from the **annulospiral endings** of muscle spindles (Ia) and from **Golgi tendon organs** (Ib). These fibres are myelinated and have a diameter of about 9–18 μm in humans. They lead signals with a velocity of 50–100 m s^{-1}.
2. The **Aβ fibres (II)** conduct **discrete touch and fine pressure signals** from cutaneous tactile receptors. These fibres have a diameter of about 5–12 μm, and a conduction velocity of 30–70 m s^{-1}.
3. The **Aγ efferents (II)** to the muscle spindles have their origin in the spinal cord, the diameter is about 3–6 μm and the conduction velocity in these fibres is 18–36 m s^{-1}.
4. The **Aδ fibres (III)** transfer pain sensations and temperature as well as crude, passive touch and deep pressure. The diameter of the fibres is about 1–5 μm, corresponding to a conduction velocity of 4–20 m s^{-1} in humans.
5. The **B fibres (III) are autonomic preganglionic** with a diameter less than 3 μm and a conduction velocity of 3–12 m s^{-1}.
6. The **C fibres (IV)** are unmyelinated and lead **pain, touch and signals from heat receptors** from the skin. The C fibres have no myelin sheath, a diameter of about 0.5–1 μm, and a conduction velocity of 1–2 m s^{-1}.

In what ways can we code information in the nervous system?

The visual system is an example of a **specific information line** for a certain modality of sensation. The neurons in the retina, the lateral geniculate nucleus, and the visual cortex describe just such a dedicated neuronal pathway. The way in which a suprathreshold stimulus is perceived is determined by the specific information line through which the signal is conducted, e.g. pressure applied on the eye will be perceived as light.

The auditory system also forms a **specific or labelled line** all the way from receptor to cortex. In all cases the modality of sensation is determined by the specific point in the cerebral cortex at which the nerve fibres end.

Now, where is the sense interpretation localized and what is its intensity?

1. **Coding in the sense organ** is **peripheral analysis**. External energy is transformed to a **receptor potential** that triggers APs in afferent nerve fibres. Peripheral analysis depends upon the special structure and sensitivity of the receptor, and upon its location. The pattern of firing of APs is the only possible variable for coding information in a single neuron. Examples of firing patterns are **on–off patterns** with mean frequencies, **off–on patterns, transient patterns** or **adaptation**, long-lasting patterns, firing with latency, etc.
2. **Central location coding in the CNS** is **central analysis**. A **neural map** is an array of neurons that receive signals from corresponding areas of the skin. These signals result in generation of motor signals that move specific muscles. The sensory–motor system, the signals from the skin to the cortical sensory homunculus, and signals from the motor homunculus make up **somatotopic maps**. Similar examples are the retinotopic map in the visual system, and the tonotopic map from the organ of Corti in the auditory system. The sense impression, its intensity and the response depend upon where the APs terminate in the brain.

What is the blood–brain and the blood-cerebrospinal fluid barrier?

The **blood–brain barrier** consists of **tight junctions** between the endothelial cells of the capillaries in the CNS. The tight junctions only allow extremely small molecules to pass into the brain tissue.

Many large molecules cannot pass from the blood to the CSF across the choroid plexus, a tight junction barrier that is called the **blood–CSF barrier**. The blood–brain and the blood–CSF barriers exist in all areas of the brain, except in the so-called **circumventricular organs** (hypothalamus, the pineal gland, and the area postrema). These discrete structures have highly fenestrated capillaries that can be easily penetrated by large and small molecules as well as ions. The circumventricular organs are located close to essential control centres regulating respiration, blood [glucose], and ECF osmolality.

The two barriers are almost impermeable to large molecules such as plasma proteins, but highly permeable to CO_2, oxygen, water, alcohol, anaesthetics, hallucinogens, and other lipophilic substances. The blood–brain barrier is almost completely impermeable to H^+, whereas CO_2 passes through the barrier (see Chapter 38).

How is the cerebrospinal fluid (CSF) produced and absorbed?

Most of the daily 500 ml CSF is produced in the **choroid plexuses** in the four brain ventricles, and the remaining is produced across the blood–brain barrier. The secretion of fluid by the choroid plexus depends on the active Na^+-transport across the cells into the CSF. The electrical gradient pulls along Cl^-, and both ions drag water by osmosis. The CSF has lower [K^+], [glucose], and [protein] than blood plasma, and higher concentrations of Na^+ and Cl^-. The production of CSF in the choroid plexuses is an active secretory process, and not directly dependent on the arterial blood pressure. The CSF is separated from the brain cells by the extremely thin **pia mater**. All natural substances that enter the CSF can easily diffuse into the brain interstitial fluid.

CSF leaves the ventricles through the roof of the 4th ventricle, traverses the subarachnoid space, and is reabsorbed into the blood of the venous sinuses via the arachnoidal villi. The **absorption** here is directly related to the pressure in the cranial cavity (CSF). Large holes through the endothelial cells allow proteins to enter the blood.

Can a neuron be replaced?

Severe injury to nervous tissue causes cellular death. Neurons are postmitotic cells. This means that once a neuron is lost, it cannot be replaced.

However, regeneration of axons in the peripheral nervous system is more likely than in the CNS. Both growth and maintenance of axons require the **nerve growth factor (NGF)**. NGF is an essential survival factor for neurons outside the CNS – in particular sensory neurons. NFG binds to receptors belonging to the **insulin receptor family (tyrosine kinase family)** – see Chapter 5.

How can an axon be replaced?

When a motor axon has been severed, the cell body undergoes **chromatolysis**. This is a neuronal reaction, where the **rough endoplasmic reticulum (the Nissl bodies)** becomes active. The Nissl bodies accumulate proteins required for repair of the axon. The **axonal reaction** is an attempt to repair the fibre by production of new protein structures.

Therefore, the rough endoplasmic reticulum becomes distended by proteins. The axon and the myelin sheath distal to the injury die and are phagocytosed. The neuroglial Schwann cells that had formed the myelin remain alive. This is the so-called **wallerian degeneration** named after Waller. The Schwann cells proliferate and form long rows along the pathway previously occupied by the dead axon. The severed axon regenerates along this pathway, and **growth cones** may eventually reinnervate the target organ.

What is axonal transport?

Fast axonal transport of organelles in the cytosol occurs as rapidly as 0.25 m per day. At this rate synaptic vesicles can travel along the motor axon from the spinal cord to a patient's foot within 3 days.

Slow axonal transport occurs as diffusion of cytosolic proteins. This occurs at a rate 100 times more slowly than fast axonal transport.

Axonal transport can be **anterograde**, when it occurs in the direction from the soma to the axonal terminals. Axonal transport can also be **retrograde**, when it occurs in the opposite direction. Here vesicles are degraded by lysosomes, when returned to the soma.

Compare and contrast the organization of the neuromuscular junction to the CNS

1. A single AP is insufficient to trigger an AP in the CNS. **Summation of a large number of APs is required to generate an AP in the CNS.** In the neuromuscular junction, **each signal** is more than sufficient to trigger muscular contraction.
2. In the CNS, the neurotransmission can be **both inhibitory and excitatory**. In the neuromuscular junction, the neurotransmission is always excitatory.
3. In the neuromuscular junction, acetylcholine is the only neurotransmitter, whereas in the CNS there is a **large variety of neurotransmitters**. Gammaaminobutyric acid (GABA) is the main inhibitory transmitter in the CNS, whereas glycine and glutamine are the main excitatory neurotransmitters found in the spinal cord and brain, respectively. The main categories of neurotransmitters in the CNS are: amino acids, amines, peptides and acetylcholine (see Chapter 4).

Further reading

Berne, R.M. and M.N. Levy (1993) *Physiology*, 3rd edn. Mosby Year Book, St Louis.

Paxinos, G. (Ed.) (1990) *The Human Nervous System.* Academic Press, San Diego.

Shephard, G.M. (Ed.) (1990) *The Synaptic Organization of the Brain*, 3rd edn. Oxford University Press, New York.

8. Multiple Choice Questions

Each of the following statements have True/False options:

A. The Nissl bodies are stacks of rough endoplasmic reticulum.

B. Taste, heat and angular acceleration follow transfer functions, so the interpreted stimulus strength decreases with the rise in actual stimulus strength.

C. Sensory receptor systems are biological transducers with a dynamic range up to 10^{12}.

D. The B fibres are autonomic preganglionic axons with a diameter less than 3 μm and a conduction velocity of 3–12 m s^{-1}.

E. The CSF has higher [K^+], [glucose], and [protein] than blood plasma, and lower concentrations of Na^+ and Cl^-.

Try to solve the problems before looking up the answers in Chapter 74.

CHAPTER 9.
THE SOMATOSENSORY SYSTEM

This system transmits signals from **sensory nerve receptors in the somatic system**. The nerve receptors are located in the skin, muscles, tendons, joints and viscera. The signals are transferred to the CNS by a pathway of first-, second-, third-, and higher-order neurons. The third- and higher-order neurons are located in the **thalamus and the cortex**. The cell body of the first-order afferent neuron is located in the dorsal root or in the cranial nerve ganglia. The signals pass through the spinal cord, the brainstem, and the thalamus before reaching the cerebral cortex.

Define adequate stimulus

The adequate stimulus is the stimulus for which the receptor has a lower energy threshold than for any other stimuli, i.e. the stimulus to which the receptor is most sensitive. The **adequate stimulus** of the receptors is mechanical deformation of the receptor or change of its temperature. The **sense impression** depends on the site in the brain which receives the sensory signal (i.e. **central analysis**) and on the receptor localization (i.e. **peripheral analysis**). This is how different neurons transmit different types of sensations, even though they may transmit the same electrical signals (see Chapter 8).

More than 99% of all incoming signals are discarded by the CNS as irrelevant.

What is the function of the somatosensory thalamus?

1. Several sensory tracts and pathways synapse in the **nuclei of the thalamus**, e.g. the **spinothalamic tract**. This is the **somatosensory thalamus**, which is a relay station for most sensory modalities. The sensory inputs are processed in excitatory and inhibitory somatotopic areas of the thalamus, and are then transferred to appropriate cortical areas. The **somatotopic organization** (see Chapter 8) is maintained all the way to the cortex.
2. The **reticular activating system (RAS)** of the brainstem is involved in arousal acting in concert with the thalamus (see Chapter 10).

How are sensory receptors classified?

Sensory receptors in the nervous system are classified as **exteroceptors** (located on the body surface), **proprioceptors** (located in muscles, tendons and joint capsules), **interoceptors** (located in the viscera), and **telereceptors** (stimulated by events far from the person).

Cutaneous receptors are fast and slowly adapting exteroceptors. The Pacinian corpuscles are fast adapting exteroreceptors, located in the cutis, in the deeper cutaneous layers, and in deeper tissues (see below). The Paccinian corpuscles are **vibration detectors (150–300 Hz)**, with a high temporal resolution. Merkel and Ruffini corpuscles are static exteroreceptors. The adequate stimulus for the cutaneous receptor is deformation of the receptor.

Thermoreceptors are also exteroreceptors. We have cold receptors just below the skin surface (200 nm deep). Cold receptors respond to changes in temperature. Heat receptors are also located in the skin. The location of certain heat and cold points in the skin was determined by bringing a thin hot or cold object in contact with the skin.

Proprioreceptors are also mechanoreceptors (muscle spindles, Golgi-receptors, Pacinian and Ruffini corpuscles, and free nerve endings). The Ruffini mechanoreceptors are also called **joint receptors**, because they are located in ligaments, tendons and **articular capsules**. They provide information for the CNS concerning articular movements, movement velocity and joint position. Joint receptors of the proximal joints are particularly sensitive. These receptors enable us to sense the position of the joint with great accuracy.

How do the cutaneous mechanoreceptors act?

Cutaneous mechanoreceptors include **hair follicle receptors, Meissner corpuscles, and deeply located Pacinian corpuscles** – all of these are rapidly adapting. **Merkel's cell endings and the Ruffini corpuscles** are, on the other hand, slowly adapting. The adequate stimulus of all these receptors is deformation of the receptor (stroking the skin, pressure). The nerves carrying signals from these receptors have **myelinated axons** ($A\beta$ fibres for active touch), but the **C fibre mechanoreceptors**, which register passive touch, such as slow strokes with a piece of cotton, are unmyelinated.

What is special about thermoreceptors?

Thermoreceptors react to temperature changes. **Cold receptors** and **heat receptors** are located close to the surface. Both types of receptors are located in the skin, in the deep tissue and in the CNS. Both types of receptors **discharge spontaneously** at normal temperature, **dynamically** when skin temperature is changing rapidly, and **adapt slowly**.

What characterizes stretch receptors in muscles and joints?

Stretch receptors are rapidly adapting mechanoreceptors located in muscle spindles and in Golgi's tendon organs. Muscle spindles signal the **length** of the muscle, whereas Golgi tendon organs are concerned with the **force** of the contraction. Ligaments and joint capsules contain **Golgi tendon organs, Paccinian and Ruffini corpuscles, and free nerve endings**. The static and dynamic receptors inform the CNS about the position and movement of the joint, respectively.

What are nociceptors?

Nociceptors or nocireceptors are responsive to stimuli that could potentially cause injury or pain. Nociceptors are probably free nerve endings of mainly two types: **the fast adapting mechanoreceptors,** consisting of finely myelinated afferent $A\delta$ fibres (III), and the **slowly adapting C-polymodal nociceptors,** consisting of unmyelinated afferent fibres that end in the superficial layers of the skin (see Chapter 8). The fast type of receptor responds to mechanical stimuli, whereas the second type responds to various chemical, thermal and mechanical stimuli.

When nociceptors become sensitized, their thresholds are reduced, thus causing **hyperalgesia** (hypersensitivity to pain). Nociceptors are **sensitized** by bradykinin, histamine, leucotrienes, prostaglandins, serotonin and K^+ which are often released near damaged nociceptors. K^+ **activates** the nociceptors. Substance P is also released through an axon reflex from nociceptor terminals, causing vasodilatation and increased capillary permeability. Glutamate may be co-released with **substance P** from polymodal C-fibre terminals involved in pain perception.

What is the most important sensory pathway for pain and temperature?

The **spinothalamic tract** conveys pain and temperature (lateral tract), and also crude passive touch (ventral tract). The first-order neurons are **afferent Aδ fibres (III)** which have cell bodies in the spinal ganglia. Second-order neurons cross immediately to the opposite side of the spinal cord, and ascend in the lateral and ventral spinal tract conveying touch.

Pain and temperature reach the thalamus in the lateral spinothalamic tract (in the lateral funiculus). The third-order neurons pass from the **somatosensory thalamus** via the **thalamocortical fasciculus** to the **somatosensory cortex** or the **primary sensory cortex** (somatic sensory area I, or **area 1, 2, 3** in Fig. 11-1) with the **sensory homunculus**. Some third-order neurons also pass to the somatic sensory area II of both hemispheres.

The gyrus cinguli (cingulate gyrus in Fig. 11-2) has the highest density of **central opiate receptors**. Pyramidal cells also have **opiate secreting interneurons** that inhibit the pain signals arriving at the gyrus cinguli.

What is the most important sensory pathway for proprioception and fine tactile sensations?

Proprioception and active tactile signals are transmitted through sensory nerve fibres to the spinal cord. Primary afferent fibres ascend in the **dorsal columns** all the way to the medulla oblongata. These primary axons synapses with second-order neurons in the **gracile and the cuneate nuclei**. These second-order neurons cross the midline in the medulla, and ascend in the medial lemniscus to end in the **somatosensory thalamus**. The **medial lemniscus pathway** transmits proprioception and fine tactile senses.

How do amino acids and peptides act as neurotransmitters?

1. The major excitatory transmitters are the **amino acids** glutamate and aspartate (EAA in Chapter 4). The major inhibitory transmitters are GABA and glycine. ACh, Ad, NA, dopamine and serotonin serve as fast neurotransmitters in the CNS.

   The **amino acids** and **monoamines** are water soluble, and bind to membrane receptors that regulate ion channels or release second messengers into the cytosol.
2. **Neuroactive peptides** (angiotensin, cholecystokinin, opiates, oxytoxin, substance P, TRH, VIP and vasopressin) have slow excitatory or inhibitory transmitter actions. Peptides are water soluble, and act through second messengers (Chapters 5 and 63).

How can glutamate antagonists and NSAIDs act as analgesics?

Glutamate is the main excitatory transmitter in the CNS, whereas **GABA** is the dominant inhibitory transmitter.

The **glutamate receptors** are also channel proteins. Activation of glutamine receptors increases Na^+ conductance which depolarizes the postsynaptic membrane. Hence, glutamate antagonists can reduce the sensation of pain.

Acetylsalicylic acid (ASA) and **prostaglandin-inhibitors** are called **non-steroid anti-inflammatory drugs (NSAIDs)**. Such drugs inhibit the synthesis of arachidonic acid derivatives (prostaglandins and leucotrienes) in peripheral nerves. Arachidonic acid derivatives and bradykinins cause pain by acting on nociceptors.

Describe the different types of central opiate receptors

Opiate substances have been found in the nervous system. They are derived from three protein molecules: pro-opiomelanocortin (POMC), proencephalin, and prodynorphin. Analgesic areas of the CNS form a **pain control system** consisting of three components: the **periaqueductal area**, the **raphe magnus nucleus** and the **dorsal horn of the spinal cord.**

We have at least three types of **central opiate receptors:** μ **for morphine**, and δ **and K for encephalin and dynorphin**, respectively. Opiate receptors are especially concentrated in the analgesia areas of the CNS. Met-encephalin and leu-encephalin are found in the **pain control system**. Encephalin inhibits both type C and type Aδ pain fibres presynaptically in the dorsal horns. **Encephalin** is the endogenous ligand for the δ opiate receptors. Dynorphin is much stronger than morphine and is only found in small quantities close to the **K** opiate receptors. β-Endorphin is present in the hypothalamo-hypophysary system. There is a 4th receptor known as the σ receptors. These receptors are non-selective opiate receptors and are also the site of psychotomimetic drugs, such as nalorphine, pentazocine and cyclazocine. It is believed that the σ receptors are associated with glutamate-activated ion channels, and may account for the dysphoria produced by some opiates. These receptors are also responsible for the dilatation of the pupil caused by some opiates.

Presynaptic transmission of opiates inhibits depolarization of nerve cell membranes, by increasing K^+ conductance – see also Chapter 64.

What is hyperalgesia?

Hyperalgesia means **hypersensitivity to pain**. Hyperalgesia is caused by either hypersensitive pain receptors (sunburned skin), or by facilitated transmission. Facilitated transmission is due to abnormal stimulation of peripheral nerve fibres, and neurons of the spinal cord or of the thalamus.

A special type of hyperalgesia is present when **herpes virus** infects one or more dorsal root ganglia. The virus excites the neurons and causes pain in the dermatomal segment subserving the ganglion. The segmental pain circles halfway around the truncus on the affected side. The virus is also transported by axonal flow to the cutaneous terminals, where it causes a characteristic rash of the dermatome. The disease is called **herpes zoster**.

What is causalgia?

Causalgia is **hyperalgesia and heat–cold sensations** accompanied by sweat secretion in a region with nerve lesion. The hyperactivity in sympathetic efferent neurons and in nociceptive afferents running along arteries is unexplained.

What is trigeminal neuralgia?

Even the lightest touch at sensitized **trigger areas** releases, within seconds, severe lancinating pain throughout the affected branch of the trigeminal nerve. The cause is unknown, and therapy is usually unsuccessful.

What is referred pain?

Referred pain is pain that originates from deep organs, is poorly localized and often referred to as coming from superficial structures. This may be explained by the fact that pain signals from viscera are transmitted through some of the neurons that transmit pain signals from a specific area of the skin. Pain due to **myocardial ischaemia** (angina pectoris) is commonly described as pain in the inner side of the left arm, and termed **referred pain.**

What is central pain?

Central pain is a sensation of pain in the absence of peripheral nociceptive stimuli. Central pain is believed to be processed in the **cortical pain areas**, and caused by lesions of the nociceptive pathways (peripheral nerves, the spino-thalamo-cortical tract and the thalamus).

What is headache?

Headache is pain on the surface of the head, which is actually due to anomalies in intracranial or extracranial structures.

Intracranial headache is released, when the dura or the tentorium is stretched or damaged, or when blood vessels (the venous sinuses and the meningeal arteries) are stretched or dilated. The pain caused by stimulating supratentorial nociceptors are felt in (referred to) the **frontal area** via the 5th cranial nerve as **frontal head ache**. Stimulation of subtentorial nociceptors causes pain in the **occipital region** through the 2nd cervical nerve (**occipital headache**).

Meningitis headache is accompanied by contraction of the neck muscles (stiff neck). The dura and the venous sinuses are inflamed, and the headache is severe. **Migraine headache** begins with prodromal nausea and vision disturbances, often occurring about 1 hour prior to the headache. The pain is located on one side of the head in classical cases. The mechanism of migraine is unknown. One hypothesis is that prolonged emotional stress in sensitive individuals causes **reflex vasoconstriction** of intra- and extracranial arteries. The brain ischaemia explains the prodromal phenomena, and leads to accumulation of vasodilatating substances such as adenosine, ADP, NO etc. The **vessels then dilate and pulsate**. The excessive stretching in the arterial walls is a possible cause of **migraine headache**.

There is evidence for genetic etiology of migraine: an autosomal dominant inheritance with reduced **penetrance** (expression only in permissive environments). **Extracranial headache** is common and caused either by spasms of the muscles of the head, or by inflamed mucous membranes of the sinuses. The pain is felt directly over the frontal or the maxillary sinuses in the case of sinusitis.

What is the thalamic syndrome?

The thalamic syndrome is frequently caused by **thrombotic blockade of bloodflow to the somatosensory thalamus**. The destruction of thalamic neurons in one hemisphere leads to **ataxia** and **loss of sensations** from the opposite side of the body. After a few months different types of sensations return, but they are accompanied by pain.

What is phantom limb pain?

Amputation of a limb is sometimes followed by **phantom limb pain**. The patient suffers from severe pains, and the sensation is projected to the amputated limb. It is not known whether the mechanism is central or peripheral.

Describe the Brown–Sequard syndrome

The Brown–Sequard syndrome or paresis includes all effects of **transection of only one half of the spinal cord at a certain level**. All motor functions on the side of the lesion are blocked in the segments below the level (paresis, spasticity, and loss of vasoconstrictor tone). Sensations of pain and temperature from all lower dermatomes on the opposite side of the body are lost, because of transection of the contralateral spinothalamic tract. The only sensation left on the side of transection is crude touch, because it is transmitted in the opposite ventral spinothalamic tract. The total sensory loss is therefore termed **dissociated anaesthesia.**

Taste and smell: how are taste sensations transmitted into the CNS?

The sensations from the anterior two-thirds of the tongue travel with the trigeminal nerve fibres, through the **chorda tympani** into the facial nerve (VIIth), and eventually reach the **solitary tract** of the **brainstem**. Taste signals from the back of the tongue and surrounding tissues are transmitted through the glossopharyngeal nerve (IXth) into the tractus solitarius. All taste fibres synapse in the **nuclei of the solitary tract** and the axons of these neurons project to the **thalamus**. From the thalamus third-order neurons reach the lower part of the **primary sensory cortex** in the postcentral gyrus (somatosensory area I = area 1 in Fig. 11-1).

Sourness is evoked by acids, because H^+ stimulates special **H^+-receptors** in the taste buds. **Saltiness** is produced by the anions of inorganic salts. The **Cl^--receptor** is particularly effective in registering saltiness. Our taste buds at the base of the tongue also have **bitter-receptors** stimulated by many long-chain organic compounds. Many alkaloids (quinine, caffeine, nicotine) also taste bitter. **Sweet-receptors** are stimulated by sucrose, glucose, lactose, maltose, glycerol, alcohols, aldehydes, ketones, and organic chemicals.

How are smell sensations transmitted into the CNS?

In the upper nasal cavity the mucous membrane is yellow and termed the **olfactory membrane**. It contains 100 million bipolar neurons called **olfactory cells**. They contain hairs or **olfactory cilia**. The olfactory cells are **smell receptors**. They work as telereceptors, and the smell pathways do not include the thalamic relay station and a neocortical projection area. Instead, olfactory signals end in the **limbic system** (Fig. 11-2), where olfactory information is correlated with feeding behaviour and emotional–motivational behaviour.

Further reading

Russell, M.B. and J. Olesen (1993) The genetics of migraine without aura and migraine with aura. *Cephalalgia* **13**(4): 245–8.

Udin, S.B. and J.W. Fawcett (1988) Formation of topographic maps. *Ann. Rev. Neurosci.* **11**: 289.

Klood, P.L. and S. Iyengar (1988) Central actions of opiates and opioid peptides. In: Pasternak, G.W. (ed.) The opiate receptors. Humana Press, Clifton, N.J.

9. Case History

A 32-year-old female is admitted to a neurosurgical ward with a discrete lesion in the spinal cord caused by a traffic accident. Her vital functions are unaffected. The most important signs are a complete lack of cutaneous temperature sensibility and pain sensibility in the left leg and the lower left side of the trunk below the umbilicus (the navel).

Where is the lesion in her spinal cord localized?

9. Multiple Choice Questions

Each of the following statements have True/False options:

A. The somatosensory thalamus is a relay station for most sensory modalities.

B. Glutamate is the main inhibitory transmitter in the CNS, whereas GABA is the dominant excitatory transmitter.

C. Arachidonic acid normally inactivates the pain nerve terminals.

D. Presynaptic transmission of opiates inhibits depolarization of nerve cell membranes, because the K^+ conductance is increased.

E. The adequate stimulus of the cutaneous mechanoreceptors is deformation of the receptor.

Try to solve the problems before looking up the answers in Chapter 74.

CHAPTER 10. CONSCIOUSNESS AND SLEEP

What is the reticular activating system and how does this system achieve arousal?

A large region of the reticular formation of the brainstem is termed the **reticular activating system (RAS)**. Stimulation of this system causes arousal and an **arousal reaction** in the electroencephalogram (EEG). **Arousal** is a high level of consciousness also called **alertness**, and the EEG arousal reaction is a high frequency–low voltage shift. Arousal is often accompanied by the **orientation reflex**, which is a fundamental change of behaviour following an external stimulus. The eyes, head and body are turned toward the external stimulus.

The **RAS** transmits facilitatory signals to the thalamus. The thalamus excites the cortex, and the cortex then excites the thalamus in a reverberating circuit. Such a positive feedback loop is what wakes us up in the morning. During the day the different activity levels are balanced by external stimuli and internal factors (inhibitory interneurons).

What is the source of circadian rhythm?

Circadian periodicities are changes in biological variables that occur daily. The circadian controller is the so-called **biological clock**, probably located in the **suprachiasmatic nucleus of the hypothalamus**. The biological clock receives many projections from sense organs including projections from the retina signalling light and darkness. These signals are transmitted further to the **pineal gland** according to one hypothesis. One possibility is that darkness stimulates melatonin secretion by the pineal gland, which inhibits the secretion of gonadotropic hormones from the anterior pituitary, and thus reduces sexual drive. Melatonin secretion may be related to ageing.

Destruction of the biological clock disrupts many biological rhythms, such as body temperature, other vegetative functions and the sleep–wake cycle.

The astronomic 24-hour cycle is shorter than the biological sleep–wake cycle (normally 25 hours). When flying east the astronomic cycle is shortened further, acerbating the discrepancy between the two cycles. This increases the problems with the circadian systems, which often require a week to regain their normal phase relation to the biological clock. Problems caused by changes of biological rhythm are summarized in the term **jet lag**.

What is an electroencephalogram?

The EEG is a **rhythmic electrical activity recorded from the surface of the skull**. In humans, the EEG is recorded from a grid of standard leads. During neurosurgery the electrical activity is recorded from the surface of the cortex as an **electrocorticogram**. In normal adult persons, the dominating frequencies are 8–13 Hz (**α-rhythm**) over the parietal and occipital lobes, as long as the subject is awake and relaxed with his or her eyes closed. With open eyes, the EEG becomes **desynchronized** with low amplitude (10 μV) and the dominant frequency increases to 50 Hz. The **theta-** (3–7 Hz) and **delta-rhythms** (0.5–2 Hz) are observed during light and deep sleep, respectively.

What is the basis for the electroencephalographic recording?

When the brain is not exposed to external stimuli there is a **thalamocortical rhythm** which produces co-ordinated extracellular currents, which can be detected on the surface of the skull as a distinct pattern – the EEG. The EEG recording is due to **large synaptic potentials** by whole groups of mainly pyramidal cells. The EEG pattern is **desynchronized** by sensory inputs through the thalamus. The EEG pattern is also modified by the level of alertness (RAS).

Each pyramidal cell – as with each Purkinje cell in the cerebellum – possesses an extraordinarily **large number of synapses** (10^6). The potentials recorded on the surface of the skull are 50–100 μV. The large pyramidal cells form a dipole with one pole directed toward the surface of the cortex, and the other toward the white matter.

Describe the term evoked potentials

When an external stimulus evokes an EEG change, the change is termed a **cortical evoked potential**. The large numbers of synaptic potentials in the cortical region is summated to form an evoked potential recorded on the skull by an electrode placed over the associated cortical area. However, evoked potentials are small, so the size has to be enhanced by **repeated stimulation and signal averaging**. The evoked potentials over the auditory, visual, and somatosensory cortex areas (I and II) are used clinically to assess the integrity of the respective sensory pathway (Fig. 11-1).

Describe the sleep–wake cycle

The endogenous circadian periodicity of the sleep–wake cycle is normally 25 hours – see above. Sleep is divided into four stages based on EEG. The relaxed individual with eyes closed has 8–13 Hz α-rhythm. As they fall asleep, they pass through the four stages of sleep. During these stages the muscles are relaxed, all vital functions are decreased, and the gastrointestinal motility is increased.

Stage 1 is a very light sleep where α-rhythm is interspersed with theta-rhythm.

Stage 2 is a somewhat deeper sleep dominated by slow waves and by **sleep spindles** (periodic spindle-shaped bursts of α-rhythm) and by large, irregular **K-complexes**.

Stage 3 is characterized by **delta waves** and by occasional sleep spindles.

Stage 4 is recognized by the very slow delta waves with frequencies around 0.5–1 Hz. The subject is difficult to wake up.

A different form of sleep with complete loss of muscle tone occurs periodically every 90 min during stage 1 sleep. This is termed **rapid eye movement sleep or REM sleep**. REM sleep is characterized by eye movement artefacts and a **desynchronized EEG** (low voltage, fast activity as in the arousal reaction in awake persons). The subject is difficult to wake up, so the condition is therefore also termed **paradoxical sleep**.

Spontaneous penal erection occurs during REM sleep, and an irregular heart rate and respiration are often observed. Dreams occurring during REM sleep are often recalled by the person when awake.

Children and young adults have all **four stages of sleep** and several periods of REM

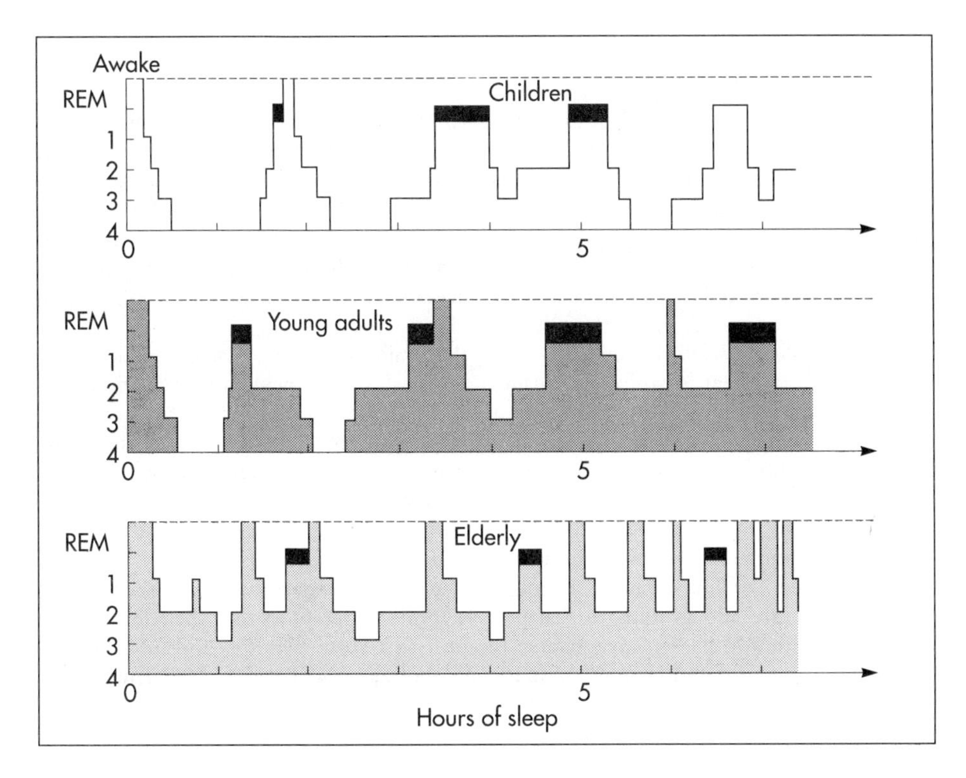

Fig. 10-1. Differences in sleeping pattern between three age groups. The REM sleep is shown by dark zones.

sleep (Fig. 10-1). The depth of the non-REM sleep diminishes through the night and the REM periods increase in duration (Fig. 10-1).

Stage 4 sleep disappears with age, and stage 3 sleep decreases in duration (Fig. 10-1). The REM sleep is also reduced, and an increasing number of awake periods occur. This is why elderly people believe that they do not sleep sufficiently.

What is the physiological basis for sleep?

Sleep was previously thought to be caused by reduced activity in RAS (i.e. the passive theory of sleep). However, transecting the brainstem in the midpontile region produces an animal that never goes to sleep. **Stimulation at the nucleus of the solitary tract can induce sleep**, suggesting that sleep is an **active process** related to centres below the midpontile level.

The question is difficult to address. An educated guess is that sleep is an active, energy-saving condition, preferable to most animals. The metabolic rate during sleep falls to 75% of the **basal metabolic rate**.

What is epilepsy?

Epileptic seizures may be partial or general.

Partial seizures can be caused by an **epileptic focus** anywhere in the cortex. This causes involuntary muscular contractions on the contralateral side. Foci in the somatosensory cortex produce sensory hallucinations called an epileptic **aura.** These hallucinations precede the epileptic seizure. The aura varies and is particular for a certain patient. Epileptic foci in the visual cortex cause visual auras, while epileptic foci in the vestibular cortex produce an aura-feeling of spinning. **Psychomotor epilepsy** originates in the limbic system and causes emotional hallucinations and muscle contractions. Focal seizures are characterized by high **epileptic spikes** in the EEG.

General epileptic seizures involve most of the brain and imply loss of consciousness. **Petit mal** is a transient loss of consciousness, which does not include the motoric cortex and is characterized by **spike-dome** waves in the EEG. **Grand mal** is characterized by an extreme and widely distributed electrical activity, with tonic convulsions of the entire body. Presumably, a basic neuronal circuit activates the cortex of both hemispheres. The increased cortical excitability, with **high voltage–high frequency discharge** over the entire cortex, is not explained. However, the **hyperactive neurons release K^+ and excitatory amino acids or EAAs (glutamate and aspartate).** Epileptic seizure activity is either initiated or propagated through ***N*-methyl-D-aspartate-(NMDA)-receptors binding aspartate and glutamate** (Chapter 4). NMDA-receptors and their ionic pores often work with a Mg^{2+}-sensitive glycine receptor as a co-transmitter. NMDA-receptors are the only **ligand-gated** channels that are also **voltage-gated** and **Ca^{2+}-permeable.**

What is insomnia?

Insomnia is **subjective sleep deficiency.** The patient complains that they sleep too little, or have the impression that they cannot sleep. Such patients sleep more than they think, when studied in sleep laboratories, and their health is not impaired. There is a natural decline of the sleep duration with age, and the use of drugs should be restrained. Monotonous sounds, such as music or the sounds from ocean waves, have proven to be an optimal 'sleeping drug' for many individuals.

Further reading

Daw, N.W., P.S. Stein and K. Fox (1993) The role of NMDA receptors in information processing. *Ann. Rev. Neurosci.* **16**: 207–22.

Klink, M.E., R. Dodge and S.F. Quan (1994) The relation of sleep complaints to respiratory symptoms in a general population. *Chest* **105** (1): 151–4.

Steriade, M. and R.W. Carley (1990) *Brainstem Control of Wakefulness and Sleep.* Plenum, New York.

Thomson, R.F. (1993) *The Brain. A Neuroscience Primer,* 2nd edn. Freeman, San Francisco.

10. Multiple Choice Questions

Each of the following five statements have True/False options:

A. The EEG arousal reaction is a low frequency–high voltage shift.

B. Circadian periodicities are changes in biological variables occurring once a day.

C. NMDA-receptors bind aspartate, dopamine and glutamate.

D. Dreams occur during REM sleep, and they are always reproduced by the person when awake.

E. Dominating EEG frequencies of 8–25 Hz are characteristic of light sleep.

Try to solve the problems before looking up the answers in Chapter 74.

CHAPTER 11.
HIGHER BRAIN FUNCTIONS

Each hemisphere consists of the following four lobes: the frontal, occipital, parietal and temporal.

What are the main functions of the frontal lobe?

The frontal lobe is located in front of sulcus centralis (central fissure) and involved in **motor behaviour**. The frontal lobe contains the motor (area 4), premotor (area 6), and supplementary motor areas (frontal eye areas 8 and 9 of Fig. 11-1). These cortex areas are responsible for planning and execution of voluntary movements.

The motor speech areas (44 and 45 or Broca's area) are located close to the motor cortex, on the **inferior frontal gyrus** of the **dominant hemisphere** in humans (the left hemisphere is controlling the expressed language in most people). Lesions here cause **motor aphasia** (difficulties with speech and writing). Patients with lesion of Broca's area (in the dominant hemisphere) frequently suffer from paralysis of the opposite side (right) of the body.

The frontal cortex is also involved in **personality and emotional behaviour** including attention, intellectual and social behaviour.

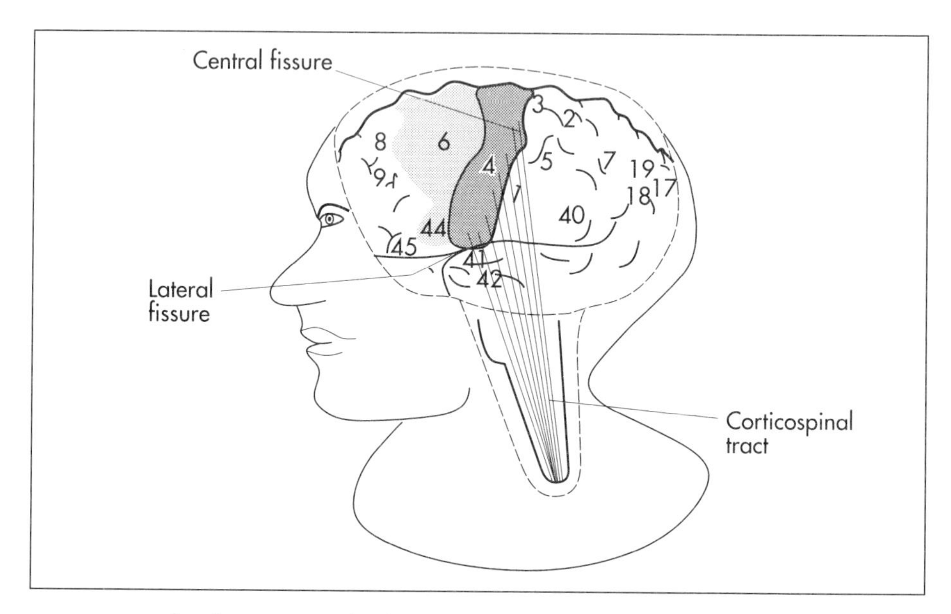

Fig. 11-1. The human cerebral cortex with distinct Brodmann areas and the corticospinal tract from area 4.

The occipital lobe?

The occipital lobe is located behind the parietal and temporal lobe, and involved in **visual processing and visual perception**. Adjustments for near vision are controlled by the primary visual cortex in **area 17** and in the cortex around the calcarine fissure occipital lobe. The conscious visual perception takes place in the primary visual cortex. The secondary visual cortex is in areas 18 and 19, where visual impressions are compared, interpreted and stored (Fig. 11-1).

The parietal lobe?

The important **somatosensory area I** is located on the postcentral gyrus (areas 1,2 and 3 in Fig. 11-1). There is a distinct spatial representation of the different areas of the body in the postcentral gyrus (the **sensory homunculus**). The **somatosensory area II** is located in the rostral part of area 40, close to the postcentral gyrus (Fig. 11-1). The **somatic association or interpretation areas** (areas 5 and 7) are located in the parietal cortex just behind the somatosensory area I (Fig. 11-1).

Each side of the cortex receives information exclusively from the opposite side of the body.

Widespread damage to the **somatosensory area I** causes loss of sensory judgement including the shapes of objects (**astereognosis**).

The temporal lobe?

Auditory and vestibular signals are processed and perceived by the superior temporal gyrus (area 41 in Fig. 11-1). Area 42 is the secondary auditory centre, where auditive signals are interpreted and stored.

The medial temporal gyrus helps control **emotional behaviour** in the limbic system and all the functions of the **autonomic nervous system**. The **hippocampus** is involved in **learning and memory**.

Signals from the auditive (area 42), visual (areas 18 and 19) and somatic (areas 7 and 40) **interpretative areas** are integrated in the posterior part of the superior temporal gyrus. This large **gnostic area** is specially developed in the dominant hemisphere, where it is called the **general interpretative or language comprehension area (Wernicke's area)**. Damage in Wernicke's area causes **sensory aphasia** (i.e. difficulties in understanding written or spoken language, although single words can be heard).

What is meant by the dominant hemisphere?

The dominant hemisphere is the hemisphere that **controls the expressed language**. Lesions of the left hemisphere produce deficits in language function of most people. These deficits are called **aphasia**. The left planum temporale in the floor of the lateral fissure of Sylvii is larger than that of the right hemisphere in most people – not only right handed. The right hemisphere is dominant for functions related to language (intonation, body language), and to mathematically related functions. Each hemisphere controls the contralateral side of the body.

How do the two hemispheres communicate?

Information between the two hemispheres is tranferred through the **anterior commissure and the corpus callosum.** The language centres on the left hemisphere cannot influence the right hemisphere unless the corpus callosum is intact. The two hemispheres can operate relatively independently with language. One hemisphere can express itself through spoken language. The other communicates non-verbally.

If an animal with an intact corpus callosum and optic chiasm learns a visual discrimination task with one eye closed, the task can still be performed with the untrained eye alone, even when the optic chiasm is transected before the animal is trained. Therefore, visual information is transferred as long as the **corpus callosum** is intact.

What can be learned from patients with split brain?

Surgical transection of the corpus callosum have been performed to prevent epilepsy from spreading. When such a patient fixates his or her vision on a point on a screen, it is possible to stimulate only one hemisphere by showing an object to one side of the visual field. Similar objects (key, ring, nail, fork, etc.) can be manipulated (but not seen) through an opening below the screen. Healthy persons can locate the correct object with either hand. Split brain patients, with the picture of the object transferred to the right hemisphere, can locate the correct object **with the left hand** (i.e. right hemisphere), not with their otherwise preferred right hand. Jigsaw puzzles are solved with such manipulo-spatial capabilities. Right-handed patients with split brain can solve three-dimensional puzzles, if the visual signals can reach the motor cortex for the hand to explore. The **visual and motor cortex are connected to each other only in the same hemisphere**, when the corpus callosum is cut.

Describe different types of memory

Memory research has characterized three temporal stages in human memory processes.

1. An **immediate memory** holds sensory information for a few hundred milliseconds to seconds for analysis and further processing. The immediate memory is erased by new incoming signals, so we can only remember a few new telephone numbers at a time. Accumulation of Ca^{2+} in the presynaptic terminals with each signal, possibly causes prolonged release of neurotransmitter at the synapse (**synaptic potentiation**).
2. The **short-term memory** is covering seconds to a few minutes, and the short-term memory receives selected information from the immediate memory. Information is erased as new items displace old data. If a person sees a rapid succession of slides, it is the last slide that remains in the **short-term memory**. We store recent events in the short-term memory, by a neural activity with improved synaptic efficacy that lasts for seconds to minutes. The improved synaptic efficacy is possibly due to **synaptic potentiation, presynaptic facilitation**, or impulses circulating in **neuronal circuits** for a restricted period.
3. The **long-term memory** is a large and permanent memory. The long-term memory receives information from the immediate and the short-term memory. Recycling of information through the short-term memory is termed **rehearsal**. The likelihood of a successful storage in the long-term memory increases with the number of cycles. When the long-term memory is searched for certain information, it may take

minutes to recall the memory. The **long-term memory** is subdivided into the **intermediate long-term memory**, which lasts for days or weeks and can be disrupted, and the **long lasting long-term memory**, which lasts for years. The **long lasting long-term memory** is the storage in the brain of highly overlearned information as one's own name and address. This memory is difficult to disrupt, and it is seldomly affected in retrograde amnesia (see below). The long-term memory and consolidation of memory relate to **effector protein synthesis** at the synapses. Electron-microscopy suggests an increased number of vesicular release sites in the presynaptic terminals.

What is retrograde amnesia?

Retrograde amnesia is a term used for a condition where the patient cannot recall information from the **immediate and short-term memory**. The mild form of retrograde amnesia is typical following head lesion with loss of consciousness (**cerebral commotion**). The short-term memories have only been rehearsed a few times and probably stored only discretely.

The long-term memories are widespread in the cortex as structurally maintained modifications of the synapses after many rehearsals. Only in severe cases is the long-term memory involved.

Describe the limbic system and its function

The limbic system is the neuronal network that controls emotional and motivational behaviour. Motivational behaviour includes control of vegetative functions such as body temperature, respiration, circulation, osmolality of body fluids, sexual behaviour, smell, thirst, appetite and body weight.

The hypothalamus constitutes the major part of the limbic system, and is located in the middle of the other limbic elements.

The limbic cortex begins in the frontal lobe as the **orbitofrontal cortex**, extends upward as the **subcallosal gyrus**, over the corpus callosum and into the **cingulate gyrus** (Fig. 11-2). The limbic cortex finally passes caudal to the corpus callosum down towards the **hippocampus, parahippocampal gyrus** and **uncus** at the medial surface of the temporal lobe (Fig. 11-2). The fornix connects the hippocampus to the mamillary body. The mamillothalamic tract connects the mamillary body to the anterior nucleus of the thalamus. Thalamus connects to the cingulate gyrus, and its cortex is associated with the hippocampus. Stria terminalis connects the amygdaloid body to the midbrain septum and to the mamillary body (Fig. 11-2). The limbic paleocortex links the subcortical limbic structures to the neocortex. Hereby, the limbic system relates behaviour and emotions to the intellectual cortex functions.

Another important pathway is the **medial forebrain bundle**, which connects the limbic system to the autonomic control functions of the brainstem reticular formation.

What is the function of the hippocampus?

The hippocampus connects with the cerebral cortex, the midbrain septum, the hypothalamus, the amygdaloid and the mammillary bodies (Fig. 11-2). The hippocampus is the **decision maker**, determining the importance of incoming signals. The hippocampus becomes habituated to indifferent signals, but learns from signals that cause either reward

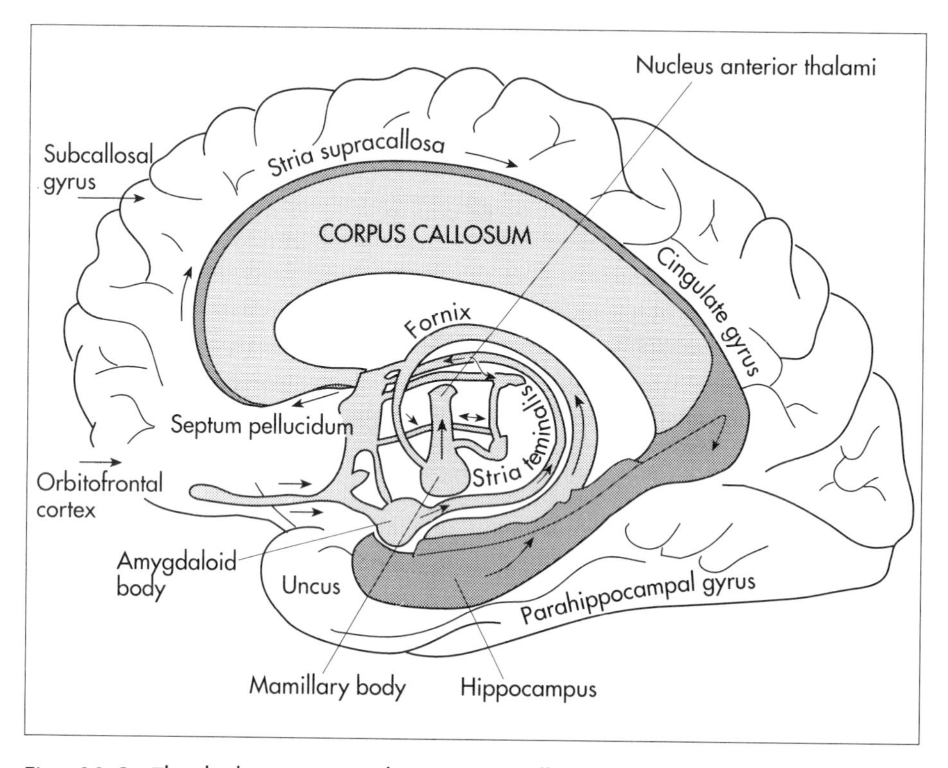

Fig. 11-2. The limbic system. The corpus callosum is transected, and we are looking at the medial aspects of the right hemisphere.

(pleasure) or punishment. It helps the brain to store new signals into the **long lasting long-term memory**. The signal molecule, nitric oxide (NO), modulates NMDA responses related to hippocampal long-term potentiation.

Bilateral removal of the hippocampi in epileptic patients permanently disrupts the ability to learn anything new (**anterograde amnesia**). Other lesions of the hippocampi cause a defect in previously learned memory material (**retrograde amnesia** – see above). Long-term alterations imply a rise in the number of synapses. Cholinergic synapses in the midbrain septum are essential to our memory, and these neurons are dependent upon the nervous growth factor (**NGF**). Repeated activation of a sensory pathway increases the reaction of pyramidal cells. This is called **long-term potentiation**. Such a reaction may last for weeks in the hippocampus and be involved in storage and retrieval of new information in the **long-term memory**.

Describe two types of learning processes

1. **Non-associative learning** means that repeated stimuli cause a reaction that gradually diminishes, as does the amount of neurotransmitters. This process includes **habituation. Sensitization** is the opposite type of non-associative learning, where a strong stimulus becomes more and more threatening to the subject, so that he or

she develops escape behaviour (the **reward and punishment hypothesis** related to the function of the hippocampus).

In the snail aplysia a **facilitating interneuron** releases serotonin on to the **presynaptic terminal neuron**. This stimulates adenylcyclase and the formation of intracellular cAMP in the presynaptic terminal neuron. The resulting protein kinase activation causes phosphorylation and blockage of K^+-outflux. The K^+-outflux is necessary for recovery from the AP. Lack of K^+-outflux prolongs the presynaptic AP considerably. This causes a prolonged Ca^{2+}-influx into the presynaptic terminal with increased release of neurotransmitter and facilitated synaptic transmission. This is probably involved in the long-term memory of snails.

2. **Associative learning** is the process of learning by associations between stimuli. **Classical conditioning** implies a temporal association between a conditioned stimulus and an unconditioned stimulus that triggers an unlearned reaction. In **operant conditioning** the response is associated with **reinforcement**, which changes the propability of the response. **Positive and negative reinforcement** increases the probability of the response, whereas **punishment** reduces its probability.

   The free radical nitric oxide (NO) modulates learning.

What is the basis for strategic or motivated behaviour?

Strategic behaviour is the basis for our social life. Strategic or motivated behaviour is related to homeostasis in general (defence, reproduction, temperature and appetite control). Previously, strategic behaviour was explained by negative feedback with the purpose as a fixpoint, and with the human brain playing a minor role. Today it is generally accepted that the **cerebral drive** is a dominant determinant for strategic or motivated behaviour. The drive that arouses individuals from inactivity originates in the limbic system (including the hypothalamus), which is acting in close relation to the thalamus and the cerebral cortex. The limbic system is connected to the autonomic control functions of the brainstem reticular formation by the medial forebrain bundle. These functions are thermocontrol, appetite control and sexual behaviour.

Describe the thermocontrol in the hypothalamus

Thermoreceptors can initiate generalized reactions to heat and cold. The signals from both superficial and deep thermoreceptors must act through the hypothalamus to arouse appropriate, generalized reactions.

Cooling or heating the denervated lower extremities of paralysed patients, as caused by spinal injury, evoked **vasoconstriction and shivering** or **vasodilatation and sweating** of the innervated upper body shortly after cooled or warmed arterial blood reached the brain. The anterior hypothalamus is responsible for sensing blood temperature variations. The anterior hypothalamus, in particular the **preoptic area**, has been shown to contain numerous heat-sensitive cells and fewer cold-sensitive receptors. Such central thermoreceptors are also found at other levels of the CNS. After destruction of the hypothalamus, the midbrain reticular formation takes over the temperature control. Sections eliminating both the hypothalamus and the mesencephalon leave the medulla and spinal cord to control temperature. The posterior hypothalamus does not contain thermoreceptors. For further details of thermocontrol see Chapters 9 and 44.

Describe the appetite control in the hypothalamus

Bilateral destruction of the ventromedial hypothalamic nuclei leads to hyperphagia and failure of body weight control. Such animals become obese, and they have high plasma [insulin].

Bilateral lesions in the lateral hypothalamic regions cause a temporary hypophagia. The cells of the ventromedial nuclei have a special affinity for glucose, and these cells are responsible for **insulin secretion** from the pancreatic β-cells. Signals from the **dorsal motor nucleus of the vagal nerve** increase insulin secretion, and sympathetic stimulation inhibits the release of insulin. The ventromedial nuclei seem to function like a **glucostat.**

Stimulations and ablations of the limbic system affect food intake. It is obvious from clinical practice that psychological factors, emotional disturbances, motivations and conditioned behaviour all affect our drive for food intake.

For further information on the control of food intake see also Chapters 43, 72 and 73.

Describe the neural and hormonal control of sexual behaviour

Hypothalamic and other limbic system co-operation is responsible for a wide variety of autonomic and somatic phenomena associated with emotions. Stimulation of the midbrain septum yields pleasurable sensations and sexual drive in patients. The **dorsomedial nucleus** of the hypothalamus is probably a major **sex centre** responsible for the complete sexual act. Stimulation of the ventromedial and preoptic regions releases sexual activities. See also Chapter 69.

Is the immune defence system under the influence of higher brain centres?

Internal and external stress affects the prefrontal cortex, whereby the limbic system with the hypothalamus is activated. Hypothalamic nuclei release **corticotropin releasing hormone (CRH)** to the portal blood. The blood reaches the adenohypophysis, where CRH triggers the release of adenocorticotropic hormone (ACTH), endorphins and met-encephalin. ACTH works through different pathways in order to protect the body. ACTH stimulates the adrenal cortex to release corticosteroids, which produce immunosuppression. Immunosuppression reduces the number of inflammatory effector cells, including **helper T cells** and **killer cells** (Chapter 68).

Cancer therapists believe that a relaxed lifestyle and positive reinforcement may stimulate the immune defence in some patients with malignant diseases, and thus explain miraculous remissions. Higher brain centres may even affect the reticuloendothelial production of killer cells through the peripheral nerves to the lymph nodes and bone marrow. See also Chapter 68.

What is Alzheimer's disease?

Alzheimer's disease is premature ageing of the brain (dementia), which rapidly progresses to complete loss of mental powers, in particular loss of memory and normal emotional behaviour. Alzheimer's disease is probably caused by neuronal **degeneration in the nucleus basalis** close to the globus pallidus, and possibly also to **lack of somatostatin and substance P** in deep brain centres.

Normally, cholinergic axons from the nucleus basalis project to the cortex, and their functions relate to memory and to the limbic system functions.

How to explain mental depression?

This is probably an impossible task, because of all the different types of **mental depression psychosis**. One type of mental depression is related to reduced formation of NA in the locus ceruleus, and of serotonin in the midline raphe nuclei of the brainstem, which seriously damages the limbic system. Medical drugs that inhibit the production of noradrenaline and serotonin often cause depression.

Monoamine oxidase inhibitors, which inhibit the destruction of monoamines, and **tricyclic antidepressants**, which block reuptake of monoamines, are effective in the treatment of depressive patients.

What is manic-depressive psychosis?

Manic-depressive psychotic patients often alternate between manic and depressive phases. Therapeutics (such as lithium compounds) that inhibit the action of noradrenaline or serotonin are effective prophylactic agents against manic phases. In this model mania is caused by overproduction of monoamines, and depression by reduced formation of monoamines in the brain nuclei mentioned above.

Further reading

Bekklers, J.M. and C.F. Stevens. (1990) Presynaptic mechanism for long-term potentiation in the hippocampus. *Nature* **346**: 724.

Kato, K. and C.F. Zorumski (1993) Nitric Oxide Inhibitors facilitate the induction of hippocampal long-term potentiation by modulating NMDA responses. *J. Neurophysiol.* 70 (5): 1260–3.

Schuman, E.M. and D.V. Madison (1994) Nitric oxide and synaptic function. *Annu. Rev. Neurosci.* **17**: 153–83.

Squire, L.R. (1987) *Memory and Brain*. Oxford University Press, New York.

11. Case History

An outstanding Russian composer, 63 years of age, recovered from a cerebral insult. However, he could no longer understand spoken or written language, although his speech was fluent. The composer also maintained his ability to compose excellent music.

1. *What is the name of this deficit in language function?*
2. *Where in the brain is the lesion localized and in what side of the brain?*

11. Multiple Choice Questions

Each of the following five statements have True/False options:

A. The frontal cortex is involved in motor and emotional behaviour.

B. The somatic association or interpretation areas (areas 5 and 7) are located in the temporal cortex.

C. Recycling of information through the primary memory is termed rehearsal.

D. Retrograde amnesia following brain commotion is a loss of the short-term memory.

E. The limbic system relates behaviour and emotions to the intellectual cortex functions.

Try to solve the problems before looking up the answers in Chapter 74.

CHAPTER 12.
MOTOR CONTROL

Motor activity can be voluntary or involuntary. **Voluntary movements** are planned and started by **feedforward control**, and when maintained for a while they are regulated by **feedback loops. Involuntary movements** comprise reflexes, such as the stretch reflexes, and autonomic functions, such as the respiratory muscle movements. We have motor centres in the cerebral cortex, the brainstem, the spinal cord, the cerebellum, and the basal ganglia. Motor centres all receive sensory information in an organized neural structure termed a **somatotopic map** (see the motor homunculus).

Here, the motor control system is presented in three sections: I. Spinal organization; II. Descending motor pathways; and III. Motor brain centres.

I. SPINAL ORGANIZATION

Describe three different types of motor units

We have 200 different skeletal muscles, which are controlled by more than 300 000 motor units.

A **motor unit** is comprised of an α-motor neuron, all its axon terminals, and the skeletal muscle fibres it innervates. The number of muscle fibres in a motor unit varies from two in highly regulated eye muscles to 2 000 in the quadriceps femoris muscle. The motor unit is the **final common pathway**, because all muscle fibres of the unit contract when a motor unit is activated. Adjacent motor units interdigitate, so they can support each other. The muscle power is increased by recruitment of more motor units and by increased frequency of discharge in each unit.

We have three types of motor units (α-motor neurons) in a mixed muscle such as the gastrocnemius.

1. **Fast fatiguable (FF) motor units** have type IIB twitch muscle fibres which have few mitochondria and small amounts of myoglobin (white fibres), and depend on glycolysis (high anaerobic metabolism) for energy. They have only small amounts of glycogen (fast glycolytic = FG). FF motor units produce short contractions (fast twitch), and fatigue easily. The FF motor neuron is larger, the axon is thick and it branches more, so that the FF motor unit innervates more muscle fibres. This is why FF motor units are capable of powerful contractions. The FF units are recruited last.
2. **Slow (S) motor units** have type I twitch fibres, which have many mitochondria and myoglobin (red fibres). They depend on aerobic metabolism and have lots of glycogen (slow oxidative = SO). S motor units have weak but long-lasting contractions (slow twitch) because they are fatigue resistant. They are small and are first to be recruited.
3. **Fast fatiguable resistant (FR) motor units** have type IIA twitch fibres with an intermediate content of mitochondria, myoglobin and glycogen. These intermediary

fibres also rely upon oxidative metabolism (fast oxidative glycolytic = FOG). They have contractions of intermediate force and duration, and they resist fatigue. FOG fibres are of intermediate size, and they are recruited before the FF fibres. This is in accordance with the **size recruitment principle**: Small motor neurons are easier to activate by excitatory postsynaptic potentials (EPSP) than large neurons.

What is the myotatic stretch reflex?

A spinal reflex is a stereotyped motor reaction to an input signal. The **myotatic stretch reflex** is the most crucial monosynaptic reflex for the maintenance of the erect body posture in humans.

The reflex has two components. First, the primary annulospiral endings (group Ia) of the muscle spindles trigger the **phasic stretch reflex**. Second, both primary and secondary endings elicit the **tonic stretch reflex**.

1. **The phasic stretch reflex** is elicited in the clinic by a light tap on a muscle tendon. When the patellar tendon from the quadriceps muscle is stretched quickly by the tap, a discharge is elicited in the afferent fibres (Ia) from the primary endings of the muscle spindle (Fig. 12-1). This is the **phasic myotatic stretch reflex** or the so-called **patellar reflex**. These Ia fibres synapse directly (**monosynaptically**) on α-motor neurons that supply the extensor muscles of the knee (E+ in Fig. 12-1). The response elicited is a brief contraction of the latter. Of all the presynaptic terminals arriving to the **motor neuron**, up to 90% are located on the surface of the dendrites. The remaining 10% synapse on the soma of the motor neuron.

   The Ia afferent fibres also synapse with small **group Ia inhibitory interneurons** in the grey matter of the spinal cord, as the one synapsing with the upper α-motor neuron in Fig. 12-1. This neuron innervates the semitendinosus muscle, which flexes the knee joint (F- in Fig. 12-1). The reflex inhibition of antagonist muscles when

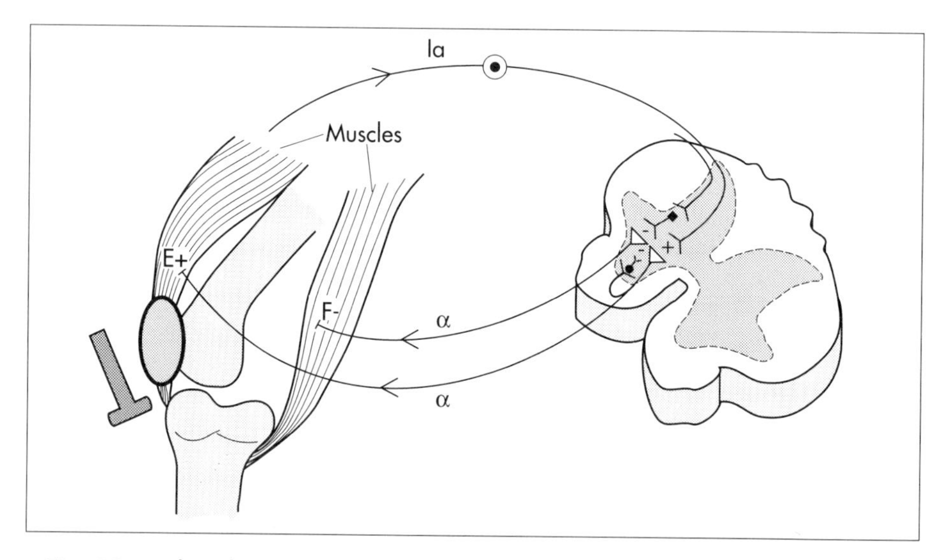

Fig. 12-1: The phasic myotatic stretch reflex and reciprocal innervation (F-).

synergistic muscles are contracted is called **reciprocal innervation.** In pathological conditions, the phasic stretch reflexes may be depressed or hyperexcitable.

2. **The tonic stretch reflex** is triggered by passive bending of a joint. This elicits a discharge in both group Ia and II afferents from the muscle spindle. The tonic stretch reflex contributes to the **erect body posture** and helps maintain posture by increasing the **tone of the physiological extensor muscles (i.e. antigravity muscles).**

Compare and contrast Renshaw inhibition and presynaptic inhibition

Renshaw inhibition. Cajal found that the α-motor axons give off thin recurrent (antidromal) collaterals in the grey matter of the spinal cord (Fig. 12-1). These collaterals synapse with Renshaw interneurons in the ventral horn (Fig. 12-1). The Renshaw cells synapse with α-motor neurons of synergistic muscles, and thus **inhibit monosynaptic reflexes (postsynaptic inhibition).** Descending signals from the brain can either amplify the postsynaptic inhibition or reduce its effect. Renshaw cells make it possible for the higher brain centres to influence spinal reflexes by central inhibition or facilitation.

Presynaptic inhibition. Presynaptic terminals contain a large number of voltage-gated Ca^{2+}-channels. Ca^{2+} must enter the presynaptic terminal from the extracellular space before the vesicles can release their neurotransmitter at the synapse.

Presynaptic inhibition takes place at presynaptic contact sites on the presynaptic terminals. Activation of these sites closes many Ca^{2+}-channels, and thus inhibits transmitter release.

Describe the function of the Golgi tendon organ

The Golgi tendon organs are the serially located terminals of **group Ib fibres** wrapped around bundles of collagen fibres in the tendons. Golgi tendon organs monitor the **force** in the tendon; they are activated either by stretch or by contraction of the muscle. The adequate stimulus is the force developed in the tendon.

What is the inverse myotatic stretch reflex?

The inverse stretch reflex or the **Golgi tendon reflex** completes the **stretch reflex** by a force-controlling feedback. The Golgi tendon organs monitor force in the tendons. Golgi tendon organs are in **series** with the muscle fibres – not parallel as the muscle spindles. If the extensor muscles of the thigh are fatigued, as during standing, the force in their tendons begin to decrease. This reduces the discharge of the Golgi tendon organs. This acts as a compensating feedback which excites the α-motor neurons and increases the force of contraction. The inverse stretch reflex helps maintain the force of muscular contraction and posture during standing. During the rapid contraction of the myotatic stretch reflex, the inverse stretch reflex reduces the force of contraction. The stretch reflexes regulate the length of the muscle, and provide a **length–force feedback** to the CNS.

Describe the function of the muscle spindle

The muscle spindle monitors **muscle length** and **rate of change of length** (velocity); they are particularly abundant in muscles that are capable of fine movements and in large muscles that are dominated by slow twitch fibres. The organ is shaped like a spindle, which lies in parallel to the large, regular, extrafusal muscle fibres. Each organ contains two main types of intrafusal muscle fibres: **Nuclear bag fibres** which swell in the equatorial region due to all the nuclei located there, and **thin nuclear chain fibres** which have central nuclei arranged in line (Fig. 12-2). The **primary afferent fibres (Ia)** twine around the equatorial regions of **both the bag and chain fibres** like a corkscrew or **annulospiral**; the **annulospiral nerve endings** signal **length and velocity** (Fig. 12-2). The **secondary afferent fibres** originate mainly from the nuclear chain fibres and with a few branches originating from the nuclear bag fibres (Fig. 12-2). They monitor only the length of the muscle.

Two types of γ-motor neurons innervate the muscle spindle. The dynamic γ-motor axons form **plate endings** (P_2) on the nuclear bag fibres (Fig. 12-2), while static γ-motor axons form **creeping trail endings** on nuclear chain fibres (Fig. 12-2). The intrafusal fibres receive Aβ-motor fibres, which terminate with P_1 plate endings on both extra- and intrafusal muscle fibres (Fig. 12-2). The Aβ-motor fibres may be involved in α-γ-coactivation.

When the extrafusal fibres contract, the muscle spindles shorten, whereby the discharge rate of their afferents decreases.

Activity of the γ-motor neurons causes the polar spindle regions to contract on either end. This elongates the equatorial regions so that muscle spindles can adjust to stretch (Fig. 12-2).

Descending commands from the brain often cause contraction of both extrafusal and intrafusal fibres simultaneously so that the muscle spindle is sensitive to stretch at all muscle lengths. When the muscle is stretched, the muscle spindles are simultaneously stretched with it, and the discharge rate of the afferents is increased.

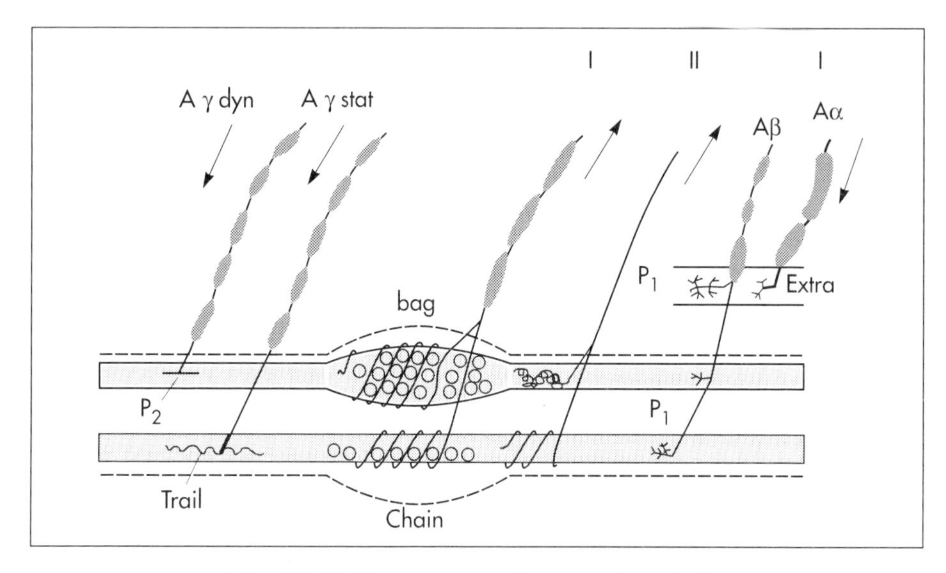

Fig. 12-2: The structure of a muscle spindle.

What are the flexion reflexes?

The flexion reflexes are triggered by various **flexion reflex afferents** including nociceptors. The flexion reflexes have a long latency, because it involves interneurons (is polysynaptic). The afferent discharge causes excitatory interneurons to activate α-motor neurons that innervate ipsilateral flexor muscles. The afferent discharge also causes inhibitory interneurons to inhibit α-motor neurons, supplying the ipsilateral extensor antagonists.

The **flexor withdrawal reflex** is crucial. This reflex is also called a nociceptive reflex or a **pain reflex**, and involves all the muscles of a limb in flexor withdrawal in order to protect from further damage. In addition, the reflex can activate the extensor muscles of the opposite limb. This contralateral activity is termed the **crossed extension reflex** by **reciprocal innervation**.

Flexion reflexes involved in locomotion are controlled by the **locomotor pattern generator**.

Severe visceral disease can trigger contraction of the chest and abdominal muscles which reduces pain by limiting movement of the body. When examining the abdomen of such a patient it will be observed that the muscles are tense. This sign is called **defence musculaire** which is a viscero-somatic protective reflex.

How is limb movement co-ordinated?

We possess pattern generators or neural circuits in the spinal cord, for every limb and for respiration, chewing, etc. The commands are organized by the **midbrain locomotor centre** via the reticular formation and through the reticulospinal tracts. Such spinal pattern generators also account for other movement patterns like scratching, dancing, etc.

II. DESCENDING MOTOR PATHWAYS

What is understood by descending motor pathways?

Clinical dichotomy, traditionally subdivides the descending fibres into the **pyramidal and the extrapyramidal pathways**; this is based on the fact that the corticospinal tract passes through the **medullary pyramids**. Therefore, interruption of the corticospinal or pyramidal tract was supposed to cause **pyramidal tract disease** (see later). The problem, however, is that the loss of the corticospinal tract does not explain all the classical signs of pyramidal tract disease.

The concept of extrapyramidal pathways raises other problems. The concept of **extrapyramidal tract diseases** is generally used to designate one or more **disorders of the basal ganglia**. While extrapyramidal pathways do play a role in basal ganglia diseases (as in cerebellar disease), the main motor pathway involved in basal ganglia diseases is the corticospinal tract.

The descending motor pathways can also be dichotomized based on their endpoint in the spinal cord, and hence which muscles they control and how. Pathways ending in the lateral horn of the spinal cord (on motor neurons or interneurons) are called the **lateral descending motor system** (the rubrospinal tract and the lateral corticospinal tract), and pathways ending on the medial ventral horn interneurons are termed the **medial descending motor system** (containing reticulo-, tecto-, and ventriculospinal tracts).

Describe the lateral descending system

The lateral corticospinal, the corticobulbar (to the facial motor and hypoglossal nucleus) and the rubrospinal tracts control the manipulative movements of the limbs and the lower face and tongue muscles. The corticospinal and corticobulbar tracts originate from areas 4, 6, 8, 9, and somatosensory area I (areas 1, 2, 3 in Fig. 11-1). The large and small pyramidal cells, and the giant pyramidal cells of Betz, are the cells of origin of these tracts. The corticospinal tract descends through the internal capsule and brainstem. At the medullary pyramid, 80% of the fibres cross to the opposite side and descend in the **dorsal lateral funiculus** as the **lateral corticospinal tract**. The fibres of this tract end on motor neurons and interneurons in the lateral horn of the spinal cord. These motor neurons innervate distal muscle groups. Interruption of the lateral corticospinal tract implies loss of the fine control of the digits. Interruption of the corticobulbar tract to the facial motor and hypoglossal nucleus implies loss of voluntary movements of the lower face and tongue. Interruption of the rubrospinal tract from the red nucleus combined with corticospinal lesions give rise to difficulty in separating finger, hand and arm movements. The red nucleus is closely linked to the deep cerebellar nuclei.

The lateral or dorsolateral descending system allows the primary motor cortex to modify the reflexes and pattern movements at the level of the spinal cord.

Describe the medial descending system

The medial or ventromedial descending system involves the ventral corticospinal tract and much of the corticobulbar tract ending in the **medial group of brainstem and spinal cord interneurons**. The ventral corticospinal tract continues caudally in the ventral funiculus on the same side and ends bilaterally on the medial interneurons. They control the axial muscles and bilateral activity including chewing and wrinkling of the eye brows.

Other medial system pathways originate in the brainstem:

1. **The lateral vestibulospinal tract** excites motor neurons that innervate proximal postural muscles. It receives input from all compartments of the vestibular apparatus and from cerebellum to the lateral vestibular nucleus.
2. **The medial vestibular tract** receives signals from the semicircular ducts and from the cerebellum, and excites motor neurons in cervical and thoracic segments. Thus, it controls the head position in response to angular accelerations of the head.
3. **The pontine reticulospinal tract** excites motor neurons to the proximal extensor muscles to support posture.
4. **The medullary reticulospinal tracts** have mainly inhibitory effects on many spinal reflexes.
5. **The tectospinal tract** from the superior colliculus causes contralateral movements of the head in response to touch and auditory stimuli. This tract allows the integration of hearing and vision with motor performance.
6. **Pathways from the solitary nucleus and the interstitial nucleus of Cajal** are involved in the pharyngeal stage of swallowing. The solitary nucleus receives all sensory signals from the mouth including taste, and is involved in cardiovascular and respiratory control.

The ventromedial system is important for the normal muscle tone and body posture.

Do we have monoaminergic descending pathways?

1. The neurons of the **pontine locus coeruleus and nucleus subcoeruleus** contain NA. These nuclei project to and inhibit interneurons and motor neurons of the spinal cord through the **lateral funiculi.**
2. The neurons of the **raphe nuclei** in the medulla, which are connected to the limbic system, also contain serotonin. The serotonergic nuclei project to and inhibit dorsal horn interneurons **reducing pain transmission**, and they also project to and excite ventral horn motor neurons of the spinal cord, thereby **enhancing motor activity**.
3. There is also a descending dopamine pathway.

The three monoaminergic pathways function as **motor system amplifiers**.

What are the signs and symptoms of capsular stroke?

Rupture of an arteriosclerotic brain artery or an aneurysm causes bleeding. Bleeding can interrupt the corticospinal tract as it traverses the internal capsule. Such a block of the excitatory pathways to the spinal cord results in **severe contralateral paresis, weakness of the finger muscles with loss of fine movements, and loss of superficial reflexes (the abdominal and cremasteric reflexes)**. Moreover, a series of **release signs** are found in the stroke patient after the spinal shock. They are the **positive sign of Babinski, spasticity, foot clonus, and flexion reflexes**. The positive sign of Babinski is a slow dorsiflexion of the big toe and fanning of the other toes, when the sole of the foot is stroked laterally from the heel and forward. **Spasticity** is a motor condition dominated by increased tonic and phasic stretch reflexes.

This is a typical result of interruption of the lateral descending system, and often termed the **upper motor neuron disease** or **the pyramidal tract syndrome**. The capsula interna damage interferes with other cortical efferents, e.g. to the basal ganglia, to the thalamus, and to the pons. Therefore, the symptoms and signs are much broader than those after injury of the corticospinal system only.

The stroke patient can slide into deep unconsciousness termed **coma**. Coma is the deepest stage of unconsciousness. The comatose patient is completely without reactions to even the strongest stimulus. The EEG is dominated by delta-waves. When coma proceeds into **brain death**, the EEG trace shows no electrical activity. Glutamate is released during cerebral ischaemia, such as the ischaemia occurring after a stroke. In animals, ischaemia-induced neurodegeneration can be prevented by ***N*-methyl-D-aspartate-receptor (NMDA)** antagonists.

What characterizes a pure lesion of the medullary pyramid?

The control of fractionated finger movements is absent. There is a positive sign of Babinski. Flexion reflexes are not found, and neither is spasticity. On the contrary, muscle tone is decreased. In summary, a pure interruption of the corticospinal tract alone does not show the same signs as capsular stroke.

What characterizes lesions of the medial descending pathways?

The main defects caused by **medial lesions** are **reduced muscle tone in the physiological extensors**, loss of balance during walking and standing, and loss of rightening reflexes (they tend to restore head and body position). However, fine finger movements are quite normal.

Describe two types of abnormal muscle tone

Spasticity is used in clinical neurology to describe muscles resisting fast, passive movements of the limbs, especially in extreme articular positions. When the limbs are moved in extreme articular positions, the increased muscle resistance suddenly disappears. Spasticity includes hyperactive stress reflexes and foot clonus. The resistance dominates in the physiological extensors (antigravity muscles).

Rigidity is muscle stiffness caused by prolonged activity in the motor units. The muscle resistance is increased toward passive movements of the limbs in any direction (**lead pipe rigidity**). This condition is found in Parkinson's disease (see later in this Chapter).

What are the signs of a spinal transection syndrome?

The **spinal shock** is immediately recognized by several characteristic symptoms: **flaccid paralysis** with loss of stretch reflexes, areflexia, loss of autonomic functions, and loss of all sensation below the level of transection. After a few weeks the spinal shock fades away and the reflexes return and become hyperactive (foot clonus), including mass reflexes and flexion reflexes. The flaccid paralysis is replaced by a **spastic paralysis or paresis**.

III. Motor Control By The Brain

The two most important EAA-receptors are the glutamate receptors, NMDA-receptors, and special glutamate receptors called AMPA-receptors (see Chapter 5). NMDA activated ion channels are only active, when the membrane is depolarized, and they increase the Ca^{2+}- and Na^{+}-outflux, and the K^{+}-influx. AMPA-activated ion channels increase the Na^{+}-outflux, and the K^{+}-influx to the cell.

What is the structure of the motor cortex?

The primary motor cortex (area 4) on the precentral gyrus controls distal muscles of the extremities. Area 4 is organized parallel to the somatosensory cortex. The face is represented laterally near the Sylvian lateral fissure, and the legs on the medial part of the hemisphere. The cortical representation is somatotopic and disharmonic, as indicated by the **motor homunculus**.

The premotor cortex helps control proximal and axial muscles.

The supplementary motor cortex is involved in motor planning and in co-ordination of movements. **The frontal eye fields** initiate saccadic eye movements (Chapter 14).

Corticospinal neurons discharge before voluntary muscle contraction, and the size of the discharge is related to the size of the contractile force. The somatosensory cortex and the posterior parietal association cortex receive feedback from the sensory nervous system, which helps correct motor feedforward commands.

What are the signs and symptoms of cerebellar disease?

1. **Damage to the flocculonodular lobe** causes nystagmus and difficulties in gait and balance (i.e. resembling lesion of the vestibular apparatus).
2. Damage to the vermis or the intermediate region and hemisphere results in motor disturbances of the trunk and limbs, respectively.

Cerebellar disorders include **cerebellar incoordination, dysequilibrium, and loss of muscle tone.**

Cerebellar incoordination comprises **ataxic gait**, as seen in alcohol intoxication and in disseminated sclerosis. Another type of ataxia is **dysmetria**, where there is an inability to move the limbs to the desired position. Many patients manifest their ataxia as **dysdiadochokinesis**, which is a disturbance of the normal ability to make repeated supinations and pronations of the lower arms. Complicated muscle function is stepwise, not smooth. **Intention tremor** is seen when the patient is asked to touch a target. Speech is slow and slurred, a defect termed **dysarthria or scanning speech**.

Dysequilibrium results in balance problems, and the patient falls to the affected side. Gyratoric vertigo is a genuine **rotational or merry-go-round** vertigo with the associated loss of equilibrium. This cerebellar vertigo is similar to that following lesion of the vestibular apparatus.

Loss of muscle tone is called **hypotonia**. The hypotonic lack of damping causes the leg to swing back and forth, when the patellar reflex is triggered – so-called **pendular knee jerk**.

Cerebellar nystagmus is involuntary movements of the eyeballs around their natural position – often accompanied by rotational vertigo, when the flocculonodular lobe is damaged.

What is the role of the cerebellum?

The **little brain**, also termed the **motor auto-pilot**, helps regulate movements and posture and influences muscle tone, eye movements and balance.

The cerebellum is particularly concerned about the timing of rapid muscular activities including the interplay between agonist and antagonist muscle groups. **Motor learning** is programmed in the cerebellum. The cerebellum compares the proprioceptive input from the actual movements, with the movements intended by the motor control areas of the brain. It controls the sequence of movements, and makes corrective adjustments – i.e. it acts as an **auto-pilot**.

The incoming pathways to the cerebellum end mostly as mossy fibres. The input signals evoke **simple spikes in Purkinje cells**. Projections from the inferior olive terminate as **climbing fibres**. They produce repetitive or complex discharges in Purkinje cells. **Complex spikes** of long duration are involved in the cerebellar programming of motor learning.

This is the basis for cerebellar co-ordination and fine, rapid adjustments of complex movements. The cerebellar hemisphere affects movements on the same side of the body, because of its crossed connection to the motor system. The motor system projects contralaterally.

Discrete electrical stimulation of cerebellum does not cause movements or sense impressions, so it is also termed the **silent brain**.

The **vestibulocerebellum** projects to the vestibulospinal and reticulospinal tracts which co-ordinate balance and eye movements. The **vestibulo-ocular reflex** produces conjugate eye movements in the direction opposite to that of the head movement. The **vestibulocollic reflex** increases the neck muscle tone damping the induced movement.

The **spinocerebellum** receives proprioceptive input from the spinal cord (the spinocerebellar tracts). The spinocerebellum controls the **axial muscles** through the **medial descending motor system,** and the **proximal limb muscles** through the **rubrospinal tract of the dorsolateral system.**

The **pontocerebellum** receives **decision signals** and **motor control signals** from the cerebral cortex by way of **pontine nuclei.** The pontocerebellum is involved in motor planning, and controls the **distal muscles** through the **lateral corticospinal tract.**

What is the function of the basal ganglia?

The main function of the basal ganglia is to initiate and stop movements. The basal ganglia inhibits the thalamus, and thus reduces the thalamic stimulation of the motor cortex.

It also contributes to cognitive (i.e. intelligence, knowledge, and motor learning) and affective (i.e. emotional) functions.

The basal ganglia includes the **globus pallidus** and the **striatum.** The **striatum consists of the nucleus caudatus and the putamen.** These deep brain nuclei function in collaboration with several thalamic nuclei, substantia nigra and the subthalamic nucleus (Fig. 12-3).

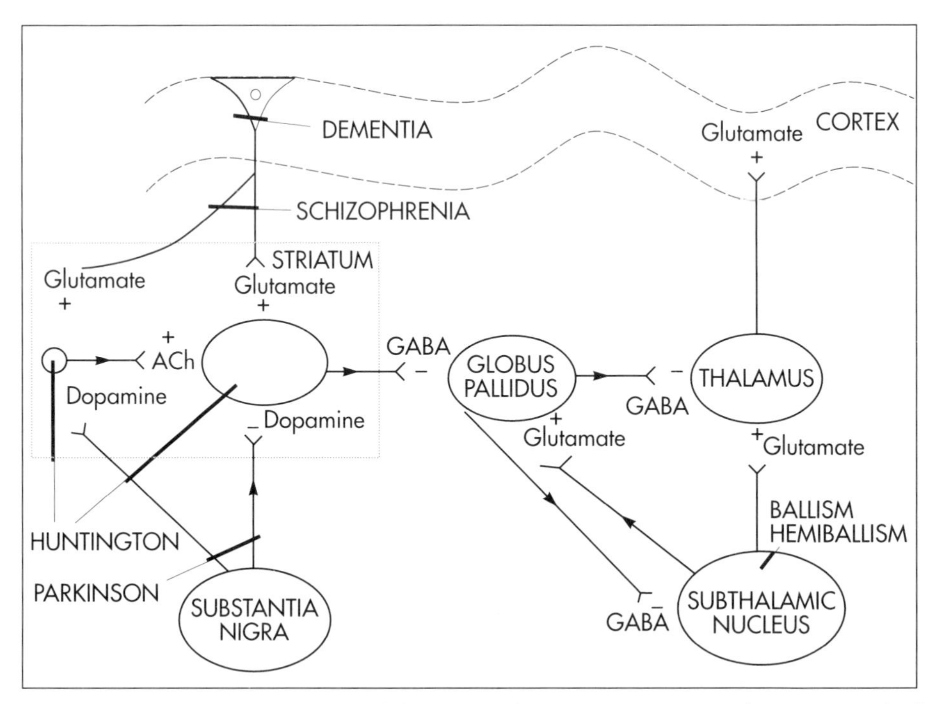

Fig. 12-3: The basal ganglia and their interplay. Transmitter stimulation is marked by +, and inhibition by −. The affected cell bodies or axons at disease states are marked with bars.

The **striatum** receives afferent fibres from the cortex (glutamate + = glutaminergic excitatory fibres), and dopaminergic (inhibitory) fibres from substantia nigra (dopamine). The striatum projects to the globus pallidus and to the substantia nigra. These connections are GABAergic and inhibitory (GABA – in Fig. 12-3). The globus pallidus receives afferent GABAergic fibres from the striatum, and projects to the thalamus with GABAergic efferents (Fig. 12-3). In the striatum, there are excitatory cholinergic pathways (Fig. 12-3).

What is schizophrenia?

Schizophrenia is a **psychosis with hallucinations, dissociations of ideas, intense fear, and paranoia.**

Schizophrenia is possibly caused by hypersecretion of **dopamine** or by blockage of the **glutamate-producing neurons** from the cortex to the striatum (Fig. 12-3). The balance between these two neurotransmitters in the striatum is seriously disturbed.

What is Parkinson's disease?

This common disorder is characterized by **pill-rolling** tremor at rest, **akinesia** with **plastic rigidity** (high tonus) and **cogwheel-movements**. Akinesia or hypokinesia means inability to initiate normal movements. Dementia or cognitive disturbances are present in some patients. Dementia is ageing of the brain and resulting loss of mental powers.

The main pathological mechanism is **degeneration of dopaminergic neurons in the substantia nigra**. The consequence is a severe lack of dopamine in the **striatum** (Fig. 12-3). Lack of dopamine results in an **overactivity of the GABA pathways to the thalamus**, which activates the motor cortex neurons (Fig. 12-3). This increases the discharge of γ-motor neurons in the spinal cord.

L-DOPA (dihydroxy-phenylalanine) is a precursor of dopamine that is capable of crossing the blood–brain barrier. Administration of this drug, which is transformed to dopamine in the brain, relieves much of the rigidity and akinesia by inhibiting the striatum.

Transplantation of dopaminergic neurons into the striatum has been explored.

A balance between the effects of dopamine and glutamate is a necessity for the normal functioning striatum (Fig. 12-3).

Parkinson patients who are treated with an overdose of L-DOPA will develop **schizophrenic psychosis**, because excess dopamine causes schizophrenia or schizophrenic symptoms and signs. Hence, antischizophrenic drugs (chlorpromazine and haloperidol) decrease the dopaminergic effects.

Parkinson patients develop **DOPA resistance** following years of dopaminergic medication. The use of antagonists of the NMDA subtype of glutamate receptor in Parkinson patients decreases the activity of the glutamate pathway to the cortex (Fig. 12-3), and reverses the akinesia and rigidity.

What is Huntington's chorea?

Huntington's chorea is an **autosomal dominant genetic defect on chromosome 4**. Its characteristic disorders are **hypotonia, dementia** and **involuntary movements**. Huntington's chorea is due to a defect in **GABAergic and acetylcholinergic interneurons of the striatum and the cerebral cortex** (Fig. 12-3). In these interneurons, the concentration of the enzymes, glutamic acid decarboxylase and acetylcholine transferase, required for the synthesis of both transmitters is markedly reduced.

GABA receptors are usually inhibitory. When the globus pallidus is no longer inhibited by GABA from the striatum, this leads to a stronger inhibition of the thalamus (Fig. 12-3), which is probably the cause of the **involitional choreiform movements.** Chorea is opposed to rigidity. Low doses of a dopamine agonist may reduce the choreiform movements of patients with Huntington's disease.

The dementia of the Huntington's chorea is caused, as are most types of dementia, by **cortical degeneration** (Fig. 12-3).

What is ballism and hemiballism?

Ballism means ballistic or flailing movements of the limbs. **Hemiballism** is a unilateral uncontrolled hyperkinesia (Saint Vitus dance). The lesion is located in the contralateral hemisphere. These disorders are caused by partial lesion of the **subthalamic nucleus** (Fig. 12-3), and the lesion is often due to thrombosis.

Further reading

Greenamyre, J.T. and C.F. O'Brien (1991) *N*-methyl-D-aspartate antagonists in the treatment of Parkinson's disease. *Arch. Neurol.* **48** (9): 977–81.

Kandel, E.R. (1991) *Principles of Neural Sciences*, 3rd edn. Elsevier, New York.

McIntosh, A.R., C.L. Grady, L.G. Ungerleider, J.V. Haxby, S.I. Rapoport and B. Horwitz (1994) Network analysis of cortical visual pathways mapped with position emission tomography. *J. Neurosci.* **14** (2): 655–66.

12. Case History

A 65-year-old male suddenly falls and is found in a deep coma by the doctor. There is a left sided hemiplegia with short arm–long leg as a flexion reflex. The paralysis and areflexia turn into spastic hemiparesis with a positive sign of Babinski. The deep stretch reflexes (patellar- and achilles-tendon reflexes) are enhanced. There is loss of superficial reflexes (the abdominal and cremasteric reflexes). When the achilles-tendon reflex is triggered it releases foot clonus. When the patient wakes from the coma his facial nerve paresis is examined. He can knit his brows and turn his eyes upwards.

1. *What is the pathophysiological basis for this condition?*
2. *What are spasticity and foot clonus?*
3. *Is the facial nerve paresis central or peripheral?*

12. Multiple Choice Questions

Each of the following five statements have True/False options:

A. Fast fatiguable motor units consist of type IIB twitch fibres with few mitochondria and small amounts of myoglobin.

B. The Renshaw cells synapse with α-motor neurons of antagonistic muscles, and thus inhibit monosynaptic reflexes.

C. The cerebellar hemisphere affects movements on the opposite side of the body.

D. The Golgi tendon organs are the serially located terminals of group Ib fibres wrapped around bundles of collagen fibres in the tendons.

E. The ventromedial descending system involves the ventral corticospinal tract and much of the corticobulbar tract ending in the medial group of brainstem and spinal cord interneurons.

Try to solve the problems before looking up the answers in Chapter 74.

Chapter 13.
The Visual System

This system detects, transmits and interprets photic stimuli. Photic stimuli are electromagnetic waves with wavelengths between 400 and 725 nm. This is **visible light** or the adequate (effective) stimulus for the eye.

The eyes can distinguish brightness and colour. The photoreceptors are rods and cones located in a specialized epithelium called the retina. The **retina** contains about 6 million cones and 120 million rods. In the peripheral region of the retina both rods and cones converge on **bipolar cells**. The bipolar cells converge on ganglion cells giving rise to the **1 million nerve fibres** in each optic nerve. In addition, there are **horizontal cells** and **amacrine cells** in the retina. They conduct impulses laterally.

Rods are most sensitive in the dark (scotopic vision). More than a hundred rods converge on each ganglion cell. There are no rods at all in the fovea.

Cones operate best in light (photopic vision). Cones have a **high resolution capacity** and hence a **high visual acuity**, because the light is focused on the fovea, where the cones are concentrated. The high resolution is also due to the small **convergence** of cones to bipolar cells in the fovea (approximately a 1:1 relationship). Cones are responsible for colour vision. Cones are surrounded by pigment, except where the light enters.

The eye contains chamber fluid, which is produced by filtration and secretion in the ciliary processes. The intraocular pressure is normally 1.3–2.6 kPa (10–20 mmHg or Torr). Increased resistance to fluid outflow at the iridocorneal junction leads to **increased intraocular pressure** – or **glaucoma**. In this condition, the retinal artery is compressed at the optic disc, where it enters the eye. This causes retinal and optic nerve atrophy which results in blindness.

What is a diopter?

A **diopter** is the unit for the refractive power of a lens. The **diopter (D) equals the reciprocal value of the focal length of the lens in metres (m).**

What is visual accommodation?

Visual accommodation is the **rise in the refractive power of the lens**, obtained when the lens rounds up, because of contraction of the ciliary muscle and relaxation of the zonule fibres.

Each object we look at has a special target point (the **fixation point**), from which **light passes unrefracted** through the **nodal point** of the eye and focuses on the fovea, creating the sharpest possible image. The nodal point in the eye is precisely the point through which a light beam passes unrefracted. The **far point (F or punctum remotum)** for the eye is the fixation point in the unaccommodated eye (Fig. 13-1). The **near point (N or punctum proximum)** of the eye is the fixation point for a maximally accommodated eye (i.e. when the lens is in its most spherical configuration). The refractive power of the lens can vary between 12 and 26 D. The **accommodative power of the eye** is the rise in refractive power from unaccommodated to the maximally accommodated condition: **Acc. power=1/F−1/N**. A child of 10 years has 12–14 D, a 20-year-old person 10 D, and a 60-year-old person 1 D.

The optical distance convention defines all distances measured from the light source to be positive. Thus, all distances from the light source are negative. Hence, the distance from the **nodal point** of the eye to a point in front of the eye is negative. Convex refractive media converge in falling light and thus have a positive diopter. Concave lenses have refractive powers with negative diopters, because the focal point is in front of the lens.

Convergence or near vision occurs when the eye focuses on an object closer than 6 m from the eye. Near vision – even with only one eye – triggers **accommodation** and **pupillary constriction**. The ciliary muscle and the pupillary sphincter muscle are innervated by the parasympathetic oculomotor nerve, and the two muscles contract simultaneously for near vision.

The visual field of both eyes is perceived as only one continuous visual field (the Cyclops eye effect). This is **fusion**, or the illusion that we are looking at the world with only one eye.

What is the refractive power of the healthy, human eye?

In a healthy eye, the light from an object in the visual scenery is focused sharply on the retina by the cornea and the lens. Both of these refract (bend) light. The cornea has a **refractive power of 43 D**, and the healthy lens has a **refractive power which varies between 12 and 26 D**. Thus the **total refractive power is 56–69 D**. The lens allows the eye to accomodate, so that both near and distant objects can be focused on the retinal fovea (**foveated**), and thus can be clearly seen. When we look at distant objects with normal eyes and relaxed ciliary muscles, the object is automatically foveated. However, when we look at nearby objects, the light is initially focused behind the retina. The lens then rounds up, by contraction of the ciliary muscles and relaxation of the zonule fibres (i.e. accommodation), to focus the image on the fovea.

What characterizes the normotropic or emmetropic eye?

The normotropic eye has the ideal refractive power. **Parallel light** from the far point (F in the upper part of Fig. 13-1) is **foveated on the retina** in the unaccommodated eye. Light from the near point (N in Fig. 13-1) in the totally accommodated eye is also foveated.

The shaded areas of the illustration are that fraction of the three-dimensional space that can be focused on the retina for a given visual axis (Fig. 13-1).

What is myopia?

Near-sighted (myopic) patients usually have **elongated eyeballs**. More rarely, myopia can be caused by too much refractive power in the lens system. Myopic persons can only foveate diverging light waves – both from F and N (Fig. 13-1). The images of distant objects are focused in front of the retina, and the image is blurred on the retina. Both F and N are located in front of the eye. Correction is accomplished by **concave lenses** (−D).

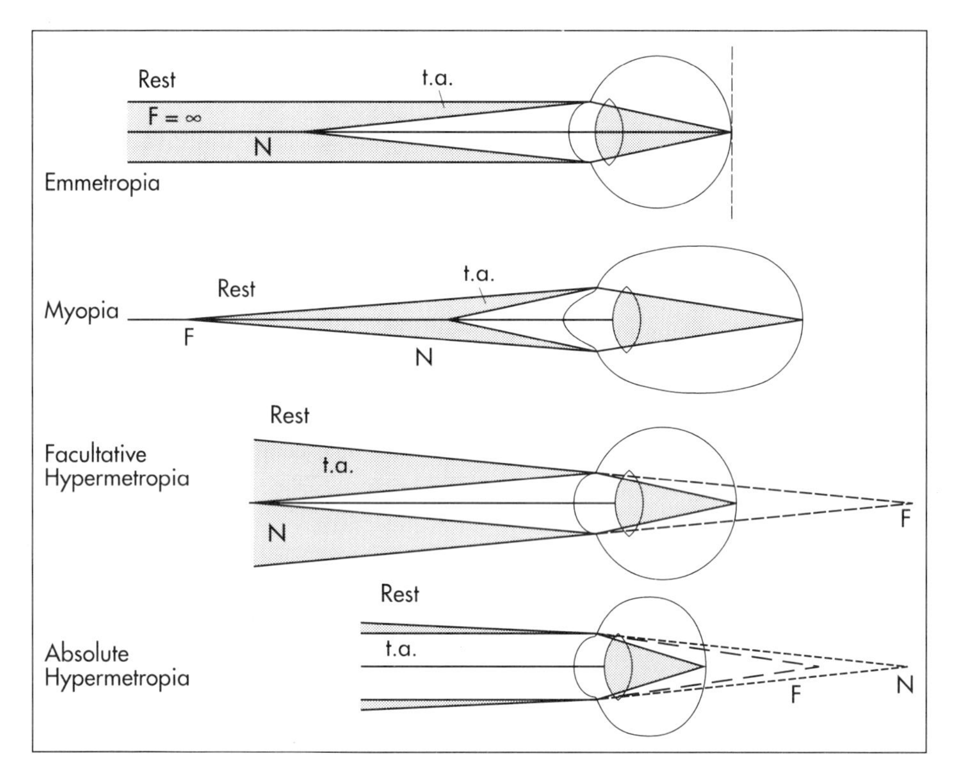

Fig. 13-1: Hypothetical light rays for emmetropic, myopic, facultative and absolute hypermetropic eyes (t.a.=total accommodation; rest=no accommodation).

What is hypermetropia?

Hypermetropic or **far-sighted persons** usually have **shortened eyeballs**. In rare cases it can also be caused by insufficient refractive power in the unaccommodated eye. The **absolute hypermetropic eye** can only focus images of distant objects behind the retina (Fig. 13-1). The **facultative hypermetropic eye** can focus converging light beams on the retina without accommodation (rest in Fig. 13-1). This patient can read the Snellen letters without problems; they also foveate diverging light by accommodation (Fig. 13-1). Thus, the patient gets eyestrain due to fatigued ciliary muscles. Hypermetropia is corrected by **convex lenses** (+D).

What is astigmatism?

Astigmatism is a refractive disorder of the eye, in which the **curvatures of the cornea or lens are different along different meridians**. The different meridians therefore have different focal distances. Therefore, astigmatism can be corrected with **cylinder lenses** that correct the curvature differences.

What is cataract?

Cataract is an eye disease, where the **vision is blurred by an opaque lens**. Precipitation of lens proteins can occur in several ways. It is often due to oxidative processes. The lens needs oxygen, but strong sun light or radiation can **oxidize lens proteins** in unprotected eyes. The oxidation is enhanced by hyperbaric oxygen therapy and by high blood [glucose] in diabetics. Oxidants in the food may be the cause in some patients. Antioxidants, such as vitamins A and D, seem to protect against the loss of transparency in long-term studies.

What is presbyopia?

Presbyopia is called **old man's sight**. The far point remains where it is, so the unaccommodated refraction is unaltered. The ability to accommodate is changed, so that N approaches F.

With age, the lenses of most people lose their elasticity, and hence their ability to assume spherical shape. The lens is the organ in our body with the highest protein concentration. Alterations of lens proteins probably cause **progressively increasing stiffness of the lens**. The accommodative power decreases from 14 D in a child to less than 2 D at the age of 50. The patient's eye becomes incapable of accommodation for near vision and reading. Presbyopia is corrected by **convex lenses**.

What is understood by conjugate eye movements?

Conjugate movements are **movements of both eyes in the same direction and magnitude, so that the relation between the visual axes is maintained**. When focusing on far-away objects, the parallel axes are maintained during conjugate movements. Likewise, conjugate eye movements maintain the convergence angles of the eye required for focusing on nearby objects.

Saccadian or jumping movements are rapid eye movements. Saccadian eye movement is an instantaneous reposition of the eye that occurs e.g. when reading or when focusing on a flash of light in the peripheral visual field. The velocity of the movement is up to $500^\circ\,s^{-1}$. The latency period is 250 ms, and the contraction time is 50 ms. The compensatory eye movement involving the vestibular system occurs when the head rotates. This is also an example of saccadian eye movement. In contrast, **pursuit movements** are smooth eye movements that allow the eye to track a moving object. They have a velocity of up to $30^\circ\,s^{-1}$.

These two movements work together in **optokinetic nystagmus**. This is a shift between smooth pursuit movements and correcting jumps. The direction of nystagmus is by convention indicated by the rapid correcting phase.

What are miniature eye movements?

Even during foveation of an object the eyes are not totally still. The eyes are continuously performing **miniature eye movements which occur at a rate of 3 microsaccades per second, with a mean amplitude of 0.1°**.

Describe the photoreceptors of the retina

The number of photoreceptors in a human eye is estimated to be 110–130 million rods and 5–7 million cones.

Each **photoreceptor cell** includes an outer and an inner segment, which are united by a thin cilium. The outer segments are directed towards the pigment epithelium of the peripheral retina, and contain stacks of **discs** that are rich in photopigment molecules. The inner segments contain the cell nucleus and numerous mitochondria. The **rods** are predominant outside the fovea, and they contain much more pigment (10^8 rhodopsin or molecules per rod) than do cones. Rods are so sensitive that a single photon can trigger a rod response. Rods are therefore well suited for night vision. Rhodopsin or **visual purple** has two absorption maxima: 350 and 500 nm. The spectral extinction curve for rods corresponds to that of rhodopsin, suggesting that rhodopsin is the chemopigment in rods. Rhodopsin consists of a glycoprotein (opsin) and a chromophore group (11-*cis*-retinal). Retinal is the aldehyde of vitamin A_1 (retinol).

The fovea only contains **cones**. Cones function in the daytime, with maximal visual acuity and colour vision. The human eye posesses **three types of cones**, each with a specific pigment related to the three basic colors: **red (erythrolab), green (chlorolab) and blue (cyanolab)**. The cones in the fovea do not contain cyanolab.

Describe the photoreceptor mechanism

When the human eye is fully adapted to darkness, its rods have **open Na^+-channels**, and the resulting influx of Na^+ maintains depolarized rods with a resting membrane potential of **−40 mV**. The rod cell synapses with bipolar and horizontal cells, and releases **glutamate** as long as the dark depolarization is maintained. Na^+ is continuously removed from the rod by the **Na^+-K^+ pump**.

Inside the rod a special amplification takes place. Light absorption by a **single rhodopsin molecule** activates **thousands of G-protein molecules (transducin)**, which then activate large quantities of **cGMP phosphodiesterase in the discs**. Each of these enzyme molecules catalyses the hydrolysis of **cGMP to 5′-GMP at a rate of thousands per second**. The **reduction in [cGMP] closes the Na^+-channels**, and causes the cell to hyperpolarize. The amplification mechanism is probably why the eye is capable of detecting a single photon.

A similar cascade of reactions takes place in cones, when they are stimulated. Cones are so small that the hyperpolarization occurs rapidly.

Describe the three types of retinal ganglion cells

Each **ganglion cell has a receptive field in the retina** that is comprised of a number of photoreceptors.

The **fraction of a receptive field belonging to each photoreceptor** is added to **neighbouring areas** in order to obtain the **receptive area of a bipolar or a horizontal cell**. An on-bipolar cell is depolarized by white light, whereas an off-bipolar cell is hyperpolarized. Signals are transmitted from the photoreceptors to the ganglion cells as a graded response. These small receptive areas are summated to form a **circular receptive field for each ganglion cell** (Fig. 13-2). Ganglion cells can generate APs and transmit signals to the brain.

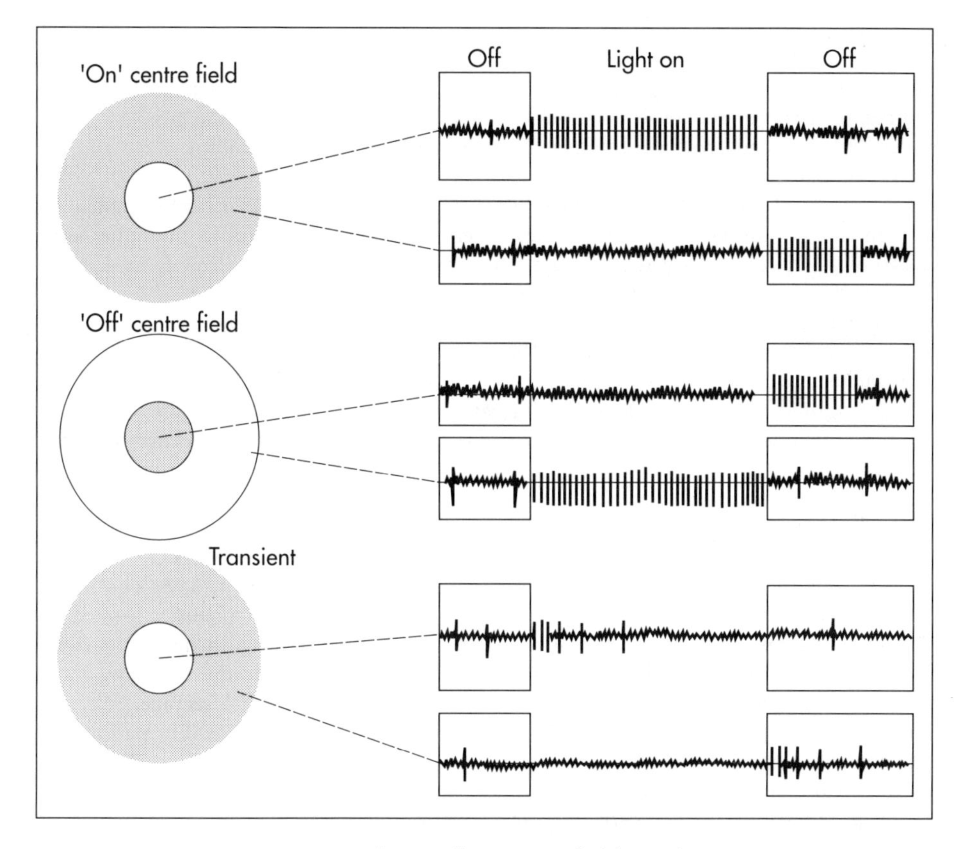

Fig. 13-2: Ganglion cell receptor fields in the retina.

1. One type of ganglion cell has a centrally located excitatory area, surrounded by an inhibitory annular area (Fig. 13-2). Together these form an **on-centre off-surround receptive field.** Here, an on-response is triggered in the bipolar cell that is connected to the on-ganglion cell.
2. Another type of ganglion cell has a centrally located inhibitory area (inhibited by light), surrounded by an excitatory annular area (Fig. 13-2). These form an **off-centre on-surrounding receptive field.**
3. A third type of ganglion cell is connected to both on- and off-bipolar cells, so its centre is both stimulated and inhibited by white light.

The ganglion cells can also produce **transient or sustained** reactions. These reactions are due to adaptation to light (decreased sensitivity with exposure) and lack of adaptation, respectively (Fig. 13-2).

Ganglion cells in the fovea are connected to few or only one cone. Some ganglion cells are excited by blue light and inhibited by its opponent colour yellow; other cells are excited by green and inhibited by the opponent colour red. This mechanism is the so-called **colour contrast analysis** of the retinal ganglion cells. Colour opponent neurons are found not only in the ganglion cells but also in the lateral geniculate nuclei.

What is the function of the lateral geniculate nuclei?

Retinal signals pass through the main visual pathway: the optic nerve, the **lateral geniculate nucleus,** the optic radiations (the geniculostriate tracts), the primary visual cortex, the pretectal area, the Edinger–Westphal nucleus, the oculomotor nerve and the ciliary muscle.

Each point of the retina has a corresponding location in the **dorsal lateral geniculate nucleus and in the visual, striate cortex (area 17).** The nerve fibres in the optic nerve are arranged in such a way that the upper quadrants of the retina are represented in the upper half of the nerve, and the lower quadrants in the lower half.

Such a **retinotopic map** is present in the **lateral geniculate nucleus** and maintained throughout the visual pathways and in the visual cortex. The receptive field in the retina is maintained all the way to the cortex. This is the basis of fusion. The consequence is that the **right striate cortex** receives information about objects located in the left side of the visual field, and the **striate cortex in the left hemisphere** receives information about the right side of the field of vision. In general, each hemisphere of the brain is connected to sensory and motor activity of the opposite side.

The **lateral geniculate nucleus** has three different pairs of neuronal layers (1–2, 3–4, 5–6). Ganglion cells from the ipsilateral (same side) eye projects to **layers 2, 4 and 6,** whereas ganglion cells from the contralateral eye projects to **layers 1, 3, 5.** The lateral geniculate nucleus is involved in integration and registration of pictures formed in corresponding areas of the retinal surfaces. Some neurons react to white light (with circular receptor fields), while other neurons react to opponent colours. When we jump from one highlight to another in the visual field, each jump is called a saccade. Selection of visual stimuli may be located in the lateral geniculate neurons (possibly performing gate control).

Most of the neurons in the geniculate nucleus projects to the striate cortex by way of the **optic radiations (geniculostriate tract).** Neurons in a certain column of the lateral geniculate nucleus project to precisely the same part of the striate cortex (area 17). The lateral geniculate nucleus also receives information from the cortex (in particular the visual cortex) that is essential for selection of signals of particular interest.

How are the cells of the visual cortex organized?

The **striate cortex (area striata, area 17)** is located around the **calcarine fissure** on the medial side of each occipital lobe. The optic radiation ends mainly in synaptic contact with simple cells in layer 4 of the striate cortex. Simple cells have on- and off-fields. Complex cells receive inputs from several simple cells, and hypercomplex cells receive inputs from several complex cells.

Axons from one eye terminate in millions of functional units called **ocular dominance columns** consisting of about 10^3 neurons. Cortical neurons are arranged in **orientation columns** showing orientation selectivity for lines, edges or bars. Other cortical neurons are arranged in **direction columns** showing direction selectivity. **Colour blops** are interspersed among the other columns (see later). The macula is represented by a large area at the occipital pole, and the upper and lower half of the visual field is represented below and above the calcarine fissure. The upper layers of the superior colliculus perform visual processing. The deep layers produce eye movements.

The **cortical area V4** contains colour-sensitive neurons, and the **visual association areas 18 and 19** (Fig. 11-1) contain many cells with complex functions.

What is the normal sensitivity to light?

The absolute sensitivity depends upon the adaptative condition of the retina, the pupillary diameter, and the source of light (spectral composition, exposure time, and light source dimensions). The threshold for the **completely dark-adapted eye is** $(7 \cdot 10^{-11})$ **W** m^{-2}. Light adaptation is a decrease in visual sensitivity during constant stimulation. This occurs rapidly because the rhodopsin bleaches readily. Hence, in daylight (photopic cone vision) we are dependent on cones for vision. Night vision (scotopic rod vision) is extremely sensitive to light, because of dark adaptation. It takes at least 20 min in dark surroundings before the rods become fully adapted. In a dark cinema we have scotopic vision with low visual acuity and colour blindness. As soon as the film is projected we experience partial light adaptation, so that the photopic cone vision is resumed.

What is the basis for colour vision?

The trichomacy theory postulates that any colour can be produced by an appropriate mixture of the three basic colours: red, green and blue. The three types of cone pigments have different opsins that differ from that in rhodopsin.

Groups of cortical neurons called cortical **colour blops** respond specifically to colour signals, and also receive signals from adjacent columns of the visual cortex. **Cortical colour blops** are probably the primary station for perception of colour, and they are found both in the primary and the secondary visual cortex areas. Perception of spectral opponent colour pairs is located in discrete colour blops of the visual cortex.

The three cone pigments are: erythrolab for red (maximal sensitivity at 555 nm), chlorolab for green (525 nm), and cyanolab for blue (450 nm). The absorption spectra of the photopigments overlap considerably. The three cone types are uniformly distributed in the retina, except in the fovea. **The fovea has no cyanolab cones** and no rods. This gives the fovea **partial physiological scotoma**, i.e. no blue vision and no scotopic dark vision. The real **physiological scotoma** is the **dark spot** corresponding to the optic papilla. Inhibition of neighbour ganglion cells from **on-centre field ganglion cells** is called **lateral inhibition**; it occurs also in the lateral geniculate nucleus or in the visual cortex. Lateral inhibition provides simultaneous contrasts and enhancement. Each **colour-contrast neuron** is excited by one colour and inhibited by the opponent colour. **Opponent colours** are red–green, yellow–blue, and green–purple.

Contrast analysis begins already in the retina and is elaborated centrally in the lateral geniculate nucleus, the thalamus and the visual cortex. If there is a multilevel neural system for the analysis of colour mixing, we also need to assume the existence of a neural system for colour brightness, depending upon the intensity of the light. Healthy people are trichromats, because they have all three cone pigments.

What are the causes of colour blindness?

The three colour genes are located on an **X chromosome**. Females have two X chromosomes, and colour blindness is rare among females. Colour blindness is inherited from the **father** – via the daughter – to her **son**. The trait is recessive and sex linked. The total incidence is about 8% of the male and 0.5% of the female population. **Monochromats** lack all three or two cone pigments, an extremely rare disorder. **Dichromats** lack one of the three cone pigments. Proteus is the first or red component, so **protanopic people** are blind for the red part of the spectrum. They cannot separate red and yellow signals in traffic. **Deuteranopic patients** are blind for the second or green colour, and **tritanopics** are blind for blue – the third basic colour. **Abnormal trichromats** have a reduced amount of one cone pigment: **protanomalous trichromats** lack erythrolab, **deuteranomalous (the most frequent type)** lack chlorolab, and **tritanomalous** lack cyanolab.

What is meant by spatial and temporal visual resolution?

Spatial resolution or **minimum separability** is the capacity of the eye to see two stimulated retinal areas as separated. In healthy young humans the spatial resolution is about 1/60 degree, depending upon luminosity, exposure time, patterns and opponent colours in the visual scenery. The most important factor limiting this capacity is the cerebral integration.

Temporal, visual resolution is the capacity of the eye to see **consecutive light stimuli as separate.** Intensity is directly related to duration of perception of light. Contrast further decreases temporal resolution, e.g. a flash of light in the dark is perceived for longer than in bright surroundings. Temporal resolution is also determined by the wavelength of light. The eye is maximally sensitive at the absorption maxima of the three cone pigments and rhodopsin.

What is meant by the positive and negative afterpicture?

The **positive afterpicture** is a visual impression lasting longer than the stimulus. It is visible on a dark background following exposure of the eye to intense light. The **negative afterpicture** follows the positive afterpicture as a dark shadow or as the opponent colour. The negative afterpicture is due to adaption of the area in the retina related to the picture.

What is the critical flicker fusion frequency?

A **flickering source of light** liberates **successive flashes so rapidly that they fuse, and appear to be continuous.** In the darkness of a cinema we do not sense the flickering frequency of 24–48 frames s^{-1}, or those of a television screen with 50–60 frames s^{-1}. With increasing intensity of illumination the critical fusion frequency increases abruptly. This is why young persons can look directly into a neon light and see its flickering character even with 60–100 flashes s^{-1}. Accordingly, the cones of the healthy human eye have a **critical fusion frequency** around **60–100 flashes s^{-1} with optimal illumination.** The photopic cones are much more sensitive to rapid alterations of light intensity than the rods.

What is hemianopsia and localized blindness?

Visual field defects are caused by interruptions of the visual pathways.

Hemianopsia means loss of vision in half of the visual field of both eyes. The loss of vision refers to the visual field, and thus to the contralateral half of each retina (Fig. 13-3).

Homonymous hemianopsia means that the same side of the visual field for each eye is defective. Corresponding halves of each retina have lost vision.

Homonymous hemianopsia occurs from lesions of the entire optic tract, the lateral geniculate body, the optic radiation, or the entire visual cortex of the contralateral hemisphere (Fig. 13-3). A lesion of the striate cortex often spares the large macular area at the occipital pole. This results in a disorder termed **homonymous hemianopsia with**

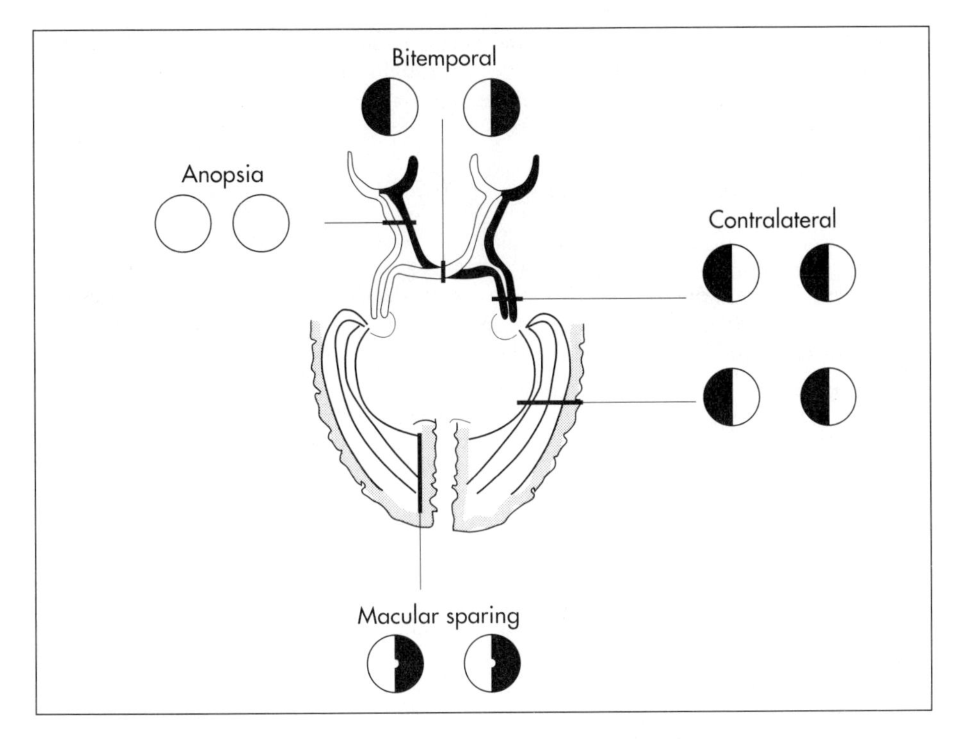

Fig. 13-3: Visual field defects. Lesions are shown with bold bars.

macular sparing (Fig. 13-3). Partial lesions may cause **quadrant-anopsia.**

Heteronymous hemianopsia can be bitemporal or binasal (Fig. 13-3). Bitemporal hemianopsia results from damage of the optic nerve fibres as they cross the optic chiasm (Fig. 13-3).

An expansively growing pituitary tumour, perhaps related to acromegaly, can damage crossing fibres, originating from ganglion cells in the nasal halves of each retina. Expansion of the tissues surrounding both carotid arteries is a rarity, which can damage nerve fibres from the temporal halves of each retina, and cause binasal hemianopsia.

As long as the patient sees with both eyes, he or she may not experience any visual defect caused by damage to non-corresponding areas of the retina.

Localized blindness or scotoma is caused by a lesion of the retina in one eye, or by partial interruption of the optic nerve. Interruption of the entire optic nerve results in complete blindness or anopsia (Fig. 13-3).

Describe the lesions responsible for visual agnosia

Bilateral temporal lobe lesions can lead to the Klüver–Bucy syndrome. In this condition, both the temporal cortex, hippocampus and the amygdaloid body are damaged. The Klüver–Bucy syndrome includes **mental blindness (visual agnosia)**. Mental blindness is the inability to recognize objects seen. Besides mental blindness, the syndrome consists of loss of short-term memory, and hypersexual behaviour incompatible with normal social adaptation. This hypersexual behaviour is related to the visual agnosia.

Damage to visual areas of the temporal cortex alone causes isolated **visual agnosia.** **Visual agnosia** or mental blindness, is lack of the ability to combine the seen object into a concept. This visual agnosia can be **colour-agnosia (acromat-agnosia)** or **face-agnosia (prosop-agnosia).**

How do we discriminate movements, depth and distance in a visual impression?

Movements in the visual scenery are depicted as opposite movements on the retina. Convergent inputs from the two eyes result in **depth perception (i.e. stereopsis or stereoscopic vision).** Stereopsis depends upon the **medial, longitudinal fasciculus and the corpus callosum.** They co-ordinate the movements of the two eyes. The two eyes are 7–8 cm apart, which causes slight disparities between their retinal images. The **perception of depth** is probably exhibited by disparate receptive fields, and thus **excitation of specific cells in the secondary visual cortex.** Distance evaluation requires high visual acuity and experience with objects of known size.

Describe the development of vision from birth

Essential for the development of the baby's brain is human milk proteins and long chain fatty acids in the mother's milk. Protein deficiency from birth reduces formation of brain neurons and thus limits brain development including the development of visual capacity. Many vitamins and key proteins have hormonal and transmitter function in the brain, and lack of such substances in the critical growth period just after birth results in irreversible damage. The action of endogenous **nerve growth factor (NGF)** is necessary for the normal functional and anatomical development of the visual system. In the critical period of visual development, which is the first 2 years of life, the child must be exposed to a **multitude of visual stimuli.** This is necessary for the development of neurons and key substances that can record future visual stimuli. The ability to fuse the two optic fields is a process that has to be practised. This fact is an important basis for the treatment of cross-eyedness (strabismus).

What is strabismus?

Cross-eyedness or squint (strabismus) is an eye disease, where the visual axes of the two eyes do not converge on the fixation point of the object simultaneously. Thus the retinal images do not fuse on corresponding areas on the two retinas. Since the fixation line only foveates in one eye, the patient can learn to suppress the other picture in the brain. Hereby, **double vision is avoided** at the expense of visual acuity.

Further reading

Bahill, A.T. and T.M. Hamm (1989) Using open-loop experiments to study physiological systems, with examples from the human eye-movement systems. *News Physiol. Sci.* **4**: 104.

Berardi, N., A. Cellerino, L. Dominici, M. Fagiolini, T. Pizzorusso, A. Cattaneo and L. Maffei (1994) Monoclonal antibodies to nerve growth factor affect the postnatal development of the visual system. **Proc. Natl Sci., USA, 91** (2): 684–8.

Livingstone, M. and D. Hubel (1988) Segregation of form, color, movement and depth: Anatomy, physiology and perception. *Science* **240**: 740.

Stryer, L. (1986) Cyclic GMP cascade of vision. *Annu. Rev. Neurosci.* **9**: 87.

13. Case History

A 30-year-old female complains of eyestrain and frontal headache during reading – sometimes followed by nausea and vomiting.

The patient is placed 6 m from the Snellen test chart. She is able to read line 6, which is the letter size read by a normotropic eye. Now thin convex lenses are placed in front of her eyes, but she can still read line 6. The diopter of the strongest convex lens with which she can still read line 6 is +4 D for both eyes. The converging light must be directed against the far-point (F), and F must be located behind the eyes at a distance of 0.25 m. Examination with concave lenses reveals the strongest concave lens by which she can read line 6 to be –3 D or N = –0.33 m (in front of each eye).

1. *What is the refractive anomaly of the patient?*
2. *A normotropic person aged 30 years has an accommodative power of 7 D. Compare the accommodative power of the female patient to that of the normal person.*
3. *Is it possible for the patient to read a fine text 0.2 m in front of her?*
4. *The saggital diameter of the patient's eyes is typical. Describe its characteristics.*
5. *This patient has a reduced outflow of chamber fluid at the iridochorneal junction. She has an increased risk of developing an eye disease, which is the most common cause of blindness in the world. Describe the most likely condition and the relation to her eye anatomy.*

Try to solve the problems before looking up the answers in Chapter 74.

CHAPTER 14. THE AUDITORY AND VESTIBULAR SYSTEMS

What characterizes the auditory and the vestibular system?

The two systems share the **labyrinth**, and transmit signals to the brain through the **8th cranial nerve**. The two systems record **fluid movements** and use the so-called **hair cells** as **mechanical transducers**.

How is sensitivity to sounds and tones measured?

Sounds are **sense impressions** that consist of complex mixtures of compression and decompression waves that can be broken down to pure tones by Fourier analysis. Pure tones are **sinusoidal waves of a specific frequency (cycles s^{-1} or Herz=Hz) and amplitude.**

Sinusoidal waves can change **phases**. The normal human ear is sensitive to pure tones with frequencies between **10 and 30 000 Hz**, in a young person. As people age, their capacity to hear high tones declines. This condition is termed **presbyacusis**.

Sound propagates at **343 m s^{-1} in air at 20°C**, although each single air molecule only moves a few μm in the direction of propagation. The unit of sound pressure (p) is the Pascal (Pa). According to international convention the **sound pressure level (SPL)** is expressed in decibel (dB):

$$\text{dB, SPL} = 20 \log p/p_o.$$

The actual pressure is p, and the threshold for sound pressure is p_o . The threshold for sound pressure is **20 μPa** in air (p_o) at **1000 Hz** in a soundproof chamber for a healthy person. This pressure corresponds to a **sound effect** of **10^{-12} W m^{-2}**. Any rise in the SPL of 10 dB implies a rise in sound pressure by a factor of 3, since the log of 3 is 0.5:

$$10 \text{ dB} = 20 \log 3.$$

Speech has an intensity of **60–65 dB**, and sounds that exceed 100 dB can damage the ear. A constant sound stimulation only results in minor adaptation. The human ear has the largest sensitivity **around 1000–4000 Hz**, the range for normal speech.

What is the function of the middle ear?

The sound pressure waves in air are converted into sound pressure waves in the fluid column within the cochlea. The pressure wave in the air is transmitted via the tympanic membrane and the ossicles (malleus, incus and stapes), to the fluid of the cochlea. The footplate of the stapes inserts in the oval window, and separates the middle ear from the fluid of the cochlea. The ratio of the effective surface area of the tympanic membrane to that of the oval window is 14:1, and the pressure is increased further by the differing lengths of the lever arms in the chain of ossicles. By this area–pressure amplification, hearing is improved by more than 25 dB. When the external ear is filled with water during diving, hearing is seriously reduced.

Two muscles are found in the middle ear. They **dampen movements of the ossicular chain** when the ear is exposed to extremely high pitched sounds that can be anticipated. These muscles are the **tensor tympani muscle** supplied by the trigeminal nerve, and the **stapedius muscle** supplied by the facial nerve. Exposure to sounds above 90 dB elicits **reflex contractions.**

How is sound transducted in the cochlea?

The cochlea is composed of three tube systems coiled together to form a pyramid: **scala vestibuli, scala media and scala tympani** (Fig. 14-1). The part of cochlea beneath the oval window is called scala vestibuli, and it is filled with a fluid column termed perilymph.

The perilymph conducts the pressure wave to the basilar membrane, which is displaced within the endolymph together with the whole organ of Corti which contains the hair cells.

Each hair cell has 40–100 hairs (**stereocilia**). The hairs have different heights, and when the pressure wave displaces the hairs towards the tallest hair, the hair cells are depolarized. When the basilar membrane moves upward towards the scala media, the reticular lamina shifts upward and inward (Fig. 14-1), causing the hair cells to depolarize by opening many K^+-channels. Downward movement of the basilar membrane towards the scala tympani moves the reticular lamina downward and outward (Fig. 14-1). This movement closes the K^+-channels and the hair cell membrane hyperpolarizes.

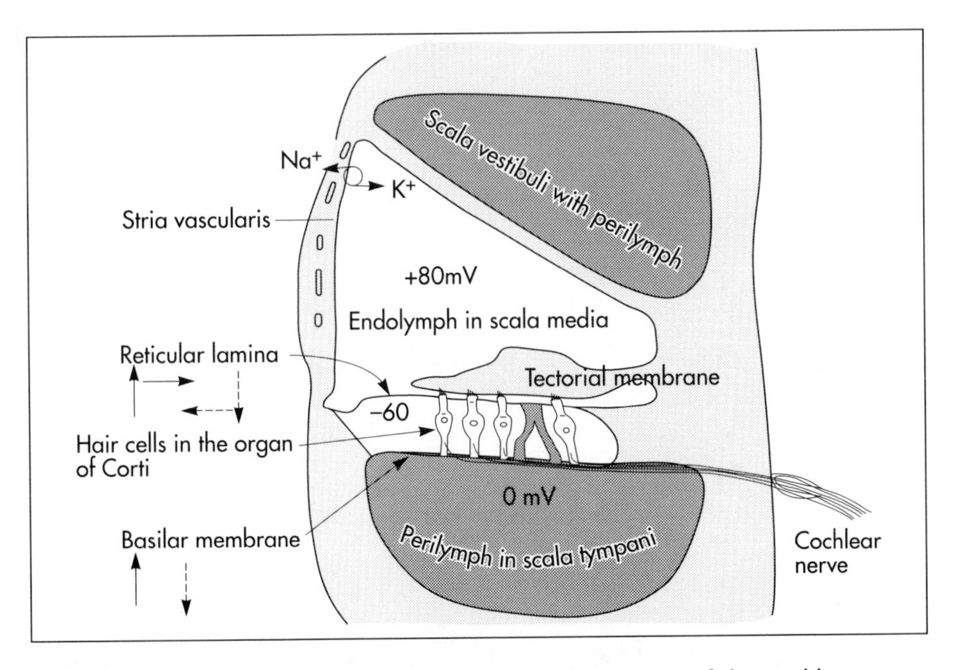

Fig. 14-1: A cross-section through one of the turns of the cochlea.

Describe the transduction process at the level of the hair cells

The endolymph in the scala media has a potential difference of +80 mV with the perilymph as reference. The inside of the hair cell is –60 mV compared to the perilymph, a RMP about the same size as in most neurons. Thus the **total potential difference** between the inside of the hair cell and the endolymph in the scala media is –140 mV. This RMP is maintained by Na^+-K^+-pumps in the stria vascularis (Fig. 14-1).

Bending of the hairs change the conductance of K^+-ions through the apical hairy membrane, and this is how the RMP is changed. A current flow is produced through the hair cell from apex to base which is resting on the **basilar membrane** (Fig. 14-1). This current flow or **receptor potential** can be recorded extracellularly with macroelectrodes as the **cochlear microphonic potential** (i.e. the sum of receptor potentials from many hair cells). This potential has the same frequency as the acoustic stimulus, and the potential is analogous to the output voltage of a microphone. The cochlear microphonic potential follows the sound stimulus without latency, without measurable threshold, and without fatigue in contrast to neuronal action potentials.

When stimulated the hair cells release neurotransmitters (glutamate, aspartate) that excite the cochlear nerve fibres. Thus, the **propagating APs** are generated in the **cochlear nerve fibres**.

Describe the travelling wave theory

A **high frequency tone** produces travelling waves along the basilar membrane. High tones travel only a **short distance** from the stapes along the basilar membrane to their **resonant point**, where the displacement amplitude of the basilar membrane is maximal (Fig. 14-2). **Low frequency tones travel all the way to the apex of the cochlea** (Fig. 14-2). The higher the tone frequency, the more basally located in the cochlea is the resonant point and its potential.

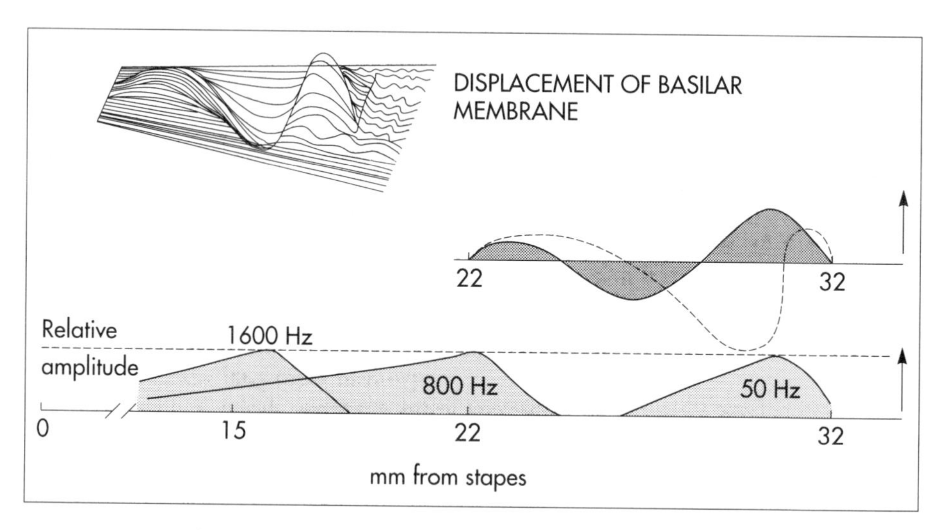

Fig. 14-2: Displacement of the basilar membrane illustrating the travelling wave theory (von Bekesy).

The existence of such a maximum of the travelling wave is termed **frequency dispersion**. Since different frequencies excite differently located hair cells the argument is called the **place analysis theory**. The brain also utilizes the temporal structure of the sound stimulus. This is the so-called **periodicity analysis**.

How do the auditory signals reach the brain?

The **receptor potentials** generate **action potentials in the cochlear nerve** (8th cranial nerve) that travel to the cochlear nuclei. Secondary neurons transfer the signals from here to the superior olivary nuclei that co-ordinates the two ears, or directly to the inferior colliculus through the **lateral lemniscus** (representing both ears). Axons from the inferior colliculus ascend to the medial geniculate nucleus of the thalamus. Axons from this thalamic nucleus form the **auditory radiation**, which terminates in the **auditory cortex in the superior temporal gyrus** (areas 41 and 42 in Fig. 11-1). High frequencies are projected to the **rostral auditory cortex**, and low tones to the caudal section.

The duration of a sound stimulus is encoded in the duration of the neural signal, and its intensity by the level of neural activity.

Projections from the auditory cortex also descend to the **medial geniculate nucleus** and the inferior colliculus. The oligocochlear bundle controls several sound impressions. Efferent stimulation through these pathways inhibits the sensitivity of these nuclei for sounds, while increasing their **tone selectivity**. This phenomenon, and a high degree of motivation, explains how a mother can hear her baby cry in spite of noise, and also how we can hear an individual in a crowd (the **cocktail party effect**).

How is sound localized?

Localization of a sound source depends upon the **difference in time** between the arrival of a low frequency sound signal to the left and right ears (time delay). The source of low frequency sounds (below 2000 Hz) is localized by this **time delay**. The source of high frequency sounds is localized by the difference in **sound amplitude** arriving at each ear caused by the dampening of the sound intensity. Sounds in the region of 2000 Hz cannot be detected by either mechanism. On average, the distance between the two organs of Corti is about 0.16 m. Then, a sinusoidal wave or pure tone with exactly the same wavelength coming from one side of the head are in phase when they reach the ears. This wavelength corresponds to the frequency of 2144 Hz (343 m s^{-1} divided by 0.16 m). In this instance the subject will be unable to determine the source of the sound.

Describe two types of deafness

Nerve deafness is caused by impairment of the cochlea or the auditory nerve or nucleus. The cochlea can be damaged by chloramphenicol, kinin and streptomycin. These drugs can cause hearing loss or deafness for **all sound frequencies**. Deafness to specific frequencies is caused by localized damage of the basilar membrane. This is typical for rock and beat musicians, soldiers, and aircraft pilots. The nerve deaf patient has a hearing loss when tested both by **air conduction** through the middle ear, and **bone conduction** through surrounding bone structures. A certain type of nerve deafness for high tones develops among older persons (**presbyacusis**).

Conduction deafness is caused by impairment of the mechanical conduction of sound into the cochlea. A hereditary disease called otosclerosis is due to fixation of the faceplate of the stapes to the oval window. Otosclerosis, blockade of the external ear with ear wax, otitis media, damage to the tympanic membrane, and to the ossicles, all cause conduction damage to hearing. Persons with conduction damage have normal bone conduction.

Describe the function of the vestibular system

The vestibular system detects if the body is in **balance**. The sensory unit of the auditory–vestibular system is the **membraneous labyrinth**, located in the petrous portion of the bony labyrinth. The membraneous labyrinth contains endolymph and is surrounded by perilymph; it is composed of the **auditory cochlear duct or scala media**, and the **balance regulating vestibular system.**

The vestibular system consists of **three semicircular ducts** and **two otolith chambers** (the utricle and the saccule). Each semicircular duct has a swelling termed an **ampulla**.

The semicircular ducts consist of a horizontal duct, a superior and a posterior duct at right angles to each other, so that they cover all three planes in space (Fig. 14-3). The semicircular ducts all communicate with the utricle. The utricle joins the saccule, which receives new endolymph from the cochlear duct.

The sensory organ of each utricle and saccule is called a **macula**. The sensory organ of each semicircular ampulla is the **crista ampularis.**

Each macula contains thousands of hair cells. Vestibular hair cells have many **stereocilia (hairs)** on their apical surface just as cochlear hair cells do; however, they also have a **large stereocilium called a kinocilium**. The hairs are embedded in a gelatinous substance, the **otolithic membrane**, that also contains **earstones or otoliths**. These otoliths increase the specific gravity of the otolithic membrane to twice that of the endolymph. Thus their hair cells are sensitive to **linear acceleration such as gravity and to static equilibrium control**, but not to angular accelerations of the head. The macula of

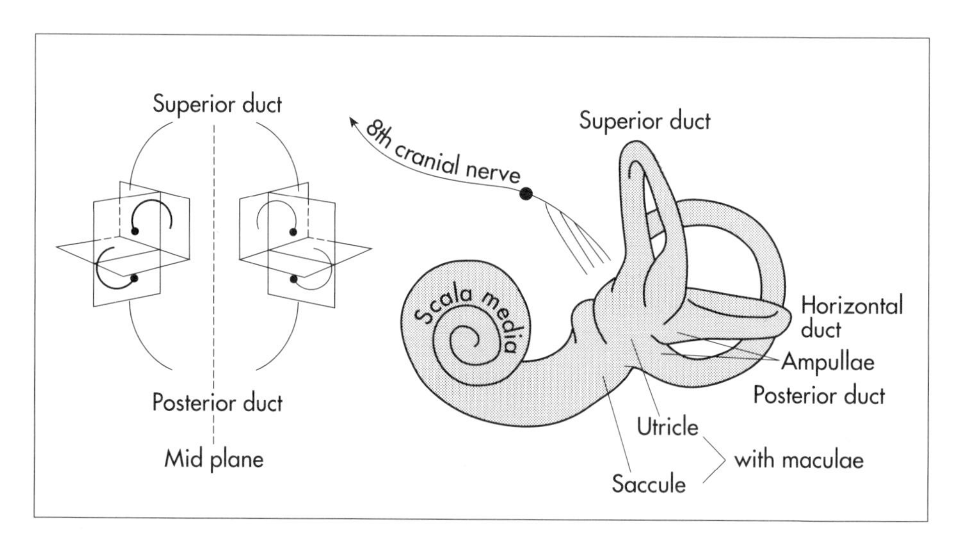

Fig. 14-3: The spatial orientation of the three semicircular ducts in the upright person (left). The horizontal duct is not drawn. The membraneous labyrinth is shown to the right.

the utricle is located in the horizontal plane, and the macula of the saccule in the vertical plane.

Each **crista ampularis** consists of many hair cells. Here the hairs are embedded in a large gelatinous substance termed a **cupula**. The cupula occludes the lumen of the ampulla completely, and its material has the same specific gravity as the endolymph. The cupula is concerned with **equilibrium control during motion and with angular acceleration** (rotation of the head), but is unaffected by linear acceleration.

When the **stereocilia are bent toward the kinocilium**, the conductance of the apical cell membrane increases for positive ions, and the hair cell becomes depolarized. Bending the stereocilia in the opposite direction hyperpolarizes the cell. The depolarized hair cell releases glutamate or aspartate and increases the discharge rate of the nerve fibre with which it synapses.

How do the utricle and saccule function?

The utricles and saccules are sensitive to **linear accelerations**. When we suddenly thrust our body forward, the otolithic membranes fall backwards on the cilia of the hair cells until the thrust stops. Then, the otolithic membranes fall forwards. The signals to the brain make us feel as if we were falling backwards. Therefore, we lean forward until the otolithic membranes are in balance.

What is nystagmus?

Nystagmus is a disorder with **abnormal involuntary movements of the eyeballs. Optokinetic nystagmus** occurs when travelling in a train or a car. The eyes remain fixed on an object long enough in order to gain a clear image. The semicircular ducts cause the eyes to rotate in the direction opposite to the direction of travel. Optokinetic nystagmus involves the vestibular nuclei, the **medial longitudinal fasciculus**, and the oculomotor nuclei.

Postrotatory nystagmus is observed in a person sitting in a rotating chair. This is the physiological adequate stimulus for nystagmus.

Caloric nystagmus refers to the horizontal reflex movement of the eye when the external ear is flushed with hot or cold water. The fast phase of the nystagmus is directed away from the ear flushed with cold water, and towards the ear flushed with hot water. The caloric nystagmus test is preferable to the postrotatory test for testing the nystagmus reflexes, because it examines one ear at a time, and is more convenient.

What is kinetosis or transportation sickness?

Many types of transportation which subject passengers to rapid changes in the direction of motion, elicit **kinetosis**. Kinetosis is a disorder with vertigo, nausea and vomiting. The disorder is triggered from the **vestibular system,** provided that the cerebellar function (the **flocculo-nodular lobe**) is intact. The flocculo-nodular lobes are linked to the equilibrium control of the semicircular system. Persons with destroyed semicircular canals or with destroyed flocculo-nodular lobes can be completely protected from kinetosis at the expense of lost equilibrium during motion.

Further reading

Brodal, A. (1981) *Neurological Anatomy in Relation to Clinical Medicine*, 3rd edn. Oxford University Press, New York.

Patuzzi, R. and D. Robertson (1988) Tuning in the mammalian cochlea. *Physiol. Rev.* **68**: 1009.

Von Bekesy, G. (1960) *Experiments in Hearing*. McGraw-Hill, New York.

14. Case History

A patient with a hearing loss of 26 dB is working in a power station, where the daily sound intensity is 100 dB and the air temperature is 20°C.

1. *Calculate the ratio between the sound pressure in the power house, and the sound threshold pressure for a healthy person.*
2. *Calculate the threshold pressure for the patient.*

Try to solve the problems before looking up the answers in Chapter 74.

CHAPTER 15.
THE AUTONOMIC NERVOUS SYSTEM

The autonomic nervous system directly influences **the cardiovascular system, other smooth muscles and glands through its two subdivisions,** the sympathetic and the parasympathetic system. The two subdivisions function in a dynamic balance. The autonomic nervous system also **modulates release of certain peptides and catecholamines, that affect both blood volume as well as total peripheral vascular resistance (TPVR).** The **enteric nervous system** lies within the walls of the gastrointestinal tract and includes neurons in the pancreas, liver and gallbladder, thus being an entity in itself. However, the enteric nervous system is clearly an important part of the autonomic nervous system that controls gastrointestinal motility, secretion and bloodflow.

The motor and premotor cortex, the cingulate gyrus and the hypothalamus can modulate the function of the **autonomic medullary control neurons.** Circulatory changes during exercise and in various stressful situations are influenced or governed by the cortex and deeper brain nuclei. The **cerebral cortex assimilates all inputs** of **visual, olfactory, labyrinthine, locomotor origin,** as well as from other **specialized sensors** (stretch-receptors, chemo-, baro-, osmo-, and thermoreceptors).

The integration of these inputs into an appropriate response takes place in the **hypothalamus** and in the **ponto-medullary circulatory centres.** From here the efferent signals pass to the periphery via the sympathetic and the parasympathetic pathways.

The primary afferent projections from the baroreceptors reach the **solitary tract nucleus** (STN), and from here we have connections to the important **dorsal motor nucleus of the vagus** (DMNV). A **high baroreceptor activity** stimulates the DMNV, so that the vagal inhibition of the heart is increased. More importantly, the high baroreceptor activity inhibits the sympathetic drive to the heart and vessels thus reducing blood pressure.

The hypothalamus and the other limbic structures are included in the **central autonomic system.** These central structures co-operate in situations of survival character: aggression and defence when attacked, feeding and drinking in starvation, reproduction and sexual satisfaction for continuation of life, thermoregulation in extreme temperatures and emotional behaviour in crises.

What are the four types of neurotransmitters in this system?

1. **Acetylcholine (ACh)** is the transmitter for all preganglionic fibres, and for postganglionic parasympathetic terminals.
2. **Noradrenaline (NA)** transmits signals from postganglionic sympathetic terminals.
3. **Peptides** function as co-transmitters.
4. **Purine nucleotides** (adenosine, adenine monophosphate (AMP)) are important vasodilatators.

How are the signals transmitted in the synapse of an autonomic ganglion?

ACh is the transmitter between the pre- and the postganglionic fibres, not only in the sympathetic nervous system, but also in the parasympathetic system. The cholinergic receptors of the ganglia are **nicotinic**. Nicotinic cholinergic receptors are activated by nicotine and ACh; these receptors are located in the ganglia and in the somatic motor endplate. Nicotinic receptors are linked to non-specific cation channels.

When the AP arrives at the preganglionic fibre, **ACh** is released from its terminals and diffuses across the synaptic cleft to bind to the specific nicotinic receptors on the membrane of the postganglionic neuron. This alters the Na^+- and K^+-conductance of the membrane by opening the non-specific cation channel for about 1 ms. The resulting current elicits an **excitatory postsynaptic potential (EPSP)**.

ACh is also the transmitter for the sympathetic innervation of sweat glands, and they are completely blocked by **atropine**. The ACh receptors of the sweat glands are **muscarinic**. These receptors are linked to G-proteins, second-messengers and specific channels, such as the adrenergic α- and β-receptors.

I. The Sympathetic Nervous System

The **preganglionic sympathetic nerve fibres** originate in small multipolar neurons in the lateral horn of the grey matter in the thoracic and lumbar spinal cord. The **central sympathetic outflow** converges on these preganglionic neurons. Their axons are thin myelinated fibres that leave the spinal cord through the ventral root. The preganglionic fibres then leave the spinal nerve forming myelinated **white rami communicantes**, through which they reach the nearest ganglion in the **paravertebral ganglia of the paired sympathetic trunk**. Typically, each fibre will end here forming synapses with up to 20 postganglionic neurons. A few preganglionic fibres pass the sympathetic trunk without interruption to form the splancnic nerves that reach the **three unpaired prevertebral ganglia** (coeliac = solar plexus, superior mesenteric and inferior mesenteric) of the lower intestinal and urinary organs. Most sympathetic ganglia are remote from the organ supplied. The postganglionic fibres are all unmyelinated, and they leave the sympathetic trunk through the **grey rami communicantes**, and thus reach the effectors supplied by the sympathetic system. The effectors are the smooth muscles of all organs (blood vessels, viscera, lungs, hairs, pupils), the heart and glands (sweat glands, salivary and other digestive glands). In addition, the sympathetic postganglionic fibres innervate adipocytes, hepatocytes and renal tubular cells.

Describe the sympatho-adrenergic system

This system is a functional and phylogenetic unit of the sympathetic system and the adrenal medulla. The **adrenal medulla** is a modified sympathetic ganglion. Any increase in sympathetic activity increases the secretion of adrenaline (Ad) and NA from the medulla into the circulation.

The preganglionic fibres to the adrenal medulla pass all the way to the special postganglionic cells in the adrenal medulla. The synapse is **cholinergic (nicotinic)** as it is for all preganglionic synapses.

The postganglionic cells have developed into cells filled with chromaffine granules, and are called **chromaffine cells**. These cells do not conduct signals, but synthesize adrenaline (Ad) (and NA) which is released in to the blood. Sympathetic stimulation triggers the conversion of tyrosine to dihydroxyphenylalanine (DOPA). A non-specific **decarboxylase** catalyses the conversion of DOPA to dopamine, which is taken up by the chromaffine granules in the cells. The granules contain the crucial enzyme, **dopamine β-hydroxylase.** This enzyme is activated by sympathetic stimulation, and catalyses the formation of NA from dopamine.

A few granules store NA, while the remaining granules liberate NA to the cytosol, where NA is methylated by **phenylethanolamine *N*-methyltransferase** to Ad. Ad is taken up by chromaffine granules and stored as the predominant adrenal hormone.

The **sympathetic system** is described above.

What are the main functions of the sympathetic system?

Stress is comprised of severe emotional and physical burdens (fear, pain, hypoxia, hypothermia, hypoglycaemia, hypotension, etc). Stress mobilizes the catecholamines of the adrenal medulla and of the sympathetic nervous system. The sympatho-adrenergic system gives rise to the **fright, flight or fight** reactions in acutely stressful situations.

Describe the adrenergic receptors

The sympathetic system exerts either excitatory or inhibitory actions through adrenergic receptors. Adrenergic receptors are **membrane receptors**. The dual response to adrenergic stimulation was known before Ahlquist in 1948 proposed that adrenergic receptors could be divided into two groups, α- and β-receptors, on the basis of **blocking drugs** (Table 15-1). The basic idea of Ahlquist was that NA acts predominantly on **vasoconstricting α-receptors,** and isoprenaline (Iso) predominantly on **vasodilatating β-receptors.** Both types of receptors are stimulated by **Ad**.

The **α-receptors** are blocked by **phenoxybenzamine and phentolamine. The** α_1-

Table 15-1: Adrenergic receptor subtypes. > indicates the rank order of sensitivity

	Adrenergic receptors			
	α-receptors		**β-receptors**	
Stimulated by:	NA > Ad		Iso > Ad $\geq$ NA	
Blocked by:	**Phenoxybenzamine**		**Propranolol**	
	α-receptors		**β-receptors**	
	α_1-receptors	α_2-receptors	β_1-receptors	β_2-receptors
Stimulated by:	NA > Ad	NA > Ad	Iso > AD = NA	Iso > Ad > NA Salbutamol
Blocked by:	**Prazosin**	**Yohimbine**	**Metoprolol**	**Butoxamine**

receptors are located on the surface of **target cells** (vascular smooth muscle, sphincter muscles of the gastrointestinal tract and bladder, and radial iris muscles). They are highly sensitive to NA, less sensitive to Ad, and almost insensitive to Iso (Table 15-1).

The α_1-receptors act through **phospholipase C** and through **intracellular [Ca^{2+}] elevation.** Ca^{2+} binds to calmodulin in the cytosol (Fig. 5-2). The complex activates **protein kinase,** which catalyses the phosphorylation of proteins. They become enzymatically active, and trigger vasoconstriction.

In contrast, the presynaptic α_2-receptors are located on the **presynaptic membrane (sympathetic end bulbs).** NA released into the synaptic cleft diffuses to the α_1-receptors on the targets cells, but part of the NA diffuses back to the α_2-receptors on the presynaptic nerve terminals. Here, **NA activates membrane adenylcyclase,** reducing [cAMP] in the cells (Fig. 5-1), and thus inhibiting release of more NA from the vesicles by negative feedback. Hence, a function of α_2-receptors is auto-inhibitory feedback. These receptors are also found in gastric smooth muscle cells and the β-cells of pancreatic islets. Stimulation decreases gastric motility and attenuates insulin secretion. An example of an α_2-antagonist is **yohimbine.** Although it is not used clinically, it is claimed to be an aphrodisiac.

The β-receptors are blocked by **propranolol** (Table 15-1). **The β-receptors** are located on effector cells that are **most sensitive to Iso, but less so to Ad and NA.** All β-receptors act through **activation of adenylcyclase and cAMP** (Fig. 5-1). β_1-Receptors are equally sensitive to NA and Ad, whereas β_2-receptors are more sensitive to Ad than to NA (Table 15-1).

β_1-Receptors are located in the **myocardium** – primarily on pacemaker cells. The β_1-receptors of the heart are **stimulated** by NA which increases cAMP production with increased chronotropic (increased heart rate) and inotropic effect (increased force). Cardioselective β_1-blockers such as metoprolol are used by heart patients, because metoprolol decreases cardiac arrhythmias and tachycardias.

β_2-Receptors are found primarily on **bronchiolar smooth muscle cells, vascular smooth muscle, uterine smooth muscle, salivatory glands, the intestine and the liver.** When NA binds to β_2-receptors, it causes **inhibition** of the target organ. Therefore, NA causes vasodilatation, bronchodilatation and uterine relaxation.

Similarly, sympatomimetics such as β_2-stimulators (salbutamol) increase cAMP production, resulting in bronchodilatation, increased salivary secretion, uterine relaxation and enhanced hepatic glucose output. β_2-Stimulators are used to eliminate bronchial asthma attacks. Butoxamine is a selective β_2-blocker.

What are catecholamines and how do they act?

Catecholamines are substances consisting of **catechol** (an aromatic structure with two hydroxyl groups) linked to an **amine.** The synthesis is described in Chapter 71 (Fig. 71-1).

Catecholamines increase heart rate and cardiac output ($\dot{Q}$) by stimulation of the **adrenergic β_1-receptors** in the myocardium. Catecholamines, released by the adrenal medulla, support the sympathetic system by modifying the circulation during exercise. During exercise the blood is directed to the working muscles from other parts. **Noradrenergic nerve fibres** innervate blood vessels **all over the body.** Sympathetic innervation accounts for vascular tone and vasoconstriction.

The most important exercise response in humans is a tremendous vasodilatation in the vascular bed of muscles. The vasodilatation is probably due to a decrease in the α-adrenergic tone of muscular arterioles, and to the action of Ad on β_2-receptors.

Catecholamines **dilate** the bronchial airways by stimulating **adrenergic β_2-receptors**. They increase both **tidal volume and respiratory frequency**. The result is an increased ventilation. Catecholamines acting on β_1-receptors cause increased $\mathring{Q}$. Catecholamines relax the **smooth muscles of the digestive tract** (β_2-receptors), but **contract the sphincters**. Catecholamines stimulate metabolism (by activation of the thyroid hormone, tri-iodothyronine (T_3)) and lipolysis. Ad stimulates hepatic glycogenolysis via β_1-receptors.

Finally, Ad stimulates the **ascending reticular system** (i.e. the reticular activating system or **RAS**) in the brainstem, thus keeping us alert and causing **arousal reactions** with desynchronization of the EEG (Chapter 10).

The resistance vessels of the striated muscles in hunting predators (and perhaps in humans) are also innervated by another system. This is the **cholinergic, sympathetic vasodilatator system**. It is capable of a rapid and appropriate bloodflow response during hunting.

When are catecholamines liberated?

Acute stress activates the splanchnic nerves and liberates large amounts of Ad from the medulla. Diabetics who are developing acute hypoglycaemia, secrete large amounts of catecholamines. Acute muscular activity starts a large catecholamine secretion in exercising persons.

Besides catecholamines, adenocorticotropic hormone (ACTH) is also released during stress by increasing hypothalamic signals. ACTH stimulates the glucocorticoid and to some extent the mineralocorticoid secretion through cAMP. **Small amounts of glucocorticoids** are **permissive** for the actions of catecholamines.

How are catecholamines eliminated?

Plasma catecholamines are rapidly removed from the blood and have a half-life in plasma of **less than 20 s**. This is the combined result of **rapid uptake by tissues** and **inactivation** in the liver and vascular endothelia (see Chapter 71).

Describe the sympathetic control of the heart

The postganglionic fibres from the three upper cervical ganglia of the sympathetic trunk pass to the heart as the **cardiac nerves to the cardiac plexus**. The sympathetic system innervates the sinus node, the coronary vessels and the myocardial syncytium. Each fibre ends in many terminals, and from the terminals the transmitter NA is released to the β_1-receptors of the smooth muscle cells of the coronary vessels and of the myocardium. In the example of a normal person, the contractility of the myocardium is increased, so the **end systolic volume (ESV)** falls from its usual volume of 70 ml to 40 ml, and the **end diastolic volume** increases due to increased venous return of blood from 140 to 180 ml, as an example. Hereby, the **stroke volume** is increased from **(140–70=) 70** to **(180–40=) 140 ml of blood**. Combining this with a three-fold increase in heart rate, we end up with a (2·3=) **six-fold rise in $\mathring{Q}$**.

Sympathetic stimulation depolarizes the sinoatrial (SA)-node, so that the threshold potential is reached faster than normal. Hereby, the heart rate is increased, and may reach

220 beats min^{-1} in young persons. Such a high frequency is due to a **maximal sympathetic activation** of the heart combined with a **reduction of the vagal tone.**

What are the effects of sympathetic activation?

Activation of noradrenergic fibres leads to **peripheral sympathetic vasoconstriction,** so that blood is shunted to central areas. The heart is stimulated through β_1-receptors so that its frequency and contractility is increased. Other organs are also stimulated to make the person fit for **fight or flight** in any stressful situation.

The **postganglionic sympathetic fibres** have **NA- and ATP-containing vacuoles** in their nerve terminals. Hence, they release NA and ATP. The NA is produced in the chromaffine granules of the neuron.

Liberation of NA and ATP to the blood does not only lead to **constriction of arterioles and arterial vessels,** but also **constriction of veins and venules.** Without venous constriction, the large **venous compliance** would cause an inordinate amount of blood to be stored in the veins upon sympathetic arteriolar constriction. The consequence would be decreased venous return which decreases $\mathring{Q}$ and perfusion of vital organs.

Activation of **presynaptic purine receptors** by **adenosine** inhibits Ad release from the postganglionic terminals innervating the blood vessels. This results in massive vasodilatation.

Exercise and stress demand mobilization of energy to muscles and heart. Activation of **β_2-receptors in the arteriolar wall** by circulating catecholamines from the medulla also contributes to vasodilatation in the striated muscles. The TPVR is reduced during exercise to 20–30% of resting values.

During stress the cutaneous circulation is reduced at first, but then the cutaneous bloodflow rises due to the **increased heat production.** The brain vessels are only modestly constricted by sympathetic stimulation (Chapters 20 and 25).

Could an adrenal medullary tumour cause hypertension?

Yes, some patients suffer from attacks of severe hypertension due to Ad hypersecretion. The hypertension is caused by release of large amounts of Ad from a **medullary tumour** (a **pheochromocytoma** of chromaffine cells).

II. The Parasympathetic System

The parasympathetic system has two subdivisions. The **cranial division** in the brainstem innervates the blood vessels of the head and neck and of many thoraco-abdominal viscera. The **sacral division** in the sacral cord innervates the smooth muscles of the walls of the viscera and their glands (the large intestine, liver, kidney, spleen, the bladder and the genitals).

The parasympathetic system only innervates a small percentage of the resistance vessels. Only arteries in the brain and of the penis, the clitoris and the labia minora receive parasympathetic innervation. Hence, the parasympathetic system has a **minimal effect on the arterial blood pressure.**

Parasympathetic fibres travelling in the **vagus nerve are of utmost importance in affecting the cardiac rate.** Vagal fibres innervate the SA- and the AV-nodes as well as the atrial muscle walls.

The parasympathetic system also innervates the tear and the salivary glands, and the muscles within the eye.

Excitation of the vagus decreases heart rate and atrial contractile force, increases intestinal motility, contracts the gall bladder and bronchi, and relaxes the sphincters of the gastrointestinal tract.

What is the effect of vagal stimulation on heart rate?

Vagal stimulation **decreases heart rate and the force of atrial contraction** as described above. The decrease in heart rate is due to the **rhythm shift** to **special P cells** which have a slow rate of depolarization. ACh is liberated on the cardiac cell membranes, **ACh-activated K^+-channels** are opened (via cholinergic receptors and G-regulatory proteins), and K^+ leaks out of the cells, thus opposing the **pacemaker current**. Vagal stimulation slows down the AV-conduction, causing the co-ordination of atrial and ventricular rhythm to be disrupted. Vagal stimulation can lead to death. Thus external massage of the carotid sinus can cause **collar death** by greatly increasing vagal stimulation.

Describe the cholinergic receptors

The effect of ACh released in the autonomic ganglia can be simulated by nicotine. Conversely, the effect of ACh released by parasympathetic nerve terminals at the target organs can be simulated by muscarine. These observations suggest the presence of two different types of cholinergic receptors. Cholinergic receptors are activated by ACh and by methacholine (MeCH).

The most important ganglionic blocking drug for blockade of both sympathetic and parasympathetic transmission is **hexamethonium** (Table 15-2).

Cholinergic receptors are located in **all autonomic ganglia (nicotinic type)**, in postganglionic terminals at **target organs** with parasympathetic innervation **(muscarinic type)**, and in the **motor endplate (nicotinic type)**.

Nicotinic receptors are those activated by ACh, nicotine and **nicotinic agonists** (e.g. dimethylphenylpiperazine, DMPP). Nicotine stimulates all autonomic ganglia simultaneously. Hence, sympathetic vasoconstriction in the limbs and viscera is accompanied by increased gastrointestinal activity and slowing of the heart via the vagus. Nicotinic receptors are blocked completely by ***d*-tubocurarine, and hexa- or decamethonium** (Table 15-2). The **motor endplate** has a different type of nicotinic receptor than the ganglions, since its receptors are not blocked by **hexamethonium**, but are blocked by ***d*-tubocurarine and decamethonium** (Table 15-2).

Muscarinic receptors are activated by ACh, muscarine and **muscarinic agonists** (pilocarpine and carbacholine, CCh). At least five different muscarinic receptor molecules have been identified (M_1, M_2, M_3 ...). Activation of the M_1 type is illustrated in Fig. 5-2. Activation of the M_2 type activates an inhibitory G-protein, which inhibits adenylcyclase (Fig. 5-1). Muscarinic receptor activation is linked to second-messenger systems.

Muscarinic receptors are blocked completely by **atropine**, and by antimuscarinic drugs such as homatropine and scopolamine (Table 15-2). These drugs do not block the nicotinic effect of ACh on the postganglionic neurons or on the motor endplate.

Table 15-2: Cholinergic receptor subtypes

	Cholinergic receptors				
	Nicotinic		Muscarinic		
Stimulated by:	Nicotine, ACh, MeCH, **DMPP**		Muscarine, ACh, **CCh**, MeCH		
Blocked by:	**Hexa**- and **deca**methonium *d*-tubocurarine		**Atropine**, scopolamine		
	Two types of nicotinic receptors:		Three muscarinic subtypes:		
	Ganglionic:	Neuromuscular:	M_1	M_2	M_3
Stimulated by:	Nicotine, ACh, **DMPP**	Nicotine, ACh			
Blocked by:	**Hexa**-methonium	**Deca**methonium *d*-tubocarine	Pirenzepine Atropine Dicyclomine	Atropine Gallamine	Atropine Hexahydrosiladifenol (HHSD)

What are the major differences between the sympathetic and the parasympathetic system?

1. The **sympathetic system** consists of **short** preganglionic and long postganglionic nerve fibres. The **parasympathetic system** contains **long** preganglionic and short postganglionic fibres.
2. The chemical transmitter at the target organ is **NA** in the sympathetic and **ACh** in the parasympathetic system.
3. The sympathetic system contains **adrenergic receptors** (α **and** β), whereas the parasympathetic system has **cholinergic receptors (muscarinic or muscarinergic and nicotinic or nicotinergic).**
4. Activation of the cholinergic system serves **anabolic functions** (i.e. stay and play), whereas activation of the noradrenergic system serves **catabolic functions** (i.e. fight, fright or flight).
5. Activation of α_1-receptors increases intracellular [Ca^{2+}], which leads to phosphorylation of protein kinases and thus to a response (Fig. 5-2). Activation of α_2-receptors triggers an **inhibition** of the membrane adenylcyclase, reducing [cAMP] in the cells (Fig. 5-1). β_1- and β_2-receptors activate adenylcyclase, which **increases cAMP production** in the cell (Fig. 5-1). Muscarinic receptors are completely blocked by atropine. Activation of M_1-receptors increases intracellular [Ca^{2+}] (Fig. 5-2). Activation of M_2 inhibits adenylcyclase, and through an inhibitory G-protein reduces the formation of cAMP (Fig. 5-1).

Further reading

Loewy, A.D. and K.M. Spyer (Eds) (1990) *Central Regulation of Autonomic Functions.* Oxford University Press, New York.

Shepherd, J.T. (1982) Reflex control of arterial pressure. *Cardiovascular Res.* **16**: 357–83.

15. Case History

A 19-year-old female is in hospital with a cranial lesion caused by a fall from her horse. The following clinical signs are found: (1) speech troubles and hoarseness, (2) swallowing problems and paresis of the soft palate, (3) rapid heart rate, and (4) dilatation of the stomach with vomiting.

Lesion of a certain cranial nerve can explain all symptoms and signs.

1. What is the name of the nerve?

2. What is special about this particular lesion?

Try to solve the problems before looking up the answers in Chapter 74.

SECTION III.
The Circulatory System

The human circulation is a continuous circuit. That is, the amount of blood pumped by the heart is the same as that which flows through each subdivision of the circulation. The heart consists of two pumps, the right heart that pumps the blood through the lungs, and the left heart that pumps the blood through the peripheral organs (with peripheral resistances).

CHAPTER 16.
THE BLOOD

In this chapter only blood viscosity, fibrinolysis, blood coagulation, thrombosis and pernicious anaemia are discussed. Other selected topics are treated in Chapters 17 and 27.

What is understood by the viscosity of the blood?

Viscosity is the **inner friction in the fluid**, which is due to the random movement of molecules and particles in the blood. This inner friction can be illustrated by telescope cylinders (laminae) of blood sliding against each other (Fig. 16-1). The outermost blood cylinder rests against the vessel wall, and the most central tube moves as a laminar flow with the greatest velocity (v). The velocity gradient, with the distance x from the centre of the blood vessel towards the outermost blood cylinder, is called the **shear rate** (dv/dx). The tangential force (F) between these blood cylinders depends upon the areas (A) sliding against each other, and the relation to **viscosity** (η) is given by the equation: $F/A=\eta \cdot dv/dx$. The **viscosity** (**η**) **1 Pascal second (1 Pa s)** is the tangential force, working on 1 m^2 of surface area, when dv/dx is 1 (s^{-1}).

This simplified description is valid for water, gas, and other homogenous fluids that are Newtonian fluids. **Newtonian fluids** are defined as **fluids with a viscosity that is independent of the shear rate**. The viscosity of **non-Newtonian fluids** decreases with increasing shear rate, according to the equation above.

Is blood a Newtonian fluid?

No, blood is not homogenous with a viscosity that is independent of shear rate. On the contrary, **at low shear rates (low bloodflow), the viscosity of blood can be 10-fold higher than normal**. The usual viscosity of **body warm blood is 5 centiPoise = 5 milliPascal seconds = 5 (mPa s)**. Viscosity of blood varies with temperature and with shear rate. The erythrocytes pile up centrally, so that fluid resistance becomes lowest at the vessel walls (Fig. 16-1). This tendency increases with increasing bloodflow, especially in the smallest vessels. The **increased axial fluid velocity**, and the shear rate, push the red cells towards the middle zone (Fig. 16-1).

With **increasing bloodflow** (and shear rate), an increasing fraction of red cells is being pulled into the axial stream of small vessels, so that friction is being minimized. At high shear rates, blood therefore largely behaves like a **Newtonian fluid**, with a low and almost constant viscosity, as well as a **linear ratio between bloodflow and the driving pressure**.

The **viscosity of blood apparently decreases in tubes with a diameter less than 0.5 mm (the Fåhraeus-Lindqvist phenomenon)**. This is because red cells have a tendency to accumulate in the middle stream in small tubes, square to the direction of flow. This falling viscosity reduces the work of the heart in the resistance vessels.

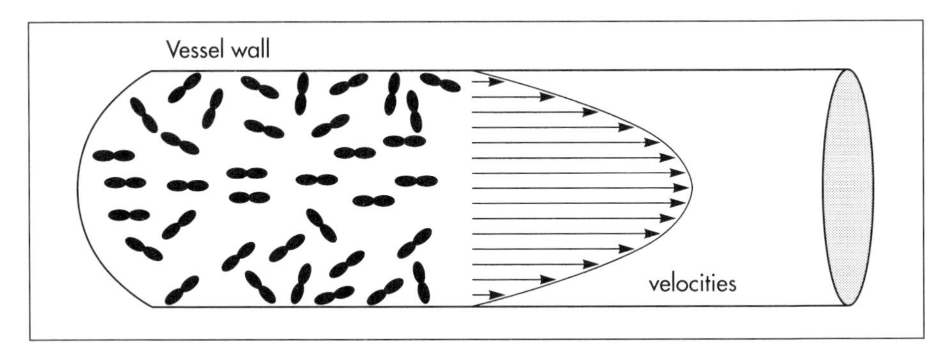

Fig. 16-1: Blood vessel with red cells and arrows showing velocities.

What is the normal function of thrombin?

Thrombin is a protease that is responsible for the formation of **fibrin monomers**, and thus for formation of a fibrin clot. Its parent molecule is **prothrombin (factor II)**. Thrombin formation from prothrombin goes through certain cleavage stages, the first of which is by **activated factor X_a (Stuart)**. These reactions are augmented by **factor IV (Ca^{2+}), factor V (proaccelerin), and phospholipid**. Thrombin initiates **blood platelet aggregation,** and disintegrates the plasma membrane of the platelets so phospholipid is provided. See the intrinsic clotting system of Fig. 16-2.

Where are all coagulation factors synthesized?

The coagulation factors are synthesized mainly in the liver. Exceptions are **factor VIII** and the **von Willebrand factor complex**. These factors are synthesized in the vascular endothelial cells and in megakaryocytes.

Describe the extrinsic clotting mechanism

The extrinsic thrombin formation is initiated by the contact of blood with injured cells. The damaged cells liberate a clot-promoting agent, **factor III or tissue thromboplastin**. Factor III interacts with a plasma protein, factor VII, to start a cascade of reactions by prothrombin activators leading to formation of thrombin within seconds (Fig. 16-2).

Clotting of blood implies conversion of a soluble plasma protein, **factor I or fibrinogen,** into an unsoluble network of **fibrin**. First, fibrinogen undergoes limited proteolysis by thrombin. The so-formed fibrin monomers polymerize into insoluble strands of **fibrin polymers** (Fig. 16-2). Finally the monomers of the fibrin strands are cross-linked by the enzyme **activated fibrin-stabilizing factor (XIIIa)**.

What is classic haemophilia?

Haemophilia A is the most frequent genetic disorder of the intrinsic clotting system, characterized by a low coagulant concentration of **antihaemophilic factor (VIII)**. This disorder is linked to the X-chromosome, and haemophilia only affects males, who

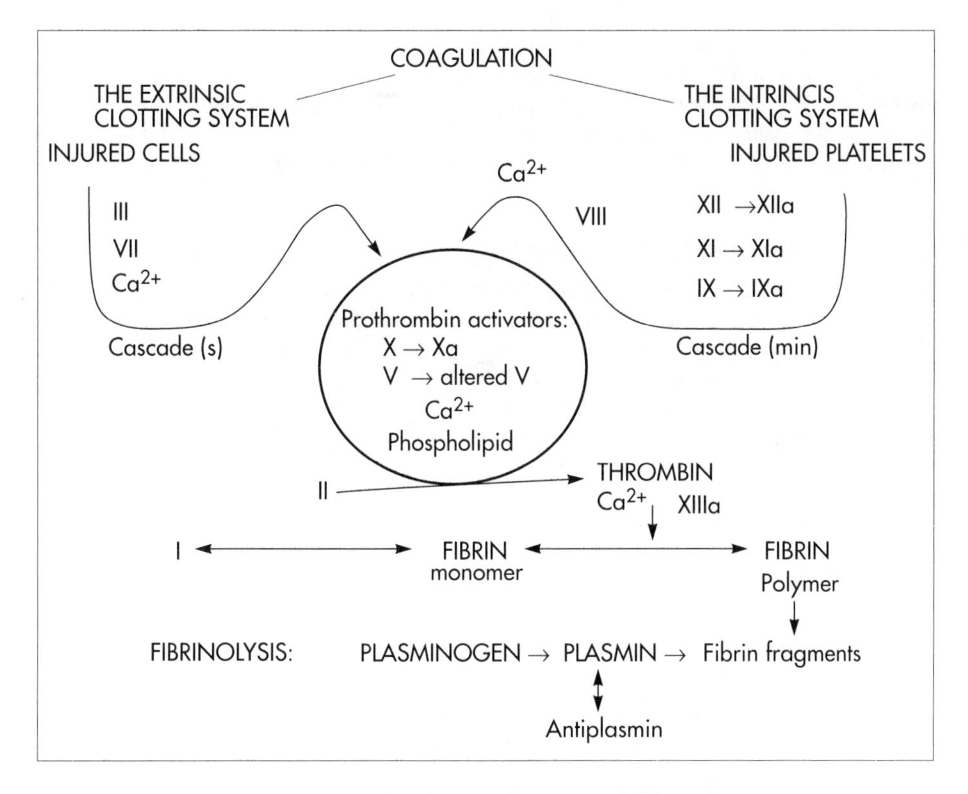

Fig. 16-2: Blood coagulation and fibrinolysis.

transfer the abnormal gene to their daughters, all of whom are carriers. The female carrier of the abnormal gene is usually without symptoms and signs of disease.

Haemophilia B (Christmas disease, Factor IX deficiency) is not as common. Most haemophiliacs suffer episodes of spontaneous bleeding. Repetitive joint bleeding (**haemarthrosis**) leads to **crippling arthritis.**

The **activated partial thromboplastin time** tests the competency of the intrinsic clotting pathway. The contact factors are maximally activated by first mixing citrated plasma with powdered glass. Then partial thromboplastins (V, cephalin, inosithin) are added. After addition of phospholipid and Ca^{2+}, the time it takes for coagulation to occur is measured. This is a preferential test of the **intrinsic clotting pathway**, because factor III (tissue thromboplastin from injured cells) is not available to trigger the **extrinsic clotting pathway** (Fig. 16-2). Normal values are 35–45 s. The time is prolonged in blood from patients with circulating anticoagulants. The time is also prolonged in haemophilia and in other disorders with defective intrinsic pathway factors.

What is von Willebrand's disease?

In most forms of this disease the plasma is deficient in **both factor VIII and von Willebrand factor**. Both sexes are affected by the disease, which is similar to mild haemophilia. The disorder is inherited as an autosomal dominant trait. The **bleeding time** tests the capacity of platelets to form plugs. A blood pressure cuff is applied to maintain venous pressure at 5.3 kPa, and a standardized incision is made on the volar surface of the forearm. Bleeding stops when a proper plug of platelets has aggregated. The incision is blotted with filter paper at 30-s intervals. The normal bleeding time is 4.5 min. The bleeding time is prolonged to **at least 10 min in von Willebrand's disease**.

Describe the intrinsic clotting mechanism

When venous blood is drawn in **silicone-coated tubes** and centrifuged for the separation of cells and plasma, the isolated plasma clots readily due to the negative surface charge of glass. **Negatively charged surfaces** generate thrombin, trigger inflammatory and immune responses and even activate fibrinolysis.

The first step is that negatively charged surfaces activate factor XII (Hageman) to XIIa, which can activate factor XI in the presence of kininogen. The **XIa** activates the vitamin-K-dependent protein, **Christmas factor (IX)**. Activated Christmas factor (IXa) converts factor X to its activated state (Stuart factor Xa). Stuart factor is a plasma proenzyme – also vitamin-K-dependent. The Xa is the enzyme immediately responsible for the release of thrombin (Fig. 16-2).

What is fibrinolysis?

Fibrinolysis is the dissolution of fibrin. The hepatic plasma glycoprotein proenzyme, plasminogen, is activated to the serine protease, **plasmin** (Fig. 16-1). Streptokinase, staphylokinase and urokinase convert plasminogen to plasmin. The tissue plasminogen activators are serine proteases. Fibrinolysis is enhanced by stress, muscular activity and emotional crises. Plasmin digests both fibrin, fibrinogen and other clotting factors. If plasmin is formed in blood plasma devoid of clots, it is irreversibly inhibited by **α_2-antiplasmin** (Fig. 16-2).

How is it possible to inhibit clotting and fibrinolysis?

Antithrombin III is the main inhibitor of thrombin and factor Xa. **Heparin** is a negatively charged mucopolysaccharide from mast cells. Heparin binds to antithrombin III forming a complex that rapidly binds **serine protesases** such as thrombin, thus functioning as a potent anticoagulant. Heparin alone does not inhibit the coagulation process significantly.

Fibrinolysis is inhibited mainly by **α_2-antiplasmin**, because plasmin combines with antiplasmin in an irreversible link (Fig 16-2).

What is the main function of platelets?

When the endothelial surface of the vascular system is disrupted, platelets adhere instantly to exposed structures (collagen and other fibres). Adherent platelets discharge ADP and other substances. Adherent platelets become spherical and send out spicules that look like the legs of a spider. The platelet plug grows and forms a firm **haemostatic plug** that stops the bleeding. Platelets provide substances that enhance thrombin production, such as **phospholipid**, the important co-factor in the clotting process.

What are thrombosis and embolism?

Thrombosis is the formation of multiple thrombi or clots within the vascular system. The cause can be damage of the vessel wall, reduced bloodflow, and hypercoagulability of the blood.

Embolism is the process through which a thrombus is dislodged from its attachment and travels with the blood until it is lodged in a blood vessel too small to allow its passage. Emboli from thrombous material in the leg veins are carried by the flowing blood to the lungs, where they become **pulmonary emboli**.

What is pernicious anaemia?

Pernicious anaemia is a disorder with an atrophic gastric mucosa. The parietal cells of the gastric glands fail to secrete HCl and **intrinsic factor**. Intrinsic factor is a glycoprotein which combines with vitamin B_{12} of the food. This combination normally makes vitamin B_{12} available for absorption in the ileum (Fig. 52-1). The site of red cell production is the red bone marrow, which is normally one of the most proliferative tissues. Lack of vitamin B_{12} in the red bone marrow turns the normal erythroblasts into abnormal megaloblasts, and the red cell production is inhibited. The cells synthesize inadequate quantities of DNA and much more RNA than normal. Also the production of leucocytes and platelets suffer, causing **leucopenia and thrombocytopenia**. Instead of normal adult erythrocytes, the megaloblasts liberate **megalocytes or macrocytes** to the circulation. Macrocytes are fragile and short-lived; the **normal average life of erythrocytes** is **120 days**, whereas that of macrocytes is about 40 days.

Megalocytic anaemia, with lack of gastric HCl and low [vitamin B_{12}] in the serum, confirms the diagnosis.

The lack of vitamin B_{12} in the liver and the red bone marrow inhibits the **methylmalonyl Co-A mutase** and spoils the **purine–pyrimidine–DNA-synthesis**. The inhibition of these two processes leads to the **neurological** and the **haematological disorders** in pernicious anaemia.

The falling red cell count leads to **falling viscosity** of the blood. The reduced viscosity can reduce the total peripheral vascular resistance (TPVR) to less than half of the resting value, which is an appropriate event in order to ease the bloodflow. A slight fall in systemic arterial pressure reduces the stimulus of the arterial baroreceptors, and causes a rise in heart rate and cardiac output. The low oxygen capacity of haemoglobin is compensated by an increased coronary bloodflow at rest. The **myocardial anoxia** results in **cardiac failure** with **oedema**, large liver, and stasis of the neck veins. **Severe anaemia** increases respiration, metabolic rate and temperature due to the large cardiac work.

16. Case History

A grey-haired male with blue eyes, 52 years old, complains of precordial pain, dyspnoea upon stair climbing, and nausea. He is depressed and suffers from frequent coughs.

The doctor observes icteric skin and eyes, ataxic walking, dysdiadochokinesis, and positive Babinski. Massive subcutaneous bleeding was found at the left hip.

Laboratory tests revealed the following abnormal results: lack of HCl in the gastric fluid during fasting and following a pentagastrin test. Haematology tests revealed large erythrocytes – many with nuclei. The red cell count was 1.4 $10^{12}\ l^{-1}$. The haematocrit was 0.21, and the blood [haemoglobin] was 4 mM. The bleeding time was 90 min and the platelet count was $50 \cdot 10^{9}\ l^{-1}$. The concentration of vitamin B_{12} in serum was 90 ng l^{-1}. The total [bilirubin] in serum was 18 mg l^{-1}, and the rise mainly due to nonconjugated bilirubin. A test with radioactive B_{12} was specific for lack of intrinsic factor production from the patient's parietal cells. Expected or 'normal' results are: Red cell count: $5.10^{12}\ l^{-1}$; haematocrit: 45%; blood [haemoglobin]: 9.2 mM; bleeding time: 3–10 min; platelet count: (150–400) $10^{9}\ l^{-1}$; serum [B_{12}]: 150 ng l^{-1}.

1. *What was the cause of this severe pancytopenia (lack of all blood cell types)?*
2. *Calculate the oxygen capacity for haemoglobin.*
3. *Why did the patient develop leucopenia and thrombocytopenia? Was the lack of leucocytes and platelets of any consequences to the patient?*
4. *Does a severe, chronic anaemia trigger physiological adaptations?*

Try to solve the problems before looking up the answers in Chapter 74.

CHAPTER 17. PRINCIPLES OF CARDIOVASCULAR PHYSIOLOGY

The principal function of the blood in the cardiovascular system is **to provide O_2 and nutrients to the tissues of the body and to remove CO_2 and waste products**. The flow of blood through the cardiovascular system follows physical laws known from fluid mechanics.

The formula for volume ($\mathring{Q}$) of a cylinder of the length $\overline{v}$, and the cross-section area (A) is: $\mathring{Q}=\overline{v} \cdot A$. This implies the following relationship: $\overline{v}=\mathring{Q}/A$. Now, the volume and the length are divided by 1 second (s), and thereby changed into the concepts: bloodflow per second ($\mathring{Q}_S$) and linear mean velocity $\overline{v}_S$. Hence, an equation is obtained for calculation of linear mean velocity in a cylindrical blood vessel.

The red cell velocity in the capillaries is observed to be approximately 1 mm in 1 s. This provides ample time for gas exchange. Since the circulating blood moves continuously, the cardiac output must pass a cross-section of all open capillaries. At rest a cardiac output of 6000 ml min^{-1} is a reasonable estimate; when changed into volume rate per second, it is: $\mathring{Q}_S=10^{-4}$ m^3 s^{-1}. Hence, it is possible to calculate the large **cross-sectional area of all open capillaries in a person at rest:**

$$A = 10^{-4}/10^{-3} = \mathbf{0.1\ m^2}.$$

These aspects will be addressed in the following chapter. See Chapter 21 for cardiac performance.

How is the cardiovascular system organized?

The heart is composed of a special type of striated muscle, called the **myocardium**. The cardiovascular system consists of **two pumps arranged in series** (Fig. 17-1). They are the right ventricle, which pumps blood into the pulmonary circulation, and the left ventricle, which pumps blood into the systemic circulation. Each of these pumps delivers blood through an efferent tube system (the arteries), and each pump receives blood through an afferent tube system (the veins). In the pulmonary system, blood is pumped from the right ventricle through the lung capillaries and is temporarily collected in the left atrium (Fig. 17-1).

The heart possesses a specialized electrical system, the **cardiac conduction system**, that ensures the optimal timing of atrial and ventricular pumping. This is what allows the heart to pump the required **cardiac output ($\mathring{Q}$ l min^{-1})**. The cardiac output is 5 l min^{-1} at rest, and the **total blood volume (TBV)** is approximately 5 l in an healthy adult.

The heart normally has a **self-firing unit**, located in the right atrium, called the **sinoatrial node (SA node)**. The electrical signal that comes from this structure has the highest frequency, and is thus the **natural pacemaker** of the heart.

The signal activates the atrial walls to contraction, and then reaches the main conduction system at the level of the **atrioventricular node (AV node)**. Here, the signal

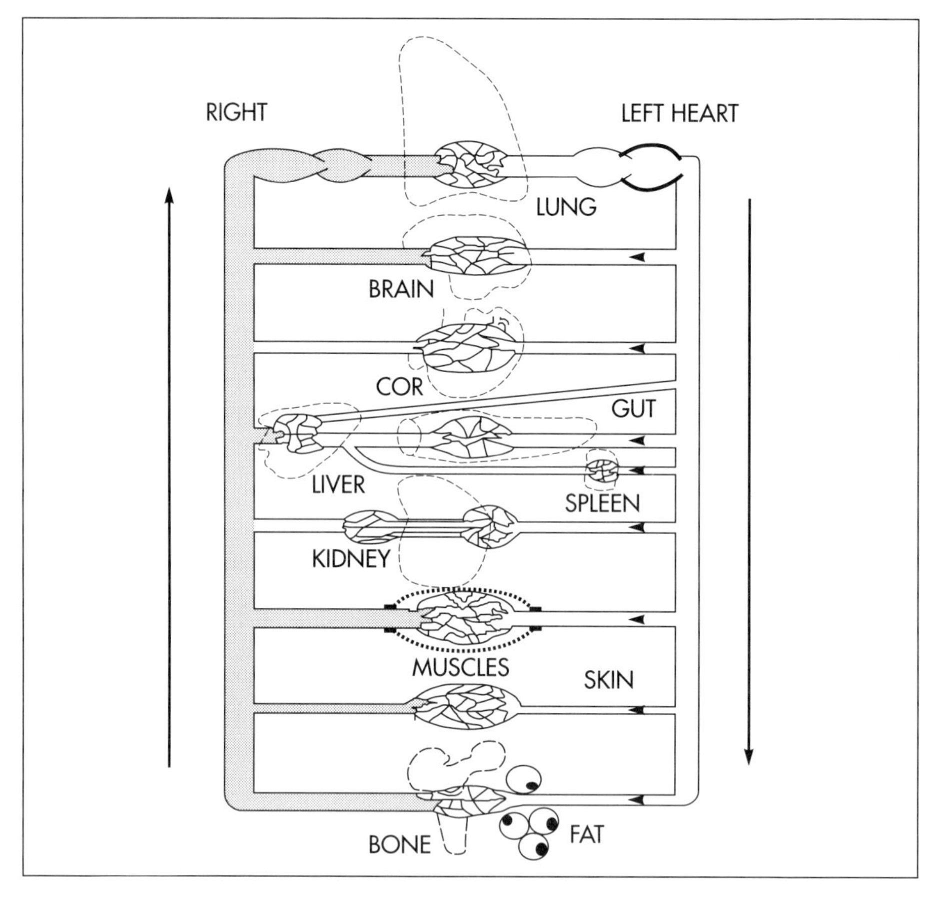

Fig. 17-1: Design of human circulation.

is transmitted down a rapid conduction pathway, composed of the **right and left bundle branches**, to stimulate the right and the left ventricle and cause them to contract.

The right atrium receives venous blood from the caval veins, and the left atrium receives oxygenated blood from the pulmonary veins. The atria function as a reservoir and conduit for blood. On average, atrial systole contributes only about 15% of the total ventricular filling, but in **cardiac insufficiency** the atrial contribution may increase importantly. The left and right ventricles provide most of the energy needed to transport the blood through the circulation. The left ventricle accelerates the blood into the **systemic or peripheral high-pressure system**, and its walls are thick in contrast to the thin, weak right ventricle which pumps blood into the **low-pressure pulmonary or central system**.

We have **four intracardiac valves**. The two **atrioventricular valves** prevent the leakage of blood backward from the ventricles into the atria. The **two-leaflet mitral valve** serves this role in the left side of the heart, and the **three-leaflet tricuspid valve** prevents retrograde bloodflow from the right ventricle to the right atrium. The other two valves are the **aortic valve** interposed between the left ventricle and aorta, and the **pulmonary valve** which lies between the pulmonary artery and the right ventricle.

The **pericardium** is a thin-walled membranous sack that surrounds the heart. It contains a small volume of fluid in the space between its two surfaces. The **coronary arteries** are the first arterial branches that arise from aorta just above the aortic valve.

What are distribution-, conductance-, resistance-, exchange-, capacitance-, and shunt vessels?

Aorta and the elastic arteries are **distribution vessels**; the muscular arteries are **conductance vessels**; the arterioles are **resistance vessels**; the capillaries are **exchange vessels**; venules and veins are **capacitance vessels**. The arterio-venous anastomoses in fingers and toes are **shunt vessels**.

How do we convert one pressure unit to another?

Pressure is **force per area unit**. The international unit is Newton per m^2 or Pascal (Pa). In the gravity field of the earth, ***G*** **equals 9.807 m** $\mathbf{s^{-2}}$ and blood has a relative density of 1033 kg m^{-3}. A 10-m high blood column resting on 1 m^2, corresponds to the following pressure: (10 m × 1033 kg m^{-3} × 9.807 m s^{-2}) = **101 300 (kg m** $\mathbf{s^{-2}}$**)** $\mathbf{m^{-2}}$**. This is 101 300 N** $\mathbf{m^{-2}}$ **or 101.3 kPa (or 1 atmosphere).**

By definition, 1 atmosphere equals 760 mmHg or 760 Torr. Hence, **1 Torr** or **1 mmHg** equals (101 300 Pa/760=) **133.3 Pa**. In this book pressures are given in Pa with Torr or mmHg in parentheses.

Is there a law relating flow rate ($\mathring{V}$), driving pressure (ΔP), and resistance (R) for fluids?

Poiseuille's law: $\mathring{V} = \Delta P/R$, is used both in the circulatory and the respiratory system. For the left ventricle, the bloodflow is actually cardiac output ($\mathring{Q}$), so the equation reads: $\mathring{Q} = \Delta P$/TPVR (l min^{-1}). The driving pressure (ΔP) is the mean arterial pressure (MAP) minus the atrial pressure, and **TPVR** is the **total peripheral vascular resistance**.

Strictly speaking, **Poiseuille's law** has validity in the circulatory system, when the bloodflow is laminar and non-pulsating, in horizontally situated cylindrical vessels of constant dimensions. In contrast to Poiseuille's conditions, the bloodflow in the human circulation is often turbulent and pulsating, and its blood vessels are not always horizontally located, cylindrical or inflexible.

Ohm's law for electrical circuits is an analogy to Poiseuille's law. Ohm's law states the connection between electrical current (*I*, analogous to $\mathring{V}$ or $\mathring{Q}$), difference of potentials (p.d., analogous to ΔP), and electrical resistance (*R*, analogous to TPVR): ***I*=p.d./*R*, analogous to:** $\mathring{V}\Delta$= ***P*/TPVR.**

How is the TPVR calculated?

TPVR can be calculated as the driving pressure, divided by bloodflow rate ($\mathring{Q}$s in ml s^{-1}): **TPVR** = $\Delta P/\mathring{Q}$**s.**

TPVR is measured in **Pressure Resistance Units (PRU)**, which are **Pascal seconds per** $\mathbf{m^3}$ **of blood** (Torr seconds/ml of blood). In the systemic or peripheral circulation the resistances in the single organs are mainly placed in parallel, and

$$1/\mathbf{TPVR} = 1R_1 + 1/R_2 + \ldots 1/R_n.$$

There are only a few serially connected elements (**portal circulations**): Spleen/liver, gut/liver, pancreas/liver and hypothalamus/pituitary.

For serial arranged resistances the formula is:

$$R_{\text{total}} = R_1 + R_2 + \ldots R_n.$$

How is the total blood volume (TBV) distributed? Is the distribution changed by a heart attack?

Eighty-eight per cent of TBV is localized in the systemic circulation at rest. Of this percentage, 50% is in the veins and venules, and the remaining blood is in the heart, the arteries, the arterioles, and the capillaries. The last 12% is found in the **pulmonary low pressure system at rest.**

During a heart attack the blood supply can no longer be passed on by the heart pump, causing accumulation of blood in the distensible vessels (the pulmonary circulation, and the distensible low pressure sections of the heart itself).

What is understood by compliance of a vessel?

Compliance is the **increase of volume per unit of transmural pressure increase** ($\Delta V/\Delta P_t$). Compliance is also called distensibility. The elastance is the reciprocal of the compliance. The **compliance of the venous system** can be 30 times as large as that of the arterial system.

Is the venous system more than a passive conductor?

Yes, the venous system can be expanded to contain **up to 70% of TBV** from 50% at rest. The veins function as **capacitance vessels,** and become very distended when blood is given in transfusions, in heart insufficiency, or during a heart attack. Severe exercise and loss of blood cause an **increase in venous tone.** During hard work the **muscular venous pump** provides up to one-third of the energy required for blood circulation (**the peripheral venous heart**). The venous system also plays an important role by its graded venous return to the heart.

Describe the arterial blood pressure recording
What is understood by the mean arterial blood pressure?

The recording shows a systolic peak pressure, a dicrotic notch as the aortic valves close, and a falling diastolic pressure. The area under the curve divided by the cardiac cycle length, is the **mean arterial blood pressure (MAP).** The area under the curve equals the area of the rectangle drawn (Fig. 17-2).

MAP is usually being defined as the **diastolic pressure plus one-third of the pulse pressure** (Fig. 17-2). The mean arterial pressure (MAP) is about 12 kPa (=90 Torr) in the arteries. The mean pressure falls to one-fifth of this in the capillaries.

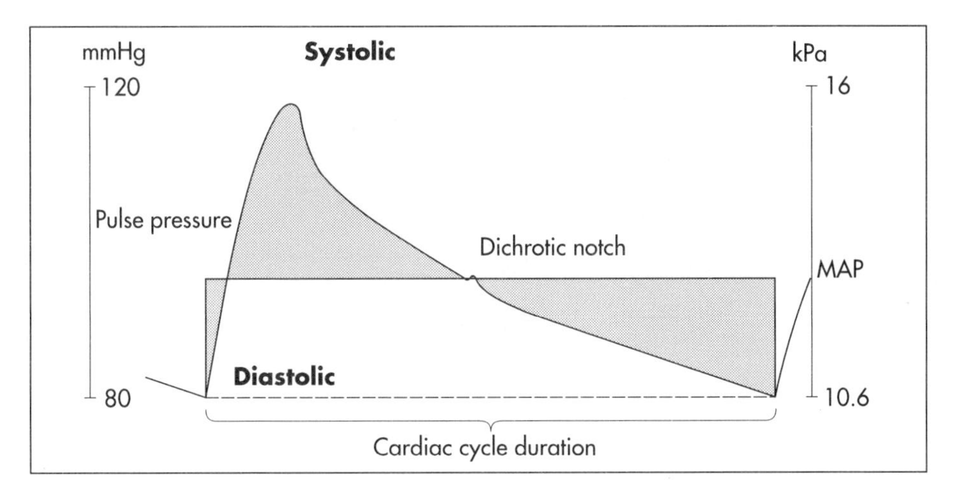

Fig. 17-2: Recording of the arterial blood pressure of a young person at rest.

Which variables determine the size of the pulse amplitude?

The **arterial pulse pressure (the pulse amplitude)** is the difference between the systolic and the diastolic arterial pressure. At a heart rate of 75 beats min^{-1}, the cardiac cycle length is 0.8 s with 0.3 s systole and 0.5 s diastole. A **stroke volume** of 80 ml is deposited in the aorta and the larger elastic arteries during systole. At the same time ($80 \times 3/8$) 30 ml of blood is streaming through the **resistance vessels**, leaving the arterial system. A young healthy subject has an **arterial compliance** of 1 ml of blood per Torr which creates a pressure rise during systole (pulse pressure amplitude) of (80–30) = 50 Torr. With a diastolic pressure of 80 Torr, this implies a systolic pressure of 130 Torr, conventionally written 130/80 (Torr or mmHg) or 17.3/10.6 kPa.

Ageing and arteriosclerosis increase the **stiffness** of the elastic arteries, causing the arterial compliance to fall to 0.5 ml of blood per Torr. In this case, a systolic volume expansion of 50 ml of blood should increase the pulse pressure amplitude to (50/0.5) **100 Torr**, and the blood pressure to 180/80 Torr. This is a likely process in an otherwise healthy person of advanced age (e.g. 75 years old).

Typically the diastolic pressure will rise as well, although it is not a law of nature.

Describe the Laplace model for the relation between transmural pressure (ΔP_t), fibre tension (T), and radius (r)

A. For a **thin-walled organ with two main radii** (*r*), Laplace assumed that the transmural pressure at equilibrium was identical with the fibre tension in the wall (*T*) divided by the two main values:

$$\text{A.} \quad \Delta P = T\left(\frac{1}{r_1} + \frac{1}{r_2}\right)$$

This model has often been used with modifications for wall thickness for the relaxed ventricle (Fig. 17-3A).

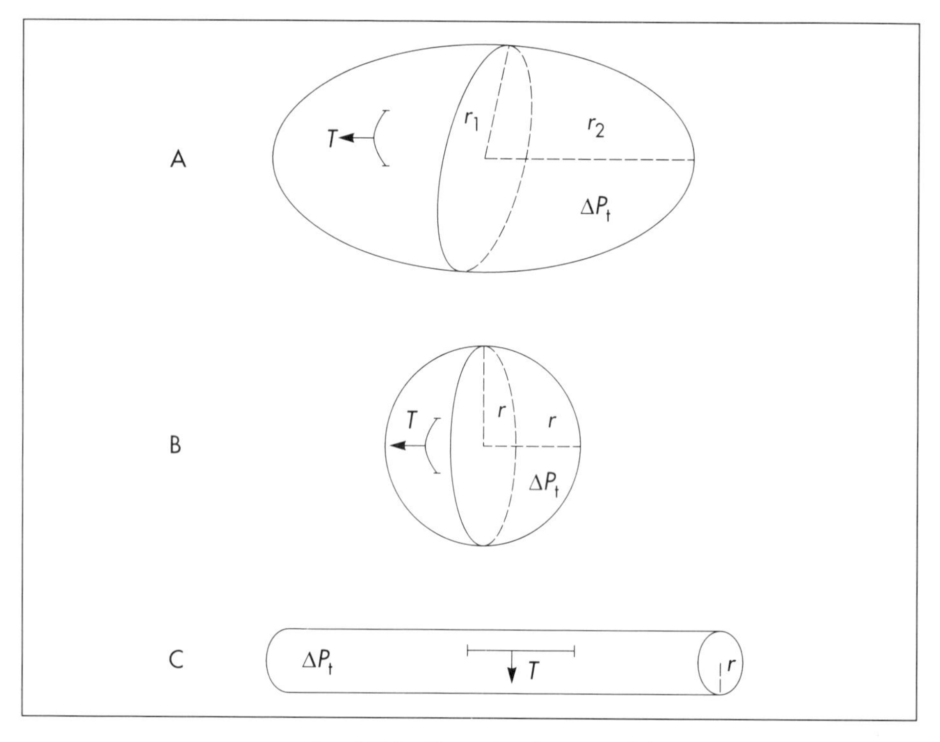

Fig. 17-3: Three Laplace models.

B. For a **thin-walled spherical organ** $r_1 = r_2 = r$, the Laplace equation can be developed from the equation above:

$$\textbf{B.} \quad \Delta P = T/(2r).$$

This model is used for both alveoli and the ventricle (Fig. 17-3B). When the left ventricle becomes spherical by diastolic filling, the T will rise with the transmural pressure, and with the increasing radius at the end-diastolic volume. The more the end-diastolic pressure and the fibre tension rises, the **more O_2 is consumed per heart beat** during contraction of the dilated ventricle.

C. For an **infinitely long thin-walled cylinder**, like a true capillary or a preferential channel, the r_2 approaches infinity and has no influence on the transmural pressure. Hence, the Laplace equation can be approximated by **equation C**:

$$\text{C.} \quad \Delta P = T/r.$$

This model is used for a thin-vessel wall (Fig. 17-3C), since T/r_2 approaches zero.

How can the thin-walled capillaries withstand a transmural pressure of 4.3 kPa (32 mmHg)?

Surprisingly enough the thin endothelial barrier (0.3 μm) easily carries a pressure of 4.3 kPa (32 Torr) or more. This is because the capillary radius is so small. According to Fig. 17-3C and **equation C**, a small radius (5–10 μm) must imply a small fibre tension (T).

Summary of main principles

The relationship between linear mean velocity ($\overline{v}_s$) and bloodflow in 1 s ($\mathring{Q}_s$) is determined by the cross-section area (A):

$$\mathring{Q}_s = \overline{v}_s A.$$

The relationship between cardiac output ($\mathring{Q}$) and driving pressure (ΔP) is determined by the TPVR, **Poiseuille's law**:

$$\mathring{Q} = \Delta P/\text{TPVR}(\text{l min}^{-1}).$$

The **driving pressure** is the **mean arterial pressure (MAP)** minus the atrial pressure. **MAP** equals diastolic pressure plus one-third pulse pressure.

For a thin-walled organ with two radii (r), the relationship between transmural pressure (ΔP) and tension (T) is determined by the radii, **Laplace's law:**

$$\Delta P = T\left(\frac{1}{r_1} + \frac{1}{r_2}\right)$$

Compliance is a change in the volume of a vessel per unit change in transmural pressure ($\Delta V/\Delta P_t$).

Further reading

Lowe, G.D.O. (1988) *Clinical Blood Rheology*, Vol. 1. CRC Press, Boca Raton, FL.

Renkin, E.M. & C.C. Michel (Ed.) (1984) *Handbook of Physiology: Sect. 2. The Cardiovascular System – Microcirculation*, Vol. 4. *Am. Physiol. Soc.*, Bethesda, MD.

17. Case History

A healthy female has a radius in the aortic ostium of 10 mm. The stroke volume of the heart is 80 ml, and is expelled from the heart within the systole of 0.3 s at rest. The viscosity of the blood is 4 mPa s. The critical linear mean velocity ($\bar{v}$) of the red cells, to be exceeded in order for turbulence to occur, is 0.4 m s^{-1}. The cross-sectional area of all open capillaries in her body is 0.1 m^2.

1. *What is the relationship between the bloodflow, the linear mean velocity of the red cells and the cross-sectional area in the aortic ostium?*
2. *Will turbulence arise in the aortic ostium of this person?*
3. *Calculate the linear mean velocity of the red cells in her open capillaries at rest.*
4. *With piezoelectrical crystals located a few centimetres from each other on the skin above the brachial artery of the patient, the pulse wave velocity was recorded as 5–10 m s^{-1}. The pulse wave velocity is inversely related to the square root of density (σ) and of distensibility (δ). The density (σ) of the arterial wall is normally 1.06 kg l^{-1}. Distensibility (δ) is also called specific compliance ($\delta = [\delta v/v]/\delta p$).*

   What happens to the pulse wave velocity, when the female shifts from supine to erect position?
5. *Calculate the wall tension in one of her capillaries with a radius of 5 μm and a transmural pressure of 2666 Pa (20 Torr).*
6. *Calculate the TPVR in the systole, when the driving pressure is 13 330 Pa (100 Torr).*

Try to solve the problems before looking up the answers in Chapter 74.

CHAPTER 18.
MICROCIRCULATION

The microcirculation is responsible for the transport of nutrients and oxygen to the tissues, and for removal of cellular waste products and CO_2. The arterioles control the flow of blood to each tissue unit, and the metabolic conditions of the tissue cells determine the diameters of the vessels. Here, the tissue unit often controls its own bloodflow by local mechanisms.

What is a microcirculatory unit?

A **microcirculatory unit** is a collection of vessels that originate from an arteriole, which is characterized by well-developed smooth musculature (Fig. 18-1). In certain tissues the arteriole branches into a **metarteriole** (with dispersed smooth muscle fibres and without nervous supply), which again branches into two types of vessels: small true capillaries, who only have a thin **hair-like endothelial cell layer**, and large capillaries termed **preferential channels**.

Smooth muscle fibres without nerve fibres, called **precapillary sphincters**, encircle the preferential channel at its origin from the metarteriole. Arterioles of the face, fingers and toes often branches into an **arteriovenous anastomose**, which functions as a **shunt vessel**, but which also can be closed completely (Fig. 18-1). The diameter of true capillaries is only 5–8 μm, barely enough for erythrocytes to squeeze through.

Important exchange vessels are thin-walled vessels with a large surface area. Exchange

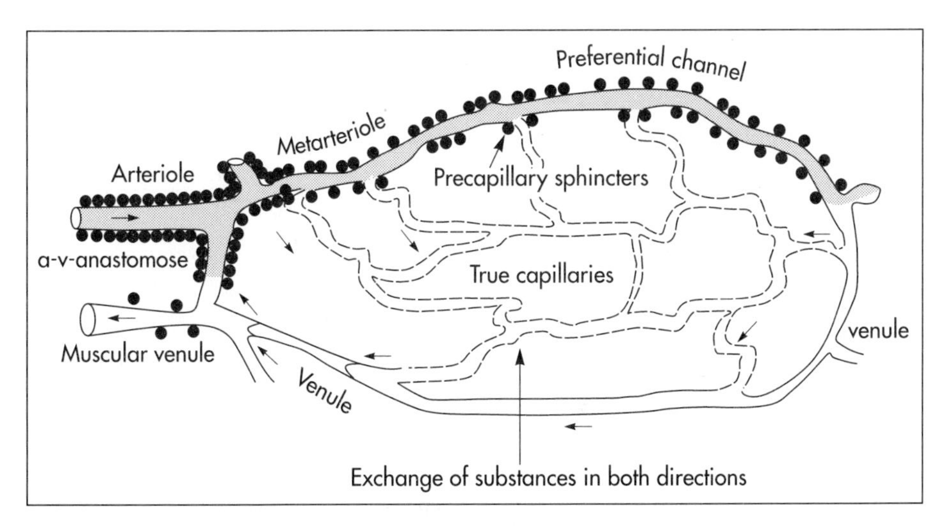

Fig. 18-1: A microcirculatory unit (modified from a previous illustration by Zweifach).

vessels comprise **true capillaries, parts of metarterioles, preferential channels, and venules** (Fig. 18-1). Exchange vessels are any blood vessels which allow **transport of substances through its wall** in both directions. The velocity of the bloodflow in capillaries varies, sometimes with rhythmic pulsations, at other times random.

At rest the intracapillary pressure varies from arteriole to venule between **3.3 and 1.6 kPa (25 and 12 Torr)**, during arteriolar **vasoconstriction between 1.6 and 1 kPa (12 and 8 Torr)**, and during **vasodilatation between 5.3 and 1.6 kPa (40 and 25 Torr)**. Arterial pressure fluctuations have been recorded even in the most distal parts of the capillaries. In venules and veins, however, the flow is smooth without fluctuations.

How is the capillary wall constructed?

The capillary wall consists of a **layer of endothelial cells** (0.1–1 μm of thickness) resting on a **basement membrane**. At least three types of capillaries exist:

1. **Continuous capillaries**, which are the most abundant. The distance between endothelial cells is 5–30 nm (Fig. 18-2). Constrictions (tight junctions) are clefts, difficult to pass for the dissolved molecules and ions. The continuous capillaries in the brain are low-permeable to ions and most hydrophilic molecules (the **blood–brain barrier**).
2. **Fenestrated capillaries** contain pores or *fenestrations (fenestrae)*, which are **water-filled channels** (50 nm in diameter). These are possibly formed by two adjacent cell membranes that have fused during removal of the lipid bilayers, so only a protein lattice is left allowing bulk flow without colloids (Fig. 18-2).

   Fenestrations are round windows found in the capillaries of organs that **transport lots of water** (the bowels, glomerular capillaries of the kidneys, pancreas and salivary glands). The protein lattice in the fenestrae is so tight, that it keeps plasma proteins back (Fig. 18-2).

   In the glomerular capillaries, water-filled fenestrations cover **20% of the surface,**

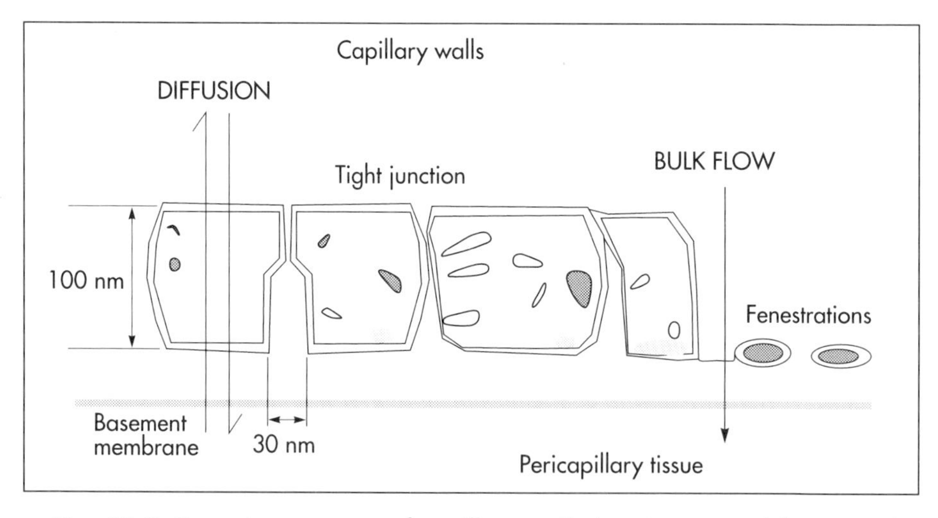

Fig. 18-2: Pores in two types of capillary walls (continuous and fenestrated).

in contrast to the continuous capillaries, where the water-filled pore surface area comprises **only 10^{-4} of the total surface.**

In the **circumventicular organs in the brain** lots of fenestrations are seen. However, most substances cannot reach the brain cells through these paths, and thus cannot bypass the blood–brain barrier. The circumventricular organs are located close to the control centres of the hypothalamus and the brainstem. Any penetration of signal molecules in the neighbourhood of these centres is probably of physiological importance.

3. **Sinusoid capillaries** have very broad openings between the endothelial cells. Sinusoid capillaries are often found in **tissues that are bathed in plasma (liver, spleen and bone marrow).**

Do macromolecules penetrate the capillary wall?

Macromolecules do penetrate the capillary wall and the content of lymph comes from plasma. Less than 0.1% of all the plasma proteins that are ejected from the heart in 24 h escape from the capillaries. The venous end of the capillaries is permeated by **pores of 40–60 nm**. Here **macromolecules can pass by filtration** in a pressure-determined fluid transport, as well as a **whole plasma portion** (convective transport of fluid or **bulk flow**). The **solvent transport** can also draw substances with it by **solvent drag**. When large amounts of lymph is being produced, **solvent drag** dominates over diffusion. At low lymph production, half of the protein transport is caused by diffusion. Fluids pass through the cell by **pinocytosis** (i.e. formation of extremely small vesicles containing fluid). Large particles, such as bacteria and cells, are destroyed in the cell by **phagocytosis** (i.e. ingestion of large particles in a vesicular membrane).

What determines the transcapillary exchange?

Starling hypothesized that the fluid exchange across the capillary wall was determined by the **hydrostatic pressure (P_c) and the colloid osmotic pressure** (π_c) in the capillary (Fig. 18-3).

The flux of substances (J) over the capillary membrane area is determined by ($P\Delta C$). Actually, this is **Fick's law for transport of matter across a membrane** of area (A): $J = -D\mathrm{d}c/\mathrm{d}x$.

Fluid moves out of the arterial end of the capillary by **filtration** because the net hydrostatic pressure ($35 - 5 = 30$ mmHg) is higher than the colloid osmotic pressure ($\pi_c = 24$–26 mmHg), and most of the fluid passes again into the blood by **reabsorption** in the venous end (Fig. 18-3). Here, the colloid osmotic pressure (26 mmHg) supersedes the hydrostatic pressure ($15 - 1 = 14$ mmHg).

What forces determine the net movement of water across the capillary wall?

The net diffusion of water molecules across the capillary wall is approximately zero. Instead, the transvascular exchange is caused by a combination of **an outward ultrafiltration and an inward colloid osmotic force**. Ultrafiltration is caused by a hydrostatic pressure gradient created by the heart. The hydrostatic pressure gradient is a net outward force, moving water through pores in the capillary wall. Plasma contains dissolved protein which cannot readily pass the small pores in capillary walls. The plasma proteins create a **colloid osmotic pressure** of about 3.3–3.7 kPa (25–28 Torr). This

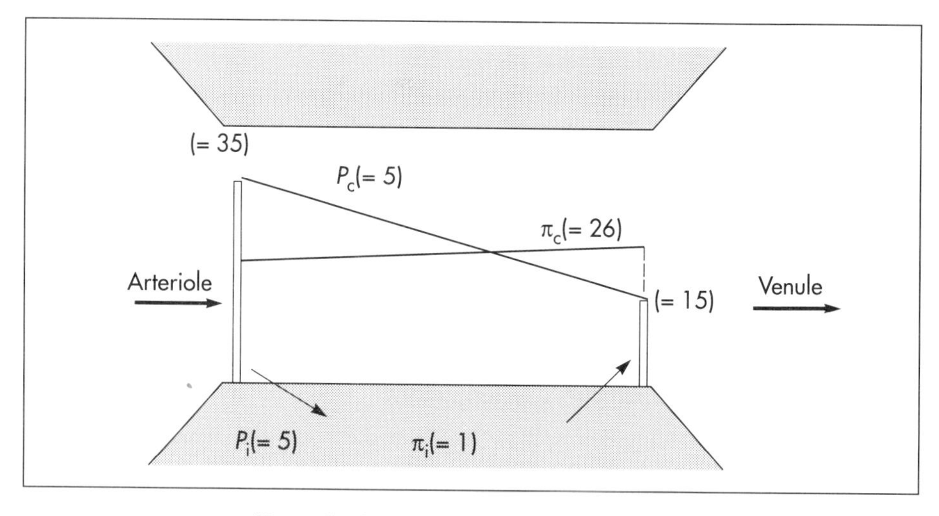

Fig. 18-3: Transcapilllary fluid exchange (Starling) over a capillary wall, with pressures in mmHg.

pressure is larger than the **interstitial colloid osmotic pressure,** so that the colloid osmotic gradient across the capillary wall is a net inward force, which draws water into the capillaries.

The transvascular water flow (J_f, volume per minute per 100 g of tissue) determined by the combined effect of these forces was described as early as in 1896 by Starling in the equation:

$$J_f = \mathbf{Cap_f}[(P_c - P_t) - \sigma(\pi_c - \pi_t)].$$

Cap$_f$ is the **capillary filtration coefficient** (in ml min^{-1} kPa^{-1} per 100 g of tissue). The driving forces are the so-called **Starling forces.** In the equation, P_c is the **capillary hydrostatic pressure,** P_t is the **tissue hydrostatic pressure** (approximately zero), π_c is the **capillary colloid osmotic pressure** (3.6 kPa or 27 Torr), π_t is the **tissue colloid osmotic pressure** (0.5 kPa), and σ is the **capillary protein reflection coefficient.** The σ is the fraction of plasma protein molecules reflected off the capillary wall. The protein reflection coefficient is 0.9–1.0 for many capillaries, expressing **that the colloid osmotic pressure gradient** is not reduced over time by diffusion of proteins over the capillary wall. **Cap**$_f$ **is represented by the product of the surface area (the length and diameter of the capillaries) and the water permeability. Cap**$_f$ is 0.075 ml min^{-1} kPa^{-1} per 100 g of tissue in the legs. The combined pressures in the Starling equation ($[(\boldsymbol{P_c} - \boldsymbol{P_t}) - \sigma(\pi_c - \pi_t)]$) determine if there is a **net pressure for water movement** across the capillary wall.

In conclusion, water moves out of the arterial end of the capillary by **filtration,** and near the venule end, water moves into the blood by **reabsorption.** Normally there is a net filtration of water and some protein into the interstitial space. This water and protein returns to the blood via the lymphatic system. The lymph volume amounts to approximately 3–5 l day^{-1}, and is mainly produced in the liver and intestine. Starling presumed that proteins were unable to leave the blood in the capillaries (Fig. 18-4: A).

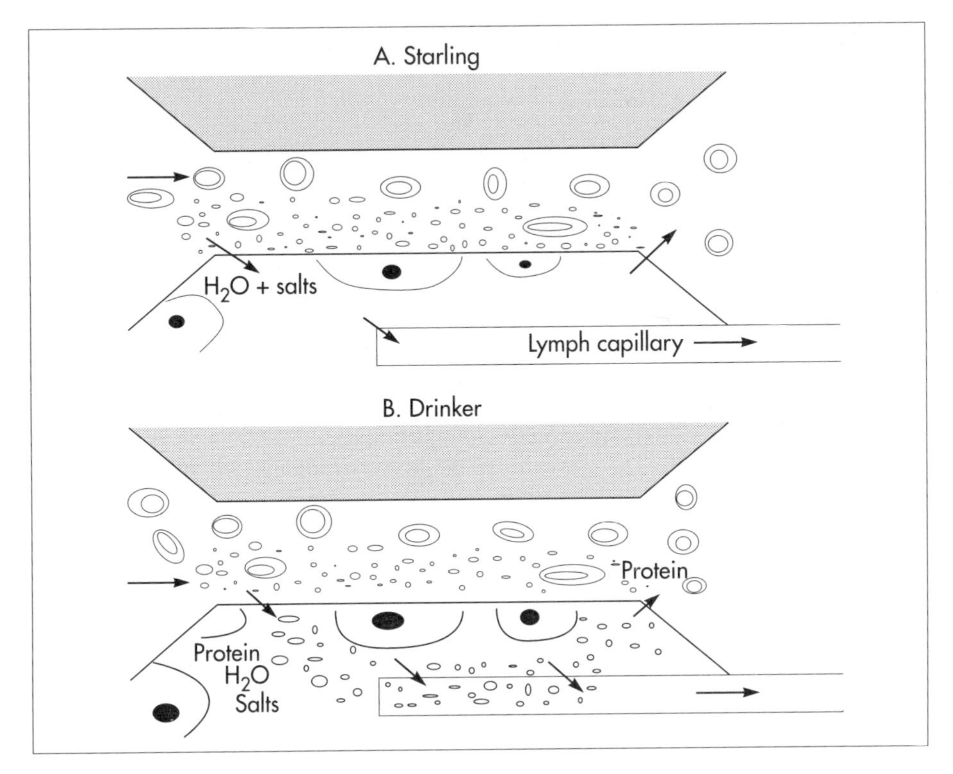

Fig. 18-4: Two models of transcapillary fluid exchange.

This assumption is wrong. The capillaries are almost universally permeable to proteins and macromolecules that resemble proteins. The physiologist Drinker found protein in lymphatic fluid. Drinker presumed that capillaries to a variable degree were **permeable to proteins** (Fig. 18-4: B).

Let us assume that the heart is pumping out **9000 l of blood every day**. Fifty-five per cent of this is plasma, which means 4950 l of plasma, with 6% proteins or in total 297 kg of protein. If less than 0.1% (e.g. 1/1440) of this protein is filtered into the interstitial fluid and lymph, it amounts to 206 g of protein daily. This amount of protein leaves the blood in the capillaries, and returns almost completely to the blood through the lymph and not the veins (Fig. 18-4: B). Hence, **Starling's paracapillary circulation** obviously plays a dominating role in the transport of **crystalloids** (small molecules of nourishment and wasteproducts) through the capillary wall.

What forces determine the capillary hydrostatic pressure?

The **capillary hydrostatic pressure (P_c)** is low in the lungs and intestine (1 kPa) and particularly high in the renal glomerular capillaries (6–8 kPa). In resting skeletal muscle capillaries, the pressure is 4.3 kPa (32 Torr) at the arterial end and 1.6 kPa (12 Torr) at the venous end. In general, P_c increases whenever the **mean arterial pressure (MAP)** increases, **venule pressure (P_v) or resistance (R_v)** increases, or when **arteriolar resistance (R_a)** decreases, according to the following formula:

$P_c = [(R_v/R_a)$ **MAP** $+ P_v]$. Normally, R_v/R_a is approximately 1/10. Thus P_c is protected from large changes in MAP, but is sensitive to changes in venous pressure including the **central venous pressure (CVP).**

Describe the lymphatic drainage

Capillary filtration predominates over capillary reabsorption resulting in an overshoot (a net filtration) of interstitial fluid. Most of the net filtration is reabsorbed into the blood of end-capillaries or venules. **The lymphatic vessels drain the remaining filtered fluid.** The lymphatics are composed of endothelium-lined vessels. Some lymphatics are equipped with one-way valves. They originate as **blind-ended sacs close to the blood capillaries.** The lymphatic drainage is particularly important for transporting **chylomicrons** absorbed from the intestine, and to **return plasma proteins** that leak from several blood capillary systems. Lung tissue has no lymphatics, because the lymphatic vessels end at the terminal bronchioli. The lymph from the liver provide us with 50% of the daily lymph produced.

What is oedema?

Oedema is an abnormal clinical state characterized by accumulation of interstitial or tissue fluid. Cutaneous oedemas can be diagnosed by the simple test: **pitting on pressure.** Theoretically, oedemas are caused by three different mechanisms:

1. A **hydrostatic pressure gradient, which is too great** (so-called **high pressure oedema** at heart failure with increased venous and central venous pressure).
2. A **colloid–osmotic pressure gradient, which is too low and caused by too low concentrations of plasma proteins** (so-called hunger oedema and renal oedema).
3. **Leakage** in the capillary endothelium (so-called **permeability oedema** with too much protein in the oedemal fluid). Increased capillary permeability for proteins is caused by burns, by infections or by allergy.

What are the possible causes for lymphatic oedema?

Lymphatic oedema can be **congenital or acquired.** A child born with insufficient development of the lymphatic system, will suffer from gradual swelling of the affected body part as a result of accumulation of interstitial fluid. Surgical destruction of lymphatic vessels can result in **acquired, lymphatic oedema,** e.g. following mastectomy.

Lymphatic vessels also can be obstructed by inflammation, cancer cells or filarias (**elephantiasis**).

How large is the protein concentration in lymphatic fluid?

Lymphatic fluids from liver and kidney have a protein concentration equal to that of plasma (6–8 g per 100 ml), and lymphatic fluid from the bronchial tree has a similar concentration of protein.

Lymphatic fluids from skin and muscles contain only 2% protein, and brain lymph contains no protein at all.

Further reading

Feng, Q. and T. Hedner. (1990) Endothelium-derived relaxing factor (EDRF) and nitric oxide. II. Physiology, pharmacology and pathophysiological implications. *Clin. Physiol.* **10**: 503.

Krogh, A. (1959) *The Anatomy and Physiology of Capillaries.* HAFNER, New York.

Starling, E.H. (1986) On the absorption of fluids from the connective tissue spaces. *J. Physiol.*, **19**: 312–20.

18. Case History

In a healthy 20-year-old male, with a mean cardiac output ($\dot{Q}$) of 7 l min^{-1} and a haematocrite of 45%, 20 l of fluid are filtered per day in the capillaries. The concentration of protein in the fluid is 5 g l^{-1}.

A daily volume of 3 l of fluid passes into the lymphatic vessels and is returned to the blood as lymphatic fluid. The rest of the filtered fluid is absorbed by the capillaries, supposedly together with a small amount of protein (10 g).

1. *Of the total amount of plasma reaching the capillaries every day, what fraction will be filtered?*
2. *Calculate the mean protein concentration in the lymphatic system.*

18. Multiple Choice Question

Each of the following five statements have True/False options:

A. Solutes are exchanged in capillaries and small venules, because of the large surface area and the thin endothelial vessel walls with many pores.
B. Oxygen diffuses from the blood to the interstitial fluid mainly across the total surface of the endothelial cell walls.
C. Systemic oedema is caused by a small increase in mean arterial pressure.
D. The bloodflow through the capillaries is regulated by arteriolar tone.
E. Oxygen is a water-soluble gas.

Try to solve the problems before looking up the answers in Chapter 74.

CHAPTER 19. OXYGEN SUPPLY TO THE TISSUES

Oxygen diffuses into the blood in the pulmonary capillaries to be carried to all parts of the body. In the tissues, oxygen diffuses from the capillary blood into the interstitial fluid and on to the cells.

How does oxygen pass the capillary wall?

Oxygen is lipophilic. Since almost the entire capillary surface is identical to the lipid-containing plasma membrane of the endothelial cells, oxygen is able to use the total capillary surface for diffusion. The transport of lipophilic molecules is **perfusion limited.**

Oxygen diffuses so easily over the capillary endothelium, that there is tension equilibrium between blood and tissues already at the **arterial capillary** end. With rising perfusion the equilibrium point is shifted towards the venous part. Due to the haemoglobin, the **O_2 tension can be maintained** through all of the capillary. The oxygen tension varies in the tissues. There is a **longitudinal tension drop** towards the venous end of the capillary, and a **radial tension drop** in the tissue itself. In brain tissue, the O_2 tension can vary from an arterial level in certain small areas (P_{aO_2} of 13.3 kPa or 100 Torr) towards zero, when bloodflow is insufficient.

How does one calculate the magnitude of oxygen transport?

Fick's first law of diffusion: The flux (J) of O_2 is equal to the **diffusion coefficient of oxygen (D is 10^{-9} m^2 s^{-1})** multiplied with the concentration gradient (dC) per distance unit (dx) through a given area (A).

Fick's first law is written: $J = (-D\mathrm{d}C)A/\mathrm{d}x$ (mol/time unit) with a diffusion gradient (dC) through A.

In international symbol language the oxygen uptake flux (J) is called $\mathring{V}_{O_2}$. Notice that D/dx is a **permeability coefficient** (m s^{-1}). The first law can also be written as shown:

$$\mathring{V}_{O_2} = (D\Delta PA)/\mathrm{d}x.$$

Marie Krogh defined the individual lung diffusion capacity as the volume of gas-uptake per minute and per Torr in a given person (see Chapter 35).

A mean value for the lung diffusion capacity for CO (D_{LCO}) for healthy persons at rest is 3.1 ml STPD per second per kPa (25 ml CO per min and per mmHg). This mean value implies a mean lung diffusion capacity for oxygen (D_{LO_2}) of **3.6 (29) of these units,** and a mean **D_{LCO_2} of 70 ml s^{-1} kPa^{-1} (565 ml min^{-1} Torr^{-1}) at rest.** During exercise the **average D_{LO_2} increases to 9 ml STPD s^{-1} kPa^{-1}** (72 ml STPD min^{-1} Torr^{-1}).

For oxygen Fick's first law can be simplified to $\mathring{V}_{O_2} = (D_{LO_2} \cdot \Delta P_{O_2})$. At rest D_{LO_2} is 3.6 ml s^{-1} kPa^{-1} (see above), and the ΔP_{O_2} is 1.16 kPa (=8 Torr). These values correspond to an oxygen uptake of (3.6 1.16 60) **250 ml STPD min^{-1}**, which is a typical normal value for an adult at rest. **STPD** defines the conditions of a gas volume: **standard temperature and pressure, dry.**

Which circumstances can ease oxygen-diffusion in the tissues?

The factors that ease O_2-diffusion and delivery are:

1. **Myoglobin** releases O_2 during muscular contraction, when blood supply is blocked.
2. **Heat energy** releases O_2 during work, since increasing heat energy equals increasing movement of O_2.
3. **Carbon dioxide**: With rising P_{CO_2}, oxygen binding to haemoglobin decreases (Bohr–effect).
4. **Liberation of 2,3 - DPG (diphosphoglycerate) from haemoglobin** eases release of O_2.
5. **Mitochondria** (located close to capillaries) have reduced diffusion pathway.
6. **Fine capillary networks** reduce the diffusion distance.

Describe the dynamic characteristics of myoglobin as an O_2 store

The **standard affinity of myoglobin** towards O_2 is much higher than that of haemoglobin. Myoglobin is important as an O_2 store, although it is not totally saturated with O_2. During muscular contraction the bloodflow is blocked, and tissue O_2 tension falls drastically. Myoglobin then gives off O_2 to the cell. Bloodflow is re-established during muscular relaxation, and **myoglobin is rapidly reloaded** even when there is only a small rise in O_2 tension.

What is understood by capillary intermittence?

Krogh presumed that **tissue capillaries shift between a closed and an open state**. The capillary diameter varies with the oxygen tension. In well-perfused tissue, **high O_2 tension will cause vasoconstriction** and thus tend to reduce its perfusion.

Which tissues are most sensitive to anoxia?

Brain and heart tissues are extremely sensitive to low P_{O_2}.

Brain tissue is found in the nerve cells of the retina. These nerve cells are deprived of oxygen in 4.5 s (occurrence of **black out**). This can be verified by pressure on the upper eye lid. Consciousness is lost (**grey out**) a few seconds after cardiac arrest occurs. After 90 s, the brain interstitial fluid $[K^+]_{ISF}$ increases drastically from 3 to 60–70 mM, and both AP and synapse transmissions are eliminated. There is ion equilibrium over the cell membranes. Intracellular $[Na^+]$ also increases drastically and intracellular brain oedema develops. A high extracellular $[K^+]$ is life-threatening. The EEG of an anoxic brain is recognizable as a **straight EEG trace** (no electrical activity) indicating brain death. Because $[Ca^{2+}]$ rises in the nerve cell, this increases the K^+ conductance, so that more K^+ leaks out into the ISF.

Where are the peripheral arterial chemoreceptors located?

See Chapter 38.

Is the coronary bloodflow special in any way?

Catheterization of the venous sinus of the heart in healthy subjects at rest reveals a venous saturation fraction of 0.30. Hence, 0.7 parts of the oxygen concentration (C_{aO_2}) is desaturated. Thus, arterial blood with a normal C_{aO_2} of 200 ml l^{-1} liberates (200 · 0.7) 140 ml of oxygen per litre of blood to the myocardium. Variations in the arteriovenous O_2 content difference at the **steep part of the O_2–haemoglobin dissociation curve**, can only change the myocardial O_2 tension modestly. The **extremely high O_2 content difference** of the heart at rest, implies that the **rise in coronary bloodflow** must be the main source of extra O_2 to the heart during exercise.

Most of the blood entering the coronary circulation is delivered during the **systolic phase**.

What is cardiac cramp?

Cardiac cramp or angina pectoris is pain beneath the upper sternum or in the precordial area, often referred to the left arm and shoulder, the neck and the jaw. The pain is provoked, when the patient is exerted, experiences a strong emotional stress, or has a high metabolic rate due to a cold environment. Cardiac pain attack lasts for 1–2 min, and is relieved by rest. If an angina patient places a small tablet of nitroglycerin under his or her tongue, the pain subsides in a matter of minutes.

Endothelium-derived relaxing factor (EDRF identical with nitric oxide or NO) is produced in the ischaemic heart and increases coronary bloodflow (see Chapter 5). Acidosis during myocardial ischaemia mediates EDRF release.

The **α-adrenergic blocking agents** prevent sympathetic stimulation of the metabolic demands of the body including the myocardium. These drugs are useful in order to decrease the oxygen demands of the heart.

The cause of angina pectoris is **coronary artery stenosis**. The reduced calibre (stenosis) of the coronary arteries is caused by **arteriosclerotic plaques** in the walls.

What is arteriosclerosis?

Arteriosclerosis is a disorder in the arterial wall, where large quantities of lipids are deposited in the intima as arteriosclerotic plaques. These fatty deposits become invaded by fibroblasts and eventually become calcified.

When arteriosclerotic plaques are present in a coronary artery, the vessel wall suffers and eventually bursts, which causes occlusion of the vessel lumen. Coronary arteries are mainly end-arteries with insufficient collaterals. Following **occlusion of a coronary artery**, part of the myocardium will suffer from lack of blood supply. Adenosine liberated from deteriorated cells to surrounding blood vessels will reduce the damage. Still, the cardiac muscle contraction is sustained in the dead myocardial cells (**myocardial infarction**).

What is aortic–coronary bypass and coronary artery angioplasty?

Aortic–coronary bypass is a surgical therapy, where vein grafts (from the superficial saphenous vein) are anastomosed to the aorta and to the sides of the coronary vessels beyond the coronary occlusion.

Coronary artery angioplasty is a therapy, where a balloon-tipped catheter is placed in a coronary artery with stenosis, and pushed through the partial occlusion. In this position the balloon is inflated. With a maintained increase in vessel calibre, the bloodflow increases, and the patient is relieved of ischaemic phenomena for some time.

Is the low renal arteriovenous O_2 content difference of importance?

The kidneys receive 25% of $\dot{Q}$ at rest (1200 ml min^{-1} containing 200 ml O_2 l^{-1}), and the kidneys only use 15 ml O_2 min^{-1}. The kidneys have the **lowest arteriovenous O_2 content difference** of all the larger organs in our body. **The large safety margin is important for this vital organ during bleeding or falling renal bloodflow.**

Renal hypoxia causes increased erythrogenesis, whereas renal artery stenosis does not. Why?

Hypoxia is a condition with insufficient oxygen supply or oxygen utility in the tissues. See Chapter 62.

Further reading

Krogh, A. (1919) The number and distribution of capillaries in muscles with calculation of the oxygen pressure head necessary for supplying the tissue. *J. Physiol.* **52**: 409–19.

Vogel, J.H.K. and S.B. King, III (1989) *Interventional Cardiology: Future Directions*. C.V. Mosby, St Louis.

Zweifach, B.W. and H.H. Lipowsky. (1984) Pressure–flow relations in blood and lymph microcirculation. In: *Handbook of Physiology*. Sect. 2. *The Cardiovascular System – Microcirculation*, Vol. IV. Am. Physiol. Soc., Bethesda, MD.

19. Case History

An adult male has a lung diffusion capacity for oxygen of 22 ml STPD min^{-1} Torr^{-1}, and a mean alveolar O_2 tension gradient of 12 Torr. His oxygen concentration in the arterial blood (C_{aO_2}) is 200 ml l^{-1} and the renal bloodflow (RBF) is 1200 ml min^{-1}. The renal O_2 consumption is 15 ml min^{-1}.

1. *Calculate his oxygen uptake ($\dot{V}_{O_2}$) in ml STPD min^{-1}.*
2. *How is it possible for this person to increase $\dot{V}_{O_2}$ to 4900 ml STPD min^{-1}?*
3. *Calculate the arteriovenous oxygen content difference in the kidneys.*

Try to solve the problems before looking up the answers in Chapter 74.

CHAPTER 20.
RESISTANCE VESSELS

Essential among the haemodynamic features of the cardiovascular system are those determined by the relations between the driving blood pressure, the bloodflow and the vascular resistance. These factors are related to the ability of each tissue to control its own bloodflow in accordance with its needs. The control of local vascular resistance is a combination of neural and metabolic factors affecting a basal smooth muscle tone.

What is understood by resistance vessels and by vascular tone?

Arterioles, which range from 10 to 150 μm in diameter, control the distribution of blood to different tissues. A few of the small arterioles serve as **thorough-fare channels** or **metarterioles** that shunt the blood directly into the venules bypassing the capillary beds (Fig. 18-1). The largest pressure fall occurs in the arterioles and they thus exhibit the largest resistance in the system, although the arteries also cause some fall in driving pressure. Smaller arterioles exhibit vasomotion, which is the rhythmic changes in the arteriole diameter that causes bloodflow to fluctuate. **Arteriolar calibre** is determined by the contractile activity of its smooth muscle cells and by the transmural arteriolar pressure.

Vasomotion is often brought about by active changes in the tension of vascular smooth muscles. The arteriole can relax completely and then close completely. The arteriolar tone is both myogenic and neural (see below). The **myogenic tone** is increased by stretching the vascular smooth muscles, because stretching activates the **contractile mechanism**, which automatically stabilizes bloodflow.

Describe the neural control of the vascular resistance

The sympathetic and the parasympathetic division of the autonomic nervous system **opposing actions control the tone** of the resistance vessels.

1. Almost all blood vessels receive **efferent nerve fibres from the sympathetic nerve system** to their smooth muscles. True capillaries do not contain smooth muscles and do not receive autonomic nerve supply. Metarterioles and capillary sphincters do not receive nerve fibres at all. The sympathetic vasoconstrictor fibres and circulating catecholamines control both **arteriolar, venous and venule tone**. The vessels are innervated by postganglionic neurons from the paravertebral sympathetic trunk. The noradrenergic control releases NA and ATP. The **transmitter transport is axonal.** NA binds to **α-adrenergic constricting receptors**. Ad binds to both **α-adrenergic constricting receptors** and to **β-adrenergic dilating receptors**. Consequently, Ad elicits vasoconstriction in arterioles where α-receptors predominate and vasodilatation where β-adrenergic receptors predominate. Adenosine is a potent vasodilatator because it inhibits release of NA possibly via presynaptic purine receptors. In the synapse, the neurotransmitter is eliminated by **re-uptake**, by enzymatic breakdown and by diffusion. The arterioles of the skeletal muscles, the skin, the kidneys and the splanchnic region are densely innervated.

Hunting predators are claimed to have sympathetic **vasodilatator fibres** to the skeletal muscle vessels.

2. The **cholinergic system** is parasympathetic. The vessels of the head, neck and thoracoabdominal organs are innervated by parasympathetic fibres (the 3rd, 7th, 9th and 10th cranial nerves). The large intestine, bladder and genital organs are innervated by parasympathetic fibres from the sacral segments 3–5. The fibres to the external genitals are activated during sexual excitation. ACh is the vasodilatating transmitter for muscarinic and nicotinic cholinergic receptors.

   Purinergic receptors use vasodilatating transmitters as ATP, AMP and the potent **adenosine.**

Describe the metabolic control of vascular resistance

Metabolic control is the sum of all metabolic factors that match the oxygen supply to the energy requirement.

A decrease in the oxygen partial pressure of the arterial blood (P_{aO_2}) elicits an important reaction, vasodilatation. A high P_{aO_2} and a high tissue P_{O_2} seem to close the metarterioles and the precapillary sphincters of a tissue unit, and they remain closed until the excess oxygen is used. When the P_{O_2} is low, the sphincters open again to begin a new rhythmic fluctuation. Krogh called this phenomenon **capillary intermittence.**

As the rate of ATP utilization increases with increasing work rate, its breakdown products accumulate, including ADP, AMP, adenosine, K^+, H^+, CO_2, lactate, pyruvate, etc. The concentration of most of these substances in the blood is directly proportional to the rate of ATP utilization and the degree of vasodilatation. **Reactive hyperaemia** (i.e. increased bloodflow following experimental vascular interruption) is probably explained by the **metabolic control theory**.

Are hormones involved in the control of local bloodflow?

Hormonal vasodilatators and factors liberated in the tissues are likely candidates. **Endothelium-derived relaxation factor (EDRF)** is a potent vasodilatator that is released from vascular endothelial cells in response to **hypoxia, ADP, and serotonin secretion. EDRF** is identical with **nitric oxide (NO)**. NO is also released when perfusion pressure or bloodflow increases, thereby stabilizing the bloodflow. NO is most potent in veins. The vasodilatator actions of NO are mediated by cGMP (Fig. 5-3).

Prostaglandins such as prostacyclin (PG_2) are vasodilatators that also inhibit the thrombocyte agglutination in the capillary bed.

Substance P is released from sensitive nerve endings by cutaneous axon-reflexes following irritation of the skin (i.e. **antidromic stimulation**). Substance P released by antidromic stimulation explains the redness and oedema in the vicinity of a skin irritation.

Inflammatory hyperaemia with accumulation of leucocytes is caused by **leucotrienes.**

VIP is vasoactive intestinal polypeptide from the intestine, the salivary glands and the penile cavernous bodies. VIP is a potent vasodilatator (Chapter 46).

Vasopressin is antidiuretic hormone (ADH) from the hypophyseal posterior lobe. ADH acts as a **vasoconstrictor** (Chapter 64).

What is autoregulation?

Autoregulation is an automatic control phenomenon, that **aims at maintaining a constant bloodflow when driving pressure is changed.**

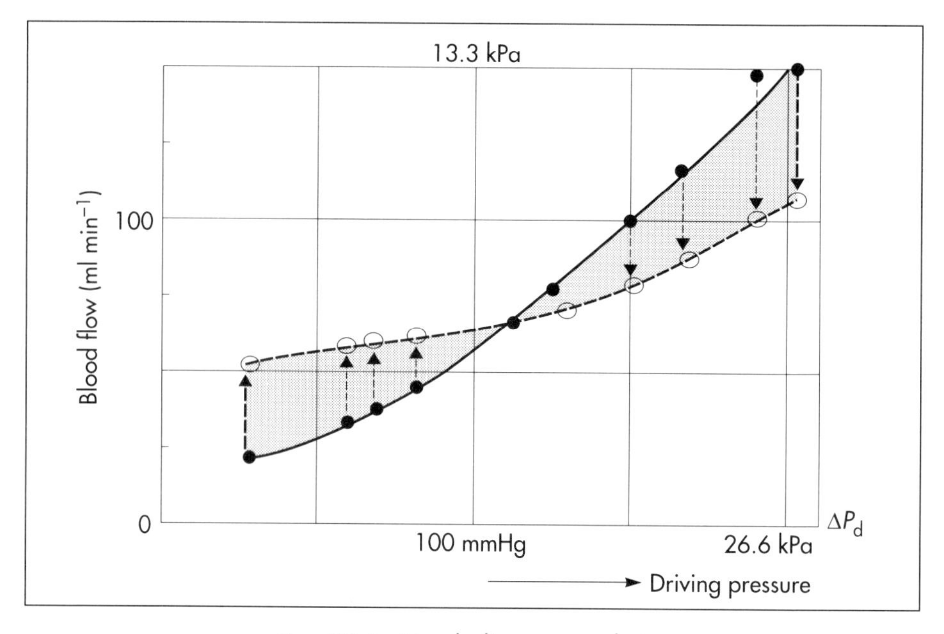

Fig. 20-1: Metabolic autoregulation.

The resistance vessels of the coronary system tend to **diminish any changes in the bloodflow** in the coronary vessels that are triggered by changes in the driving pressure (ΔP_d). Increases or reductions in ΔP_d are followed by similar changes in coronary bloodflow. However, the resistance of the vessels is then changed – metabolically and mechanically – so that the coronary bloodflow is maintained at control levels at all times (arrows in Fig. 20-1).

Autoregulation has been explained by at least two theories: (1) the **myogenic theory** considers autoregulation as a **myogenic response,** which is an intrinsic property of vascular smooth muscle. Increased stretch of the smooth muscle elicits contraction, whereas diminished stretch elicits vasodilatation; (2) the **metabolic control theory** (see above) is especially concerned with a potent vasodilatator, **adenosine.**

Adenosine is continuously produced by breakdown of ATP. **Adenosine** is the most likely candidate for the role of **metabolic mediator**, because it is such a potent vasodilatator and because it diffuses readily across the cell membranes. Adenosine may work via **presynaptic inhibition of sympathetic nerve fibres** to the smooth muscles of the resistance vessels. Falling perfusion pressure leads to diminished rate of adenosine washout and thus to local vasodilatation. Adenosine dilates the vessels and causes increased coronary bloodflow (see Chapter 25). Increased perfusion pressure washes out adenosine, which leads to vasoconstriction and local decrease in bloodflow.

How is the arterial blood pressure regulated?

The three types of regulators involved are classified as **rapid, intermediate and long-term regulators.**

1. **Rapid regulators** are the **arterial baroreceptor reflexes** originating from the carotid

sinuses and the aortic arch. These arterial presso-receptors work within seconds following dynamic changes in blood pressure, but they do not regulate chronic blood pressure changes. A drastic fall in arterial blood pressure will release **large amounts of vasopressin** in order to constrict the resistance and capacitance vessels.

2. The **intermediate regulators** begin their function within minutes. These mechanisms include **transcapillary capillary volume shifts** in response to changes in capillary blood pressure, and **vascular stress relaxation**, which refers to a particular smooth muscle phenomenon well developed in veins. When veins are stressed by increased pressure, they slowly expand so that the blood pressure decreases. Conversely, when the intravascular volume decreases, the opposite occurs. A third intermediate regulator is the **renin–angiotensin–aldosterone cascade**, which is activated by a fall in arterial pressure (see below).
3. **Long-term regulators** comprise different mechanisms for **renal regulation of the body fluid volume**. When arterial pressure rises, more urine is excreted. Hereby, the plasma and interstitial volume is reduced. The diminished plasma volume decreases venous return to the heart, reducing $\mathring{Q}$, so that elevated arterial blood pressure is brought back toward normal. A **decrease in arterial pressure** elicts the opposite reaction: **aldosterone from the adrenal cortex** promotes Na^+ reabsorption and K^+ secretion from the renal tubules. The reabsorbed Na^+ augment water retention, as does also increased **vasopressin (ADH) secretion** from the posterior hypophysis. A falling arterial pressure also diminishes the release of atrial, natriuretic factor (ANF) and its Na^+ and water excreting actions diminish.

How does the renin–angiotensin–aldosterone cascade work?

See Chapters 59 and 61.

Further reading

Calver, A., J. Collier and P. Vallance (1993) Nitric oxide and cardiovascular control. *Exp. Physiol.* **78**: 303–26.

Chien, S. (Ed.) (1988) Vascular endothelium in health and disease. Plenum, New York.

Opie, L.H. (1991) *Heart: Physiology and Metabolism*, 2nd edn. Raven Press, New York.

Schuman, E.M. and D.V. Madison (1994) Nitric oxide and synaptic function. *Annu. Rev. Neurosci.* **17**: 153–83.

Shepherd, J.T. (1982) Reflex control of arterial pressure. *Cardiovascular Res.* **16**: 357–83.

White, C.W., R.F. Wilson and M.L. Marcus (1988) Methods of measuring myocardial blood flow in humans. *Prog. Cardiovasc. Dis.* **31**: 79.

20. Case History

A female, age 22 years, is sitting on a bicycle ergometer with her calf muscles 0.9 m below heart level. She is at rest and the venous pressure is 10 Torr (1.3 kPa) at the level of the heart. The oxygen uptake ($\dot{V}_{O_2}$) is 0.247 l STPD min^{-1} and the muscle bloodflow is 3 ml min^{-1} per 100 g of tissue (3 Flow Units, FU). The total weight of all her skeletal muscles is 30 kg.

Following a 5-min rest, she starts cycling, hereby increasing $\dot{V}_{O_2}$ to 4.5 l STPD min^{-1}, and her muscular arterioles dilate to reach a three-fold increase in inner radius. During exercise the arterio-venous O_2 content difference is 170 ml STPD l^{-1}, and the oxygen uptake in the skeletal muscles increases from 1 to 100 ml STPD min^{-1} kg^{-1}.

1. *Calculate the venous pressure in the calf muscles at rest.*
2. *Calculate the relative alteration of the muscular vascular resistance during exercise.*
3. *Calculate the driving blood pressure over the working muscles during exercise, where the arterial blood pressure is 170/70 mmHg (22.7/9.3 kPa) and the venous pressure in the calf muscles is reduced to 20 mmHg (2.6 kPa).*
4. *Calculate the rise in muscle bloodflow during exercise.*

Try to solve the problems before looking up the answers in Chapter 74.

Chapter 21. Cardiac Performance

The heart is a four-chambered double pump. Every 24 h the heart ejects more than 10^4 l of blood, and contracts more than 10^5 times. The total amount of work performed over the lifetime of a person is enormous. The order of size is calculated at the end of this chapter.

Describe the left ventricular performance in a cardiac cycle

The cardiac cycle describes volume, pressure, and electric phenomena in the left ventricle as a function of time, and one heart beat is shown below (Fig. 21-1). The solid clock-shaped curve is the **intraventricular pressure** (Fig. 21-1). The left ventricle is closed to the aorta in diastole and blood flows from the atrium to the left ventricle (Fig. 21-1). The QRS complex in the ECG reflects ventricular depolarization (see Chapter 23). Contraction or **ventricular systole** results in closure of the mitral valve (Fig. 21-1).

Systole is **isovolumetric** until the intraventricular pressure exceeds the aortic pressure. Then the aortic valve opens and ventricular ejection occurs (see thick upward arrow in Fig. 21-1). **Bulging of the cuspidal mitral valve** into the left atrium during **isovolumetric contraction** causes a rise in left atrial pressure (see the **c-wave** in the thin **a-c-v-curve** of Fig. 21-1). The intraventricular pressure reaches a plateau around 15–16 kPa and then begins to decrease. The aortic valve closes when the intraventricular pressure falls below aortic pressure (see small arrow indicating **retrograde flow in aorta** in Fig. 21-1). This is the end of the ejection phase or the **left ventricular ejection time (LVET)** (Fig. 21-1).

The **ECG** does not reflect the mechanical performance of the heart, but **closure of the aortic valve** corresponds in time with the **end of the T-wave in ECG.**

The **left atrial pressure** rises slowly during ventricular systole as blood reaches the atrium by means of the pulmonary veins. The atrial pressure starts to fall when the cuspidal mitral valve opens. This produces the **v-wave** in the thin **atrial pressure curve** (Fig. 21-1). Filling of the atrial chambers with blood is aided by the **great atrial compliance**. This is why the atrial pressure only rises modestly.

At the end of the ventricular diastole the left atrial pressure increases, because the atrial systole begins just before the ventricular systole. This **early atrial contraction** forces extra blood into the ventricle just before the mitral valve closes, and gives rise to the **a-wave** on the thin atrial pressure curve (Fig. 21-1).

Describe the intraventricular volume and pressure changes in a cardiac cycle

The **intraventricular pressure–volume loop** is a time-independent representation of the cardiac cycle. The instantaneous intraventricular pressure and volume is plotted (Fig. 21-2). In diastole from B to C, the ventricle receives blood from the left atrium. The small

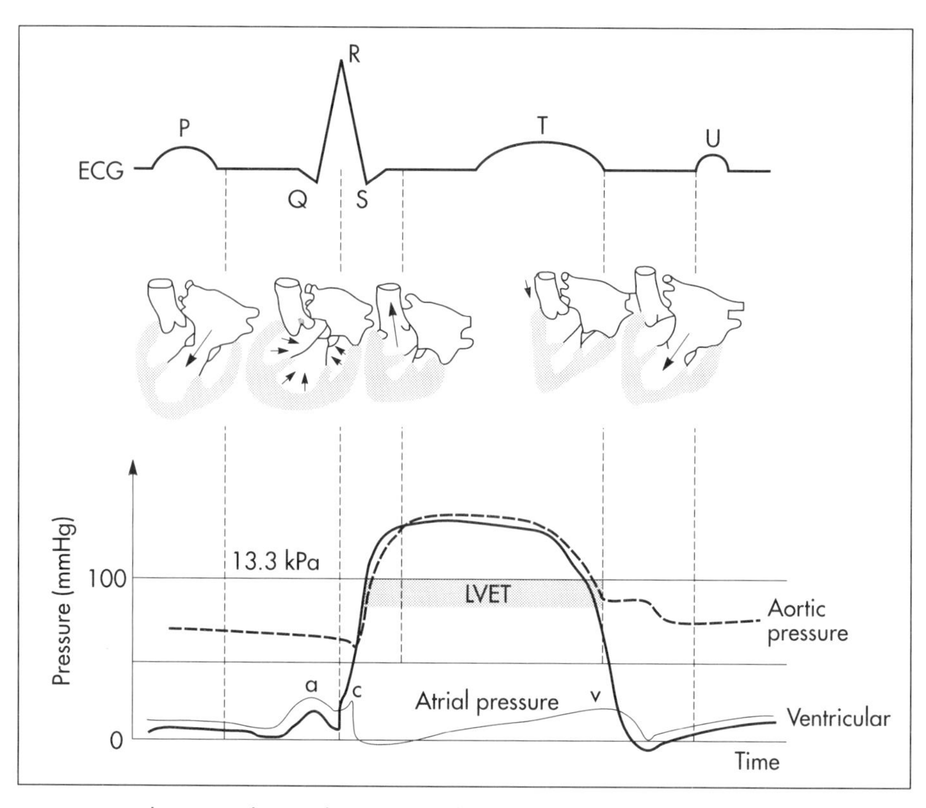

Fig. 21-1: Electromechanical events in the cardiac cycle. The thin atrial pressure curve has a, c, and v waves. For ECG see Chapter 23.

increase in ventricular pressure reflects passive expansion and elastance of the myocardial wall. Pressure and volume increase with a slope that is called **ventricular stiffness** or **contractility**. The line marked **min.** is **minimal contractility**. The line starts from the ventricular **dead volume**, that is a virtual minimal volume of blood, which can never be ejected (50 ml in Fig. 21-2). A steep rise in pressure occurs from C to D, with no change in ventricular volume, the **isovolumetric contraction**. At D the aortic orifice opens, because the end-diastolic pressure in the aorta is passed. During the rapid ejection phase, the fall in ventricular blood volume is accompanied by a continuous increase in pressure. During ejection the volume falls by a size equal to stroke volume, pressure rises and falls until the **residual ventricular volume** is attained (about 80 ml in Fig 21-2). The last ejection phase is slow, because the pressure decreases towards A, where the aortic orifice closes. The last event from A to B is the **isovolumetric relaxation** with a sharp drop in pressure at constant volume (Fig. 21-2). The maximum slope of the curve is the maximum chamber stiffness or **maximal contractility** (line **max.** in Fig. 21-2). The **reverse (dV/dP) is chamber compliance.**

The **area of the loop represents the pressure–volume work** on the **stroke blood volume** performed by the **ventricular contractile elements** during ejection. The pressure in C is the **end-diastolic intraventricular pressure (EDIP)** or the so-called **preload**. The force against which the ventricle contracts is termed **afterload**. A good index of the

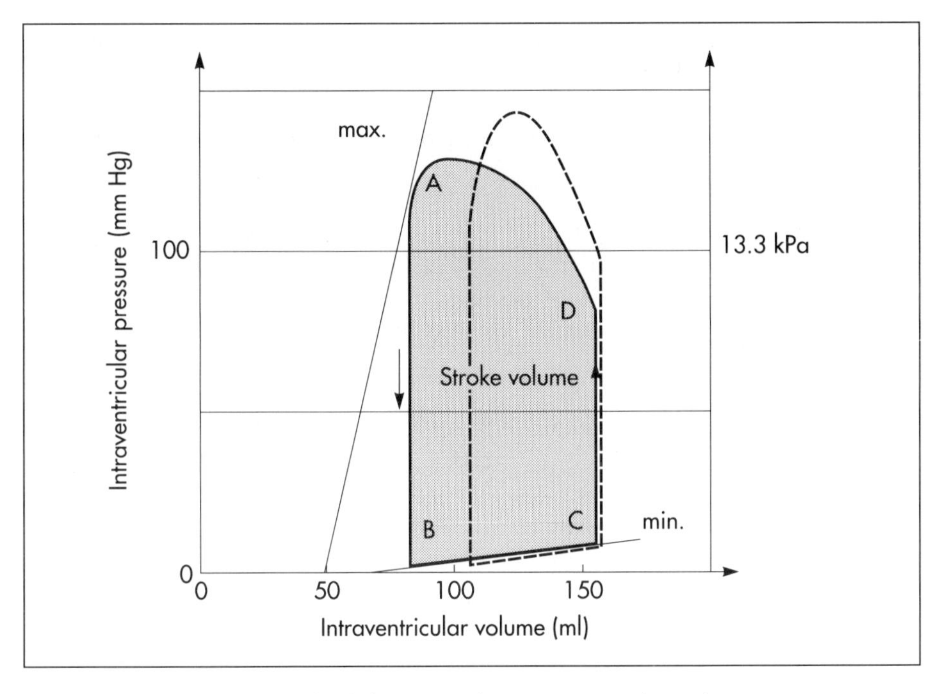

Fig. 21-2: The left ventricular pressure-volume loop.

afterload is the **peak aortic or intraventricular pressure during systole**, equal to the highest pressure shown in Fig. 21-2. The afterload is almost equal to the **peak systolic pressure** in the arterial tree. When the afterload is increased at constant end-diastolic pressure and volume, a greater ventricular pressure develops in order to expel blood (the dashed curve from D in Fig. 21-2). The result is a smaller stroke volume (and hence a greater residual ventricular volume), because of the high aortic and intraventricular pressure (Fig. 21-2).

What is the Frank–Starling law of the heart?

The Frank–Starling relationship states that increasing left ventricular end-diastolic volume increases the stroke volume of the next heart beat. During isovolumetric contraction, the **EDIP** or the **preload** increases. Increasing end-diastolic volume increases the ventricular filling pressure and thus stroke volume and stroke work.

An increase in afterload (theoretically defined as the maximal tension in the left ventricular wall during ejection) occurs when the aortic pressure increases. Such an increase causes a decrease in stroke volume. Hereby, the EDIP and the end-diastolic volume increases, so the **cardiac stroke work** increases concomitantly.

Calculate the cardiac stroke work

The **ventricular stroke work** is the **sum of the pressure–volume work and the kinetic work:**

$$\textbf{Stroke work rate} = [(\boldsymbol{P} \cdot \boldsymbol{V}) + \tfrac{1}{2}\, \boldsymbol{m} \cdot \boldsymbol{v}^2].$$

Both the pressure–volume work and the kinetic work are work per stroke duration or time unit, that is comparable to work-rate or **effect** in Watts.

What happens to the stroke work of the heart when the preload is increased?

Preload is the end-diastolic filling pressure of the ventricle just before contraction (C and E in Fig. 21-3).

When the left ventricle expands from C to E, by receiving more blood than before, the end-diastolic volume is increased. The greater diastolic filling results in a larger stroke volume according to Starling's law of the heart.

The stroke work of the heart on the blood is the area A, B, E, F. Accordingly, the stroke work is increased.

What are the consequences of increased contractility with constant preload?

A sympathetic increase in contractility without a change in end-diastolic pressure (**preload**), results in an increased intraventricular pressure (Fig. 21-4). The max. line through G illustrates the **increased contractility**. The larger stroke volume, smaller residual volume, and larger stroke work on the blood (area C, D, G, H in Fig. 21-4) are also shown.

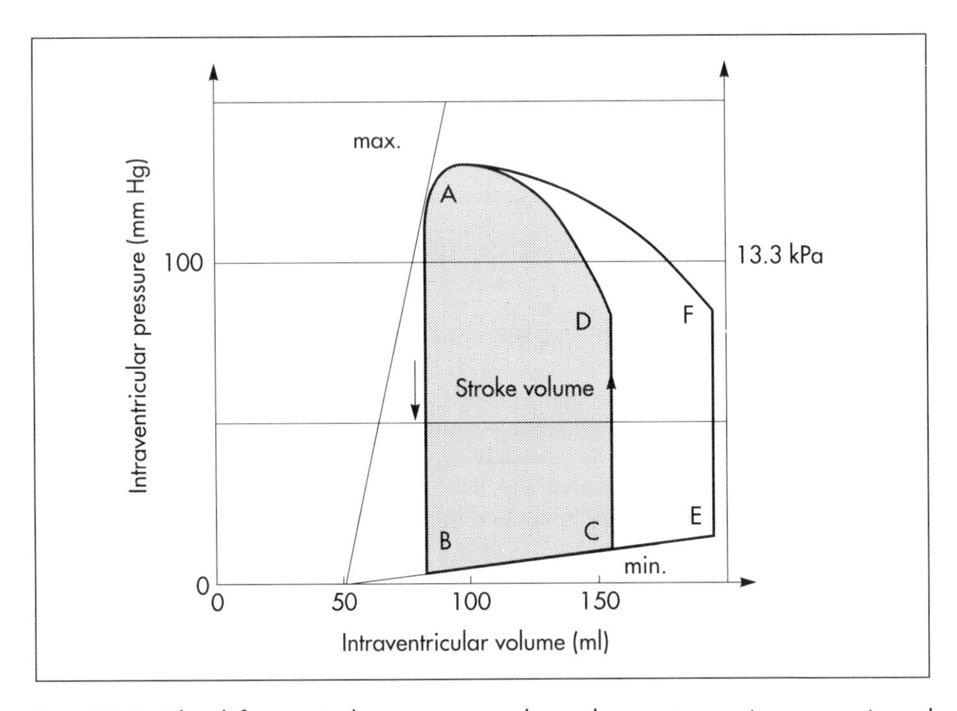

Fig. 21-3: The left ventricular pressure–volume loop at rest (grey area) and following an increase in end-diastolic volume (E–F).

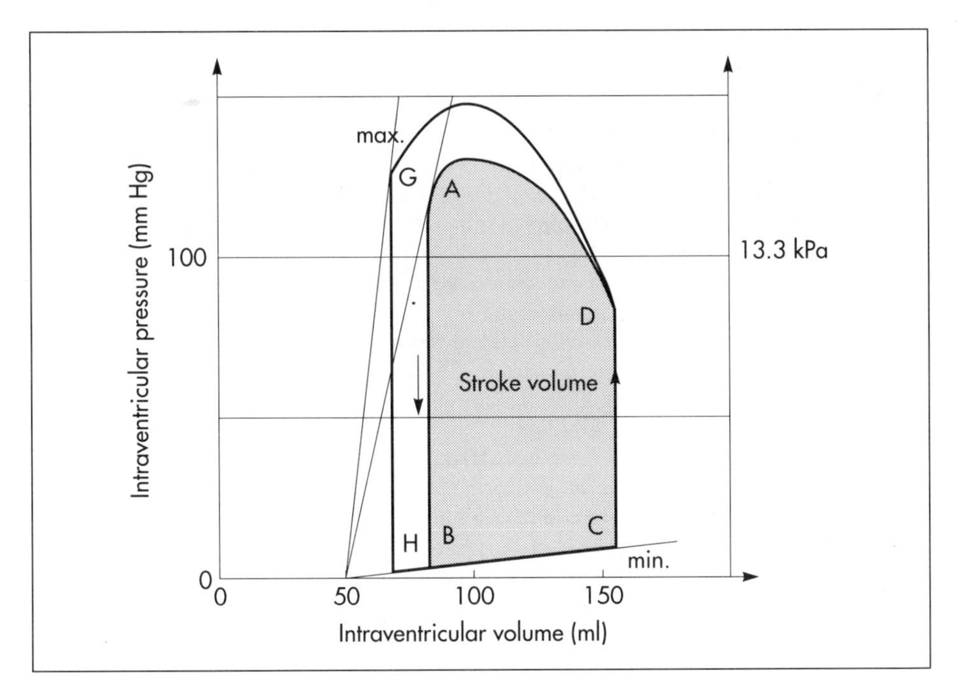

Fig. 21-4: The left ventricular pressure–volume loop at rest (grey area) and following an increase in contractility (G–H).

Is the law of Laplace acceptable as a model of ventricular contraction?

The transmural pressure rises during ventricular contraction even at constant fibre tension. This is because the short ventricular radius is reduced to the same extent:

$$\Delta P = T\left(\frac{1}{r_1} + \frac{1}{r_2}\right)$$

Details of the model are given in relation to Fig. 17-3A.

Echocardiography shows that the short ventricular radius is reduced by 5–6 mm during systole at rest in healthy people. The Laplace law is acceptable in such a situation.

If a cardiac chamber of a heart patient increases its diameter to double, and a spherical chamber is assumed, it implies a **four-fold greater fibre tension** (Fig. 17-3B). To the patient, this means an enormous energy requirement in the myocardium to maintain the necessary pressure. This disorder is called cardiac stress or **cardiac failure (insufficiency)** – see Chapter 22.

Further reading

Bruetsaert, D.L. and S.U. Sys (1989) Relaxation and diastole of the heart. *Physiol. Rev.* **69**: 1228–47.

Sheu, S.S. and M.P. Blaustein (1991) Sodium/calcium exchange and control of cell calcium and contractility in cardiac muscle and vascular smooth muscle. In: H.A. Fozzard *et al.* (Ed.), *The Heart and Cardiovascular System.* Raven Press, New York.

21. Case History

A male, 20 years of age, has an aortic mean pressure of 13.30 kPa, a pressure in the pulmonary trunk of 2.67 kPa, a heart frequency of 60 and a stroke volume (sv) of 68 ml. The atrial pressure is zero. The cross-sectional area of the aortic ostium is estimated to $3.8 \cdot 10^{-4}$ m^2. The mechanic efficiency of his ventricles is only 10%.

1. *Calculate the pressure–volume work rate (in watts) for an average cardiac cycle in the right and the left ventricle.*
2. *Is this man at rest or exercising during these measurements?*
3. *During heavy exercise the $\mathring{Q}$ rises to 25 l min^{-1} and the aortic mean pressure rises to 20 kPa. Calculate the pressure–volume work rate for the left ventricle during this exercise.*
4. *Calculate the pressure–volume work rate on the blood during the life span of this man, assuming a life duration of exactly 63 years (2×10^9 s), performing the heavy exercise above for 1/1000 of the time and spending the remaining part of his life at rest.*
5. *Calculate the energy requirements of the heart necessary to perform such a task.*
6. *Calculate the kinetic work rate at rest with a 40% ejection time. Calculate the kinetic work rate during the exercise above, with an ejection time of 55%. The relative density of his blood is assumed to be 1.06 kg l^{-1}.*

Try to solve the problems before looking up the answers in Chapter 74.

CHAPTER 22. THE VENOUS SYSTEM

Veins are highly distensible or **compliant vessels** (i.e. large d*V*/d*P*) that have **one-way valves.**

The venous system (veins, venules, and venous sinuses) is used to control the amount of blood that is translocated from the venous to the arterial side of the circulation.

The venous system is an important determinant of the **cardiac output ($\mathring{Q}$)**. It does so by a strong nervous influence from the **sympathetic nervous system on its smooth muscle fibres.** During normal conditions the venous system contains two-thirds of the systemic blood volume. Only half of the total blood volume is located in the exchangeable pool of the small veins. Any increase in the venomotor tone decreases the venous capacity and redistributes blood volume, thus increasing $\mathring{Q}$. Any decrease in venomotor tone increases venous capacity, causing $\mathring{Q}$ to decrease. Studies of cardiovascular system models by Guyton explain the venous return to the heart in the following manner. The model contains an arterial and a venous compliance and an arteriolar resistance. A so-called **venous return curve** is compared to the **cardiac output curve** from a denervated heart with the right atrial pressure as the x-axis (Fig. 22-1).

The steep part of the $\mathring{Q}$ curve shows that $\mathring{Q}$ can double following a rise in pressure of only 0.27 kPa (2 Torr). The mean systemic filling pressure depends upon the blood volume and the compliance of the systemic circulation. The right atrial pressure is strongly influenced by body posture, and the pressure is often termed the **central venous pressure (CVP).**

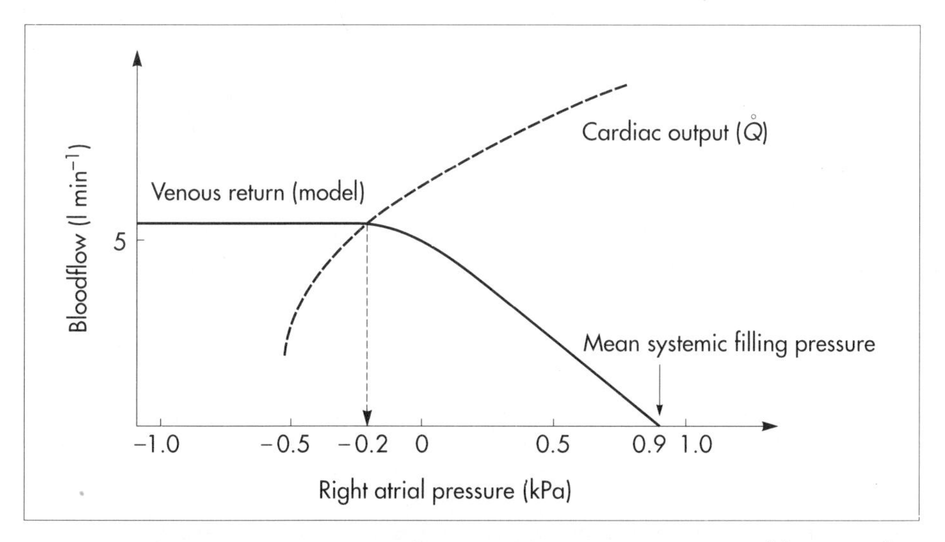

Fig. 22-1: The venous return curve from model experiments, crossed by a cardiac output curve from a denervated heart.

As the right atrial pressure of the circulatory model increases, it exerts a back pressure on the venous system to impede the bloodflow into the right atrium, the **venous return** (Fig. 22-1). At the normal right atrial pressure about −0.2 kPa (−1.5 Torr) the venous return curve crosses the $\mathring{Q}$ curve, and both flows are 5.5 l min^{-1}. If the atrial pressure is suddenly increased to 0.9 kPa (7 Torr), equal to the so-called **mean systemic filling pressure** in all parts of the systemic circulation, then all flow of blood is stopped (Fig. 22-1). This relatively low pressure is mainly due to the very distensible venous system.

The right atrial pressure has only increased slightly, but enough to decrease the venous return to zero and thus the $\mathring{Q}$ is zero (Fig. 22-1). The more the atrial pressure falls below the venous pressure, the more the venous return will rise up to a certain level at an atrial pressure of −0.2 kPa (almost zero). Negative atrial pressures have the same venous return. This is because negative transmural pressures in the central, thoracic veins imply collapse.

The venous return is therefore constant, regardless of further fall in right atrial pressure (Fig. 22-1).

$\mathring{Q}$ must equal venous return in the steady state. Thus the $\mathring{Q}$ curve for the left ventricle must cross the venous return curve in one point (Fig. 22-1).

What is the importance of the central venous pressure (CVP)?

CVP is the **pressure measured in the large thoracic veins** at the level of the heart or in the right atrium.

The **driving pressure** for the systemic circulation is the **mean aortic pressure (MAP)** minus **CVP**. The relationship to $\mathring{Q}$ and total peripheral vascular resistance (TPVR) is given by the formula: $\mathring{Q}$ = (**MAP** – **CVP**)/**TPVR**. A small fraction of TPVR is found in the venous system.

The venous return is expressed by the equation:

venous return = (**venule pressure** – **CVP**)/**venous resistance**.

Small variations in CVP alter the volume of blood considerably in the venous system. A normal value for venule pressure is 1.3 kPa (10 Torr) and for CVP about zero. Since venous return must equal $\mathring{Q}$ in steady state, the venous resistance is only about 10% of the TPVR.

Describe the three venous pressure waves of the right atrium

The **a-wave** occurs during the **contraction of the right atrium** by which it squeezes out extra blood just before ventricular systole. As the atrium relaxes, the CVP falls (Fig. 21-1).

The next wave is the **c-wave**, and this wave is caused by closure of the **cuspidal valves** and by the right ventricular contraction, because the increased ventricular pressure is transmitted backwards to the right atrium and large veins (Fig. 21-1). The third wave is the **v-wave**. Throughout ventricular systole and isovolumetric relaxation, **venous blood** returns to the heart, but the tricuspid valve is closed, so that the central veins and right atrium are distended. The coinciding pressure buildup is relieved, when the tricuspid valve opens at the start of diastole, where the pressure falls (Fig. 21-1). The increase in **right atrial pressure** is transmitted backwards to the large veins near the heart. Prominent waves can often be seen in the neck veins when supine.

Describe the function of the venous pump

The **venous pump** is defined as **all local external pressures that facilitate venous return to the heart**. Two important pump mechanisms are involved:

1. **The skeletal muscle pump**. The deep veins of the arms and legs are affected by pressure exerted by exercising skeletal muscles. The veins are compressed by muscle contraction, and the one-way valves prevent the blood from flowing backward, and thus secure the transfer of blood toward the heart. Even the superficial veins are compressed during contraction.

   As soon as a venous segment is emptied of blood, its **transmural pressure** is so low that the **filling pressure** from more peripheral veins can fill the empty segment with blood.

   This **skeletal muscular venous pump** is also called the **peripheral venous heart**, because its force must be equal to or larger than that of the heart. In the erect position the **peripheral venous heart** must also overcome the force of gravity. During muscle rest there is an added hydrostatic pressure load of 13.3 kPa (100 Torr) in the dependent limb. With a MAP of also 13.3 kPa the total pressure in a foot artery is 26.6 kPa, and in the dependent vein just above 13.3 kPa. During muscle contraction the venous pressure rises driving blood into the heart, and just after muscle contraction the venous pressure falls again.
2. **The thoraco-abdominal pump**. The large veins are also affected by the positive intra-abdominal pressure and by the negative pressure in the thoracic cavity. The inferior vena cava returns blood from lower regions to the heart. During inspiration the intrathoracic pressure becomes more negative, and **blood is sucked into the large thoracic veins facilitating venous return to the heart**. The inspiratory contraction of the diaphragm increases abdominal pressure favouring venous return. During expiration the intra-abdominal pressure decreases and intrathoracic pressure increases but remains negative, so that the venous return is maintained. **Intrinsic cardiac mechanisms**, including the **length–tension relation** (Chapter 6), allow the heart to increase stroke volume **beat-by-beat**, when the venous return increases.

What are varicose veins?

Varicose veins have incomplete valves. The skeletal muscle pump maintains the venous bloodflow towards the heart. Patients with this disorder can develop venous pooling or **stasis** and **ankle oedema**.

Why is the capillary pressure much more dependent of venous than of arterial pressure changes?

The venous resistance is minimal, whereas the arteriolar resistance is the dominant portion of the total TPVR. Thus, any given **venous pressure change** will affect the capillary pressure most.

What is cardiac failure?

Cardiac failure or insufficiency is a disorder, where the heart cannot pump enough blood to satisfy the nutritive needs of the body. Cardiac insufficiency is manifested by a consequential decrease in $\mathring{Q}$ (lower output failure) or by an increase in $\mathring{Q}$ (higher output failure). The cardiac failure can be acute or chronic.

Damming of blood in the vessels behind the insufficient heart pump is typical. In chronic cardiac failure the venous system is expanded by blood, and the left ventricle as well (Fig. 22-2).

What is low output cardiac failure?

Acute myocardial infarction, and also slow cardiac deterioration with chronic cardiac failure and cardiac oedema, render the heart incapable of pumping the minimal blood volume required to transfer sufficient oxygen to the mitochondria. All tissues deteriorate, the patient is extremely tired, and even the work of breathing is a heavy task. The condition is also called **cardiogenic shock** (Fig. 22-2) and it is a case of **general ischaemic hypoxia**.

Is it possible to develop cardiac failure with high $\mathring{Q}$?

Patients with a $\mathring{Q}$ much higher than normal can develop cardiac failure. The venous return is much too high, and after some time with an overexpansion of the heart, the cardiac pump fails to eject the same blood volume, and an increasing blood volume is accumulated behind the insufficient ventricle.

The rise in CVP and venous pressure leads to pulmonary oedema and eventually to peripheral oedema.

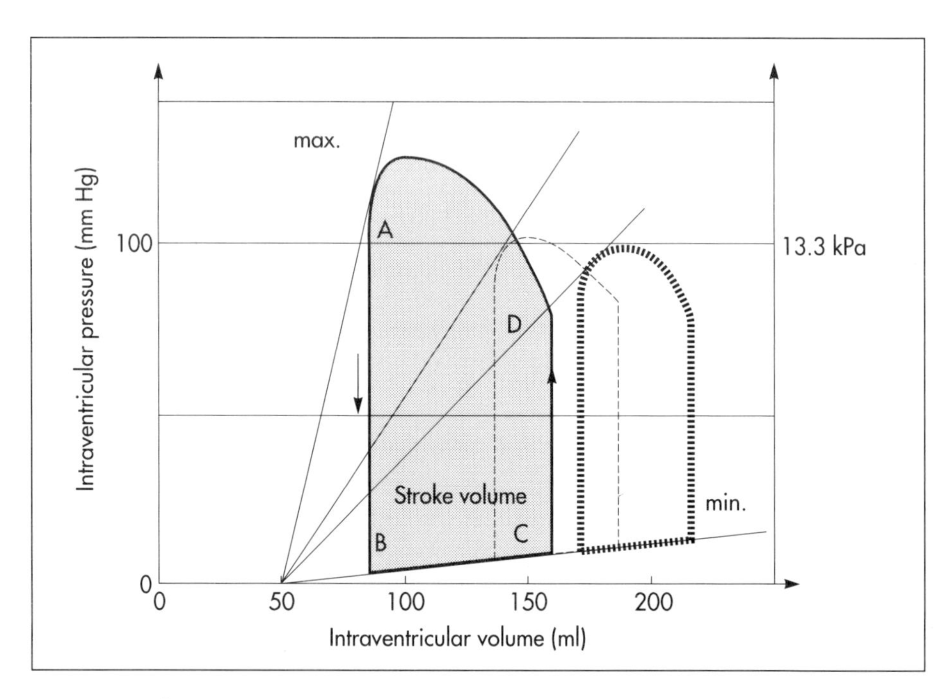

Fig. 22-2: The ventricular pressure–volume loop in a healthy person (solid loop), and in persons with acute cardiac failure (dotted curve) or chronic cardiac failure (dashed curve with higher contractility).

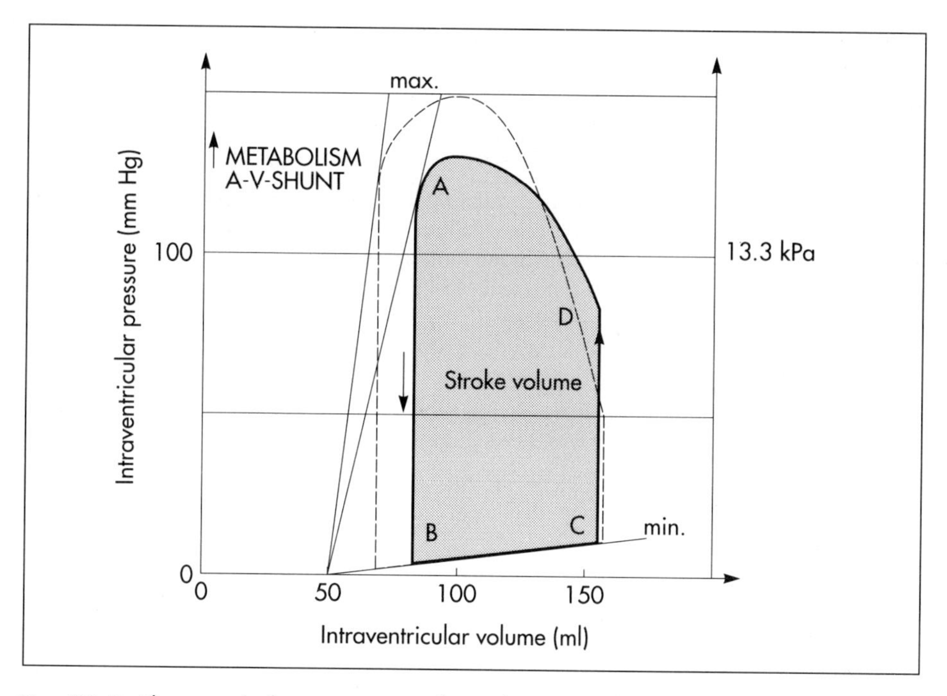

Fig. 22-3: The ventricular pressure–volume loop in a healthy person (solid loop), and in persons with high metabolic rate or with arteriovenous shunts.

Examples of this condition are cardiovascular disorders with a drastic reduction of the TPVR. The low TPVR is illustrated by the low opening pressure in the left ventricle being equal to a low end-diastolic aortic pressure (Fig. 22-3).

In **hyperthyroidism** (with an abnormally high metabolic rate), all the vessels in the systemic circulation dilate, and the venous return overloads the heart. On the other hand, short term administration of L-thyroxine to patients with chronic heart failure improves cardiac and exercise performance.

Any major **arteriovenous shunt** leads a large fraction of the arterial blood directly into the veins. This greatly increases venous return and overloads the heart.

Are the kidneys involved in cardiac oedema?

Cardiac oedema develops during chronic cardiac failure, because the kidneys retain fluid. The accumulated fluid increases MAP and thus venous return, which in turn elevates the right atrial pressure. The rising atrial pressure elevates the venous and the capillary pressure. This causes loss of fluid into the interstitial fluid volume. Accumulation of abnormal volumes of fluid here is the definition of **oedema**.

The low $\mathring{Q}$ and blood pressure causes an increased sympathetic tone with constriction of the afferent renal arterioles to the glomeruli. As a consequence, the **glomerular filtration rate** decreases. The renin–angiotensin–aldosterone cascade is activated. Angiotensin is a strong vasoconstrictor, which further decreases the renal bloodflow, and aldosterone promotes the reabsorption of NaCl and water from the distal renal system. Cardiac failure with excessive increase of the atrial pressure leads to a rise in the **atrial natriuretic hormone ([ANH]) in the blood**, whereby a part of the retained NaCl and water is excreted.

What happens to $\mathring{Q}$ and venous pressure in chronic heart failure?

The venous system contains two-thirds of the total blood volume. Thus, venoconstriction shifts significant quantities of blood from the peripheral to the central circulation. Since **CVP varies inversely with TPVR**, it is possible to maintain $\mathring{Q}$ in resting patients with chronic heart failure (insufficient contractile force) at the expense of increased CVP, by reduction of TPVR:

$$\mathring{Q} = (\mathbf{MAP} - \mathbf{CVP})/\mathbf{TPVR}.$$

When $\mathring{Q}$ decreases more and more during development of chronic heart failure, the compensation fails, and both CVP and end-diastolic ventricular pressure and volume rises further (Fig. 22-2).

Further reading

Milnor, W.R. (1982) *Hemodynamics.* Williams & Wilkins, Baltimore, MD.

Moruzzi, P., E. Doria, P.G. Agostoni, V. Cappacione, and P. Scanzerla (1994) Usefulness of L-thyroxine to improve cardiac and exercise performance in idiopathic dilated cardiomyopathy. *Am. J. Cardiol.* **73**(5): 374–8.

Smith, J.J. (Ed.) (1990) Circulatory Response to the Upright Posture. CRC Press, Boca Raton, FL.

22. Case History

A 30-year-old male is lying supine on a tilt table. The mean capillary pressure of his leg muscle tissues is 3.1 kPa. The capillary filtration coefficient (Cap_f) is 0.075 ml min^{-1} kPa^{-1} per 100 g of tissue. The tissue hydrostatic pressure is negligible (−0.13 kPa or −1 Torr). The protein concentration of the blood plasma is 70 g l^{-1} and its colloid osmotic pressure is 3.3 kPa. The protein concentration of the interstitial fluid is 10 g l^{-1}, and its colloid osmotic pressure is 1/7 of that of plasma. The protein reflection coefficient is 0.9. The total weight of both legs is 30 kg.

1. *Calculate the capillary waterflow from 100 g of his leg tissues per minute.*
2. *Calculate the capillary waterflow from all of his leg capillaries per 24 h (30 kg tissue).*
3. *The male subject is tilted to an upright position without standing, and the pressure is measured in a foot capillary 1.2 m below the level of the heart. Calculate the expected value which is measured in kPa.*
4. *Now the male subject leaves the tilt table and stands erect with his foot capillary 1.2 m below heart level. The high capillary pressure in the foot is eventually partially compensated to a lower level measured to be 9.8 kPa. Explain this compensation by at least three physiological mechanisms.*

Try to solve the problems before looking up the answers in Chapter 74.

CHAPTER 23.
THE CARDIAC ACTION POTENTIAL AND ECG

Describe the neural regulation of the heart function

A large number of sympathetic (S) and vagal (X) motor nerve fibres end close to the sinoatrial node (SA node in Fig. 23-1).

Sympathetic stimulation speeds up the SA node (**sinus tachycardia**) and vagal activity slows the SA node (**sinus bradycardia**). Increased concentration of the sympathetic transmitter noradrenaline (NA), and of adrenaline (Ad) from the adrenal glands, cause **positive inotropia** (increased contractility), **positive chronotropia** (increased frequency), **positive dromotropia** (increased conduction velocity), and **positive bathmotropia** (increased irritability) on the heart. NA activates α-adrenergic constrictor receptors in the coronary vessels, whereas Ad activates β-adrenergic vasodilatator receptors.

The neurotransmitter acetylcholine (ACh activating muscarinic receptors) and vagal stimulation caused **negative ino-, chrono-, dromo-, and bathmotropia.**

Describe the action potential during a contractile cycle of a ventricular fibre

Across the ventricular cell membrane there is a steady potential difference of the same size as the equilibrium potential for K^+ (−94 mV), that is almost −90 mV (Fig. 23-2).

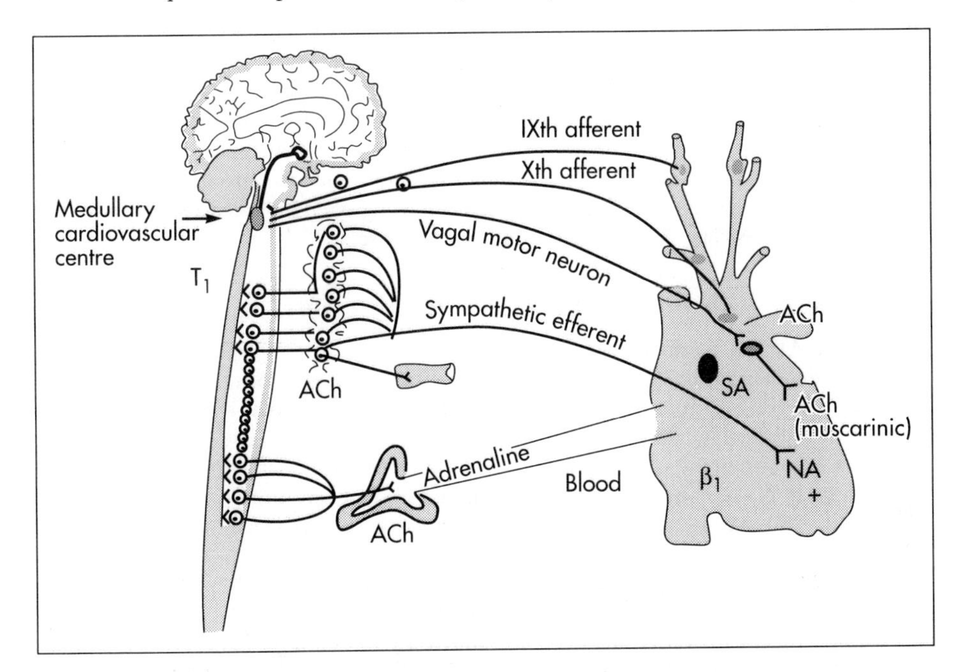

Fig. 23-1: The neural control of the heart.

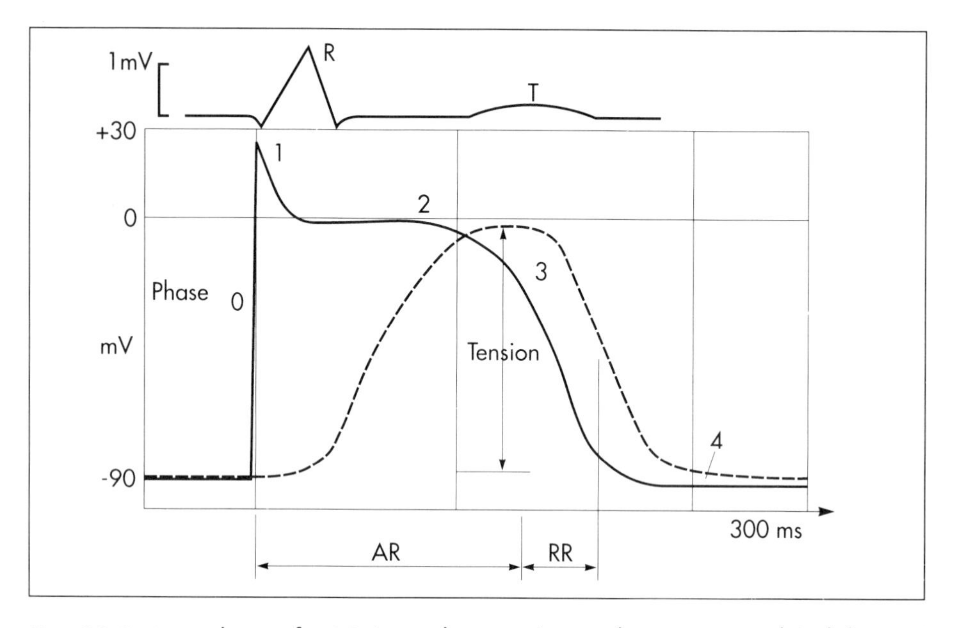

Fig. 23-2: Recordings of ECG (R and T wave), membrane potential (solid curve) and contraction (stippled curve) of one heart cycle in a ventricular fibre.

This negative potential is referred to as the **resting membrane potential (RMP)**, because it represents the potential difference across the cell membrane (inside negative) between successive APs.

Depolarization occurs by a small, abrupt influx of Na^+ from neighbouring regions favoured by electrochemical forces. Hereby, the potential difference approaches the equilibrium potential for Na^+ (+60 mV). The threshold potential for an AP is a rise of 25–30 mV, where the **fast Na^+-channels** begin to swing open for only a short time, and then abruptly close. The AP is an **all-or-none response**.

What is the ionic basis for the AP in a ventricular fibre?

The **cardiac AP** can be divided into **five phases**:

The fast depolarization (phase 0) is shown by the abrupt upstroke, which is related to the rapid entry of Na^+ into the cell through **fast voltage-gated Na^+-channels**, suddenly allowing the electrostatic and chemical forces to work (Fig. 23-2). The **fast Na^+ current** across the cell membrane causes **phase 0 of atrial, ventricular and Purkinje APs** (Fig. 23-2). The **fast Na^+-channels** are both **voltage-dependent and time-dependent**. Phase 0 stops about +30 mV, because the fast Na^+-channels become voltage-inactivated.

Phase 1 is the early repolarization from the upstroke. Here, the **slow Ca^{2+}-Na^+-channels** operate, so there is a steady influx of Ca^{2+} and Na^+. The K^+-channels are also voltage-gated and here the K^+-permeability of the membrane is small. The early limited repolarization in phase 1 is related to the sudden arrest of Na^+-influx.

Phase 2 is the plateau of the AP, where the **slow Ca^{2+}-Na^+-channels** remain open for up to 300 ms. The net influx of Ca^{2+} and Na^+ is almost balanced by a net efflux of K^+, so the **balance is forming the plateau** (Fig. 23-2). Ca^{2+} activates the muscle contractile process. When the slow Ca^{2+}-Na^{+}-channels close, the potential is optimal for the voltage-gated K^{+}-channels, and the permeability for K^{+} increases rapidly.

Phase 3 is the terminal repolarization. With all the **K^{+}-channels open**, large amounts of K^{+} diffuse out of the ventricular fibre. The equilibrium potential for K^{+} (−94 mV) and the RMP is rapidly approached.

Phase 4 is recognized by the RMP of −90 mV (Fig. 23-2). The **Na^{+}-K^{+} pump** restores ionic concentrations by exchanging Na^{+} for K^{+} in a ratio of 3:2.

Why is it not possible to bring ventricles into smooth tetanus?

Phase 3 covers the **relative refractory period (RR)**, and the T-wave in the ECG. The long **absolute refractory period (AR)** of the ventricular cells, AR, covers the whole shortening phase of the contraction (Fig. 23-2 stippled curve), where all the fast Na^{+}-channels are voltage-inactivated. As a consequence, no stimulus can be released regardless of size. In RR enough of the fast Na^{+}-channels are working, so that a sufficiently large stimulus can break through and produce an AP smaller than normal.

The long AR protects the cardiac pump, as it is not possible to bring ventricles into smooth tetanus.

Why does the ventricular fibre have an AP-plateau?

See the **slow Ca^{2+}-Na^{+}-channels** and phase 2 described above. Skeletal muscle fibres have no plateau, because they do not have **slow Ca^{2+}-Na^{+}-channels** open for such a long time.

What is the role of Ca^{2+} *in the cardiac excitation–contraction coupling?*

When a wave of excitation spreads along the myocardial sarcolemma from cell to cell via electrically **conducting gap junctions**, the wave also spreads into the cell interior via the T-tubules. During the phase 2 plateau of the AP (Fig. 23-2), Ca^{2+} permeability increases, and Ca^{2+} flows down its electrochemical gradient. Ca^{2+} enters the cell through the **slow Ca^{2+}-channels** and the large T-tubules fill with mucopolysaccharides. The **channel proteins** are phosphorylated by a **cAMP-dependent protein kinase A**. The small Ca^{2+} influx is called **trigger-Ca^{2+}**, because it releases large amounts of Ca^{2+} from the sarcoplasmic reticulum. Hence the cytoplasmic [Ca^{2+}] increases from the resting level of 10^{-7} mol l^{-1} by a factor of 10–100 during excitation. The free Ca^{2+} binds to troponin C, and the complex interacts with tropomyosin to activate sites between actin and myosin filaments. This is the basis for **crossbridge cycling** and thus for contraction of myofibrils.

Catecholamines and increasing extracellular [Ca^{2+}], raise the cytoplasmic [Ca^{2+}] and thus the developed force of contraction.

Cardiac glycosides poison the **Na^{+}-K^{+}-pump**. This blockage increases $[Na^{+}]_{in}$, to the extent that less Ca^{2+} is removed from the cell. This – and any – form of elevated cytoplasmic [Ca^{2+}] enhances contractile force. Increased [K^{+}] in the extracellular fluid reduces the force of contraction, so the heart becomes dilated and flaccid.

What parts of the heart can depolarize spontaneously? How does the AP of a pacemaker cell look?

The rhythm of the heart is initiated by a complex flow of electrical signals which are called **action potentials or APs**. The APs generated in the SA node display **automaticity** (i.e. they undergo spontaneous and rhythmic depolarization without external stimuli). SA is the **primary pacemaker**, because it has the highest frequency. The automaticity is associated with small **pale cells**. Pale cells are all potential pacemakers. The **secondary pacemakers** are situated in the AV node, the bundle of His and in the Purkinje network.

The **membrane potential** of pacemaker cells (about −55 to −60 mV) is **never constant** (Fig. 23-3). At this potential almost all the fast Na^+-channels are **voltage-inactivated or blocked** by the inactivation gates on the inside of the channel and they remain blocked. Only the **slow Ca^{2+}-Na^+-channels** can open and thereby cause the AP. The AP is therefore slower to develop. After a while (100–300 ms) the **slow Ca^{2+}-Na^+-channels** close, and a large number of K^+-channels open. Thus the influx of Ca^{2+} and Na^+ ceases, while lots of K^+ diffuse out of the pacemaker cell during the relatively slow repolarization.

The K^+-channels remain open during the repolarization, so the membrane potential approaches −60 mV at the end of the AP. The rising membrane potential between two heart beats is caused by the inherent leakiness of the pacemaker membrane to Na^+. When the membrane potential reaches the threshold potential (TP about −40 mV), the slow Ca^{2+}-Na^+-channels open and the cycle starts again continuing for a lifetime (Fig. 23-3).

How is the cardiac propagating wave conducted?

The heart beat is self-initiating, and normally the propagating wave (impulse) originates in the SA node (see above). The impulse propagates from the SA node via **three bundles**

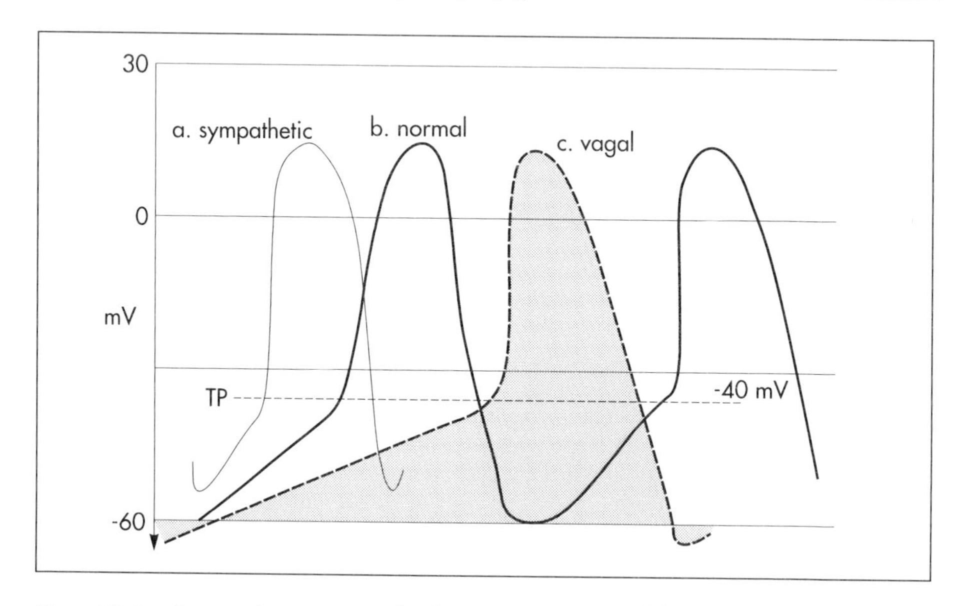

Fig. 23-3: Pacemaker potentials from a sinus nodal fibre: a. Sympathetic stimulation; b. normal heart rate; c. vagal stimulation.

of internodal syncytial cells, through the left and right atrial wall to the **atrioventricular (AV) node**. This point is typically reached within 40 ms. After passing through the AV node (with a so-called **AV delay** of 100 ms), the propagating wave (excitation) reaches the **bundle of His** and specialized conduction fibres, which activate almost synchronously all the ventricular tissue and thus impart maximal thrust to the blood. The propagation velocity in these large **Purkinje fibres** is 1 m s^{-1}, which is the fastest velocity possible in the heart. The Purkinje system terminates just under the endocardial surface on **gap junctions** in the myocardial cells. The **AV delay** provides time for the atrial systole to pass extra blood to the ventricles before the ventricular systole occurs.

How was the electrocardiograph developed?

In **1903 Einthoven began a systematic study, with a string galvanometer** (the first electrocardiograph), of the potential differences between electrodes (four **bath tub** electrodes) on the skin surface of humans during heart beats. As the fluids of the body conduct electricity quite easily, the body conducts electrical signals uniformly in all directions (an almost **uniform volume conductor** increasing the potential electrical field around the **heart generator**).

An electrocardiogram (ECG) is a curve showing the potential variations against time in the whole body stemming from the heart, which is an **electrochemical generator** suspended in a conductive medium. The myocardial cell membranes have separate charges, and small ion gradients produce electrochemical gradients. Small ion fluxes occur only during depolarization and repolarization, where potential differences are produced between polarized and depolarized cell regions in the heart. An **electrical current flows from polarized (plus outside the cell) to depolarized (minus outside the cell) regions.** In contrast, the direction of the **propagation wave of the impulse** goes from outside negative (depolarized) to outside positive (polarized) regions.

Einthoven arranged the ECG equipment so that the **direction of the propagating wave toward the positive pole produces an upright deflection**. If it is directed towards the negative pole a downward deflection is recorded in bipolar leads (Fig. 23-4). In **unipolar recordings, upright deflection indicates movement of the propagating wave toward the positive, exploring electrode**, and downward deflection indicates that the propagation wave is moving away (Fig. 23-4).

Einthoven's law states that any two of the three bipolar limb leads determine the third one with mathematical precision. The potential differences recorded over time in an ECG can be estimated using the rules of **vectorial projection.**

The QRS deflections in two of the three standard leads can be drawn graphically in a triangle and their resultant is the **mean QRS-axis of the heart** (Fig. 23-4). The atrial depolarization (the three arrows passing the upper part of the heart in the lower part of Fig. 23-4) and the P-waves are shown from three different electrodes positions. The **electrical field generated** by definition **moves in the same direction as the charge from positive to negative.**

Describe the normal ECG and its interpretation

Each heart cycle has a fixed pattern of waves (**P,Q,R,S,T,U**). The SA node is a minimal muscle mass, and there is no potential difference (wave in the ECG) before the atria depolarize with a P-wave. When the propagating wave is directed towards the electrode (as in lead II) the atrial depolarization will produce a **positive P-wave** (Fig. 23-5). The P-

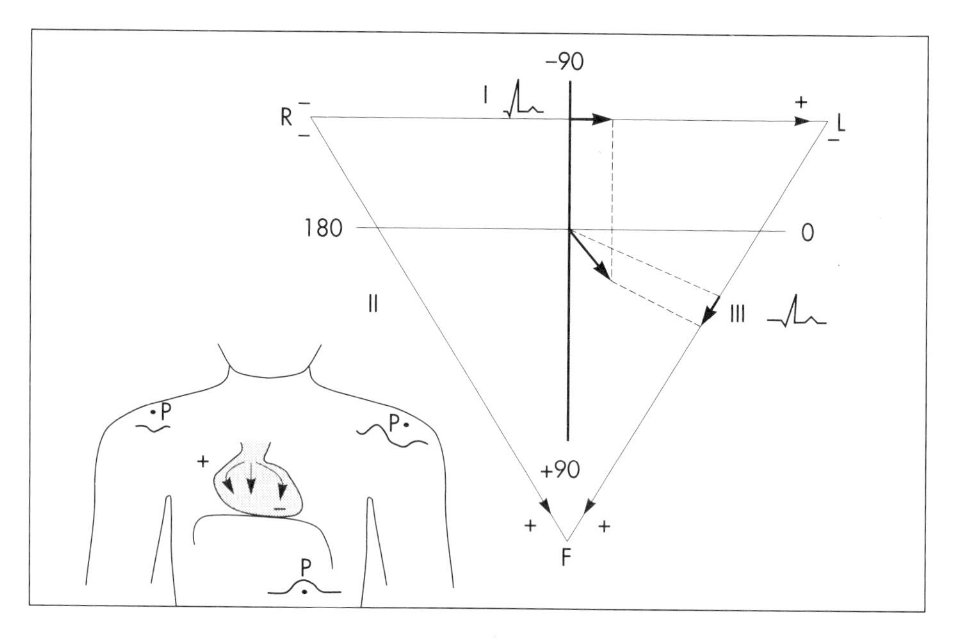

Fig. 23-4: Einthoven's triangle.

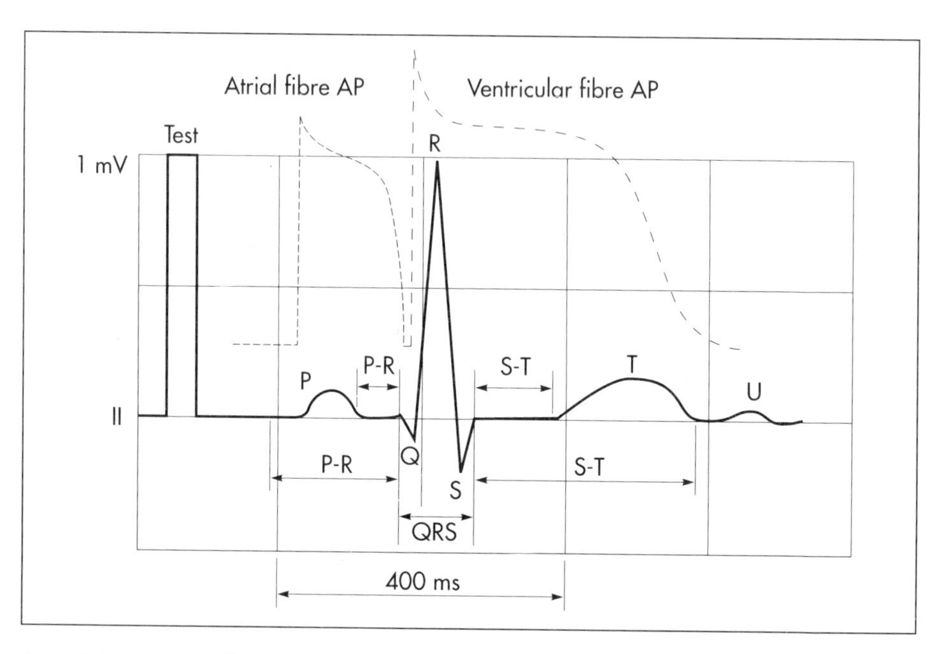

Fig. 23-5: Normal ECG (II. lead). The action potentials from an atrial (dotted curve) and a ventricular fibre (dashed curve) are shown above.

waves correspond to the impulse distribution in the atria, and the QRS-complex origin from depolarization of the strong ventricular myocardium (Fig. 23-5).

Since the activation of the septum is from left towards right, the progagating wave moves away from the exploring electrode, and the **Q-wave becomes negative**, whereas the dominating transfer of the propagating wave towards the apical electrodes provides the **large, positive R-wave** in almost all leads. The small propagating wave moving away from the electrode at the apex and to the right to reach the right ventricle, is responsible for the small, negative **S-wave**.

The **T-wave** is positive in most leads, and due to the apically directed repolarization of the ventricular AP (Fig. 23-5). The U-wave is a rare phenonenon seen in children and young people.

How is ECG recorded?

Einthoven used three standard bipolar leads placed in the frontal plane (Fig. 23-6): Lead **I**: records the potential difference between the right and left arms; lead **II**: between the right arm and left leg; and lead **III**: between the left arm and left leg. The right leg is used to ground the patient (Fig. 23-6; observe the positive and negative signs).

Unipolar, precordial leads assume that **one exploring electrode** is the actual recording electrode, detecting changes in the local potential relative to zero. The exploring electrodes are defined as the three standard extremity leads connected to one **indifferent electrode** (Fig. 23-6). Conventionally, the six Wilson leads are recorded from the precordium (V_1–V_6). The QRS complexes of the normal heart are mainly negative in

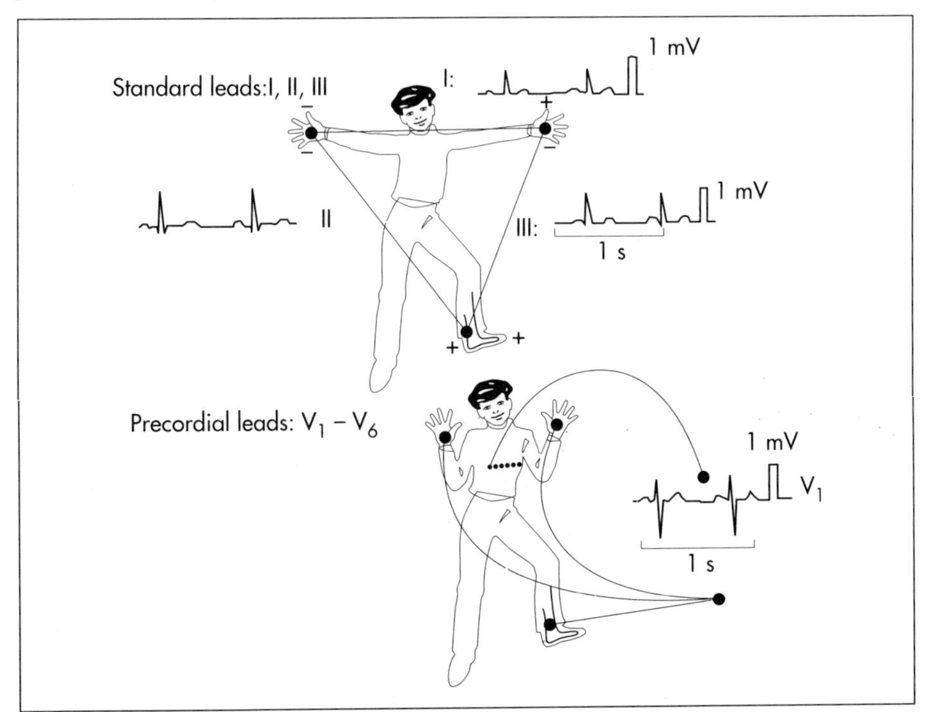

Fig. 23-6: Standard and precordial ECG leads.

leads V_1 to V_2 because the exploring electrodes are more influenced by the base than by the apex, and the opposite is the case in leads V_4 to V_6.

What is atrial fibrillation?

Atrial fibrillation is a condition in which the **sinus node** no longer controls the rhythm. The fibrillation is characterized by a tumultuous rapid twitching of atrial muscle fibres, and a total irregularity of ventricular contractions. An excitation wave with 400–600 cycles per minute, courses continuously through the atrial wall over a circular pathway about the origin of the great veins (the **circus motion theory**). Because of the refractoriness of the AV-bundle, only 80–150 ventricular beats are recorded each minute. The many P-waves in the ECG are characteristic for atrial fibrillation.

What is Adam–Stokes syndrome?

The **Adam–Stokes syndrome** is a clinical disorder caused by a partial AV-block, with a long P–Q interval and a wide QRS complex in the ECG, suddenly becoming a **total bundle block**. The condition results in **unconsciousness and cramps caused by brain**

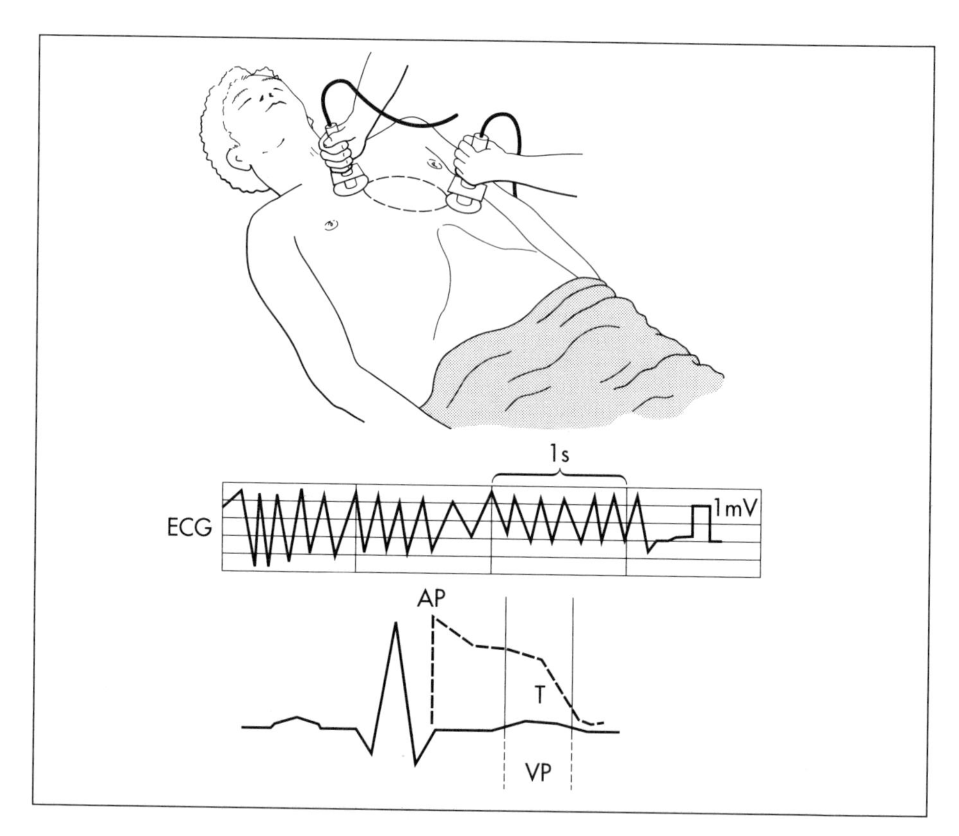

Fig. 23-7: Ventricular defibrillation.

hypoxia and sometimes resulting in **universal cramps (grand mal)** due to violent activity in the motor cortex. The **key hole** is the **AV node and the bundle of His**. Disease processes here elicit the Adam–Stokes syndrome.

Therapy is to provoke sinus rhythm by a few forceful strokes in the precordial area of the thorax, accompanied by **mouth-to-nose-resuscitation** and **external heart massage**. A sympathomimetic drug can be injected if necessary, even into the heart directly. The patient must be brought immediately to hospital for special intensive care. **Permanent pacemaker treatment** may become necessary.

What is ventricular fibrillation?

Ventricular fibrillation is a tumultuous twitching of ventricular muscle fibres, which are ineffectual in expelling blood, and thus incompatible with life. The irregular ventricular rate is 200–600 twitches min^{-1} (Fig. 23-7). Without co-ordination the force is used frustraneous. Negligible amounts of blood leave the heart, so within 5 s unconsciousness occurs, because of lack of blood to the brain. In patients with coronary artery disease, ventricular fibrillation is a cause of **sudden death**. The trigger is **anoxia**, with an ineffective Na^+-K^+-pump, or electrical shock (electrocution).

When is electrical defibrillation performed?

Ventricular fibrillation is the most serious cardiac arrhythmia. It must be converted to sinus rhythm at once by the application of a **large electrical shock** to the heart (ventricular defibrillation) or the patient will die. Alternating current is applied for 100 ms, or 1000 V direct current is applied for a few milliseconds.

The vulnerable period (VP in Fig. 23-7 is actually **phase 3** and represented in the ECG as the T-wave) **is dangerous**, because an electrical shock, when given during this period, will cause in itself **ventricular fibrillation**.

Further reading

Noble, D. (1984) The surprising heart: A review of recent progress in cardiac electrophysiology. *J. Physiol. (London)* **353**: 1–9.

Zipes, D.P. and J. Jalife. (1990) *Cardiac Electrophysiology: From Cell to Bedside.* W.B. Saunders, Philadelphia.

23. Multiple Choice Questions

Each of the following five statements have True/False options:

Stem statement: The ventricular action potential is

A. Initiated by rapid entry of Na^+.

B. Characterized by slow Ca^{2+}-Na^+-channels.

C. Characterized by closed K^+-channels in phase 3.

D. Dependent upon Ca^{2+} influx.

E. Independent of the Na^+-K^+-pump in phase 4.

Try to solve the problems before looking up the answers in Chapter 74.

CHAPTER 24. BLOODFLOW MEASUREMENTS

When the radius of a vascular bed is doubled, its bloodflow may increase by as much as 16 times. However, when the driving blood pressure is doubled, the bloodflow rises only slightly due to **autoregulation**. Hence, bloodflow is much more sensitive to changes of the vascular dimensions, than changes in driving blood pressure, as resistance varies with the 4th power of the radius.

What is mass balance and how is it used for bloodflow measurements?

According to the **law of conservation of matter**, mass or energy can neither be created nor destroyed (the principle of mass balance). **Adolph Fick** applied natural occurring indicators like O_2 and CO_2. Using O_2 as an indicator and the law of mass balance, he expressed in an equation that the O_2 flux, taken up by the lung blood ($\mathring{V}_{O_2}$), plus the O_2 flux which arrives to the lungs with the venous blood ($\mathring{Q}C_{\bar{v}O_2}$), must be equal to the O_2 flux, which leaves the lungs in the oxygenized blood ($\mathring{Q}C_{aO_2}$): $\mathring{V}_{O_2} = \mathring{Q}C_{aO_2} - \mathring{Q}C_{\bar{v}O_2}$. Thereby, Fick proposed that the $\mathring{Q}$ can be calculated as follows:

$$\mathring{Q} = \mathring{V}_{O_2}/(C_{aO_2} - C_{\bar{v}O_2}).$$

A classical example of the usefulness of Fick's principle is to consider the data of a healthy male at rest. The typical data for such a person are an **O_2 uptake of 250 ml STPD min ^{1}** and an **arteriovenous O_2 content difference of 50 ml STPD l^{-1}**. According to Fick's principle, a man can only satisfy his O_2 demands, if 5 l of blood is oxygenated in his lungs every minute. Thus 5 l min^{-1} is his $\mathring{Q}$.

The oxygen concentration of mixed venous blood ($C_{\bar{v}O_2}$) is usually obtained through a venous catheter inserted up the median cubital vein, through the subclavian vein, and finally into the right ventricle or pulmonary artery, where the blood is well mixed. Arterial blood is easily obtained from the radial artery (C_{aO_2}). The disappearance rate of oxygen from the respired air can be recorded in a metabolic ratemeter ($\mathring{V}_{O_2}$).

The principle of mass balance is valid only for a system in **steady state**. Steady state is a state where the indicator is administered at a constant rate, and is neither stored, mobilized, synthesized nor used by the system, and where no shunts are present.

This method has been used to measure a large increase in $\mathring{Q}$ in different patient groups. For example, patients with anaemia have been found to have higher $\mathring{Q}$ at rest.

What is anaemia?

Anaemia refers to a condition of **reduced blood [haemoglobin] and the consequential reduction in working capacity**. This occurs at concentrations below 8 mmol l^{-1} (130 g l^{-1}). **Anaemic hypoxia** is the diminished delivery of oxygen to the tissues, despite normal P_{aO_2}, that results in local vasodilatation.

Anaemia with few red cells reduces the viscosity of the blood. Hypoxia and reduced viscosity reduce the TPVR. The falling TPVR forces the heart to increase the output in order to avoid a serious fall in arterial blood pressure.

Describe the use of the dilution principle for bloodflow measurements

When an **indicator** bolus (mass or dose of **tracer** in weight or molar units) is instantaneously injected in the right side of the heart, the indicator and blood will mix. The mixture leaves the right ventricle through a **well mixed outlet,** passes the pulmonary circulation and then returns to the left side of the heart. The indicator concentration during the first passage of any peripheral artery is recorded **continuously or by multiple sampling.** The resulting curve is shown in a semilog scale (Fig. 24-1). The indicator concentration (in mol ml^{-1} of blood) reaches a peak and then decreases in a few seconds, before it again rises due to indicator recirculation with the blood (Fig. 24-1). The first decrease in concentration is assumed to be mono-exponential. Hence it is easy to extrapolate to the concentration zero, and read the so-called **first passage time,** T_1. In this case the T_1 is 9 s (Fig. 24-1). The mean concentration ($\bar{c}$ mol ml^{-1}) of indicator in the period T_1 seconds is determined by planimetry.

The average amount of indicator (in moles) leaving the left ventricle per second in 1 ml of blood is $\bar{c}$, hence $\bar{c}$ is given in mol ml^{-1} of blood. The volume of blood (V) in which the indicator dose is distributed is dose/$\bar{c}$. Since blood carries only $\bar{c}$ mol of tracer in each ml, the heart needs at least a bloodflow of V ml (dose in mol/$\bar{c}$) in order to carry the entire dose through the aortic orifice in T_1 (9) s. Accordingly, the $\mathring{Q}$ per second is **dose/($\bar{c}T_1$).** The product ($\bar{c}T_1$) is the area under the curve (Fig. 24-1). Thus the **dose/area ratio** must be equal to the bloodflow leaving the left ventricle in ml s^{-1}.

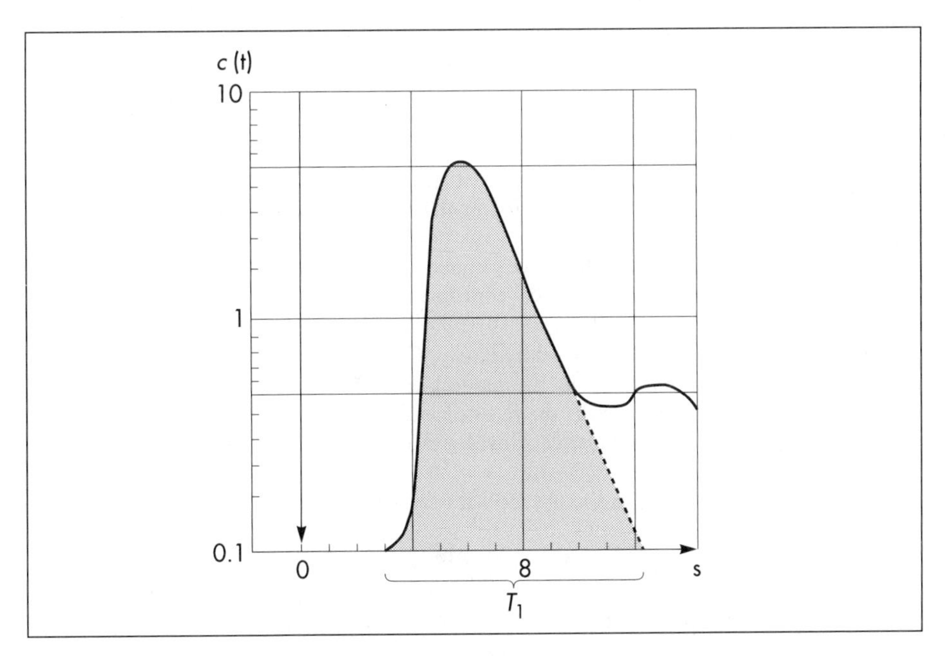

Fig. 24-1: The indicator dilution principle.

Put simply, the bloodflow Q (ml of blood per second, or more conveniently, per minute) can be measured by **dividing the dose of indicator injected upstream by the area under the downstream concentration curve.**

The bloodflow equation is also called the **dose/area equation.**

An attractive choice of indicator is to use **cold saline,** of known temperature and volume. A flexible catheter, with a thermistor located at its tip, and an opening through which cold saline can pass, is used. The catheter tip is advanced to the pulmonary artery, while the opening supplies the right atrium with saline. The thermistor records the downstream alterations in temperature as the saline bolus passes. This is the **thermodilution technique.** This technique can be repeated frequently without having harmful effects. Moreover, there is negligible recirculation, and the technique spares the patient the ordeal of an arterial puncture.

This method is widely used. For example, interesting indicator dilution studies have shown the pump effect of external cardiac massage to be modest.

What is clearance?

Clearance is a theoretical tool for estimating bloodflow in the kidney and other organs. Clearance is the **volume of blood plasma, which is totally cleared of a given indicator each minute by a specific organ** (renal clearance, see Chapter 55). The extraction (E) is the fraction of substance which is extracted from the total amount transported to the organ per minute:

$$E = \mathring{Q}(C_a - C_{\bar{v}})/(\mathring{Q} \cdot C_a) = (C_a - C_{\bar{v}})/C_a.$$

Clearance for **para-amino hippuric acid (PAH) at low plasma concentrations** is a measure of the **renal plasma flow (RPF).** The high hepatic extraction of bromsulphalein or of indocyanine is used to estimate the **splanchnic perfusion.**

Describe the isotope-wash-out-method

A homogenous muscle tissue of the weight, W_{tis}, is presumed. A lipid-soluble indicator such as ^{133}Xe dissolved in saline, is injected in the **tibialis anterior muscle** (* in Fig. 24-2).

At steady equilibrium state, the tracer concentration in the venous blood (C_v) is assumed to be the **average blood concentration,** and C_{tis} the **mean tissue concentration.**

A **distribution coefficient** is introduced: $C_{tis}/C_v = \lambda$, which is known for Xe. The decrease of the mass of indicator in the tissue per time unit (dt) must be equal to the mass supplied (which is zero) minus the mass of indicator leaving the tissue in the venous blood. The **principle of mass balance** provides the following equation:

$$(W_{tis}\, dC_{tis}) = [\text{mass supplied minus mass eliminated}].$$

$$(W_{tis}\, dC_{tis}) = (-C_v \,\text{Flow}\, dt) \text{ or } W_{tis}\, dC_{tis}/C_{tis} = (-C_v/C_{tis}\, \text{Flow}\, dt).$$

$$\mathbf{dC_{tis}/C_{tis} = -\,Flow\, d\mathit{t}/(W_{tis}\lambda).}$$

The **fractional fall in the mean tissue concentration of Xe (C_{tis}) per time unit (dt)** is

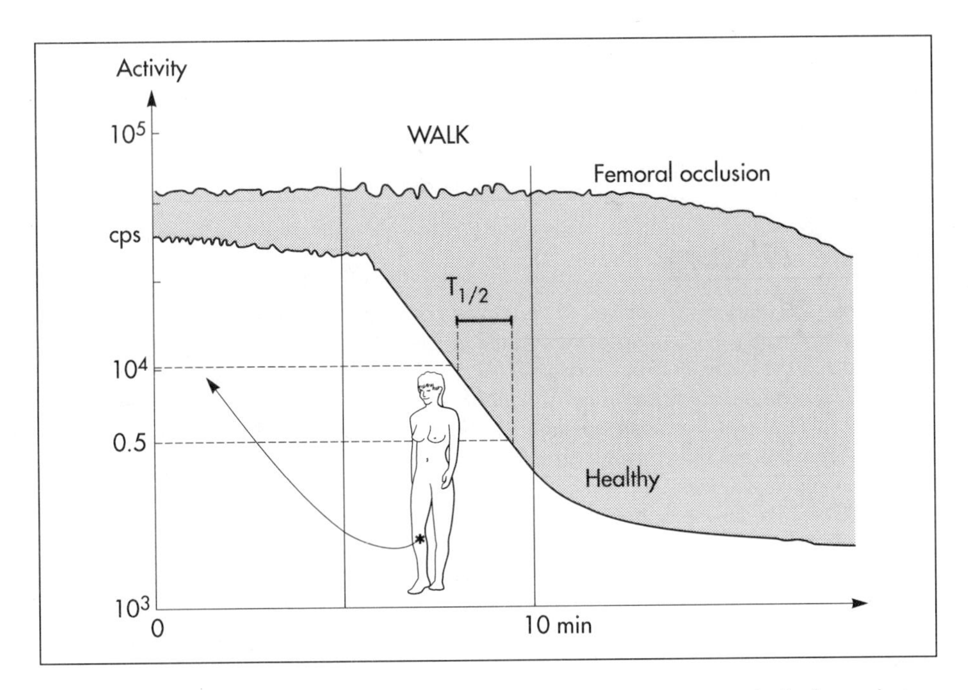

Fig. 24-2: Isotope (^{133}Xe) wash out from the gastrocnemius muscle before, during and after walk on a treadmill. Upper curve is when the femoral artery is occluded – lower curve is from the healthy leg.

constant during the whole elimination period (a **rate constant** = **ln** $\mathbf{2/T_{\frac{1}{2}}}$**).** **Flow/**$\mathbf{W_{tis}}$ is a perfusion coefficient in FU (ml of blood per minute and per 100 g tissue). The **ratio** $\mathbf{dC_{tis}/C_{tis}}$ is **measurable as** $\mathbf{T_{\frac{1}{2}}}$ **on the skin surface** above the Xe deposit in fat or muscle tissue with a scintillation detector (Fig. 24-2).

The method is used clinically to detect peripheral vascular diseases. An example is **intermittent claudication** which refers to constricting pain arising during activity of any muscle group but most commonly in the calf muscles. The hypoxic pain and cramp appear after having walked a certain distance and is promptly relieved by rest. The cause is **femoral occlusion** due to arteriosclerosis with insufficient local bloodflow and **ischaemic hypoxia** (Fig. 24-2).

What is the mean transit time?

The **mean transit time** ($\bar{t}$) for indicator particles in a system is equal to the sum of all **transit times for all single particles divided with their number**. The volume equation for a cylinder implies that the flow per second ($\mathring{Q}_s$) and its volume ($\mathring{V}$) is related by $\bar{t}$: $\bar{t} = \mathring{V}/\mathring{Q}_s$.

This concept is used in a wide variety of indicator methods.

By means of intravascular catheters it is possible to measure the **partial circulation time** through most parts of the circulation. For a healthy adult at rest the normal ranges include the following: arm–ear 8–12 s, arm–lung 5–7s, and lung–ear 3–5 s.

Further reading

Opie, L.H. (1991) *Heart: Physiology and Metabolism*, 2nd edn. Raven Press, New York.

24. Case History

A male, 21 years old, is located in the supine body position. He has a $\mathring{Q}$ of 5.4 l min^{-1} at rest, a circulating blood volume in the pulmonary circulation of 650 ml and in the systemic circulation of 4750 ml. His total mass of skeletal muscle is 35 kg. The muscular perfusion is 3 ml of blood min^{-1} per 100 g of muscle tissue (3 FU), and the mean passage time in the muscular capillaries is 5 s.

1. *Calculate the mean transit time ($\bar{t}$) for all red blood cells in the total circulatory system.*
2. *Calculate the mean transit time for the pulmonary circulation only.*
3. *Calculate the total perfusion of the skeletal muscle mass at rest (ml min^{-1}) and calculate the bloodflow per second.*
4. *Calculate the functioning capillary volume in the muscular capillaries.*

Try to solve the problems before looking up the answers in Chapter 74.

CHAPTER 25.
REGIONAL VASCULAR SYSTEMS

The myocardial metabolism is an **exclusively aerobic process** under normal conditions. It depends on oxidative phosphorylation to resynthesize ATP. The O_2 needs of the myocardium are therefore great, even at rest. Exercise can increase the needs **six-fold**; however, the myocardium cannot extract a greater fraction of the O_2 delivered, since the myocardial O_2 extraction is already close to maximum at rest. Thus the coronary bloodflow must rise importantly during exercise in order to deliver the O_2 needed.

Two main coronary arteries arise from the aorta. The **left main coronary artery** divides into two major branches: the **left anterior descending artery**, which courses down the interventricular groove towards the apex of the heart, and the **left circumflex artery** which courses leftward and posteriorly in the atrioventricular groove to the posterolateral wall of the left ventricle.

The **right coronary artery** arises from the right aortic sinus and courses rightward and posteriorly in the atrioventricular groove to reach the right atrium, and via the **posterior descending artery** to the posterior wall of the left ventricle and the lower part of the interventricular septum. Later the right coronary artery also gives off branches to the **posterolateral wall of the left ventricle.**

This arrangement of coronary vessels exists in half of the population in western countries. In 30% of the population the **posterior descending artery** arises from the **right** coronary artery, and the **posterior left ventricular branch** arises from the left circumflex artery. In another 20% of the population the **right coronary artery** is small and supplies only the right atrium and the right ventricle with blood, and all the blood supply to the left ventricle comes from the **left main coronary artery**.

The main arteries run along the epicardial surface and divide several times on the surface of the heart before they send off small penetrating vessels forming a network of intramural arteries, arterioles and capillaries on their way to the endocardium. The myocardial capillaries feed into a net of intramural venules. They drain eventually into the epicardial collecting veins. **Right ventricular venous blood** drains into the right atrium. Left ventricular venous blood drains into the **coronary sinus** that empties in the right atrium, except for a small blood volume which drains into the left ventricle. The epicardial coronary vessels contain a preponderance of constrictor receptors called **adrenergic α-receptors,** whereas the intramuscular and endocardial coronary receptors have a preponderance of dilator receptors called **adrenergic β-receptors.**

Describe the regulation of coronary bloodflow

The coronary bloodflow is mainly controlled by **local metabolic autoregulation**, and sympathetic stimulation does not always cause significant vasoconstriction. Accordingly, a moderate decrease in arterial blood pressure down to 9.3 kPa (70 Torr) does not significantly reduce the bloodflow through the myocardium. The cause of autoregulation is discussed in Chapter 20.

What is special about myocardial metabolism?

Unlike skeletal muscle tissue, the myocardium cannot function anaerobically for extended periods by building up an **oxygen debt**. Thus oxidative ATP synthesis must continuously match ATP utilization in the heart. At rest the heart produces 70% of its ATP from oxidation of fatty acids and 30% from oxidation of carbohydrates.

During exercise with lactate production by skeletal muscles, this lactate becomes an important substrate for the myocardial metabolism, entering the tricarbocyclic cycle (TCA) after conversion to pyruvate.

What is atherosclerosis and what are its consequences in the heart?

Atherosclerosis is a process of **progressive lipid accumulation and calcification of the inner arterial walls in the abdominal aorta, lower extremities and the arteries of the heart, brain and kidneys.** Atherosclerotic coronary artery disease remains a leading cause of death, and is manifested as focal narrowings in the **epicardial coronary arteries.** The gradually narrowed vessel segment can be abruptly occluded by **clot formation (thrombus)** or by vasoconstriction at the atherosclerotic lesion. When a thrombus flows along the arterial tree with the blood and occludes the vessel, it is called an **embolus.**

What happens to the coronary bloodflow when a healthy adult switches from rest to maximal exercise?

A healthy adult at rest has a $\mathring{Q}$ **of 5 l min^{-1}**, a **coronary bloodflow of 200 ml min^{-1}**, and a **low O_2 saturation** in the venous blood of the coronary sinus (30% or $C_{\bar{v}O_2}$ 60 ml l^{-1}). The **myocardial O_2 consumption** is thus (200 · 0.7 · 0.200) or **28 ml STPD min^{-1}**. With maximal exercise $\mathring{Q}$ can rise eight-fold, whereas the O_2 uptake of the whole body increases 20-fold. The myocardial part of this exercise response in $\mathring{V}_{O_2}$ is at least 20-fold, i.e. to (28 · 20) = **560 ml STPD min^{-1}**. Since the **myocardial arteriovenous O_2 content difference** can rise from 140 ml STPD l^{-1} (30%) at rest to an absolute maximum of 190 ml STPD l^{-1} (of the total C_{aO_2} 200 ml l^{-1}) during maximum exercise, the total coronary bloodflow must be increased to at least (560/190) **2.9 l of blood min^{-1}**.

What is the coronary bloodflow in patients with coronary sclerosis or occlusion?

Normally, the bloodflow in healthy persons increases during exercise in both systole and diastole (Fig. 25-1A). Exercise with a high heart rate implies a **great diastolic bloodflow** (Fig. 25-1B).

The conditions are totally different in patients with coronary occlusion causing cellular death (**infarct**) of a myocardial area. Distal to the coronary occlusion the blood pressure is low. The thin-walled subendocardial vessels receive the smallest bloodflow, which is the cause of the **frequent subendocardial infarcts**. The subendocardial myocardial infarcts are often **silent** (which means without pains; the pain relief is due to destruction of subendocardial nerve fibres). The typical infarct causes severe and long-lasting pain.

Hypoxia pains in the substernal or precardial area (with projections to the left arm, neck, jaw, etc.) are brought on by **exercise and cold** in a patient with **cardiac cramps or angina pectoris**. Hypoxia pains are transmitted by subendocardially situated nerve fibres

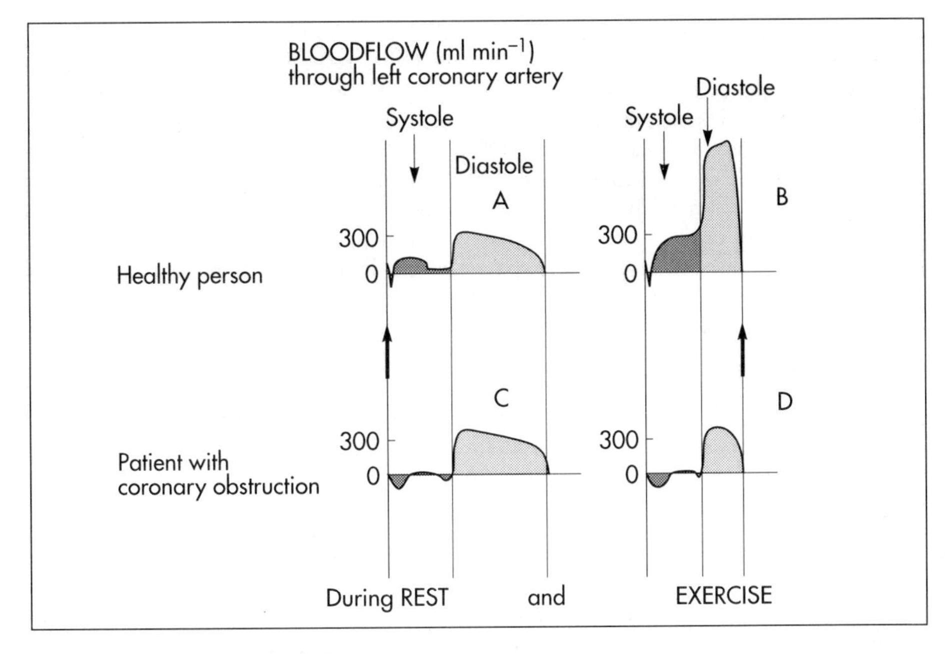

Fig. 25-1: Bloodflow through the left coronary artery at rest and during exercise in a healthy person (upper) and in a patient with coronary obstruction (lower) and cardiac cramps.

functioning during **cardiac cramps**. The **vasodilatation can be maximal** already at rest (Fig. 25-1C). **Exercise reduces the diastolic bloodflow further** (Fig. 25-1D), and the resulting hypoxia causes **cardiac cramps**. The beneficial effect of drugs such as glyceryltrinitrate in cardiac cramps has been known for more than a century. Recently it was realized that the drugs act by releasing nitric oxide (NO) in the vascular wall (see Chapter 5).

What organs besides the heart exhibit autoregulation?

The dominant control of cerebral bloodflow (CBF) is also metabolic autoregulation. The sympathetic nervous system plays a secondary role.

Some degree of autoregulation is found in many other organs including the **skeletal muscle mass, the splanchnic area, and the kidneys**.

What factors determine the brain bloodflow or CBF?

CBF reaches the brain through the internal carotid and the vertebral arteries. The vertebral arteries join to form the basilar artery, which forms the circle of Willis together with blood from the internal carotids. When brain arterioles dilate the CBF increases, and since the brain tissue within the cranium is relatively incompressible, the venous outflow must balance.

CBF is normally 55 ml of blood min^{-1} per 100 g of brain tissue (55 FU) in humans

at rest. With a normal brain weight of 1300 g this value corresponds to a total of (55·13) = **715 ml of blood min^{-1}**. The brain $\mathring{V}_{O_2}$ and this resting CBF can double during **cerebral activity**, and triple in active brain regions during an **epileptic attack** (see Chapter 10). Brain vessels are metabolically regulated. Increased P_{aCO_2}, and reduced P_{aO_2} dilate brain vessels and increase CBF. CO_2 (not H^+) passes the blood–brain barrier easily. Brain vessels show **autoregulation**. The mean arterial pressure can double without any appreciable rise in CBF. Brain vessels contract by **sympathetic stimulation**. This neurogenic control only concerns the **larger cerebral resistance vessels**. A neuropeptide released in response to transient hypotension, **calcitonin gene-related peptide**, is probably implicated in the autoregulation of brain vesssels.

How is the skin and fat bloodflow regulated?

Blood flows through the skin and subcutaneous tissues to **nourish the cells, and to regulate shell temperature**. Blood flows much faster through the arteriovenous anastomoses in the skin of the face, the fingers and toes in a cold environment. The sympathetic activity constricts the metarterioles that lead to the skin, so the blood bypasses the cutaneous circulation. Hereby, the skin blood flow can fall from **5 FU and approach zero**. The heat content of the blood returns to the body core which helps to maintain the core temperature.

In a warm environment, the sympathetic tone is minimal and the arterioles dilate, so that the **skin perfusion can rise to 70 FU**, and much energy is given off to the atmosphere. **Psychological influence** can cause one to blush or to have a white face, by **changing α-adrenergic constrictor tone** and via the effect of **local, vasoactive substances** normally found in the skin. When a **large fat combustion** is occurring (e.g. in hunger and distance running), the fat bloodflow can increase from 3 to 20 FU. **Cold and warm environments** alter the fat perfusion just like the skin perfusion. The **sympathetic regulation of the arteriolar tone** in fat and skin tissue is also similar. The sympathetic change in tone is not related to the classical baroreceptors.

Is the pulmonary circulation fit to meet metabolic demands?

This part of the circulatory system is basically a low-pressure, low-resistance, highly compliant vessel system, which is **meant to accommodate the entire $\mathring{Q}$ during gas exchange** with the alveolar air – and not meant to meet special metabolic demands as in the case of the systemic circulation.

What are the pressures in the pulmonary vascular system? Are pulmonary vessels controlled actively?

See Chapter 36.

What is special about the splanchnic circulation?

The splanchnic area is drained (1.5 l min^{-1}) via the hepatic veins at rest, so **all blood passes through the liver**. The liver receives more than 1 l of blood from the portal vein and less than 0.5 l from the hepatic artery each minute. A special characteristic for the splanchnic circulation is that two large capillary beds are partially in series with one another forming a **portal system**. The splanchnic perfusion increases after meals, and decreases during fasting and duration exercise. The sympathetic nervous system has a tonic activity on **splanchnic vessels via α-adrenergic nerve fibres**. Vagal fibres dilate the splanchnic vessels. Haemorrhagic shock can elicit a **fatal splanchnic hypoxia**.

What factors determine the exchange of gas and nutrients between maternal and foetal blood?

The **placental barrier** has an area of 10 m^2, and can be passed by low molecular substances (nutrients, gases, and waste) by diffusion. The **foetal haemoglobin (F)** has a dissociation curve which is shifted to the left. Foetal blood has a **high haemoglobin concentration**, so the foetus takes up large amounts of O_2 in placenta. This occurs even at low P_{O_2}, and the maximal value in the placenta is only 6.7 kPa or 50 Torr.

What hormones can cross the placental barrier? Essential foetal hormones?

Steroid hormones, maternal thyroid hormones, and catecholamines cross the placental barrier to the foetus. Peptide hormones cannot traverse the placental barrier, except small peptides such as **thyrotropin-releasing hormones (TRH)** and **antidiuretic hormone (ADH)**.

Foetal insulin contributes to anabolism and lipid storage. Human chorionic somatomammotropin, prolactin, and insulin-like growth factor (IGF-2) are the most important growth factors during foetal life. Foetal parathyroid hormone stimulates the transport of Ca^{2+} to the foetus.

How is the foetus perfused during normal pregnancy?

The maternal blood rich in O_2 and nutrients is injected into the intervillous spaces of the placenta via spiral arteries, and returns with CO_2 and waste to the mother via veins draining to the uterine veins. Foetal blood rich in O_2 and nutrients returns to the foetus from the placenta in the umbilical veins. The blood flowing in the umbilical veins continues in the **ductus venosus** to the inferior vena cava, or the blood enters the foetal liver.

The well-saturated blood from the umbilical veins flows through the oval communication between the right and left atria (**foramen ovale**) to reach the left ventricle (Fig. 25-2A). This shunt delivers well-saturated nutritive blood to upper-body foetal organs (brain and heart). Venous drainage from these essential organs returns to the foetal heart in the superior vena cava. The venous return is ejected by the right ventricle.

The foetal pulmonary vascular resistance is high due to the compressed inactive lungs (Fig. 25-2A), so the major part of the right ventricular output bypasses the pulmonary circulation and flows through a foetal channel (**ductus arteriosus**) between the pulmonary trunk and the descending aorta (Fig. 25-2A).

Much of the blood flowing in the descending aorta of the foetus is directed toward the placenta, so venous drainage from all organs is shunted toward the placenta, where **wastes are eliminated from foetal blood, whereas O_2 and nutrients are acquired**.

What happens to the foetal circulation at birth?

At birth P_{CO_2} increases, and the first breath reduces the intrathoracic pressure in the newborn, so placental blood is sucked into the baby (**placental transfusion**). Massive sensory stimuli of the baby caused by labour and delivery, cutaneous cooling, the rising

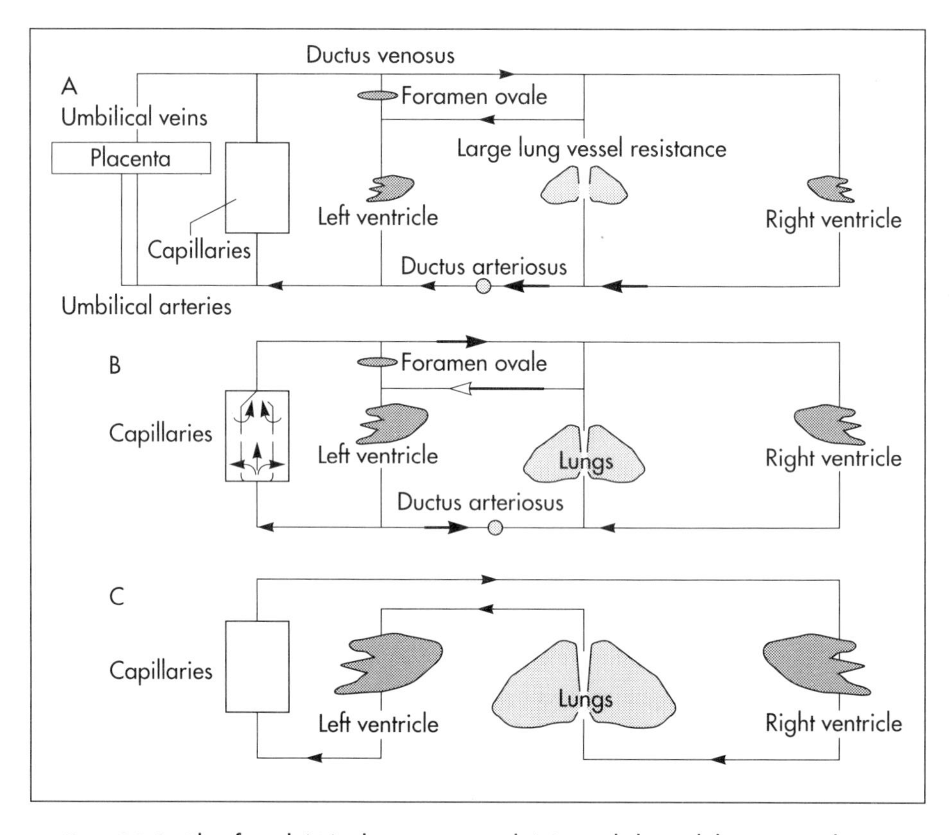

Fig. 25-2: The foetal (A), the transitional (B), and the adult (C) circulation.

P_{aCO_2}, and withdrawal of a **placentally-produced respiratory inhibitor** all add up in maintaining breathing in the newborn. The **newly established air–liquid interface** reduces pulmonary surface tension which eases the lung expansion.

Distension of the lungs also distend pulmonary vessels, so pulmonary vascular resistance **(PVR) decreases drastically**. Hereby, pulmonary arterial pressure decreases and pulmonary bloodflow increases. As a consequence, bloodflow via ductus arteriosus slows, and pulmonary venous return to the heart increases (Fig. 25-2B).

Normal arterial blood gas tensions are established by 30 min of age. Left atrial pressure increases and foramen ovale closes soon after birth. This **reverses the blood pressure gradient across the foramen ovale,** so now the left atrial pressure exceeds the right. When the umbilical cord is closed, and the placental circulation is thus eliminated, the **TPVR of the newborn increases**. The decrease in PVR and increase in TPVR means a great difference in the **size of the blood pressures in the aorta and in the pulmonary artery.**

In conclusion, the **parallel foetal circulation is transformed into a series circulation in the baby**. The foramen ovale and ductus venosus close within 3 days of birth, and the ductus arteriosus closes within 10 days (Fig. 25-2C). The **sharp increase of O_2 content (C_{aO_2}) in the baby's blood is a potent and universal vasodilatator**. The dramatic changes in gas exchange affect **cardiopulmonary and vascular regulation**, probably via local mediators such as **arachidonic acid** and **prostacyclin.**

Further reading

Adamson, S.L. (1991) Regulation of breathing at birth. *J. Developmental Physiol.* **15** (1): 45–52.

Calver, A., J. Collier and P. Vallance (1993) Nitric oxide and cardiovascular control. *Exp. Physiol.* **78**: 303–26.

Grover, R.F. *et al.* (1983) Pulmonary circulation, and Heistad, D.D. and H.A. Kontos (1983) Cerebral circulation. Both in: *Handbook of Physiology*: Sect. 2: *The Cardiovascular System – Peripheral Circulation and Blood Flow*, Vol. 3. Am. Physiol Soc., Bethesda, MD.

Hong, K.W., K.M. Pyo, W.S. Lee and B.Y. Rhim (1994) Pharmacological evidence that calcitonin gene-related peptide is implicated in cerebral autoregulation. *Am. J. Physiol.* **266**: H11–16.

25. Multiple Choice Questions

Each of the following five statements have True/False options:

A. The capillaries have the greatest cross-sectional area of the systemic circulation.
B. The systemic arterioles offer the greatest vascular resistance to bloodflow.
C. The sympathetic regulation of the arteriolar tone in fat and skin tissue depends upon the classical baroreceptors.
D. Increased compliance in the venous system means decreased venous return.
E. The lactate produced by skeletal muscles during exercise is not an important substrate for the myocardial metabolism.

Try to solve the problems before looking up the answers in Chapter 74.

CHAPTER 26.
WORK AND ENVIRONMENTAL PHYSIOLOGY

This chapter deals with **cardiovascular aspects of exercise and environmental physiology**. Respiratory aspects are described in Chapters 39 and 41.

I. EXERCISE

How much blood and oxygen passes the skeletal muscle mass at rest and during maximal exercise?

A top athlete can show a six-fold increase in cardiac output ($\mathring{Q}$) from 5 to 30 l of blood min^{-1} when going from rest to maximal dynamic exercise. However, the muscle bloodflow can rise from 3 to 75 ml min^{-1} per 100 g of muscle tissue or FU (factor 25) in a total muscle mass of 35 kg. The **muscular a-v-O_2 difference** can rise from the resting level (200 − 150 =) **50** ml STPD l^{-1} of blood to (200 − 30) = **170** ml STPD l^{-1}.

The athlete at rest has a $\mathring{V}_{O_2}$ of 250 ml STPD min^{-1}. The total muscle bloodflow at rest is (35 000/100) × 3 = **1050 ml of blood min^{-1}**. The total muscular $\mathring{V}_{O_2}$ at rest is (1050 · 50/1000) = 52.5 ml min^{-1}.

During **maximal dynamic activity** the total muscle bloodflow is: (35 000/100) · 75 = **26 250 ml min^{-1} or 26.25 l min^{-1}**. The **total muscular $\mathring{V}_{O_2}$** is increased to (170 · 26.25 l min^{-1}) = 4463 ml STPD min^{-1}.

Accordingly, the total muscle $\mathring{V}_{O_2}$ rises by a factor of (4463/52.5) **85** from rest to exercise.

How is $\mathring{Q}$ distributed during dynamic exercise?

At the start of exercise, signals from the brain and from the working muscles bombard the cardiopulmonary control centres in the brainstem. **Both $\mathring{Q}$ and ventilation increase, the α-adrenergic tone of the muscular arterioles falls abruptly**, whereas the vascular resistance increases in inactive tissues. The systolic blood pressure increases, whereas the mean arterial pressure (MAP) only rises minimally during dynamic exercise. **The total peripheral vascular resistance (TPVR) falls during exercise to one-quarter of the level at rest**, because of the massive vasodilatation in the muscular arterioles of almost 35 kg muscle mass. This is why the major portion of $\mathring{Q}$ passes through the skeletal muscles (Fig. 26-1) and why the diastolic pressure often decreases during exercise. The **coronary bloodflow** increases, and at some intensities of exercise we see increases in the skin bloodflow (Fig. 26-1).

How does the contraction affect the Na^+-K^+-exchange across the skeletal muscle membrane?

Contraction releases a great ionic leak (Na^+ influx and a K^+ efflux), which elicits the action potential. Thus the muscle cell loses K^+ and gains Na^+ during intensive exercise.

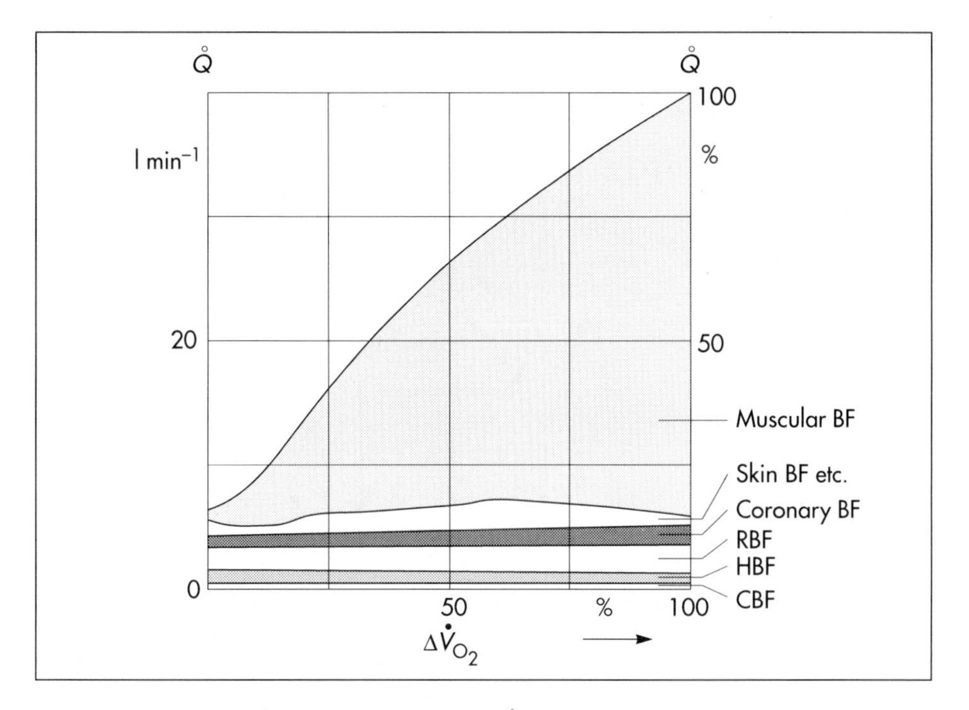

Fig. 26-1: Distribution of cardiac output ($\mathring{Q}$) during exercise. BF, bloodflow; RBF, renal bloodflow; HBF, hepatic bloodflow; CBF, cerebral bloodflow.

Contraction stimulates the **Na^+-K^+-pump** acutely, and training increases its activity. Still, at high intensity exercise the ionic leaks can exceed the capacity of the Na^+-K^+-pump for intracellular restoration. Training reduces **exercise-induced hyperkalaemia**.

What are the cardiovascular reactions to physical training?

Training improves the **capacity for oxygen transport** to the muscular mitochondria, and improves their **ability to use oxygen**. After long-term endurance training the athlete typically has a **lower resting heart rate, a greater stroke volume, and a lower TPVR than before**. The maximum oxygen uptake ($\mathring{V}_{O_2}$max) progressively increases with long-term training, and the **extraction of oxygen from the blood is increased**. The lung diffusion capacity for oxygen probably increases by **endurance training**. The capillary density of skeletal muscles, the number of mitochondria, the activity of their oxidative enzymes, ATPase activity, lipase activity and myoglobin content all increase with endurance training. **Endurance training** also produces a rise in left ventricular volume. **Strength training** (weight lifting) produces a rise in left ventricular wall thickness without any important increase in volume.

During dynamic exercise the stroke volume increases as does heart rate, and the residual ventricular volume decreases (Fig. 26-2). Although the peak ventricular pressure during systole rises considerably and thus the arterial peak pressure, the diastolic pressure falls because of the massive fall in TPVR. For dynamic exercise TPVR is typically 20–30% of that at rest. The rise in contractility is seen from the slope of the pressure–volume curve.

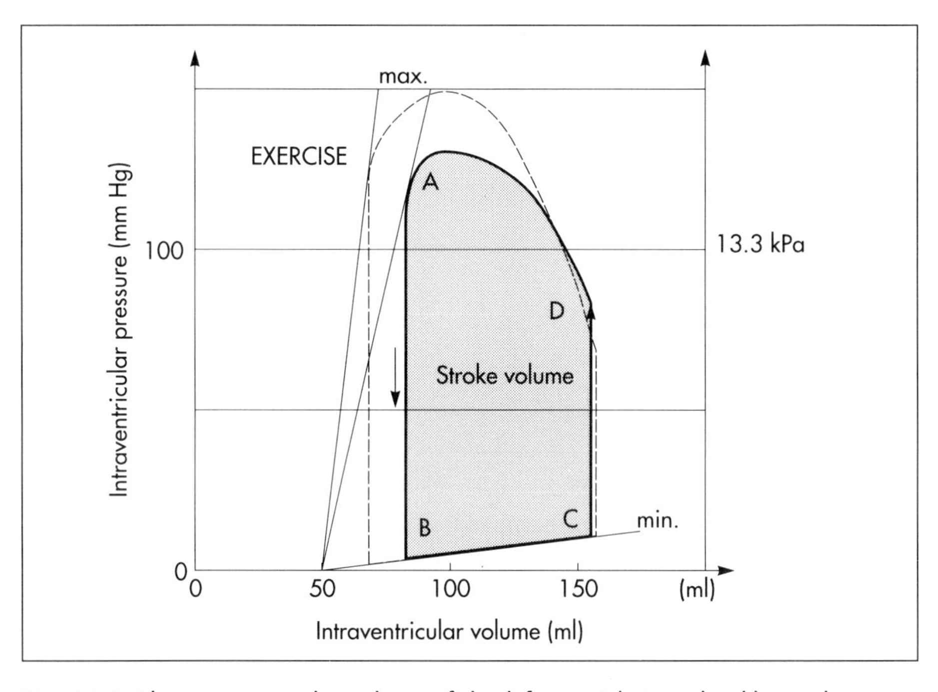

Fig. 26-2: The pressure–volume loop of the left ventricle in a healthy male at rest (solid curve) and during dynamic exercise (dashed curve).

Why is dynamic exercise beneficial and static exercise often dangerous in older people?

The rise in mean arterial pressure (MAP) during dynamic exercise is often minimal, because of the arteriolar dilatation with a large, rhythmic bloodflow through the working muscles. The TPVR is typically reduced to one-quarter of the value at rest. In contrast, static exercise often results in a doubling of the MAP, because a large muscle mass is contracting and the contraction is maintained. Static work is typically accomplished with a low $\mathring{Q}$, so the TPVR is relatively high. This is dangerous to **elderly people with known or unknown degrees of arteriosclerosis.**

What limits exercise performance?

The **rate of oxygen utilization** is one possible limitation, and another possibility is the **transport capacity for oxygen** to the mitochondria.

1. The rate of oxygen utilization is not a likely limitation in healthy people. The $\mathring{V}_{O_2}$max is not further increased, when we recruit more active muscles during an already maximal leg exercise.
2. **Limited transport capacity for oxygen** could be caused by failure of the lungs to fully oxygenate blood. This is certainly not the case in healthy people. The arterial blood is almost fully saturated with oxygen even during the most strenous exercise at sea level.

Limited transport capacity for oxygen caused by limited bloodflow is the only logical explanation.

A limited bloodflow could be caused by limitations in reducing TPVR, or by a limited pumping capacity ($\mathring{Q}$) of the heart. When work is maintained at peak $\mathring{Q}$ and $\mathring{V}_{O_2}$ max, the blood pressure falls as more vasodilatation occurs and there are no signs of even a slight relative increase in the low exercise TPVR. Thus the major limitation to exercise is the **heart's pumping capacity**.

Can smoking decrease athletic performance?

Yes, smoking has acute and deleterious effects on both the cardiovascular and the respiratory system.

CO blocks off part of the haemoglobin and limits the transport capacity for oxygen to all mitochondria, which is especially inhibitory to the heart and the skeletal muscles. **Nicotine constricts terminal bronchioli and arterioles in many vascular beds.** Nicotine also **paralyses the cilia** of the epithelial cells of the respiratory tract.

Chronic smoking leads to obstructive lung disease (chronic bronchitis and emphysema – see Chapter 40).

Can drugs increase athletic performance?

Many drugs have been reputed to increase athletic performance. However, the reputation is widely overestimated. Most of the drugs in use by doping addicts have **deleterious effects**, and no proven improvement of performance in double-blind cross-over experiments. On the contrary, such drugs are dangerous to the healthy person.

Androgens and other anabolic steroids **can increase muscle strength** for a while, and they **certainly decrease testicular function, damage the liver, are cancer suspect, and change the personality of the athlete**.

Does the plasma volume and electrolyte balance change during high intensity exercise?

During intensive exercise the osmolarity of the contracting muscle cells increases together with the capillary hydrostatic pressure. As a consequence, the **plasma volume can fall by 20% within a few minutes**. The **plasma [K$^+$] can rise to 8 mmol l^{-1} due to efflux** from the contracting muscle cells and from red blood cells into a reduced plasma volume.

Is fatal heat stroke realistic during running?

Endurance athletics in a hot and humid environment can increase the temperature of the **body core to more than 41°C**. Such a level is deleterious to the brain cells, and CNS signs and symptoms develop: headache, dizziness, nausea, confusion, staggering gait, unconsciousness and profuse sweating. When the victim suddenly faints, this is termed **heat stroke**, which can be fatal.

II. Environment

Is the heart the correct reference point when measuring vascular pressures?

The heart is not always the correct reference point for blood pressure measurements. The elastic properties of the vascular tree differ throughout the body.

Actually, the point in which the pressure does not change with change of position is approximately 5 cm beneath the diaphragm (Fig. 26-3). This is called the **hydrostatic indifference point (HIP)**. Above this horizontal level, all vascular pressures are lower in the erect than in the recumbent position. The subatmospheric intrathoracic pressure counteracts venous collapse, so the intrathoracic veins remain open and the atrial pressures are zero in the erect position. The veins of the neck and face are collapsed. The venous sinuses of the brain are kept open by attachment to the surrounding tissues, and their pressures are around −1.3 kPa (−10 Torr) in the erect position.

HIP must not be confused with the **cardiac mean equilibration pressure**, which is a pressure of 1 kPa (6 Torr) measurable in all divisions of the circulatory system just after cardiac arrest. This is also called the **cardiac mean filling pressure**, because it is a determinant of the venous return.

What happens to $\mathring{Q}$ when we move from supine to standing position?

When a supine person arises, his or her TPVR increases, the systolic blood pressure falls and the diastolic blood pressure rises. Thus the **pulse pressure (PP) falls, while the MAP**

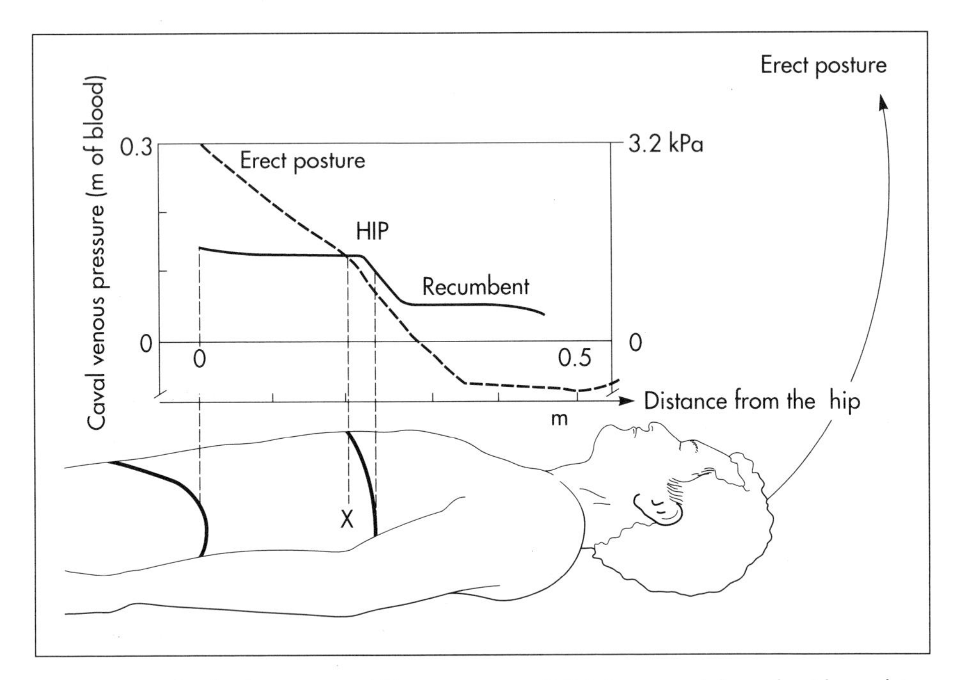

FIG. 26-3: The hydrostatic indifference point (HIP) in an adult male. The subject changes position from recumbent to erect.

is unchanged. The stroke volume falls more than the heart rate rises, so $\mathring{Q}$ will decrease in the standing position.

An elegant way of studying circulatory consequences of standing is by use of **lower-body-negative-pressure (LBNP)**. LBNP applied to a recumbent subject simulates the circulatory effects of standing.

The venous return to the heart is dependent upon the body position, and upon the total blood volume. The venous return is also dependent upon the venous compliance and upon the sympathetic tone in the venous system and in the arteriolar system.

How does the venous pressure vary with the position of the body?

When a person is located on a tilt table in a horizontal position, his or her blood pressure in a superficial vein on the foot is approximately 1.6 kPa (12 Torr) and in the femoral vein 0.8 kPa (6 Torr). When the person is turned upright towards the vertical plane, the venous pressure increases by the hydrostatic column up to the **HIP** just below the heart as long as he or she is not standing and using their skeletal muscle pump. If the tilt table is turned, so the head of the person is downward (Trendelenburg position), then the venous pressure increases in neck and head. The **Trendelenburg position** is rational during neck and head surgery. With the head upward the patient risks **air embolism** if blood vessels are cut during neck and head surgery.

Describe the circulatory consequences of G-forces on a pilot flying in a curve

A jet fighter seated normally feels the forces of acceleration along the long axis of his body when flying in a curve. Above an individual G-threshold (approximately 4 G), blood will be pooled in the lower veins, and the pilot will lose vision (**black-out**), consciousness (**grey out**) and probably his life due to insufficient cerebral bloodflow. The venous return to the heart is reduced when blood is pooled in the lower extremities. Hereby the stroke volume is reduced according to Starlings law of the heart. The pressure in the veins of the foot increases by a factor equal to the G-exposure.

For short periods forces up to 10 times the Earth's gravity have been sustained (10 G). To accomplish this an anti-G-suit is essential. This is a suit which increases the external pressure on the lower body and legs of the jet fighter. This prevents blood from being pooled in the lower regions of the body during **loops**.

What are the physiological problems of weightlessness?

The major problems for the astronaut during prolonged stay in space are: (1) space sickness; (2) hydrostatic pressures (suddenly without the effect from gravity); and (3) physical inactivity.

These problems lead to the following disorders: dizziness, nausea, vomiting, bone loss, reduced $\mathring{Q}$ capacity, reduced working capacity ($\mathring{V}_{O_2}$max), reduced blood volume, and reduced erythrocyte volume. **Space sickness** is caused by lack of gravitational signals to the vestibular system (Chapter 14). Physical inactivity and the lack of gravitational stimuli result in a tendency to faint from **postural hypotension** after return to Earth.

Further reading

Ahn, B., Y. Sakakibara, P.-E. Paulev, A. Masuda, Y. Nishibayashi, W. Nakamura and Y. Honda (1989) Circulatory and respiratory responses to lower body negative pressure in man. *Jpn. J. Physiol.* **39**: 919–30.

Fisher, A.G. and C.R. Jensen (1989) *Scientific Basis of Athletic Conditioning*, 3rd edn. Lea & Febiger, Philadelphia.

Maughan, R.J. and J.B. Leiper (1963) Whole body volume and electrolyte balance during exercise. 32. *Congress of the Internat. Union of Physiological Sciences* (IUPS), p. 61.

26. Case History

A frogman in the US Navy practises breathing at rest in a closed circuit. The circuit consists of a respiratory double valve, a rubber bag and a CO_2-absorber. The system contains 10 l STPD of oxygen, and 6 mmol of carbon monoxide (CO) as an indicator. Before the breathing practice a blood sample from the median cubital vein contains 0.17 mmol CO l^{-1} (CO saturation degree: 1.6%; $S_{CO,1}$: 0.016). The frogman is a cigarette smoker with a body weight of 60 kg.

The frogman starts breathing for 10 min in the circuit. At the end a second venous blood sample contains 1.5 mmol CO l^{-1} (CO saturation degree: 16.6%; $S_{CO,2}$: 0.166). No trace of CO was detectable in the closed circuit at the end of the 10-min breathing period. The haematocrit of both blood samples was 43%, after correction for trapped plasma by subtraction of 1%.

1. *Calculate the total blood volume (TBV) and the total erythrocyte volume (TEV) of the frogman.*
2. *Are the obtained results normal or far too small?*
3. *Calculate the total haemoglobin content (TH) of the circulating blood. Is the calculated result normal?*
4. *Describe the main sources of error for estimating the TH by the CO method.*

Try to solve the problems before looking up the answers in Chapter 74.

Chapter 27.
Shock and Haemorrhage

Loss of more than **35% of the total blood volume (TBV)** of a healthy person is a threat, if the loss is unaided by blood transfusion. If the **blood loss approaches 50%,** the person slides into **irreversible shock** and death ensues. When a healthy person delivers 500 ml of blood for transfusion the volume is replaced within an hour.

What is shock?

Shock is a clinical condition characterized by **falling arterial blood pressure, and rapid heart rate**. Respiration is also rapid and the skin is pale, moist and grey. The most common cause is **haemorrhage** with loss of large quantities of blood. The condition of the patient is dominated by **universal ischaemic hypoxia**. The progressive deterioration becomes irreversible at a blood loss of more than 50% of the TBV.

The reduced delivery of oxygen and nutrients to virtually all cells of the body is consequential: the mitochondria synthesize less ATP, the **Na^+-K^+-pump** operates insufficiently, the metabolic processing of nutrients is depressed which profoundly depresses muscular contractions, and finally digestive enzymes destruct the damaged cells. Glucose transport across the cell membranes in the liver and in the skeletal muscles is depressed including a severe inhibition of the actions of insulin and other hormones. During progressive shock the heat energy liberated in the body of the patient is reduced and the **body temperature tends to decrease**, if the patient is not kept warm.

The fall in arterial blood pressure reduces the **renal glomerular filtration pressure** below the critical level, so filtration is diminished or abolished, leading to abolished urine output (anuria). The patient loses consciousness and falls into a state of **stupor** or **coma**.

Compensatory mechanisms operate to counteract the fall in blood pressure. Baroreceptor responses and many hormonal control systems that tend to raise the falling blood pressure operate on **negative feedback** mechanisms (see baroreceptors and hormonal reactions below).

Other compensatory mechanisms exaggerate the primary fall in blood pressure. This is called **positive feedback**. A positive feedback mechanism can lead to a **vicious cycle**, if its gain is above one.

The gain of a feedback system is defined as the ratio of the response to the stimulus itself. A gain greater than one may lead to death.

How do the baroreceptors react following blood loss?

The falling arterial blood pressure during haemorrhage decreases the stimulation of the baroreceptors. Reduction of vagal tone and increase of sympathetic tone enhance myocardial contractility and heart rate. The increased sympathetic tone also produces **generalized arteriolar and venous constriction**. The latter has positive consequences resembling those of a transfusion of blood. Vasoconstriction is most severe in the muscular, cutaneous and splanchnic area. The falling cardiac output ($\mathring{Q}$) is redistributed to favour bloodflow through the brain and heart. If the arterial mean blood pressure falls below 8 kPa (60 Torr) there is no additional baroreceptor response.

How do the chemoreceptors react following blood loss?

The arterial **chemoreceptors** are stimulated by a MAP below 8 kPa (60 Torr), because of **hypoxia due to the low bloodflow** through the chemoreceptor tissues (**ischaemic hypoxia**). The hypoxia stimulus of the chemoreceptors **enhances the vasoconstriction** evoked by the baroreceptors.

What happens if shock leads to cerebral ischaemia?

Cerebral ischaemia (hypoperfusion) occurs when the MAP is below 5 kPa (38 Torr). The severe hypotension reduces cerebral bloodflow (CBF). With progressive hypotension the cardiopulmonary centres in the brainstem are depressed, because of inadequate bloodflow and ischaemic hypoxia.

Such a low arterial pressure activates the **sympathoadrenergic system**. Its high discharge leads to a pronounced vasoconstriction and increased myocardial contractility.

The terminal loss of sympathetic tone towards the heart and vessels then reduces $\mathring{Q}$ and TPVR. This leads to reduced MAP, which intensifies the falling CBF – a **classical vicious cycle**.

Finally the vagal tone is increased leading to **terminal bradycardia** (final fall in heart rate). A severe shock becomes irreversible, when the high-energy phosphate stores of the liver and heart are depleted. All of the ATP has been degraded to ADP, AMP and eventually to the even more efficient vasodilatator **adenosine**.

Adenosine is converted into hypoxanthine and uric acid, a substance that cannot re-enter the cells. Cellular **depletion of high-energy phosphate** is probably causing the final state of irreversibility.

Is there hormonal reactions to such a severe haemorrhage?

1. **Catecholamines** and encephalins are released from chromaffine granules in the adrenal medulla as part of the sympathoadrenergic response to stress. The blood [catecholamine] increases 40–50 times. Ad is released almost exclusively from the adrenal medulla. NA is derived from sympathetic nerve endings and from the adrenal medulla. Catecholamines increase the heart rate and the $\mathring{Q}$ by stimulation of the **adrenergic β_1-receptors** in the myocardium. **Catecholamines** constrict vessels **all over the body** by stimulating **α_1-receptors** located on the surface of **vascular smooth muscles**.
2. **ADH (vasopressin)** is secreted from the **posterior pituitary gland** in response to shock, because the sinoaortic baroreceptors are understimulated. Vasopressin is a modest vasoconstrictor and a strong ADH.
3. **Renin** is secreted from the juxtaglomerular apparatus, when blood pressure and renal perfusion fall drastically. Renin acts on the plasma protein, angiotensinogen, to form the powerful vasoconstrictor, **angiotensin II**. The most likely trigger of the **renin–angiotensin–aldosterone cascade** during shock is the falling NaCl concentration of the fluid at the macula densa (see Chapter 70).

   Other mechanisms are also involved in the release of renin at falling arterial pressure: (1) falling perfusion pressure to the kidney; (2) stimulation by prostaglandins; and (3) signals through the renal nerves, which stimulate renin secretion directly via **β-adrenergic receptors** on the juxta-glomerular cells. Angiotensin II is a powerful stimulator of the aldosterone secretion from the renal cortex. Aldosterone **promotes the reabsorption of Na^+ and increases the secretion of K^+ and H^+ in the distal tubular system** of the kidneys. Water follows by osmosis, so the extracellular volume is increased. The rise in normal serum-[K^+] due to the ischaemia of shock also releases aldosterone.

4. **ACTH and β-endorphins** are released into the blood from the anterior pituitary gland in response to haemorrhage or other forms of stress. ACTH and endorphins both excerbate and restrict the development of shock. These opioids depress the brainstem control centres that normally mediate autonomic responses to stress. Hence, naloxone (an opioid antagonist) improves the circulation and increases the rate of survival from life-threatening shock. On the other hand, ACTH has a small aldosterone and a strong cortisol stimulating effect (Chapter 70).

Does the heart suffer during shock?

Myocardial contractility is depressed by **hypotension** and ischaemic hypoxia. The ventricular function curves (i.e. $\mathring{Q}$ versus atrial pressure) shift to the right. This leads to a further fall in MAP. The falling blood pressure reduces both coronary blood pressure and $\mathring{Q}$. The condition is analogous to acute cardiac failure in Fig. 22-2. The fall in $\mathring{Q}$ leads to a further decline in arterial blood pressure – a **classical vicious cycle**. The low $\mathring{Q}$ also reduces tissue bloodflow with accumulation of vasodilatator metabolites (see above). This decreases TPVR and further aggravates the **hypotension**.

Is shock related to metabolic acidosis?

The low $\mathring{Q}$ and bloodflow result in **stagnant hypoxia of all mitochondria**. Hypoxia increases lactic acid liberation. Renal failure prevents excretion of excess H^+. The high [H^+] further depress the myocardium and thus the tissue bloodflow. This aggravates the metabolic acidosis – a **classical vicious cycle**.

Does the blood clot normally during blood loss?

Initially, the bleeding patient suffers from **hypercoagulability**. Thromboxane A_2 (TxA_2) aggregates thrombocytes, and the aggregate releases more TxA_2. This positive feedback prolongs the clotting tendency. In this phase anticoagulants (heparin) reduce mortality from shock (see Chapter 16).

In the later stage of haemorrhagic hypotension with shock, there is **fibrinolysis** and prolonged coagulation time (**hypocoagulability**). Hence, **heparin therapy** can be lethal.

Do we maintain our phagocytic activity during blood loss?

No, the **phagocytic activity of the reticulo-endothelial system (RES)** is depressed during shock. Endotoxins constantly enter the blood from the bacterial, intestinal flora of a healthy person. These endotoxins are normally inactivated by the RES (Chapter 68). Following loss of half the total blood volume the shock patient must have lost about 50% of his or her circulating macrophages, and control substances modulating the phagocytic activity of RES. The depressed defence mechanisms in RES result in an **endotoxic shock** which aggravates the **haemodynamic shock** – a **classical vicious cycle**.

Is shock always caused by haemorrhage?

Haemorrhage is the most frequent cause, but not the only one. Alternatives are:

1. **Hypovolaemic shock by severe burns or other denuding conditions** of the skin.
2. **Hypovolaemic shock due to intestinal obstruction.** Distention of the upper intestine causes protein-rich fluid to leak from the intestinal capillaries into the intestinal lumen. Conditions (1) and (2) reduce total plasma volume and greatly increase blood viscosity.
3. **Dehydration** is a condition caused by loss of water from any fluid compartment of the body (Chapter 54). Such a hypovolaemic shock is caused by **excessive sweating, severe diarrhoea, severe vomiting, water deprivation, hypo-aldosteronism with urinary loss of salt and water (Chapter 70),** excessive use of antidiuretics, and excessive loss of water and proteins by nephrotic kidneys.
4. **Neurogenic shock** is caused by a sudden increase in the vascular capacity. The normal blood volume is inadequate. The major cause is a **sudden loss of vasomotor tone**, resulting in **venous pooling** of the blood and insufficient venous return. A common cause is **emotional fainting (vasovagal syncope)** with a strong emotional activation of the autonomic system via the hypothalamus. Other causes are deep anaesthesia and brain damage, both with operative failure of the vasomotor centres in the CNS.
5. **Anaphylactic shock (anaphylaxis)** is a severe allergic disorder in which the $\mathring{Q}$ and the MAP fall rapidly and drastically. As soon as an antigen to which the patient is sensitive has entered the blood the **antigen–antibody reaction** triggers **release of histamine** from basophilic cells in the blood and mast cells in the tissues. Histamine dilates arterioles and all peripheral vessels. This results in falling arterial pressure and increased capillary permeability with rapid loss of plasma water into the interstitial fluid.
6. Septic shock (see below).

What is septic shock?

Septic shock or **blood poisoning** is a widespread bloodborne bacterial infection – often life-threatening. Examples are gas gangrene bacilli spreading from a **gangrenous limb,** colon bacilli with **endotoxin** spreading into the blood from infected kidneys (pyelonephritis), and fulminant peritonitis due to acute abdominal disease. Frequent causes of fulminant peritonitis are rupture of the infected gut or the uterus, and rupture of the uterine tube due to extrauterine pregnancy.

Septic shock is characterized by tremendously high fever, high $\mathring{Q}$, marked vasodilatation, red cell agglutination, disseminated intravascular coagulation with **microclots** spread all over the circulatory system. When the clotting factors are used up, internal haemorrhages occur. Overproduction of **nitric oxide (NO)** may contribute to the vasodilatation and the depressed myocardial contractility found in septic shock.

The high $\mathring{Q}$ is due to a high stroke volume and a high heart rate. The diastolic pressure is low and the systolic pressure is high until endotoxins begin to inhibit myocardial contractility seriously. Now the condition is in a vicious cycle which is often fatal.

What is the proper treatment for shock?

First of all the **cause** has to be established (see above).

1. **Head-down position** (placing the patient's head below the level of the heart) is the immediate therapy for haemorrhagic and neurogenic shock.
2. **Haemorrhagic arrest** (closing the abdominal aorta with the pressure of a fist, or blocking the bleeding from an artery with a finger) is often life saving when applied without unnecessary delay.
3. **Replacement transfusions.** The best possible therapy for haemorrhagic shock is whole blood transfusion. The best treatment of shock caused by plasma loss is plasma transfusion, and the best therapy for dehydration shock is transfusion with the appropriate solution of electrolytes.
4. **Oxygen breathing** is always helpful in shock with insufficient delivery of oxygen.
5. **Sympathomimetic drugs (NA, Ad, etc.)** are often beneficial in neurogenic and anaphylactic shock. They are seldom useful in haemorrhagic shock.
6. **Glucocorticoids** are administered to patients in severe shock. Glucocorticoids increase the force of the damaged heart, improve the glucose combustion of damaged cells, and their lysosomal membranes, preventing release of digestive enzymes from the lysosomes of the cells.

Further reading

Calver, A., J. Collier and P. Vallance (1993) Nitric oxide and cardiovascular control. *Exp. Physiol.* **78**: 303–26.

Guyton, A.C. (1991) *Textbook of Medical Physiology*, 8th edn. W.B. Saunders, Philadelphia.

Heffernan, J.J. *et al.* (1989) *Clinical Problems in Acute Care Medicine*. W.B. Saunders, Philadelphia.

27. Case History

A 64-year-old male normally has a body weight of 74 kg, a total blood volume of 5 l and a blood [haemoglobin] of 10 mmol l^{-1} (mM). One day he suddenly vomits large quantities of fresh blood. For 2 days his stools have been tarry. The last weeks have been stressful at work. The patient calls his doctor and the emergency ward at the hospital is alerted. Owing to an incompetent local ambulance service, the patient is brought to hospital without delay by taxi. Here, the MAP is below 10 kPa (75 Torr) and falling. The heart rate is above 150 beats min^{-1} and rising. The blood [haemoglobin] is 5 mM measured 1 h after the first massive blood loss.

The emergency team immediately institutes transfusion of blood. The following 8 days the patient receives three transfusions of blood and at least 10 l of physiological saline. On the second day his [haemoglobin] has increased to 7.2 mM, but on the third day it falls again to 5 mM. On the 4th day at the hospital the patient develops high fever (maximum 40.6°C), and a broad-spectrum antibiotic programme is started without delay. On the 8th day at hospital the patient has normal temperature, but he develops watery swellings of legs and lower abdomen, in spite of pronounced urination. The body weight is now 80 kg.

1. *What is the most likely cause of the haemorrhage?*
2. *Estimate the size of his blood loss.*
3. *Why did the patient develop high fever?*
4. *Why did the patient accumulate water?*

Try to solve the problems before looking up the answers in Chapter 74.

SECTION IV.
The Respiratory System

The **respiratory process** covers the exchange of air between the atmosphere and the lung alveoli, the diffusion of gases between the alveolar air and the blood, the transport of oxygen and carbon dioxide in the body to and from the cells, and the regulation of alveolar ventilation ($\mathring{V}_A$ l min^{-1}) and cardiac bloodflow or cardiac output ($\mathring{Q}$ l min^{-1}). The airways are protected by humidification with a liquid layer, which prevents dehydration of the epithelium and surrounds the epithelial cilia. The airway mucus consists mainly of polysaccharides from goblet cells. This mucus forms a gelatinous blanket on top of the liquid layer. The cilia continuously pushes the gelatinous blanket upwards. Smoking reduces **mucociliary transport**. Lung secretions contain **lysozyme and lactoferrin** with bacteriocidal effects, **interferon** produced in response to viral infection, and α_1**-antitrypsin**, which inhibits trypsin and **chymotrypsin**. These topics are presented here. Consult Chapter 75 for respiratory symbols.

CHAPTER 28. LUNG FUNCTION

How many generations of sequential branching take place in human airways?

Air passes into the airways through the nose and mouth, where it is warmed, humidified and filtered. From the trachea to the alveoli, there are 23 generations. The **first 16** (as an average) constitute the **conducting zone** which is an **anatomic dead space**, because no gas exchange takes place. Each generation of branching increases the total cross-sectional area of the airways, but reduces the radius of each airway and the velocity of air flowing through that airway. The **branching generations (17–23)** constitute the **exchange effective respiratory zone**, which comprises of the **respiratory bronchioles, alveolar ducts and alveolar sacs.**

What is the gas-exchange interface?

There are approximately **300 million alveoli** in the lungs. Each alveolus has a diameter of 100–300 μm and is surrounded by 1000 capillaries. Each capillary is in contact with several alveoli, so the capillaries present a sheet of blood to the alveolar air for gas exchange. The total contact area averages **70–140 m^2 in adult humans** – and is an extremely difficult estimate. Oxygen must cross **six zones** to enter erythrocytes in the lungs. These zones are a **surfactant-containing fluid layer, the alveolar epithelium, an interstitial space, the capillary endothelium, blood plasma, and the erythrocyte membrane and cytosol to the haemoglobin binding site.**

The first four layers represent a barrier of 0.5 μm. To this must be added the distance from the blood plasma to the haemoglobin binding site. The six zones form the ideal **gas exchanger** for **diffusion.** Haemoglobin creates a diffusion gradient owing to its high affinity to oxygen.

Does the law of ideal gases apply to real gases?

Real gases do not obey the **ideal gas equation**. The ideal gas equation reads: $PV = n\mathrm{R}\mathrm{T}$. However, the deviations are acceptable at the pressures (P) and temperatures (T) of life on Earth.

One mole of gas occupies a **volume of 22.4 l at Standard Temperature (273 K), Pressure (1 atmosphere = 101.3 kPa = 760 Torr), Dry air (STPD).** This volume (V) is easy to calculate from the **universal gas constant R = 0.08207 l atm mol^{-1} K**: $V = 1$ mol $\times$ 0.08207 $\times$ 273/(1 atm) = 22.4 l STPD.

For a fixed number of moles (n), the product $n\mathrm{R}$ is constant, and equal to PV/T. In general: $P_1V_1/\mathrm{T}_1 = P_2V_2/\mathrm{T}_2$ = **constant**.

If the temperatures are the same in two states of one mass of gas, then: $P_1V_1 = P_2V_2$ = constant. The fact that the product of the pressure and volume of a fixed mass of gas is constant at constant temperature was discovered by Robert Boyle in 1660 (Boyle's law). **Boyle's law** is not a fundamental law like Newton's laws or the law of conservation of energy, but a practical approximation for real gases.

How do we convert a gas volume from one state to another?

Consider a cylinder with a movable piston containing n moles of a gas at volume, V, at a certain pressure and temperature. At the above described **standard conditions (STPD)** the following equation applies: $V_{STPD}760/273 = n\mathbf{R}$. Now consider the same mass of gas at room temperature (t°C or $273 + t$ K), saturated with water vapour, and at actual barometric pressure (P_B). These conditions are known as ATPS (**Ambient Temperature, Pressure, Saturated** – with water vapour at tension P_{water}). The air volume rises with temperature: $V_{ATPS}(P_B - P_{water})/(273 + t) = n\mathbf{R}$.

Now consider the same amount of gas at the conditions present in the alveoli; the air is saturated with water vapour, which exerts a partial pressure of **6.3 kPa** (47 Torr) at 37°C, at ambient pressure. These conditions are known as BTPS (**Body Temperature, ambient Pressure, Saturated with water vapour**). At BTPS conditions the air volume only varies with barometric pressure: $V_{BTPS} \cdot (P_B - 47)/(273 + 37) = n\mathbf{R}$.

Since we have considered one mass of gas the products of volume and pressure divided by temperature, must in all three states equal $n\mathbf{R}$. This we can state in the following three equations which are recalculated to pressure in kPa:

$$V_{STPD}101.3/273 = V_{ATPS}(P_B - P_{water})/(273 + t)$$
$$= V_{BTPS}(P_B - 6.3)/(273 + 37) = n \cdot \mathrm{R}.$$

How is the ventilatory mass flow of gas generated?

The rib cage consists of two elastic components that work together: the lungs, which behave like an elastic **balloon** trying to collapse, and the thoracic cage, which is an elastic cage trying to expand. The following important pressures depend on the elastic properties of these two components:

1. The **atmospheric or barometric pressure** (P_B). The barometric pressure is 1 atmosphere and is frequently defined as zero. All pressures measured are given in reference to the barometric pressure, which is the pressure at the mouth or at the surface of the thoracic cage.

   By convention, all transmural pressures are expressed as the difference between the pressure measured **inside the body and P_B**. Transmural differences are essential during **static conditions** (i.e. apnoea with relaxed muscles).
2. The **intrapulmonary (alveolar) pressure** (P_{alv}). This is the pressure in the alveoli and is equal to the **static mouth pressure** when there is no air flow (i.e. during apnoea with the glottis open). Apnoea is a temporary stop in breathing – a **static** condition. The **static mouth pressure** (P_{alv}) varies with the lung volume during apnoea with the mouth and nose sealed (Fig. 29-1). The **driving pressure** for air to move is: ($P_{alv} - P_B$). The driving pressure and airway resistances are studied when air moves into and out of the lungs, and the condition is therefore called **dynamic**. The driving pressure for inspiration is a negative **alveolar pressure** (P_{alv}), when P_B is defined as zero (Fig. 29-2).
3. The **(intrapleural) intrathoracic pressure** (P_{it}). The intrathoracic pressure is the pressure in the fluid-filled pleural space between the parietal and visceral layers of pleura. The intrapleural fluid reduces the friction between the two layers. P_{it} can be measured with a **pressure-sensitive balloon transducer which is passed into the oesophagus**. Pressure changes in the intrapleural space equal the **oesophageal**

balloon pressure changes, because the oesophagus traverses the intrapleural space. P_{it} is subatmospheric owing to the opposing directions of the elastic recoil of lungs and thoracic cage.

How do we relate the partial pressure to mole fraction?

The relation is called **Dalton's law** which states that the **partial pressure or tension of a single gas in a mixture is equal to the product of the total pressure and the mole fraction** (*F*). Thus, the F_{O_2} of the alveolar air (F_{AO_2}) is: $F_{AO_2} = P_{AO_2}/(101.3 - 6.3) = P_{AO_2}/(760-47)$. With a P_{AO_2} of 13.3 kPa (or 100 Torr), the F_{AO_2} is **0.14**. Each gas acts as if the others did not exist.

How are physiological lung volumes and capacities measured?

These volumes are measured by **spirometry** (Fig. 28-1). A spirometer consists of a counterbalanced bell which is connected to a pen writing on a rotating drum. The air-filled bell is inverted over a chamber of water, so an air-tight chamber is formed. The bell moves up and down with respiration (Fig. 28-1). Volume changes are recorded on volume and time calibrated paper. If the spirometer is supplied with a **CO_2 absorber**, the device is called a **metabolic ratemeter** (Benedict–Krogh).

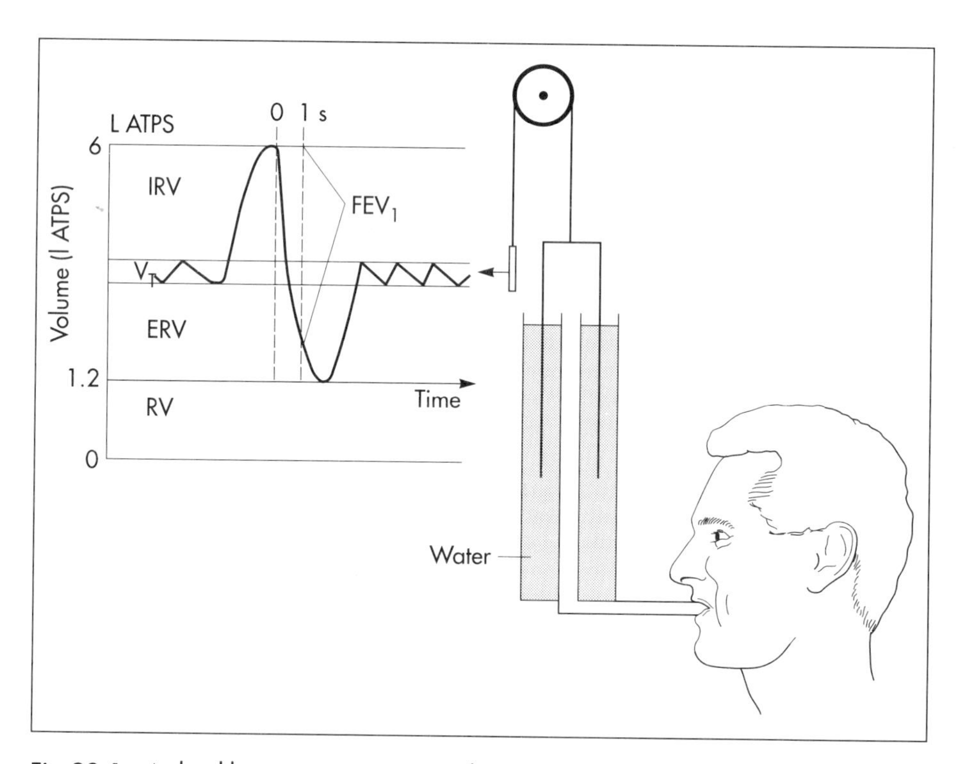

Fig.28-1: A healthy person, connected to a spirometer, is performing a vital capacity manoeuvre from maximal inspiration to maximal expiration (RV).

The **residual volume (RV)** represents the volume of air still left in the lungs after a maximal expiration (1.2 l in Fig. 28-1). The **vital capacity (VC)** is the maximum volume of air that can be exhaled after a maximal inspiration (4.8 l in Fig. 28-1). VC has three components. The first is the **inspiratory reserve volume (IRV)**, which is the quantity of air that can be inhaled from a normal end inspiratory position. The second component of VC is the **tidal volume (V_T)**, which is the volume of air inspired and expired with each breath (about 0.5 l at rest), in the average young adult human.

The third component is the **expiratory reserve volume (ERV)**, which is the amount of air that can be exhaled from the lungs from a normal end-tidal expiratory position which is characterized by a relaxed **expiratory pause** (Fig. 28-1). This position is the easiest one to reproduce in healthy individuals, and the lung volume in this position is called **functional residual capacity (FRC) (FRC = ERV + RV)**. The **total lung capacity (TLC)** is the total volume of air in the lungs, when they are maximally inflated (RV+VC) – approx. 6 l of air.

When the person in Fig. 28-1 exhales with maximal pressure, the **forced expiratory volume in 1 s (FEV_1)** can be recorded.

Are lung volumes of consequence in modern medicine?

Lung volumes can aid in differentiating between the two major functional types of lung disorders, and in quantifying the degree of the abnormality. The two types are **restrictive** and **obstructive** lung disorders. **Restrictive lung diseases** are characterized by **small lung volumes**, eg fibrosis and **obstructive lung diseases** by **increased airflow resistance**, eg asthma, emphysema and chronic bronchitis.

What is dynamic airway compression?

At a certain point in the airway the forces that expand the airway equal the forces that tend to collapse. This is **the equal pressure point.**

During a forceful expiration, the intrathoracic pressure (P_{it}) rises and causes the alveolar pressure (P_{alv}) to exceed the downstream pressure at the airway openings ($P_B = 0$). As flow resistance dissipates the driving energy along the bronchial tree, the driving pressure of the cartilagenous bronchi falls towards zero at the mouth (Fig.28-2). Beyond the equal pressure point the driving pressure falls below the **external pressure,** and the bronchi becomes narrower (Fig. 28-2). Total closure of the tubes does not occur as the bronchi have cartilagenous support. At this point the person cannot voluntarily increase the rate of expiratory airflow, because increased effort also increases the external pressure. This phenomenon is called **dynamic airway compression.**

Apply Bernouilli's law to the phenomenon of airway compression

According to **Bernouilli's law** the driving energy equals the sum of the **kinetic energy** and the laterally directed energy (i.e. the **lateral pressure** directed towards the walls). When airways are compressed during expiration, kinetic energy increases causing acceleration of the gas molecules through the narrowed region, while lateral pressure decreases.

What happens during coughing?

Coughing is momentary collapse of the tracheal wall. The airway closure occurs when the **equal pressure point** has moved to a part of the trachea that is not supported by cartilage.

Where is airway resistance maximum?

The **peak airway resistance**, where flow limitation takes place, is found in the **medium-sized segmental bronchi** around the 4th–7th generation moving peripherally as lung volume decreases. In healthy people the least resistance to air flow is found in the numerous terminal bronchioles. At low lung volumes the elastic pull in the bronchioles becomes smaller (the structures relax with falling volume) and the airways tend to collapse more easily (Fig. 28-2).

What is a pneumotachograph?

A pneumotachograph consists of a **respiratory tube** with a minimal resistance (*Res*) to airflow (Fig. 28-3). The two chambers separated by *Res*, connects to the two differential transducer chambers by thin tubes.

The forced vital capacity manoeuvre is performed with all the expiratory and accessory muscles. When we contract our strong expiratory accessory muscles, we generate a high airflow velocity at thoracic volumes near TLC (Fig. 28-4). Just following peak-expiratory flow (PEF) the airflow velocity deceases linearly with volume no matter how hard the subject tries. This is the **effort-independent airflow** (Fig. 28-4) as explained above caused

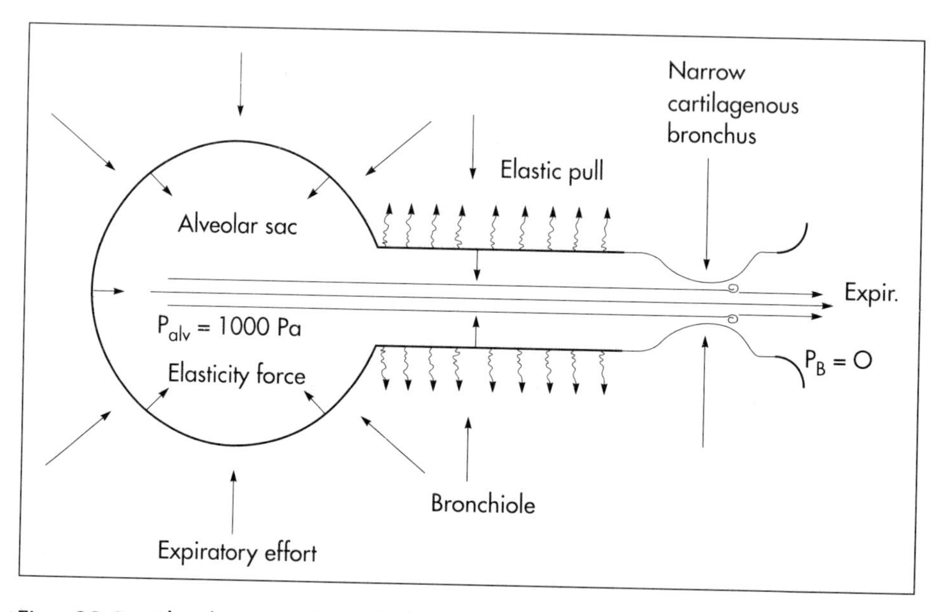

Fig. 28-2. Alveolar sac, bronchiole and a cartilagenous bronchus collapsing during expiration.

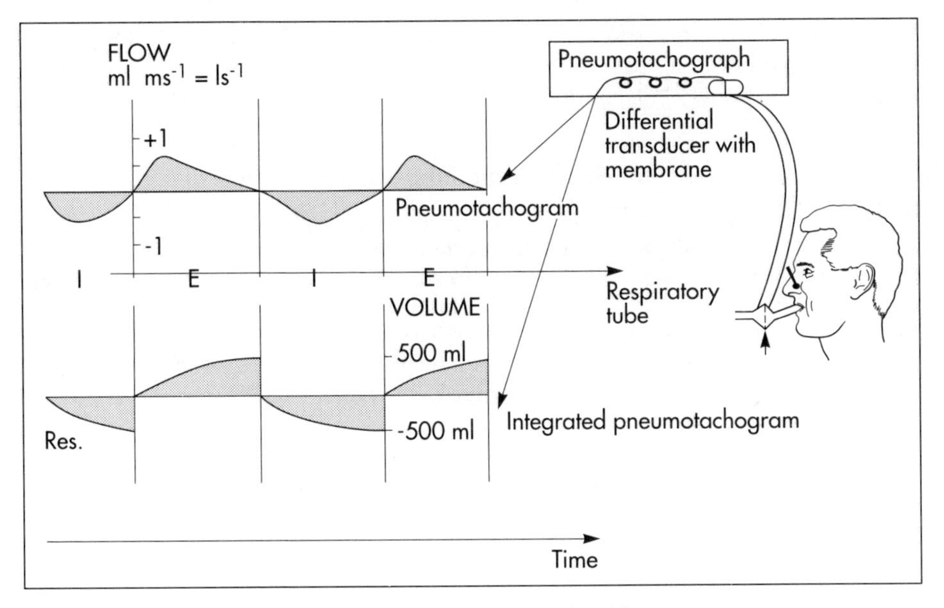

Fig. 28-3: Pneumotachograms from a healthy person at rest.

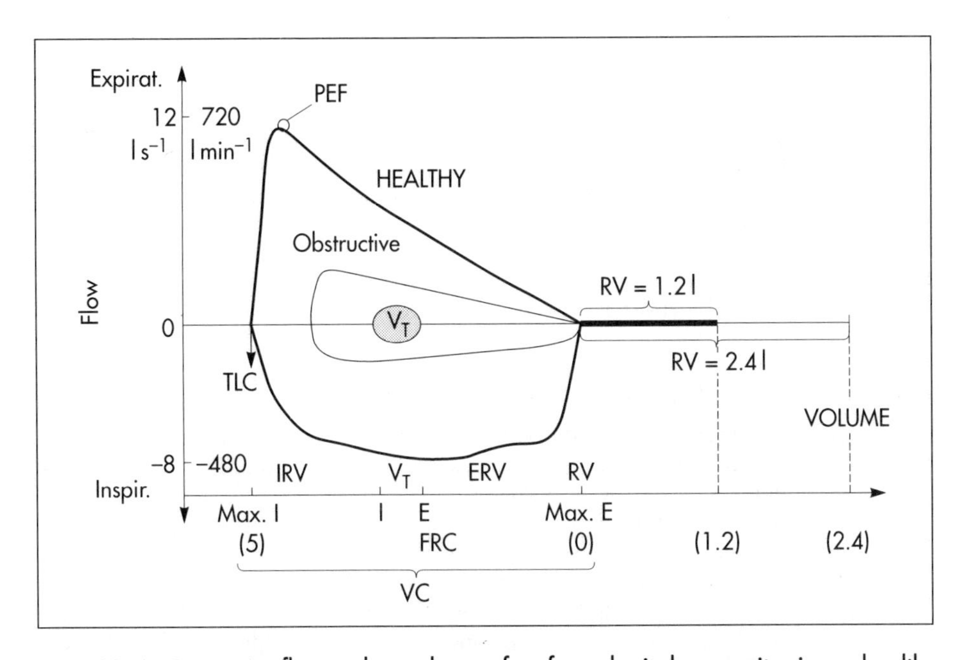

Fig. 28-4: Dynamic flow–volume loops for forced vital capacity in a healthy person (large solid loop) and in a patient with an obstructive lung disease (the thin loop with RV 2.4 l). The small grey loop is tidal breathing for both persons. All volumes are in litres.

by dynamic airway compression. During respiration through the respiratory tube, a small pressure difference ($\triangle P$) is created across the resistance, and this $\triangle P$ is directly proportional to the airflow: $\mathring{V}$ (l min^{-1}) = $\triangle P/Res$ according to Poiseuille's law.

What can be learned from the dynamic flow–volume loop?

Figure 28-4 shows a dynamic flow–volume loop generated by plotting airflow velocity measured at the mouth with a **pneumotachograph** against lung volume (integrated airflow velocity). The large, solid loop is from a healthy person performing a forced expiration from full lung inflation (TLC) to full lung deflation (RV). This is a so-called **forced vital capacity manoeuvre** (Fig. 28-4). The inspiratory airflow velocity increases rapidly, and reaches a plateau until maximal inspiration, at which point it falls rapidly to zero (Fig. 28-4). Inspiration is not flow limited as is expiration. The transmural pressure causing dynamic compression during expiration, actually causes expansion during inspiration. Inspiratory airflow is limited by the inspiratory muscles.

The **small grey loop from FRC represents the flow–volume curve** for a tidal breath both in a normal person and in obstructive lung disease (Fig. 28-4).

The **patient with obstructive lung disease** has a smaller flow–volume loop than a normal subject, when they perform a forced vital capacity manoeuvre (Fig. 28-4). The RV of the patient is 2.4 l or twice as high as that of the healthy subject (Fig. 28-4), because of **air trapping** (a large volume of trapped air).

Although the flow–volume curve in obstructive lung disease is consistently reduced in the flow direction, it is not always reduced in the volume direction as shown in Fig. 28-4.

Do airways collapse during forced expiration?

Airflow resistance is always higher during expiration, because the airways narrow during expiration. During a forced expiration – **forced vital capacity manoeuvre** – the maximal driving pressure raises alveolar pressure above the pressure at the mouth (Fig. 28-4). The main stem bronchi and the trachea are compressed, airflow velocity is increased, cross-sectional area decreases reducing lateral pressure, and this leads to further **dynamic airway compression** (see above). This limits expiratory airflow and causes the maximum expiratory airflow to be **effort-independent** (Fig. 28-4).

How are the bronchial smooth muscles innervated?

Adrenergic sympathetic activity (and sympathomimetic drugs) relax bronchial smooth muscle via **adrenergic β_2-receptors**, whereas **parasympathetic cholinergic activity (and parasympathomimetics)** constrict bronchial smooth muscles via muscarinic receptors.

Smoke, dust and other irritants (perhaps also histamine and substance P) constrict the airway smooth muscles via a reflex triggered by rapidly adapting irritant-receptors. Decreased P_{ACO_2}, thromboxane and leucotrienes also act as bronchoconstrictors.

Vasoactive intestinal peptide (VIP) can dilate airways and reduce airflow resistance. Substances that dilate airways include increased P_{ACO_2}, adrenergic α-blockers, catecholamines and atropine.

Regularly inhaled **β_2-agonists** with steroids are, in fact, beneficial in stable asthma. Cardioselective **adrenergic β_1-blockers** must be administered with care to patients with a combination of asthma and cardiac disease.

Further reading

The glossary Committee of the International Union of Physiological Sciences (IUPS) (1973) Glossary on respiration and gas exchange. *J. Appl. Physiol.* **35**: 941–61. See Chapter 75.

Heino, M. (1994) Regularly inhaled beta-agonists with steroids are not harmful in stable asthma. *J. Allergy Clin. Immunol.* **93** (1): 80–4.

Levitzky, M.G. (1991) *Pulmonary Physiology*. 3rd edn. McGraw-Hill, New York.

Staub, N.C. (1991) *Basic Respiratory Physiology*. Churchill-Livingstone, New York.

28. Case History

A 24-year-old male, with an oxygen uptake of 333 ml STPD min^{-1}, is breathing from a metabolic ratemeter containing 50 l of atmospheric air with an oxygen fraction of 0.2093. The room temperature is 293 K, the water vapour tension is 18 Torr and P_B is 760 Torr.

1. *Calculate the original STPD volume of the metabolic ratemeter.*
2. *Calculate the time period in which it was safe for him to breathe in this device. Use a safety margin of 50%.*

Try to solve the problems before looking up the answers in Chapter 74.

CHAPTER 29.
MECHANICS OF BREATHING

Normal lungs are very distensible at fundamental residual capacity (FRC), but stiffen progressively towards total lung capacity (TLC). Lung distensibility is called compliance. Compliance is an index of **expandability** of elastic organs and defined as the change in volume per unit change in pressure (dV/dP). The falling compliance during inflation near TLC is caused by an increase in the **air–liquid surface tension**, because the liquid contains tension reducing molecules (**surfactant**) that are spread further and further apart. The lungs and chest wall move together and support each other. This is what makes the **total standard compliance** of the respiratory system less than that of the lungs or rib cage alone (Fig. 29-1).

What are the forces opposing lung expansion with air flow?

Two forces are involved: the elastic recoil of the lung, and the non-elastic or airflow resistance.

1. The **overall elastic recoil (dP/dV)** is the sum of the pulmonary **elastic recoil** and its **surface tension**. These forces relate to the **elastic work** of Fig. 29-2. Traditionally,

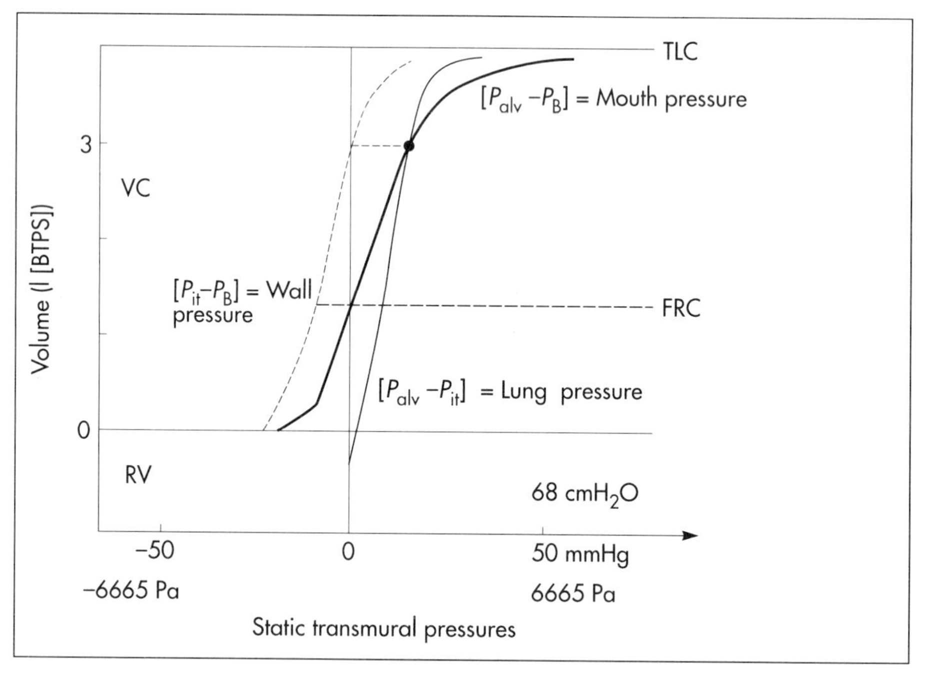

Fig. 29-1: Transmural, static pressures and lung volumes.

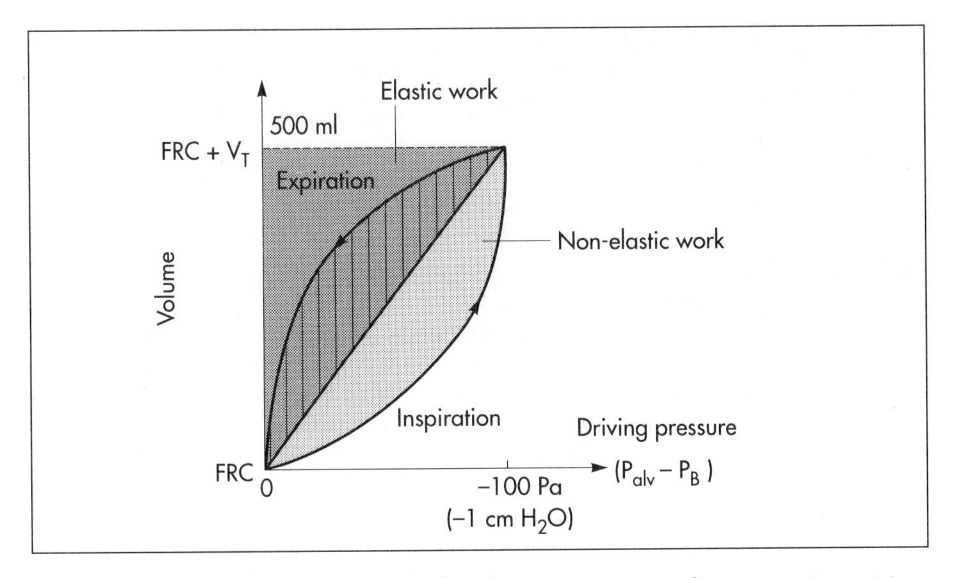

Fig. 29-2: Tidal volume (V_T) and the dynamic transmural pressure ($P_{ip}-P_B$).

the **reciprocal elasticity** or **compliance (d*V*/d*P*)** is preferred as an index of the expandability of the lungs and the rib cage. The specific standard compliance is defined at a comparable lung volume (FRC). Increased specific lung compliance (reduced elasticity) means that the lungs are easier to inflate. Reduced specific lung compliance means that the lungs are stiffer (increased elasticity) and harder to inflate (restrictive lung disease).

There is a thin fluid layer containing **surfactant** on the inner surface of the alveolus. Because of the **alveolar fluid – air interface,** a surface tension is created that tends to collapse the alveolus (just like the elastic recoil). The contribution of **surface tension** to the **overall lung elasticity** is large. **Pulmonary surfactant** is secreted into the thin air–fluid interface of the alveolar lining. It is **dipalmitoyl phosphatidylcholine, DPPC, which decreases surface tension** and thereby promotes the stability of the 300 million alveoli in both lungs.

2. The **airflow resistance forces** relate to the **non-elastic work** of Fig. 29-2. The **total airflow resistance** is the sum of all the resistances of the nose and mouth (a substantial portion) and of the 23 generations of the tracheobronchial tree. Airway resistance is created by the friction between gas molecules and between gas molecules and the walls. The **airway resistance** is important and make the sliding of lung tissue over each other (**tissue resistance**) a minor issue.

STATIC RELATIONS

When is compliance changed in the lungs?

In the lungs, the term **static compliance** is used because the volume and pressure measurements are made when there is no airflow. **Increased lung compliance** is caused by **reduced lung elasticity**, and means that lungs with elastic tissue degeneration are easier to inflate. **Reduced lung compliance** is caused by **increased lung elasticity** and means that stiff, fibrotic lungs are harder to inflate.

How is the pressure–volume curve derived for the lung and rib cage under static conditions?

The subject expires completely so that there is only a residual volume remaining in his lungs (RV in Fig. 29-1). He then inspires a fixed volume of air from a spirometer. The subject then relaxes the respiratory muscles while the glottis is open. The alveolar pressure ($P_{alv} - P_B$) can then be measured as the **mouth pressure** by a manometer (Fig. 29-1). The procedure is repeated by varying the inspired lung volume between RV and TLC, and the **solid relaxation curve** of Fig. 29-1 is recorded. This relaxation curve shows the **specific standard compliance for the total system** (defined at static conditions as the slope of the curve at FRC). In healthy persons, the **specific total standard compliance** is **1 ml per Pascal** (0.1 l BTPS per cm of water at FRC).

When lung volume increases, the alveolar surfactant molecules are spread, thus **increasing** the surface tension. Thereby the compliance of the lung is reduced. Compliance also **decreases** with age; there is a corresponding decrease in lung volume.

How can the compliance of lung tissue alone be measured?

The mouth pressure and intrathoracic pressure are measured during apnoea as described above. The **intrathoracic pressure** can be measured using an **oesophageal balloon transducer**.

The volume of the lungs before each apnoea is varied in the same way as when measuring the total lung compliance. The pressure difference between the mouth and intrathoracic pressure ($P_{alv} - P_{it}$) is equal to the intramural pressure across the lung tissue (thin curve marked **lung pressure** in Fig. 29-1). In healthy persons, the **specific lung compliance is 2 ml per Pascal** (0.2 l BTPS per cm of water) at FRC, where the thin lung curve is almost linear (dV/dP at FRC).

What about the rib cage wall compliance?

The **static transmural wall pressure** ($P_{it} - P_B$) is indirectly obtained from the two static transmural pressures measured above. With the two elastic systems in static equilibrium ($P_{it} - P_B$) must be equal to the difference between the static transmural pressure of the total system and that of the lungs: ($P_{alv} - P_B$) – ($P_{alv} - P\text{it}$). According to this equation a minimal increase in lung volume (dV) implies the following relationship:

$$d(P_{it} - P_B)/dV = d(P_{alv} - P_B)/dV - d(P_{alv} - P_{it})/dV.$$

These three values are elastancies (dP/dV). The elastancy of the thoracic cage is equal to the total elastance minus the lung elastance. The **specific compliance of the thoracic cage** (dV/dP) is the specific reciprocal elastance, and equal to **2 ml per Pascal** (0.2 l BTPS per cm of water). The **chest wall compliance curve** is constructed by using the above equation (dashed curve in Fig. 29-1).

Quote the common name for the group of lung disorders with small lung volumes

Restrictive lung diseases cover all disorders with **decreased compliance (d*V*/d*P*)** or **inhibited thoracic expansion.** The lungs cannot expand normally because of pulmonary fibrosis with connective tissue infiltration of the interstitial space, extreme obesity, hunchback or kyphoscoliosis. Consequently, the **lung volumes are small** (VC and TLC), and all volumes are often proportionally decreased.

Pulmonary compliance is also reduced, when **alveoli collapse** (atelectasis) or when left heart failure causes **alveolar oedema (pulmonary oedema).**

Quote the common name for the group of lung diseases with increased air flow resistance

Obstructive lung diseases are caused by a **narrowing of the airways due to spasm of the bronchial smooth muscle, inflammatory increase in mucus production and bronchial wall oedema.** The maximal air flow is reduced both in acute asthma, chronic bronchitis, and emphysema.

Emphysema is characterized by loss of lung tissue, and in particular of elastic fibres, septa and capillaries. Thus the emphysematous lung has increased compliance (increased d*V*/d*P*), because the elasticity is decreased. Chronic bronchitis and emphysema often co-exist, and are commonly grouped together as **chronic obstructive lung disease (COLD)** or more commonly known in the UK as **chronic obstructive airways disease (COAD).** This is often a progressive disease during years of smoking, causing impaired exercise capacity. In the **terminal stages** the patients become hypoxic and later also hypercapnic with severe disability. COLD patients have a high oxygen cost of breathing (up to 30% of their resting $\mathring{V}_{O_2}$), and even food intake is a demanding task, causing wasting. Obstruction of airflow causes uneven ventilation, and destruction of lung tissue reduces capillary blood volume. The maldistribution of bloodflow is upsetting the normal matching of ventilation and perfusion ($\mathring{V}_A/\mathring{Q}$). This is the major cause of hypoxia and later hypercapnia. Hypercapnia is signifying the inability to maintain adequate alveolar ventilation.

The hypercapnic patient has given up the control level of normal P_{aCO_2}, because it is much too costly for them in terms of energy expenditure for breathing. The patient is in a state of hypoxia with low P_{aO_2} and **malnutrition.**

DYNAMIC RELATIONS

How does one derive the dynamic PV-curve?

Respiratory volume is recorded graphically with a x,y-recorder. The tidal volume is plotted against the **driving pressure**, which is equal to the **dynamic intrapulmonic or alveolar pressure** (Fig. 29-2).

The **resistance to airflow**, and viscous resistance of lung tissue, cause the dynamic pressure–volume curve (Fig. 29-2) to deviate from the static (Fig. 29-1). The slanting straight line (diagonal) is sometimes called **dynamic compliance** for the total system (Fig. 29-2).

Integrating pressure with respect to volume gives the grey area corresponding to the **elastic work** of inspiration (Fig. 29-2). This is the work needed to overcome the elastic resistance against inspiration. The dotted area to the right of the diagonal is the extra work of inspiration called the **flow-resistive work** or alternatively **non-elastic work** (Fig. 29-2).

During expiration, the **flow-resistive work** is equal to the area with vertical bars between the diagonal line and the expiration curve (Fig. 29-2). The inspiratory and expiratory curve form a so-called **hysteresis loop**. The lack of coincidence of the curves for inspiration and expiration is known as **elastic hysteresis**. With deeper and more rapid breathing the hysteresis loop becomes larger, and the non-elastic work relatively greater.

The elastic work is performed in stretching the elastic structures of the lung and thoracic cage which includes the work performed against **surface tension in the alveoli** (Fig. 28-2). **The non-elastic work** consists of the work done against airway resistance (the **flow-resistive forces**), and the work done to overcome inertia and friction (which is negligible at rest).

Does the surface tension constitute a major portion of the total wall tension in the alveoli?

Yes, this is easily demonstrated. When the pressure–volume curves in air are contrasted to those from saline-inflated excised lungs, it is clear that the air-filled lungs are much less compliant and show a much larger **hysteresis loop** than when they are inflated with saline. **Saline-filled lungs have no air–liquid interface, and thus no surface tension.** More than half the total elastic recoil force of the lungs is caused by **surface tension.**

How does the intrapleural pressure vary in the upright position?

The lungs are suspended in a gravity field. The weight of the lung causes the change in gravity from top to bottom of the lung. Consequently, the effect of gravity on the mass of lung tissue is highest at the apex. This effect, which tends to separate the parietal and the visceral pleura, is greater at the lung apex than at its base. Thus, the **intrapleural, intrathoracic pressure (P_{it}) is more subatmospheric at the apex of the lung than at its base** (−900 to −200 Pa or −9 to −2 cm of water at FRC). Since the alveolar pressure (P_{alv}) is more or less the same throughout the 300 million communicating alveoli, it follows that the transmural pressure gradient ($P_{alv} - P_{it}$) over the lung tissue is higher at the apex than at the base of the lung. Therefore, the alveoli at the apex are always more expanded than those at the base. The expanded apical alveoli will distend less during inhalation than the small, compliant alveoli located in the middle and basal regions. Consequently, there is a continuum from the apex to the base of the upright lung, with the **lowest level of ventilation per unit lung volume at the apex and the highest at the base of the lung.**

What is the physiological importance of pulmonary surfactant?

Surfactant is a complex phospholipid that is a combination of **DPPC** and other lipids and proteins. DPPC orients perpendicular to the air–water interface, such that the charged choline base is dissolved in water (hydrophilic) and the non-polar, **hydrophobic fatty acids** project toward the alveolar air. Surfactant is secreted by the type 2 alveolar epithelial cells.

Surfactant lowers the surface tension importantly in the alveoli which increases the lung compliance.

Surfactant promotes the stability of all alveoli and helps keep them from collapsing. According to the law of Laplace, the transmural distending pressure in spherical alveoli is equal to $T/(2r)$, where T is the total wall tension (elastic recoil plus surface tension; N m^{-1}) and r is the radius. Because the distending pressure is essentially the same in all communicating alveoli, the **total wall tension** must change commensurate with variations in diameter. During expiration the diameter is reduced, surfactant molecules are packed tightly together, separating the water molecules and reducing the **total wall tension**. During inspiration the diameter is increased, the surfactant molecules are scattered, and the water molecules closer to each other so the **total alveolar wall tension** increases more and more; the lung becomes stiffer. **Surfactant helps to keep the alveoli dry.**

What is the respiratory distress syndrome of newborns?

Respiratory distress in premature infants is caused by **inadequate synthesis of surfactant by the type 2 cells**. Such infants have lungs with enormous surface tension forces and low compliance, causing collapse of alveoli (atelectasis) and oedema. Positive-pressure ventilation opposes these changes and improves gas exchange. Administration of aerosolized surfactant has been shown to be effective.

Is it possible to measure the airway resistance (R_{aw})?

The **driving pressure (ΔP) for laminar flow** (air flow, $\mathring{V}_E$) through the airway resistance is the **intrapulmonary pressure (P_{alv})** minus the external **barometric pressure, P_B**.

Poiseuille's law for laminar flow is an analogy to Ohm's law: $R_{aw} = \Delta P/\mathring{V}_E$. R_{aw} is directly related to the **air viscosity** (η) and to the length (L) of the tube, and inversely related to its radius in the 4th power: $\boldsymbol{R_{aw} = 8\ \eta\ L/r^4}$. Doubling the length of the airways only doubles the airway resistance, but halving the radius increases the resistance 16-fold. Such a reduction takes place in the small airways during **bronchiolitis**. The walls of the bronchioles are inflamed causing oedema (swelling), constriction, sloughing of epithelium, and excessive secretion. A similar reversible bronchoconstriction takes place in **hyperirritable airway disease** (asthma).

In the clinic R_{aw} is measured by one of the following two methods.

(1) A **body plethysmograph** is a technically demanding piece of equipment, so in everyday clinical practice (2) the **forced expiratory volume in 1 s (FEV_1)** from TLC is used as an indirect measure. This indirect method only requires simple, reliable and accurate spirometers. The patient is asked to expire as fast as possible from TLC by creating a maximal driving pressure. This **maximal driving pressure** is considered an **arbitrary unit**, because it is a reproducible constant for each patient. Therefore, it is essential that maximal expiratory pressure is applied. As long as a fatigued patient applies an expiratory pressure above the threshold pressure able to create dynamic airway compression, its absolute size is immaterial. The R_{aw} is obtained by dividing the **expiratory pressure unit** by the airflow velocity, FEV_1. The R_{aw} relates to the **non-elastic work** of Fig. 29-2.

Is the airway flow laminar or turbulent?

Laminar or streamline flow (the streamlines move parallel to the sides of the tubes) is limited to airways with low airflow velocities and smooth walls. Such conditions are normally present in small airways. Laminar flow is silent. However, all airways branch and there is **transitional flow** at each bifurcation. This **transitional (laminar-turbulent) flow** depends on the following **driving pressure**:

$$\Delta P = [\mathring{V}_E R_1 + \mathring{V}_E^2 R_2].$$

The second (turbulent) component is small during quiet breathing.

Turbulent flow is the agitated random movement of molecules, which accounts for the sounds heard over the chest during breathing. Turbulent flow develops at the branch points of the upper airways even in quiet breathing.

Turbulence also develops when the radius of the airways is decreased by constriction, mucus, infection, tumours, or foreign bodies. Vagal stimulation (by smoke, dust, cold air, and irritants) leads to airway constriction.

Does R_{aw} vary with relative ventilation?

With increasing lung volume the expanding lung tissue **pulls the airways open** and thereby decreases the airway resistance. There is a continuum from the top to the bottom of the upright lung, with respect to the degree of airway and alveolar distension. The greatest relative lung distension – at any lung volume – is found at the top. Consequently, the distended top of the lung has the lowest relative ventilation. Thus airway calibre is larger at the top than at the bottom of the upright lung, **causing R_{aw} to increase progressively from the top to the base of the lung.**

Further reading

Levitzky, M.G. (1991) *Pulmonary Physiology*, 3rd edn. McGraw-Hill, New York.

Rahn, H., A.B. Otis, L.E. Chadwick and W.O. Fenn (1946) The pressure–volume diagram of the thorax and lung. *Amer. J. Physiol.* **146**: 161–6.

Taylor, A.E., K. Rehder, R.E. Hyatt and J.C. Parker (1989) *Clinical Respiratory Physiology*. W.B. Saunders, Philadelphia.

West, J.B. (1990) *Respiratory Physiology*, 4th edn. Williams & Wilkens, Baltimore.

29. Case History

A male with a FRC of 2.5 l shows an intrathoracic pressure change of 3 cm of water during normal tidal breathing of 0.5 l. His chest wall compliance is 0.15 l BTPS/cm of water. He has a total alveolar wall tension force (T) of 0.07 N m^{-1} tending to collapse two alveoli with radius 0.00004 and 0.00008 m, respectively. The total alveolar wall tension consists of the surface tension plus the elastic recoil forces.

1. *Calculate the ΔP which can prevent collapse of the two alveoli.*
2. *Is this result consequential for the stability of his alveolar design?*
3. *Is there a natural solution to this problem?*
4. *Calculate the specific lung compliance of this patient and compare it to the normal value.*
5. *What pressure must be applied to supply this person with 1 l (BTPS) of air per breath under positive-pressure ventilation?*

Try to solve the problems before looking up the answers in Chapter 74.

CHAPTER 30. EXPIRATORY AND ALVEOLAR AIR

This chapter deals with the two common sense concepts **(sacred cows)** on which almost all calculations in respiratory physiology are based. The two concepts (axioms) lead us to **Bohr's equation** for determination of dead space (V_D), the **alveolar ventilation equation, the equations for calculation of $\dot{V}_{O_2}$ and $\dot{V}_{CO_2}$, and the alveolar gas equation**. The international symbols are a precise short-cut for intellectual transfer used by all physiologists (see Chapter 75 before reading on).

What is the air volume from the alveoli?

Every exhaled breath (tidal volume, V_T) passing the airway opening consists of an air volume from the **anatomic dead space (V_D)** and an **effective volume ($V_T - V_D$)** from the alveoli (Fig. 30-1). The **tidal volume, V_T**, is all the air that leaves or enters the lungs with each breath. Most lung volumes are expressed as BTPS (see Chapter 28). The anatomic V_D is all conducting airways without alveoli and gas exchange in a healthy person. It is normally equal to twice the body weight in kilograms. Multiplying V_T with the respiratory frequency (f) produces a minute-flow, the **expired minute ventilation ($\dot{V}_E$)**. The alveolar ventilation ($\dot{V}_A$) is the volume of fresh gas that reaches the alveoli each minute, that is ($V_T - V_D$) multiplied by f. The $\dot{V}_A$ is of key importance in the body,

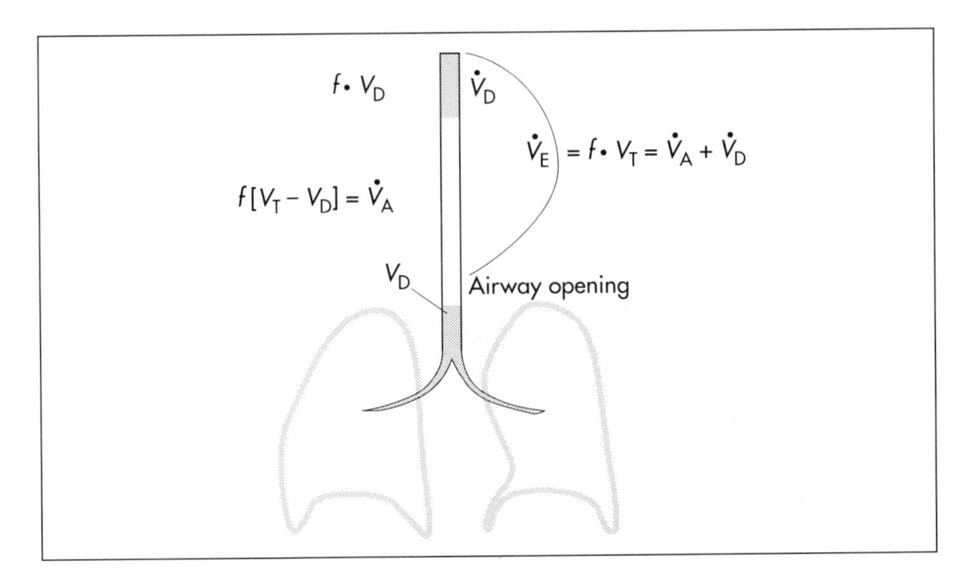

Fig. 30-1: Expiratory volume above the airway opening from dead space (V_D is the grey column) and from the alveolar space (the white column of air).

because it represents the volume of fresh air that is inspired for gas exchange. Thus it follows:

$$\mathring{V}_D + \mathring{V}_A = \mathring{V}_E \qquad \text{(Fig. 30-1 illustrated the \textbf{First Axiom})}.$$

This simple and fundamental equation has a multitude of practical applications.

What is understood by the physiological dead space?

This volume is defined as any part of the lungs and airways that is ventilated but not perfused. It represents wasted **respiratory volume**, and it is the sum of the anatomic and the alveolar dead space. The alveolar dead space is the alveolar volume ventilated but without bloodflow. The ratio of **physiological dead space** to V_T is 0.3 in normal people. In other words the **wasted ventilation** is 30%. The wasted ventilation increases in many pathological conditions, with a large alveolar dead space.

How to measure V_D *using dilution methods*

Such methods employ indicator gases.

Carbon dioxide can be used as an indicator. The **anatomic** dead space can be measured by one expiration into an empty bag. The number of expired indicator molecules is equal to the number of molecules present in the anatomic dead space plus those that are derived from the alveolar air. The fractions of CO_2 in the expired and the alveolar air are shortened as F_{ECO_2} and F_{ACO_2}, and the related partial tensions as P_{ECO_2} and P_{ACO_2}.

For a single expiration:

$$(V_T F_{ECO_2}) = (V_D F_{ICO_2}) + (V_T - V_D) F_{ACO_2}.$$

Atmospheric air contains virtually no CO_2 ($F_{ICO_2} = 0$). From the molar fraction definitions it follows that:

$$F_{ECO_2}(P_B - 47) = P_{ECO_2}; \text{ and } F_{ACO_2}(P_B - 47) = P_{ACO_2}.$$

This leads to the following equation:

$$(V_T P_{ECO_2}) = (V_T - V_D) P_{ACO_2} \text{ or}$$

$$\textbf{anatomic } V_D = V_T(P_{ACO_2} - P_{ECO_2})/P_{ACO_2}.$$

This is the **Bohr equation**, which can be solved for either V_D or for F_A.

The existence of the **alveolar dead space** implies that the partial CO_2 tension of the arterial blood (P_{aCO_2}) must be higher than P_{ACO_2}. In the absence of this **alveolar dead space** the P_{ACO_2} would be equal to P_{aCO_2}.

Hence, if one wants to measure the **physiological (anatomic plus alveolar)** dead space, one has to substitute P_{ACO_2} with P_{aCO_2}. Only P_{aCO_2} reflects the **efficiency of total gas exchange** across the alveolar barrier:

$$\textbf{physiological } V_D = V_T(P_{aCO_2} - P_{ECO_2})/P_{aCO_2}$$

The mean alveolar CO_2 fraction only reflects the **anatomic dead space**:

$$F_{ACO_2} = \mathring{V}_{CO_2}/(V_T - V_D).f$$

What is the composition of the mean alveolar air in the lungs of a resting person?

In the alveolar air F_{AO_2} is 0.15 and F_{ACO_2} is 0.056 (see symbols in Chapter 76).
Accordingly, $F_{AN_2} = 1 - (0.15 + 0.056) = 0.794$.

One can easily calculate the partial pressure of oxygen, carbon dioxide and nitrogen at body temperature and saturation, since the vapour pressure is 6.3 kPa or 47 Torr at 37°C. P_{AO_2} and P_{ACO_2} are surprisingly constant:

$$\mathbf{P_{AO_2} = (101.3 - 6.3) \times 0.15 = 14.25\ kPa}$$

$$\mathbf{P_{ACO_2} = (101.3 - 6.3) \times 0.056 = 5.3\ kPa;\ or}$$

$$\mathbf{P_{ACO_2} = (760 - 47) \times 0.056 = 40\ Torr.}$$

What is the composition of the air expired by a resting person?

$F_{EO_2} = \mathbf{0.164}$ and $F_{ECO_2} = \mathbf{0.041}$ so $F_{EN_2} = (1 - 0.205) = \mathbf{0.795}$. These are normal average values and can be used in other calculations.

How is the net oxygen intake calculated?

The intake is simply input minus output:

$$\mathring{V}_{O_2} = (\mathring{V}_I F_{IO_2}) - (\mathring{V}_E F_{EO_2}).$$

What happens to nitrogen?

The flux of nitrogen into the body is equal to the flux of nitrogen out of the body. This concept (axiom) has not – to this day – been refuted:

$$\textbf{Second axiom: } (\mathring{V}_I F_{IN_2}) = (\mathring{V}_E F_{EN_2}).$$

This equation makes it possible to calculate V_I, when V_E is already measured or vice versa.

By use of the **first axiom,** we can express the second axiom without dead space. This is equivalent to: $(\mathring{V}_{AI} F_{IN_2}) = (\mathring{V}_A F_{AN_2})$. With these equations we can eliminate the variables inspired ventilation and inspired alveolar ventilation ($\mathring{V}_I$ and $\mathring{V}_{AI}$) once and forever. In the international symbol language, $\mathring{V}_A$ is defined as the **alveolar expired ventilation,** and the **alveolar inspired ventilation ($\mathring{V}_{AI}$)** is only used once for the above stated argument.

Why does F_{EN_2} and F_{AN_2} deviate from F_{IN_2}?

$$(F_{EN_2} \mathring{V}_E) = (F_{IN_2} \mathring{V}_I) \text{ and } (F_{AN_2} \mathring{V}_A) = (F_{IN_2} \mathring{V}_{AI}).$$

All deviations between fractions are obviously caused by **volume deviations.** The two standard volumes are the same only when R is equal to one, and in this situation the two fractions are therefore also the same in both equations.

How do we measure oxygen uptake correctly by only measuring $\dot{V}_E$?

Since $\dot{V}_I = (\dot{V}_E(F_{EN_2}/F_{IN_2}))$ it is possible to avoid to measure $\dot{V}_I$, and we simply calculate the O_2 difference between inspired and expired air:

$$\dot{V}_{O_2} = \dot{V}_E[(F_{IO_2}F_{EN_2}/F_{IN_2}) - F_{EO_2}].$$

How is oxygen uptake in terms of alveolar ventilation expressed?

Eliminate V_D in the above equation by the first and second axiom:

$$\dot{V}_{O_2} = (\dot{V}_{AI}F_{IO_2}) - (\dot{V}_A F_{AO_2});$$

$$\dot{V}_{O_2} = \dot{V}_A[(F_{IO_2}F_{AN_2}/F_{IN_2}) - F_{AO_2}].$$

How is the rate of CO_2 removal from the blood calculated?

This **CO_2 output** must be equal to the **expired minus the inspired CO_2.**

$$\dot{V}_{CO_2} = \dot{V}_E F_{ECO_2} - (\dot{V}_I F_{ICO_2}).$$

$$\dot{V}_{CO_2} = \dot{V}_E[F_{ECO_2} - (F_{ICO_2} \times F_{EN_2}/F_{IN_2})].$$

Atmospheric air contains virtually no CO_2 ($F_{ICO_2} = 0.0003$). Therefore, the last expression is zero. Thus, the CO_2 fractions are merely just the ratio between two flows:

$$F_{ECO_2} = \dot{V}_{CO_2}/\dot{V}_E \text{ and } F_{ACO_2} = \dot{V}_{CO_2}/\dot{V}_A.$$

Without the dead space the equation reads:

$$\dot{V}_{CO_2} = (\dot{V}_A F_{ACO_2}).$$

This is the **alveolar ventilation equation:** $\dot{V}_A = \dot{V}_{CO_2}/F_{ACO_2}$, describing the hyperbolic relationship between $\dot{V}_A$ and F_{ACO_2} (Fig. 30-2). F_{ACO_2} is equal to $P_{ACO_2}/(101.3 - 6.3)$, so P_{ACO_2} is easily exchanged for F_{ACO_2}.

What is hypo- and hyperventilation?

Alveolar hypoventilation occurs when $\dot{V}_A$ is less than the metabolic demands, so P_{ACO_2} will increase. **Alveolar hyperventilation** arises when $\dot{V}_A$ exceeds the metabolic demands, whereby P_{ACO_2} and P_{aCO_2} decrease. These definitions were introduced by **Wallace O. Fenn** as early as 1950.

In general, $\dot{V}_A$ and $\dot{V}_{CO_2}$ do not change independently. Any increase in metabolic CO_2 production accelerates $\dot{V}_{CO_2}$. The excess CO_2 increases P_{ACO_2} and P_{aCO_2}. This will stimulate the central and peripheral chemoreceptors, which causes $\dot{V}_A$ to increase appropriately (Fig. 30-2).

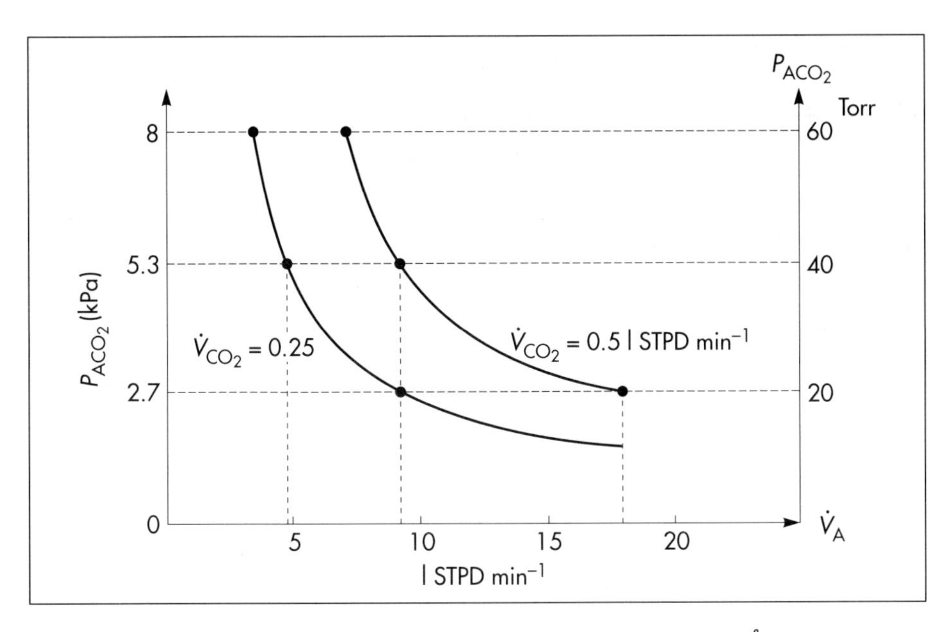

Fig. 30-2: An illustration of the hyperbolic relationship between $\mathring{V}_A$ and P_{ACO_2} at two values of $\mathring{V}_{CO_2}$.

What is the respiratory exchange quotient, R?

R equals the ratio between the alveolar $\mathring{V}_{CO_2}$ and $\mathring{V}_{O_2}$. R equals the metabolic respiratory quotient (RQ) for all body cells at steady state.

$\mathring{V}_A$ is the same in the nominator and in the denominator. Accordingly:

$$R = (F_{ACO_2} - F_{ICO_2} \times F_{AN_2}/F_{IN_2})/(F_{IO_2} \times F_{AN_2}/F_{IN_2} - F_{AO_2}).$$

This is the classical **alveolar gas equation** which is also expressed as:

$$(P_{IO_2} - P_{AO_2}) = P_{ACO_2}[F_{IO_2} + (1 - F_{IO_2})/R].$$

This equation is further simplified when $F_{ICO_2} = 0$, and $R = 1$ (implying dominant carbohydrate metabolism and equal N_2 fractions):

$$R = F_{ACO_2}/(F_{IO_2} - F_{AO_2}).$$

This is the **simplified alveolar gas equation** for $R = 1$. At a given F_{IO_2} and a constant R, P_{aO_2} and P_{aCO_2} are inversely related. The **mass balance for the body as a whole** explains why our body weight does not increase, when we live on a mixed diet, and inhale more O_2 than we exhale CO_2 ($R = 0.8$).

How do healthy people wash out alveolar nitrogen?

Inspiration of pure oxygen is started at time zero. The nitrogen fraction in **end expiratory air** at time zero is an acceptable estimate of the **mean alveolar fraction** ($^0F_{AN_2}$). Exactly at time zero the person is shifted from atmospheric air to pure O_2. At the end of the first normal inspiration of O_2 the amount of nitrogen in FRC is still: $(FRC^0 F_{AN_2})$ and the size of the alveolar volume: $(FRC + (V_T - V_D)$.

The fraction of nitrogen in this new alveolar volume is:

$$^1F_{AN_2} = [FRC^0 F_{AN_2}/(FRC + (V_T - V_D))] = {}^0F_{AN_2} \cdot k.$$

This dilution constant, k, obviously dilutes the previous nitrogen fraction with each O_2 inspiration. Accordingly:

$$^2F_{AN_2} = (k^1 F_{AN_2}) = (k^2\,{}^0F_{AN_2}).$$

Following n respirations of pure O_2: $^nF_{AN_2} = (k^n\,{}^0F_{AN_2})$.

The nitrogen fraction of the alveolar air is reduced with a **constant fraction for each respiratory cycle** (as long as FRC, V_T and V_D are constant and the lung ventilation is even in all regions).

The logarithm of F_{AN_2} as the ordinate and the respiratory number as abscissa form **one straight line**. Healthy people with ideal ventilation eliminate alveolar nitrogen **monoexponentially**. The ratio between alveolar ventilation and FRC is **k** ($\mathring{V}_A$/FRC). **Accordingly, $T_{\frac{1}{2}}$ equals (ln 2/k) or (0.693/k).**

The nitrogen fraction is also an exponential function of time, and the **nitrogen wash out curve** can be used for **FRC calculation**.

Are nitrogen wash out curves of clinical importance?

Persons with uneven ventilation or patients with lung disease eliminate alveolar nitrogen with different rates from different lung regions. Such a patient has a nitrogen wash out curve which consists of several exponential functions (several straight lines). The **oxygen test** or the **multiple breath method** are synonyms for the nitrogen wash out curve.

Further reading

Nunn, J.F. (1987) *Applied Respiratory Physiology*, 3rd edn. Butterworths, London.
West, J.B. (1989) *Respiratory Physiology*, 4th edn. Williams & Wilkins, Baltimore.

30. Multiple Choice Questions

Each of the following five answers have True/False options:
Stem statement: In a healthy person P_{aCO_2} is tightly controlled, compared with arterial P_{O_2}, because:

A. The hypercapnic ventilatory response is exponential.
B. The hypoxic ventilatory response is linear.
C. The hypercapnic ventilatory response is linear with a steep slope.
D. The hypoxic ventilatory response is blocked.
E. The hypercapnic ventilatory response is hyperbolic.

30. Case History

A healthy female, age 22 years, in respiratory steady state, has a metabolic rate of 10 000 kJ day^{-1}, a respiratory frequency (f) of 12 breaths min^{-1} and a respiratory quotient (RQ) of 1. The fraction of nitrogen in the inspired air is 0.79 and P_B is 760 Torr. During measurement of her metabolic rate the expired air is gathered in a rubber bag (Douglas bag) saturated with water vapour (20 Torr) and the P_{EO_2} is estimated to be 120 Torr. The P_{ACO_2} is 40 Torr. Following the steady state study at rest, the patient is asked to increase her ventilation while following a fixed respiratory pattern shown to her on a screen. This standardized hyperventilation is tripling her alveolar ventilation and increasing her ventilatory exchange quotient (R) to 3 with a new but constant alveolar air composition and an unchanged metabolic rate (F_{AN_2} falls to 0.78). Through a catheter in the pulmonary trunk mixed venous blood is sampled and the following values determined: $C_{\bar{v}O_2}$ 160 ml STPD l^{-1}, and $C_{\bar{v}CO_2}$ 540 ml STPD l^{-1}. In an arterial blood sample the C_{aO_2} is measured to 200 ml STPD l^{-1}. The respiratory enthalpy equivalent for oxygen is 20 kJ l^{-1} STPD.

1. *Calculate her alveolar ventilation at rest.*
2. *Calculate the anatomic dead space.*
3. *Calculate her alveolar air composition during hyperventilation.*
4. *Calculate the cardiac output ($\dot{Q}$) (see Chapter 21).*

Try to solve the problems before looking up the answers in Chapter 74.

CHAPTER 31.
THE OXYGEN TRANSPORT OF THE BLOOD

Oxygen is transported from the alveolar air to the red blood cells by **diffusion** across the alveolar–capillary barrier. The transit time for the erythrocytes through the lung capillaries is virtually always adequate for the haemoglobin to become fully saturated with the oxygen of the alveolar gas. Accordingly, diffusion-limitation is not present in healthy persons. The arterial blood gas tensions are ideal informers about the status of blood and lung gas exchange.

The second step in the transfer is the tissue diffusion from the capillary blood to the mitochondria. Here, the conditions are less favourable and **diffusion-limitation is present** in distant cells, although their mitochondria can maintain oxidative metabolism at 1 Torr (133 Pa). The first problems concern life without haemoglobin.

How to calculate cardiac output ($\mathring{Q}$)

Fick proposed that the cardiac output ($\mathring{Q}$) can be calculated:

$$\mathring{Q} = \mathring{V}_{O_2}/(C_{aO_2} - C_{\bar{v}O_2}); \text{ see Chapter 24.}$$

Standard data for a healthy person at rest are an **O_2 uptake of 250 ml STPD min^{-1}** and an **arteriovenous O_2 content difference of 50 ml STPD l^{-1}** or 25% of C_{aO_2}. This implies a $\mathring{Q}$ of 5 l min^{-1}.

How to calculate the hypothetical $\mathring{Q}$ in a person deprived of red cells

The solubility coefficient for oxygen in body warm plasma is 0.022 ml STPD per ml and per atmosphere (101.3 kPa). The person is assumed to have a normal P_{aO_2} of 13.3 kPa. The dissolved oxygen in his or her arterial plasma must be:

$$[0.022 \cdot 13.3 \cdot 1000/101.3] = 2.76 \text{ ml STPD l}^{-1}.$$

Seventy-five per cent oxygen remains in the venous blood, if 25% is utilized by the tissues as above. Thus, the **hypothetical $\mathring{Q}$** is 250/(2.76 · 0.25) = **362.3 l of blood min^{-1}!**

This is a totally unrealistic $\mathring{Q}$ for humans. Life without **haemoglobin** is not possible for humans at 1 atmosphere of air pressure (P_{aO_2} = 13.3 kPa). However, life without haemoglobin is possible for 10–20 min by **hyperbaric oxygenation** (see Chapter 41).

Describe the structure of haemoglobin using the words globin, haeme, imidazol-group, ferro-ion, α- and β-chain

Haemoglobin is an oligometric protein that consists of **four polypeptide chains and four prostetic haeme-groups.**

The four polypeptide chains form a protein-part called **globin**. The polypeptide chains are not covalently linked but are held together by **hydrophobic forces**. Each haeme-group is connected to one polypeptide chain and consists of a complex organic ring

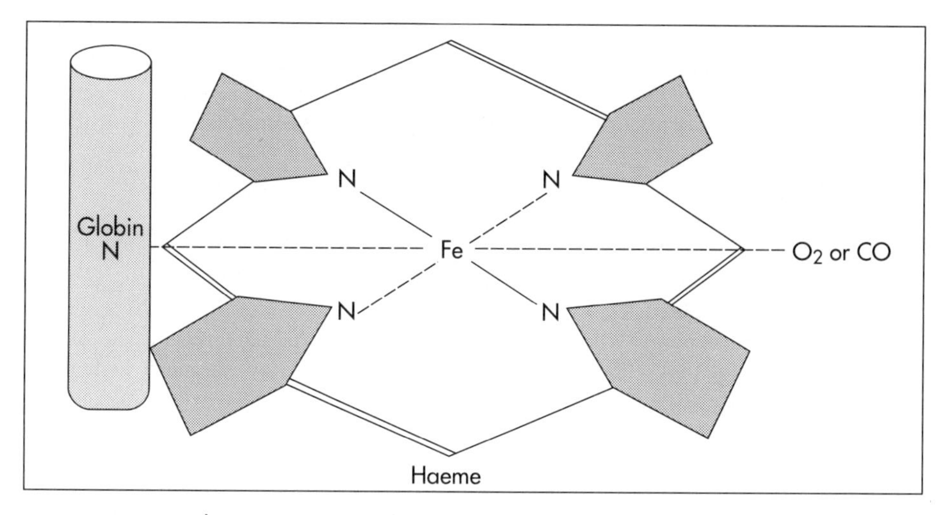

Fig. 31-1: One haeme-group with an open binding for O_2 or CO.

structure, **protoporphyrin**, which is built up of four imidazol-groups (Fig. 31-1). In the centre of the porphyrin ring there is one iron atom that is co-ordinated by six ligands, four of which bind the metal to the porphyrin chain and one to histidin on either the α- or the β-chains (Fig. 31-1), and the last is an **open binding** which is able to bind either O_2 or carbon monoxide (CO).

When haeme is bound to O_2 or CO, it has a cherry-red colour and is dark red when it is in the deoxygenated form. The breakdown of porphyrin (which entails ring-opening) produces bilirubin which is yellow in colour. This explains the yellow colour in **jaundice (icterus)** and in blue-yellow skin-spots. **Jaundice** is pigmentation of cell membranes and secretions with bile-pigments.

What is the molecular weight of haemoglobin?

Haemoglobin A (for Adult) has a molecular weight of 64 443 g mol^{-1}.

How to calculate the maximum volume of oxygen that can bind to 1 g of haemoglobin

One haemoglobin molecule binds **four oxygen molecules**. One gram of haemoglobin consists of (1/64 443) mol. One mole of any gas has a STPD volume of 22.4 l. Therefore, 1 g of haemoglobin binds:

$$((1/64\,443) \times 4 \times 22\,400 \text{ ml STPD g}^{-1}) = \mathbf{1.39 \text{ ml } O_2 \text{ STPD g}^{-1}}.$$

Haemoglobin is saturated by a O_2 partial pressure of approx. 45 kPa. Atmospheric O_2 partial pressure is only 20 kPa (150 mmHg or Torr); thus, haemoglobin is theoretically not saturated under normal atmospheric conditions.

Despite its incomplete saturation at 20 kPa, the **oxygen capacity of haemoglobin** is defined as the volume of O_2 that can be bound to 1 g of haemoglobin at the normal conditions of 20 kPa (150 mmHg): **1.34 ml O_2 STPD g^{-1}.** This value is generally accepted as **oxygen capacity**, but obviously less than the theoretical oxygen capacity of 1.39 ml g^{-1}.

How to calculate the [haemoglobin] in blood with a C_{aO_2} of 200 ml l^{-1}

The blood [haemoglobin] is **200/1.34** (ml STPD O_2 l^{-1})/(ml STPD O_2 g^{-1}) = **149 g haemoglobin l^{-1} of blood.**

Notice that when blood is saturated under normal O_2 partial pressure the **oxygen capacity is 1.34** and not 1.39 ml l^{-1}. The latter holds only for extremely high partial pressures (above 45 kPa).

Are other important substances also transported by haemoglobin?

Carbon dioxide (CO_2) and **hydrogen ions (H^+).**

Sketch the reaction between oxyhaemoglobin ($Hb(O_2)_4$), CO_2 and H^+

$$\mathbf{Hb_4(O_2)_4 + CO_2 + H^+ \rightleftharpoons Hb_4CO_2(H^+) + 4O_2.}$$

According to **Henry's law** the **concentration of physically dissolved O_2, $[O_2]$, is directly proportional to its partial pressure in the blood**. In the lungs, the **concentration is high** (due to the high partial pressure), hence the reaction is shifted to the left. This results in the release of CO_2 and the oxygenation of haemoglobin. In the muscles, the **$[O_2]$ is falling** (due to mitochondrial consumption of O_2), hence the reaction is shifted to the right which results in the release of O_2 and reduced haemoglobin.

How can the release of H^+ from haemoglobin in the lungs increase CO_2-removal from the blood?

The reaction is **shifted to the right** when the H^+-concentration rises:

$$H^+ + HCO_3^- \rightleftharpoons H_2CO_3 \rightleftharpoons H_2O + CO_2.$$

The last reaction is catalysed by **carbonic anhydrase** (CA).

Sketch the curve that shows the relationship between the oxyhaemoglobin saturation and the oxygen partial pressure

In fully oxygenated pulmonary blood, the oxygen concentration is approximately 200 ml STPD l^{-1} ($S_{O_2} = 1.0$) at an atmospheric oxygen partial pressure of 20 kPa and a P_{aO_2} of 13.3 kPa (Fig. 31-2). This fully oxygenated blood is mixed with venous blood that passes through the **physiological shunt** on its way to the left heart. Thus, the oxygen tension falls to 12.6 kPa (95 Torr or mmHg) in arterial blood from 13.3 kPa at the pulmonary capillaries and S_{aO_2} is only 0.985 (Fig. 31-2).

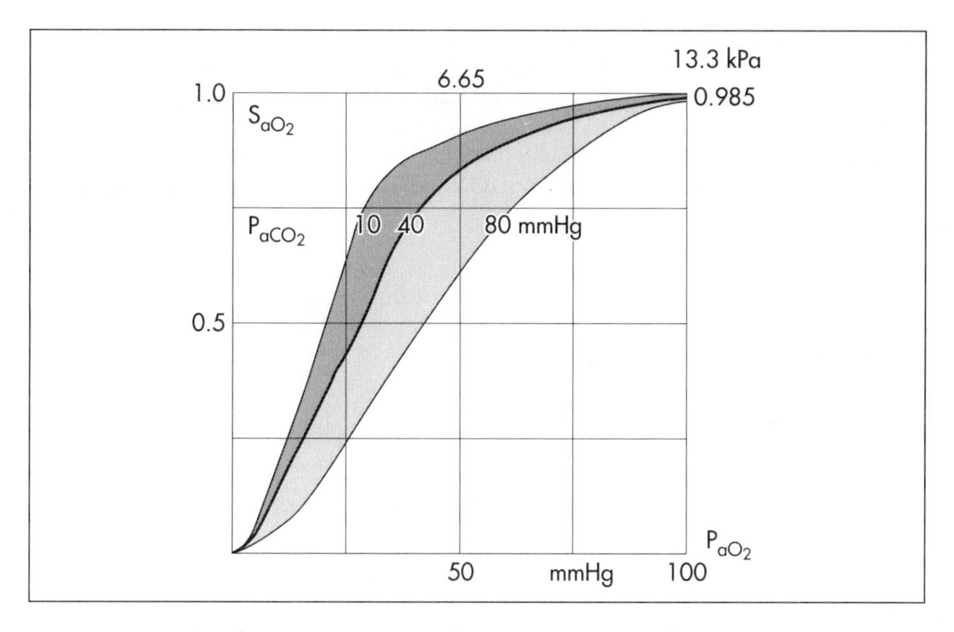

Fig. 31-2: The dissociation curve of O_2 – pressure in kPa or mmHg (Torr).

Read the arterial O_2 saturation from the dissociation curve and calculate the concentration of dissolved O_2

The solubility coefficient of oxygen is 0.022 ml STPD per ml per atmosphere. From the curve, $S_{aO_2} = 0.985$ (Fig. 31-2). The chemically bound O_2: (0.985·200) = **197** and the physical bound:

$$(0.022 \times 12.6/101.3) \times 1000 = \mathbf{2.74} \;-\; \textbf{a total of 199.75 [ml STPD l}^{-1}\textbf{]}.$$

Which factors affect the oxyhaemoglobin dissociation curve?

Standard affinity of haemoglobin for O_2 is defined as the **reaction rate when haemoglobin is 50% oxygenated**. The standard affinity of haemoglobin for oxygen is therefore the gradient of the dissociation curve when haemoglobin is 50% oxygenated. A practical expression for standard affinity is the O_2 partial pressure at 50% oxygenation – called P_{50}. A metabolite of anaerobic glycolysis called 2,3-**DPG (diphosphoglycerate)** is highly concentrated in erythrocytes (1 mol per mol of Hb_4). The highest values are found during hypoxia and during exercise. The 2,3-DPG, like H^+ and CO_2, facilitates the unloading of O_2 from haemoglobin by reversible changes of its molecular configuration.

The following factors **shift the curve to the right** (i.e. increase P_{50}): (1) increasing P_{CO_2} (the **Bohr-effect**); (2) increasing $[H^+]$; (3) increasing 2,3-DPG; and (4) increasing temperature. This rightward shift is due to the metabolic activity of the cells, and means that less O_2 is bound to haemoglobin at a given P_{aO_2} (metabolic activity facilitates unloading of O_2).

The factors that **shift the curve to the left** (low P_{50}) are: (1) increasing O_2 tension (The **Haldane-effect**); (2) decreasing $[H^+]$ and P_{aCO_2}; and (3) decreasing temperature, as it occurs in the lungs. A leftward shift is also caused by: (4) increasing CO tension; and by (5) foetal haemoglobin. The leftward shift in the lungs means that more O_2 is bound to haemoglobin at a given P_{aO_2} (facilitates the binding of O_2 to haemoglobin).

Foetal haemoglobin (F) contains two γ-chains instead of the two β-chains of adult (A) haemoglobin. **F binds less 2,3-DPG** and the **F-curve** is clearly shifted to the left (low P_{50}, since F binds O_2 strongly). Also species differences in the globin part affect the curve.

Further reading

Mines, A.H. (1986) *Respiratory Physiology*, 2nd edn. Raven Press, New York.
Siggaard-Andersen, O. (1974) The acid–base status of the blood. Munksgaard, Copenhagen.

31. Multiple Choice Questions

Each of the following five statements have True/False options:

A. Most of the carbon dioxide in the blood is transported in the form of bicarbonate in the plasma.
B. Near the terminal bronchioles the movement of air is accomplished by bulk flow.
C. When blood flows through tissue capillaries carbon dioxide is released from haemoglobin.
D. Pulmonary surfactant helps to equalize the distending pressure in alveoli of different sizes by reducing their surface tension.
E. The conducting zones of the airways constitutes the anatomic dead space.

31. Case History

A male, age 25 years, exercises at a steady state maximum with $\dot{V}_{O_2}$max of 4.5 l min^{-1}, a C_{aCO_2} of 500 ml STPD l^{-1}, C_{aO_2} of 200 ml STPD l^{-1}, and a $C_{\bar{v}CO_2}$ of 650 ml STPD l^{-1}. The C_{O_2} in his coronary sinus is measured to 30 ml STPD l^{-1}, and his myocardial oxygen consumption is 420 ml STPD min^{-1}. He is working with a RQ of 0.9.

One hour later, the male is at rest with $\dot{Q} = 5450$ ml min^{-1}, and $\dot{V}_{O_2}$ 273 ml STPD min^{-1}.

1. *Calculate his $\dot{V}_{CO_2}$ and his $\dot{Q}$ during exercise.*
2. *Calculate the coronary bloodflow during exercise.*
3. *What are the energy sources of the heart during exercise and during rest?*
4. *Calculate the concentration of oxygen in his venous blood at rest, and thus the arteriovenous oxygen difference.*

Try to solve the problems before looking up the answers in Chapter 74.

Chapter 32. Gas Exchange between Blood and Tissue

Tissue diffusion is the rate-limiting process for oxygen transport. The oxyhaemoglobin saturation curve (Fig. 31-1) shows that the P_{O_2} must fall appreciably before an important quantity of oxygen dissociates from haemoglobin. The gas tensions of the mixed venous blood ($P_{\bar{v}O_2}$; $P_{\bar{v}CO_2}$) provide the ideal information about the status of blood and tissue exchange. However, mixed venous blood is obtainable only by invasive procedures.

What factors are important for the P_{O_2} in a tissue point?

The transfer of gas is governed by **Fick's diffusion law**. This law states that the amount of gas transferred from blood to a certain point (mitochondrion) is **directly related** to: the pressure gradient (ΔP = mean systemic capillary minus mean tissue oxygen tension), area of the systemic capillary barrier (A), solubility of the gas (α). The flux of gas is **inversely related** to: the length of the diffusion pathway from the capillaries to the mitochondria (L), and the square root of the molecular weight of the gas.

Marie Krogh incorporated solubility (Bunsens α), molecular weight (mol.weight), A, and L in her **lung diffusion constant (D_L)** and by analogy we can define a **mitochondrial diffusion constant (D_m) equal to the solubility** (α) **divided by the square root of the mol.weight** (see Chapter 35).

The lung diffusion constant is reduced by lung diseases with degeneration of elastic tissues including septal capillaries (**emphysema**).

The **diffusion volume of O_2** is $\mathring{V}_{O_2}$, and this must equal ($D_m \cdot \Delta P$).

Oxygen equilibrates rapidly across the barrier, but its transport is **perfusion limited**.

The ΔP depends upon the bloodflow through each capillary, the density of open capillaries, and their degree of dilatation. The P_{aO_2} is kept high by O_2 released from oxyhaemoglobin, and the steep dissociation curve illustrates the substantial O_2 delivery to the mitochondria.

Apart from the **factors leading to increased P_{O_2}**, all **factors which imply a rise in $\mathring{V}_{O_2}$** are also important.

The force driving oxygen to the most distant mitochondria is often small.

Fortunately, mitochondria have the capacity to maintain oxidative metabolism at a tissue tension as low as **133 Pa** or **1 Torr**. Even such a low P_{O_2} increases the rate of the respiratory chain events.

Healthy lungs have a diffusion distance (L) of less than 1 μm. Are the distances similar in other tissues?

No, most other tissues have a diffusion pathway 10-fold longer or more. This is particularly **critical in brain and heart tissues**, whereas the even longer diffusion distance in resting muscle tissue is of minor consequence.

What is the most effective way to improve oxygen delivery?

The body has no choice, but to **reduce the diffusion distance** by **recruitment** of more capillaries at increasing demand. This is particularly important in skeletal muscles during heavy exercise, where the capillary density increases three-fold. Such a rise **increases the systemic, capillary surface area,** simultaneously with the decrease in diffusion distance.

What is the benefit of myoglobin during muscular contraction?

The stream of blood is more or less blocked during the contraction phase, and the tissue P_{O_2} falls toward zero. At low P_{O_2}, the gradient of the dissociation curve of myoglobin is steepest. Hence, myoglobin releases its O_2 readily. During muscular relaxation, bloodflow is restored and myoglobin is rapidly reloaded with oxygen. The myoglobin dissociation curve has a P_{50} some five-fold lower than that of haemoglobin.

The rate of CO_2 diffusion is larger than that of O_2. How much?

The **diffusion constant** is equal to α divided by the square root of the mol.weight. The square root of their molecular weights (square root of 32/44) is 0.85. This result is obtained by solving the equations for CO_2 and O_2. The ratio of their solubilities is 0.51/ 0.022 = 23.2. The **rate of CO_2 diffusion** is 20-fold (23.2 · 0.85 = 20) greater than that of O_2.

Is the shape of the oxyhaemoglobin dissociation curve of physiological importance?

Yes, the flat plateau part going towards the right (Fig. 31-1) provides a **rapidly available O_2 store**. The P_{aO_2} can fall from 13.3 to 8.7 kPa (100 to 65 Torr) with little change in S_{aO_2}. Almost the same mass of O_2 will attach to haemoglobin. The steep, middle portion of the curve allows cell mitochondria to extract relatively large quantities of O_2 from haemoglobin with relatively small changes in the P_{aO_2}. Consequently, 50% of the O_2 can be delivered to metabolically active cells when the P_{aO_2} is low, but high enough to maintain diffusive force.

Which organs have a particular high arterio-venous O_2 difference?

The **heart** uses **at rest 140 ml O_2 l^{-1}** of the 200 ml l^{-1} that is supplied by the coronary arteries. The venous oxygen saturation in the coronary sinus is therefore (200 – 140)/200 = 0.30 at rest. Thus, increases in the myocardial O_2 uptake during exercise are mainly met by an increase in bloodflow.

Intensive working skeletal muscles – just like the myocardium at rest – utilize approximately **140 ml O_2 l^{-1}**. The main part of the $\mathring{Q}$ passes the working muscles during severe work. The **arterio-venous O_2 content difference** for the total muscle mass becomes dominant, and almost equal to that of the body as a whole.

Describe classical oxygen transport deficiencies

1. **Anaemic hypoxia** is caused by anaemia or CO poisoning. The disorder is an insufficient oxygen carrying capacity of the patient's haemoglobin. For both sexes a blood [haemoglobin] below 130 g l^{-1} (8 mM) implies reduced working capacity and

thus a consequential anaemia. The normal blood [haemoglobin] is 150 g l^{-1}. Smokers expose their bodies to anaemic hypoxia and to lung cancer.

2. **Hypotonic hypoxia** is caused by insufficient oxygen uptake into the blood from the lungs. This condition is due to lung disease or it occurs in space, during flying, at altitude and during diving. Hypotonic hypoxia is defined as a P_{aO_2} of less than 7.3 kPa (55 mmHg). Below this threshold the ventilation starts to increase.
3. **Ischaemic or stagnant hypoxia** is caused by insufficient bloodflow (low $\dot{Q}$ or locally).
4. **Histotoxic hypoxia** is caused by blockade of the mitochondrial metabolism as in cyanide poisoning.

Further reading

Cherniack, N.S., M.D. Altose and S.G. Kelsen (1983) The respiratory system. In: Berne, R.M. and M.N. Levy (Eds) *Physiology*, Section VI. C.V. Mosby, St Louis.

Krogh, A. (1919) The number and distribution of capillaries in muscles with calculations of the oxygen pressure head necessary for supplying the tissue. *J. Physiol.* (London) **52**: 409.

Nunn, J.F. (1987) *Applied Respiratory Physiology*, 3rd edn. Butterworths, London.

Taube, G. (1993) Claim of higher risk for women smokers attacked. *Science* **262**: 1375.

West, J.B. (1989) *Respiratory Physiology*, 4th edn. Williams & Wilkins, Baltimore.

32. Case History

A 20-year-old person is unconscious after a traffic accident and brought to hospital. No focal lesions are found. A cardio-pulmonary screening reveals the following: metabolic rate: 10 450 kJ daily, respiratory frequency 14 min^{-1}, P_{AO_2} 4.5 kPa and P_{ACO_2} 5.3 kPa. In the mixed expiratory air P_{EO_2} is 15.6 and P_{ECO_2} 4.4 kPa.

Cardiac catheterization includes the pulmonary artery with gas concentrations: $C_{\bar{v}O_2}$ 150 and $C_{\bar{v}CO_2}$ 542 ml STPD l^{-1} of blood.

The concentrations of gases are: C_{aO_2} 200 and C_{aCO_2} 500 ml STPD l^{-1} in a blood sample from the radial artery. The dietary energetic equivalent for O_2 is here 20.6 kJ l^{-1} STPD. The barometric pressure is 101.3 kPa and the tracheal water vapour tension is 6251 Pa.

1. *Calculate the RQ.*
2. *What can be inferred from this RQ value?*
3. *Calculate the $\dot{V}_{O_2}$ of the patient.*
4. *Calculate the alveolar ventilation and the dead space.*
5. *Calculate the $\dot{Q}$ of the patient.*
6. *Is the unconsciousness caused by respiratory, cardiovascular or central nervous system malfunction?*

Try to solve the problems before looking up the answers in Chapter 74.

CHAPTER 33. CARBON DIOXIDE TRANSPORT BY THE BLOOD

Carbon dioxide is the most important final product of cellular metabolism, since it operates as a **control molecule** in essential regulatory processes linking ventilation to cardiac output and vascular resistance. Carbon dioxide is carried by the venous blood to the lungs, where it is eliminated in the expired air by ventilatory effort. Our blood contains large quantities of carbon dioxide. The [total CO_2] in arterial blood is 500 ml STPD l^{-1} (22.3 mM), and in venous blood 540 ml STPD l^{-1} (24 mM). **The solubility coefficient for CO_2 is 0.51 ml STPD per ml and per 101.3 kPa, P_{aCO_2} is 5.3 and $P_{\bar{v}CO_2}$ is 6.1 kPa.**

There are **three forms of CO_2 transport** from the cells to the lungs: **(1) physically dissolved; (2) in combination with blood proteins as carbamino compounds; and (3) as bicarbonate.**

1. How is the physically dissolved CO_2 concentration in arterial and in mixed venous blood calculated?

The concentration is equal to $(0.51 \times 1000\ \Delta P/101.3)$. One mmol equals 22.4 ml. The results are for arterial blood: **26.7 ml STPD l^{-1}** (26.7/22.4 = 1.19 mmol l^{-1}), and for mixed venous blood: **30.7 ml STPD l^{-1}** (30.7/22.4 = 1.37 mmol l^{-1}). The difference (30.7 − 26.7) equals 4 or 4/22.4 = **0.18 mM dissolved CO_2.**

Because CO_2 is 24 times more soluble than O_2, dissolved CO_2 is a more significant form of transport in the venous blood than is dissolved O_2 in the arterial blood (Fig. 33-2c).

2. How is CO_2 transported in combination with blood proteins as carbamino compounds?

Terminal amine groups, primarily on haemoglobin, react with CO_2:

$$Hb - NH_2 + CO_2 \rightleftharpoons Hb - NH - COO^- + H^+.$$

The concentration of CO_2 binding compounds in arterial blood is 1 mM and in mixed venous blood 1.4 mM, so carbamino compounds account for **0.4 mM of the CO_2 transported in blood.**

3. How is the bicarbonate part of the [total CO_2] in blood calculated?

Some of the dissolved CO_2 reacts with water, forming carbonic acid, which immediately breaks down to bicarbonate and H^+:

$$\mathbf{CO_2 + H_2O \rightleftharpoons H_2CO_3 \rightleftharpoons H^+ + HCO_3^-}.$$

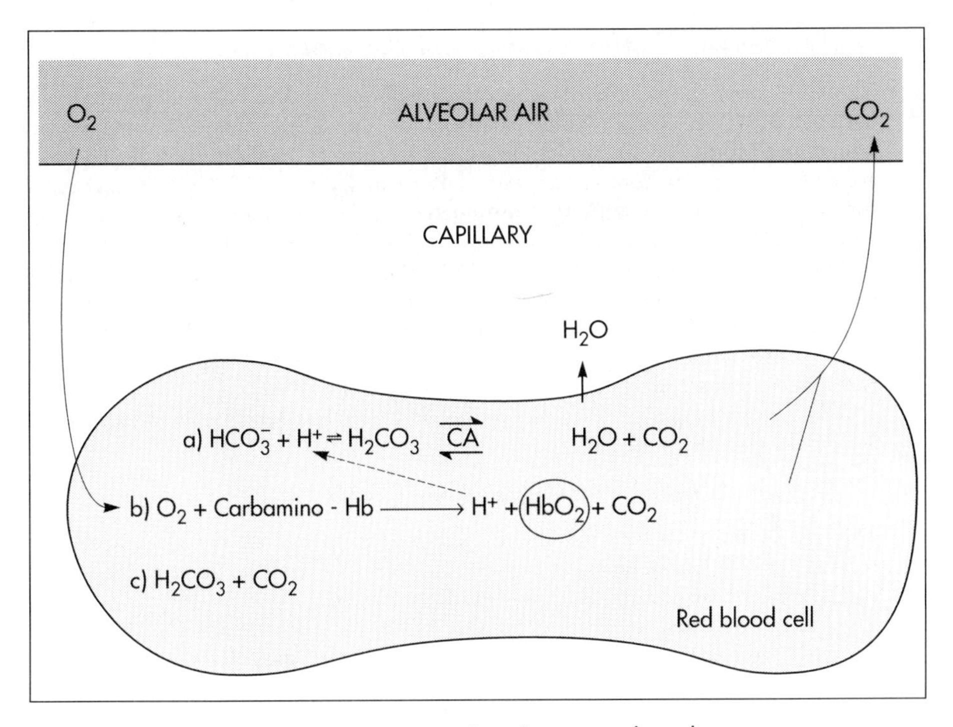

Fig. 33-1: Reactions in the plasma and erythrocytes.

This reaction is slow in plasma, but more than 10 000 times faster in the erythrocytes because of the presence of the enzyme **carbonic anhydrase** (CA), which catalyses the hydration of CO_2 to carbonic acid. The H^+ is buffered by the partially oxygenated haemoglobin. The bicarbonate diffuses out of the erythrocytes into plasma, thus preventing an accumulation, which would slow down the hydration of CO_2 according to the law of mass action. Of the total CO_2 in blood two-thirds occurs in the form of bicarbonate, and most of this is found in the **plasma sink** (Fig. 33-2) following exchange with Cl^- (Fig. 33-2).

The actual [bicarbonate] is calculated as follows:

$$[\textbf{Bicarbonate}] = [\textbf{Total CO}_2 - \textbf{carbamino CO}_2 - \textbf{dissolved CO}_2].$$

Arterial blood [bicarbonate]: 22.3 − 1 − 1.19 = **20.1 mM**.

Venous blood [bicarbonate]: 24 − 1.4 − 1.37 = **21.2 mM**.

How is CO_2 released from blood to alveolar air?

As CO_2 is removed from the blood, the equilibrium is shifted toward the formation of CO_2 from bicarbonate according to the law of mass action (Fig. 33-1). This process, plus the binding of oxygen to haemoglobin causes both CO_2 and H^+ to dissociate from haemoglobin (Fig. 33-1). The released H^+ then combines with bicarbonate to form more CO_2 for diffusion (Fig. 33-1). CA in the red cell speeds up the CO_2 formation.

What are the chloride–bicarbonate shift and the water shift?

1. The CO_2 from the cells enters the capillary blood, and is buffered primarily by haemoglobin (Fig. 33-2). Within milliseconds the erythrocyte CA catalyses its hydration to form H_2CO_3, which dissociates to form H^+ and bicarbonate. To avoid accumulation of bicarbonate, two out of three newly formed HCO_3^- is exchanged for Cl^- in plasma, through **anion antiporters** in the membrane (Fig. 33-2). In this way the membrane potential is sustained. This **bicarbonate–chloride shift** occurs within milliseconds and is vital for the CO_2 transport (Fig. 33-2). The **bicarbonate–chloride shift** occurs simultanously with, and probably independent of, the release of Na^+ and HCO_3^- from erythrocytes into plasma possibly by **Na^+-H^+ exchange.**
2. CO_2 is also rapidly bound to the amino groups of deoxyhaemoglobin to form carbamino-groups and H^+ within the erythrocyte (Fig. 33-2).

   When oxyhaemoglobin loses its O_2, it can accept H^+, which combines with the NH moiety on the imidazole ring:

   $$O_2.Fe.Hb\text{-}NH + H^+ \rightleftharpoons O_2 + Fe.Hb\text{-}NH_2^+.$$

   This allows more CO_2 to be hydrolysed and be carried in the **plasma sink** as bicarbonate (Fig. 33-2). Thus, at a given P_{CO_2} the CO_2 concentration at the venous end is greater than in arterial blood (the Haldane effect).
3. There is a small rise in $[CO_2+H_2CO_3]$ of 0.2 mM dissolved CO_2 (Fig. 33-2). Venous erythrocytes contain **more osmotically active particles** than arterial red

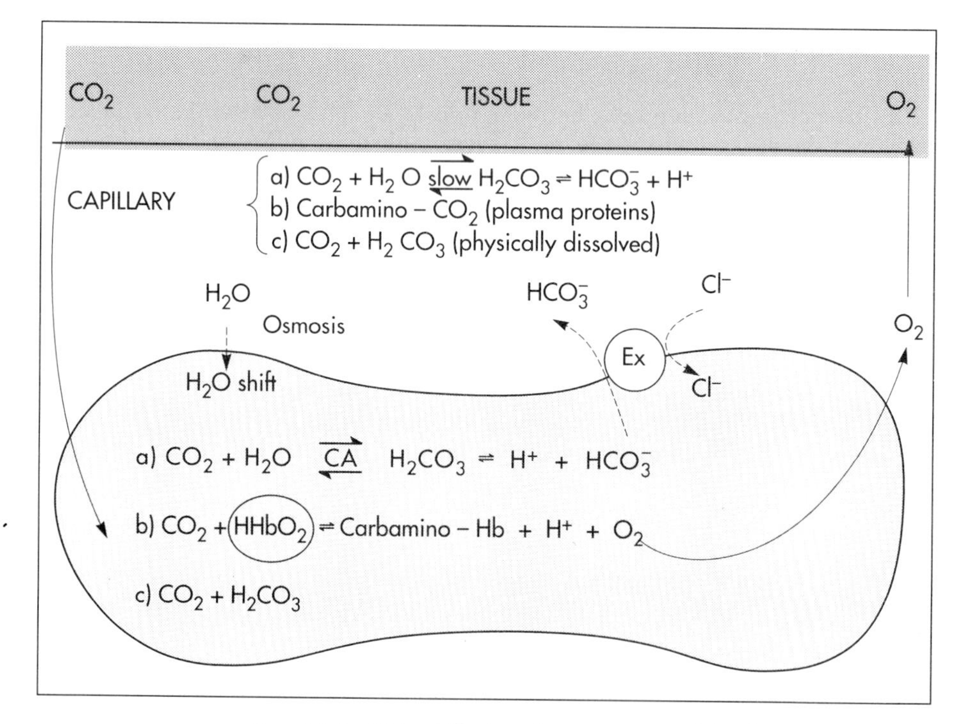

Fig. 33-2: CO_2 diffusion from tissue to blood.

cells. Consequently, they swell because **water from the plasma enters the venous erythrocytes by osmosis** (Fig. 33-2). The venous haematocrit packed cell volume (PCV) is larger than the arterial, but the advantage of this water transport is uncertain. This so-called **water shift** is almost solely due to the shift of anions, secondary to the establishment of a new Donnan equilibrium.

Is haemoglobin of vital importance for the CO_2 transport?

Without haemoglobin the transport would be minimal. With haemoglobin most of the **CO_2 transport is isohydric** (i.e. it takes place without any change in H^+ concentration). The actual rise in venous H^+ is also buffered by haemoglobin as by any buffer present. However, this function of haemoglobin is **not vital for the CO_2 transport.**

Further reading

Nunn, J.F. (1987) *Applied Respiratory Physiology.* 3rd edn. Butterworths, London.
West, J.B. (1989) *Respiratory Physiology*, 4th edn. Williams & Wilkins, Baltimore.

33. Multiple Choice Questions

Each of the following five statements have True/False options:

A. Most CO_2 in the blood is transported as bicarbonate.
B. Some CO_2 is transported dissolved in plasma and erythrocytes.
C. Some CO_2 is transported as bicarbonate in erythrocytes.
D. More than 1% of the CO_2 is transported as carbonic acid.
E. Some CO_2 is transported as carbamino compounds in plasma.

33. *Case History*

A female, 24 years of age, has a P_{AO_2} of 13.3 kPa and a F_{ACO_2} of 0.056. Her $P_{\bar{v}CO_2}$ is 46 Torr or 6.1 kPa and her $P_{\bar{v}O_2}$ is 42 Torr or 6.0 kPa. The solubility coefficients (α) are 0.022 and 0.51 ml STPD ml^{-1} for O_2 and CO_2, respectively. The barometric pressure is 101.3 kPa, and the water vapour tension in the alveolar air is 6.2 kPa.

1. *Calculate the partial pressure of CO_2 in the alveolar air.*
2. *Develop an equation showing the relation between gas concentration (C) and partial pressure (P) in a fluid (Henry's law). Does the solubility of a gas depend upon temperature in the fluid?*
3. *Calculate the concentration of O_2 and CO_2 physically dissolved in 1 litre of arterialized blood in equilibrium with the alveolar air of the above female.*
4. *Calculate the concentration of CO_2 and O_2 physically dissolved in her mixed venous blood.*

Try to solve the problems before looking up the answers in Chapter 74.

CHAPTER 34. ACID–BASE BALANCE

The H^+ concentration ($[H^+]$) of human arterial blood is 40 nM, corresponding to a pH of 7.40 (with a range of 7.36–7.44). Such values and a P_{aCO_2} of 5.3 kPa (40 Torr or mmHg) are found when an arterial blood sample is taken from a healthy person at rest. The low $[H^+]$ of the body fluids is maintained within a narrow range by the lungs and the kidneys: 16–160 nmol l^{-1} or pH 6.8–7.8.

The lungs eliminate the volatile acid (carbon dioxide), whereas the non-volatile or fixed acids are excreted renally (Chapter 60). The most important buffer system of the ECV is the **carbon dioxide–bicarbonate buffer.**

The kidneys reabsorb virtually all the HCO_3^- that is filtered through the glomerular barrier. The kidneys excrete just as much non-volatile acid as the amount produced and ingested.

Metabolic acid–base disorders are compensated partially by the lungs in a matter of minutes. **Primary pulmonary disorders** are compensated by the kidneys in a matter of days.

What is understood by the buffer equation?

$$\mathbf{pH = pK + log\ ([Base]/[Acid])}.$$

For the carbonic acid system ($CO_2 + H_2O \rightleftharpoons H_2CO_3 \rightleftharpoons H^+ + HCO_3^-$) the buffer equation is:

$$\mathbf{pH = pK + log([bicarbonate]/[dissolved\ CO_2])}.$$

This is the **Henderson–Hasselbalch Equation** – a logarithmic rearrangement of the mass action equation.

How is [dissolved CO_2] calculated?

Dissolved CO_2 (ml l^{-1} of blood) is in equilibrium with and equal to P_{CO_2} multiplied by the **solubility coefficient** (the Bunsen $\alpha = 0.51$ ml/ml per 760 Torr):

$$(0.51 \times 1000 \times P_{CO_2})/(760 \times 22.4) = (\mathbf{P_{CO_2} \times 0.03})\ \ \mathbf{mM\ (mmol\ l^{-1})}$$

with P_{CO_2} in Torr or mmHg, or with kPa as unit for P_{CO_2}:

$$0.03 \times 760/101.3 = 0.225;\ \text{equal to}:\ (\mathbf{P_{CO_2} \times 0.225})\ \mathbf{mM}.$$

Present the final form of the Henderson–Hasselbalch equation

For plasma of normal ion strength, temperature and $[H^+]$ the pK is 6.1.

$$pH = 6.1 + \log([bicarbonate]/(0.03 \cdot P_{CO_2}))\ \text{with}\ P_{CO_2}\ \text{in Torr or}$$

$$\mathbf{pH = 6.1 + log([bicarbonate]/(0.225 \cdot P_{CO_2}))\ with\ kPa}.$$

Why is our CO_2/HCO_3^- buffer system so important?

Theoretically, the **primary/secondary phosphate buffer system** should be the best, because its pK is 7.3 in contrast to 6.1 for the CO_2/HCO_3^- system. This later buffer system is important because it is abundant in the body. Moreover, the amount of buffer can be adjusted by respiration, so that P_{aCO_2} is maintained normal (the ratio $\mathring{V}_{CO_2}/\mathring{V}_A$ is equal to F_{ACO_2}). This is easily realized when comparing the addition of 1 mM of strong acid to a closed and an open system with 24 mM bicarbonate each (pH 7.4; P_{CO_2} 5.3 kPa).

In the closed system, the base concentration is reduced by 1 mM to 23, and the acid concentration is increased by 1 mM due to a high P_{CO_2}. Accordingly, the acid concentration is now: $([0.225 \cdot 5.3] + 1) = 0.225 \cdot P_{CO_2}$. Thus the new P_{CO_2} is 2.19/0.225 or 9.74 kPa. The pH of the closed system is changed to: **pH = 6.1 + log [23/(0.225 · 9.74)] or 7.12.** The **buffer capacity** is 1/(7.4 – 7.12) = 3.57 mM/pH unit.

In an open system such as the body, the ventilation simply eliminates excess carbon dioxide and the P_{CO_2} is kept constant: 5.3 kPa. Here, the new pH is: **pH = 6.1 + log [23/(0.225 · 5.3)] or 7.385.** The **buffer capacity** is 1/(7.4 – 7.385) = 67 mM/pH unit.

Draw a graph of the Henderson–Hasselbalch equation with decade equidistance (see Fig. 34-1 below).

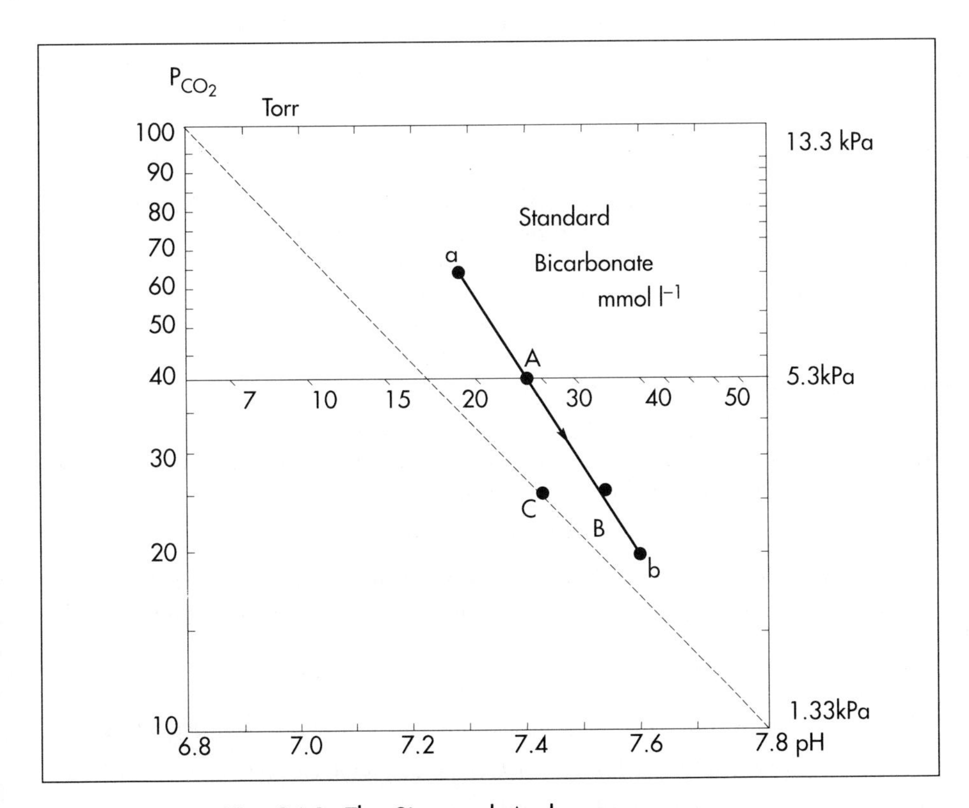

Fig. 34-1: The Siggaard–Andersen nomogram.

What is the slope for a curve showing the relation between pH and log P_{CO_2} for a 15 mM bicarbonate solution in water in Fig. 34-1?

$$\log P_{CO_2} = -\text{pH} + \text{k}.$$

Thus, y is a function of x, and the slope is **−1** (stippled line in Fig. 34-1). All lines with a slope of −1 are **isobicarbonate lines.**

The curves for plasma and whole blood (line a–b, Fig. 34-1) have a steeper slope. Why?

These fluids contain not only bicarbonate, but several other buffers. The blood plasma contains protein anions and phosphate, and whole blood also contains haemoglobin. Since the **buffering power** is the sum of the effect of all buffers present, it is clear that the buffering power must increase, and this is shown by the steeper slope of the buffer line for whole blood (Fig. 34-1).

The normal buffer base (**NBB**) concentration in plasma is **42 mM** (24 mM bicarbonate, 1 mM secondary phosphate, and 17 mM plasma proteins). The NBB concentration in whole blood is **48 mM** (20 mM bicarbonate, 0.5 mM secondary phosphate, 9.5 mM plasma proteins, and 18 mM haemoglobin anions).

What is standard-bicarbonate (SB) and how is SB measured?

SB is the [bicarbonate] in a blood or plasma sample of body temperature, which has been fully oxygenated and equilibrated at normal P_{CO_2} (5.3 kPa or 40 Torr).

Is SB clinically important?

Changes of SB from the **normal 24 mM** in plasma, show that the acid–base disorder is **metabolic**, because any respiratory disturbance is not reflected by SB.

The blood is simply equilibrated at normal standard P_{CO_2}.

The deviation from normal depicts the amount of fixed acid or base, which the bicarbonate–carbonic acid buffer has buffered in each litre of plasma.

What is base excess and base deficit in plasma?

The alternative to SB in plasma is **base excess (BE)/base deficit (BD)**, which is the buffer base (BB) of plasma minus NBB. BB is the **actual concentrations of bicarbonate, phosphate, and plasma proteins.** Accordingly:

$$\textbf{BE} = \textbf{BB} - \textbf{NBB}.$$

In healthy individuals NBB in plasma is 42 mM, of which 24 are bicarbonate. BE depicts the total surplus of base in plasma and BD depicts the total surplus of acid in plasma.

BE and BD are easy to use in clinical work and therapy. **SB depicts only a part of the deviation.**

In lung diseases with hypoventilation, the following equation is shifted towards right:

$$CO_2 + H_2O \rightleftharpoons H_2CO_3 \rightleftharpoons H^+ + HCO_3^-.$$

BBs different from HCO_3^- bind H^+ and form acids.

Thus, for each molecule of alternative BB eliminated, one molecule of HCO_3^- is produced, and the BB concentration is unaltered. Lung diseases with hyperventilation shift the equation to the left. For each HCO_3^- eliminated, alternative buffers liberate one H^+ from their acid form and produce one BB molecule. The total BB concentration is unaltered.

How is BE/BD measured?

BE/BD is the **amount of strong acid or base, which must be added to 1 l of blood to bring pH to 7.4.**

The blood sample is at body temperature, fully saturated with oxygen in an equilibrium with normal P_{CO_2} 5.3 kPa (40 Torr).

Based on the measured BE/BD, the absolute base surplus or deficit for the total extracellular fluid (ECF) can be calculated. The ECF (including the erythrocytes) is assumed to be 20% of the body weight.

What is the dietary intake of acids and bases?

Persons on a high-protein diet metabolize some **50–100 mmol fixed** or non-volatile acids daily (organic acids, sulphuric acid and phosphoric acid).

Diets rich in vegetables and fruits generate non-volatile bases (salts of organic acids such as oxalate and citrate), which must also be excreted by the kidneys. Normally, we absorb 30 mmol base daily from the gut, but more in vegetarians. In the body only amino acids are metabolized to bases (ammonia). See Chapter 60.

How is bicarbonate reabsorbed in the renal tubules?

Already in the proximal tubules we reabsorb 90% of the filtered bicarbonate (4000 mmol daily) by means of **H^+ secretion**. Most of the H^+ secreted in the proximal tubules is derived from the **Na^+–H^+ exchange** and the rest is mediated by a **H^+–ATPase**. See also Chapter 60 for the function of CA and CA-inhibitors.

What are the two other buffer systems reacting with the H^+ secreted? How do we excrete an acid surplus?

See Chapter 60.

What is acidosis and alkalosis?

Acidosis and alkalosis are abnormal conditions characterized by proton surplus and proton deficit, respectively (Fig. 34-2).

Clinicians measure pH and P_{aCO_2} in the arterial blood. **Acidaemia–alkalaemia** are conditions with blood pH below 7.35 and above 7.45, respectively.

What is metabolic acidosis?

The proton surplus reduces bicarbonate and also SB. **Metabolic** acidosis is diagnosed by a **SB below normal** (or **negative BE, a BD**). This is condition 2 in Fig. 34-2 (SB below 20 mM). The high $[H^+]$ in a sustained condition stimulates ventilation, so P_{ACO_2} and

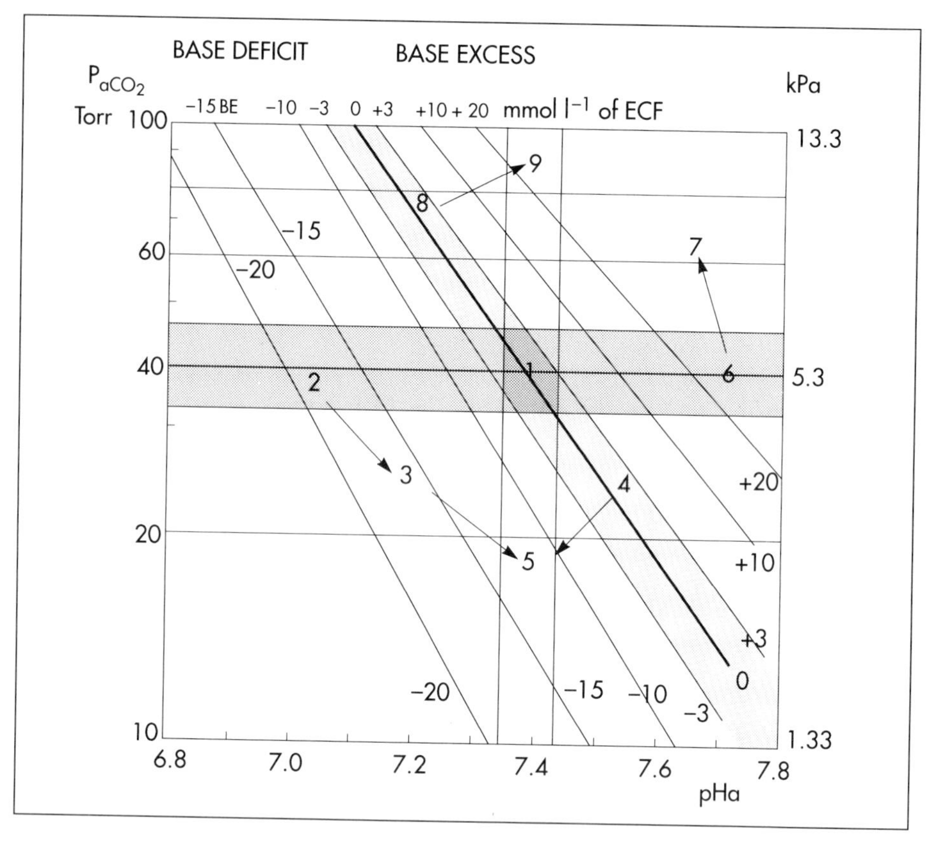

Fig. 34-2: The 9 van Slyke conditions.

P_{aCO_2} are reduced by the **hyperventilation**. This is compensated metabolic acidosis (condition 3 in Fig. 34-2). Strong acids can accumulate because of excess production (diabetic ketoacidosis, lactic acidosis) or by impaired excretion. The high ventilation with deep and frequent respirations is called **Kusmaul breathing** in diabetic coma.

Metabolic acidosis increases $\mathring{V}_{CO_2}$, R$\mathring{Q}$ (R), P_{50}, and P_{aO_2}, whereas $(P_{AO_2} - P_{aO_2})$ simultaneously decreases. The smaller oxygen tension gradient is caused by the Bohr effect and by an improved $\mathring{V}_A/\mathring{Q}$ ratio.

What is metabolic alkalosis?

This condition is caused by a **primary fall in the proton/bicarbonate** ratio in the ECV. Loss of gastric fluid (vomiting HCl-containing fluid) and use of diuretics can cause this form of alkalosis. The low $[H^+]$ reduces ventilation and increases bicarbonate and thus SB (and BE). This is condition 6 in Fig. 34-2 (SB above 27 mM and positive BE). The **hypoventilatory compensation** raises P_{aCO_2} and pH is reduced, but the compensation is never total, since the rise in P_{aCO_2} and the fall in P_{aO_2} limit the fall in ventilation. The **respiratory compensated metabolic alkalosis** is condition 7 in Fig. 34-2.

What is respiratory acidosis?

This condition is caused by a **primary fall of the alveolar ventilation** (asthma, chronical bronchitis with emphysema, tuberculosis, lung cancer) **relative to a maintained or increased $\mathring{V}_{CO_2}$. Impaired respiratory pump muscles** (neuromuscular diseases such as poliomyelitis) and **ventilatory depressing drugs** (such as narcotics and anaesthetics) cause hypoventilation and elevate P_{aCO_2} (hypercapnia). This is condition 8 in Fig. 34-2 with a P_{aCO_2} above 6.4 kPa or 48 Torr. The accumulated CO_2 increases the H^+ secretion in the renal tubules, and maximal renal compensation is reached within 5 days (often with a normal pH). This is **totally compensated respiratory acidosis** or condition 9 in Fig. 34-2. SB is normal, and BE is zero. An acute respiratory acidosis, caused by breath-holding before underwater swimming, can develop into a life-threatening condition (Chapter 41).

What is respiratory alkalosis?

The essential event is a **too high alveolar ventilation** (disproportionate to CO_2 production), causing excess clearance of CO_2, which results in reduced P_{ACO_2} (hyperventilation with hypocapnia). The standard bicarbonate is normal. This is condition 4 in Fig. 34-2 with a P_{aCO_2} below 4.4 kPa or 33 Torr. The most common causes are lung diseases and other conditions with severe hypoxaemia such as **high altitude**, pregnancy, and hysteria.

The low P_{ACO_2} reduces the tubular H^+ secretion, and after a few days the kidneys compensate completely for the respiratory alkalosis (normal pH). This is **totally compensated respiratory alkalosis**, or condition 5 in Fig. 34-2.

Is there a fixed pattern in acid–base compensations?

Primary **respiratory disturbances are compensated renally** (i.e. by changes of the H^+ excretion in the urine after a few days).

Primary **metabolic disturbances are compensated by respiratory means** (i.e. by changes of $\mathring{V}_A$ – and thus pulmonary CO_2 output – after a few minutes).

Further reading

Astrup, P. and J.W. Severinghaus (1986) *The History of Blood Gases, Acids and Bases.* Munksgaard, Copenhagen.

Eldridge, F.L., J.P. Kiley and D.E. Milhorn (1985) Respiratory responses to medullary hydrogen ion changes in cats: different effects of respiratory and metabolic acidoses. *J. Physiol.* **358**: 285–97.

Frans, A., T. Clerbaux, E. Willems, and F. Kreuzer (1993) Effect on metabolic acidosis on pulmonary gas exchange of artificially ventilated dogs. *J. Appl. Physiol.* **74** (5): 2301–8.

Siggaard-Andersen, O. (1974) *The Acid–Base Status of the Blood.* Munksgaard, Copenhagen.

34. Multiple Choice Questions

A patient who has been vomiting for several days is in hospital with the following findings in a sample of arterial blood: pH = 7.58; [standard bicarbonate] = 39 mM; and P_{aCO_2} = 51 Torr. Each of the following disorders and statements have True/False options:

A. Compensated respiratory alkalosis.
B. Metabolic alkalosis.
C. Hypoventilation.
D. Increased [standard bicarbonate].
E. CO_2 accumulation.

34. Case History

A female patient, 25 years of age, is undergoing a routine examination at an outpatient ward. Her C_{aCO_2} is 500 ml STPD l^{-1}, and P_{aCO_2} is 4.9 kPa (37 Torr). The patient is suffering from a disorder releasing acute attacks of anxiety, panic, dyspnoea and tetanic cramps.

A week later the patient is admitted to the emergency ward of the hospital with severe hyperpnoea and tetanic cramps. She has abnormal blood gas tensions: P_{aCO_2} 2.66 kPa (20 Torr), P_{aO_2} 14.6 kPa (110 Torr), and C_{aCO_2} is 21 mmol l^{-1}.

1. *Calculate the pH of the arterial blood at the routine check, and describe the normal condition of this patient.*
2. *Calculate the pH of the arterial blood during emergency conditions and describe her acid–base status.*
3. *What is the most likely diagnosis?*

Try to solve the problems before looking up the answers in Chapter 74.

CHAPTER 35.
GAS EXCHANGE IN THE LUNGS

Blood passing the pulmonary capillaries of a healthy person is rapidly equilibrating with the alveolar air. Oxygen from the air diffuses into the blood and binds reversibly with haemoglobin. The **normal oxygen capacity is 200 ml STPD l^{-1} of blood** (150 g haemoglobin l^{-1} carrying 1.34 ml STPD g^{-1}).

What are the six layers that alveolar gas must cross to enter erythrocyte binding sites?

The six layers of the **alveolar–capillary barrier** are: (1) a fluid layer containing **surfactant**; (2) the **alveolar epithelium** with its basement membrane; (3) a **fluid-filled interstitial space**; (4) the **basement membrane and capillary endothelium**; (5) the **blood plasma**; and (6) the **erythrocyte membrane.**

At the dead end of the narrow tube system with 23 branch points there are 300 million tiny blind end sacs (alveoli) in the lungs. Fortunately, the alveoli are flushed with fresh air as we breathe, and this bulk flow reaches the alveolar lining. The alveolar air would stagnate if the bulk flow stopped before, because the maximum diffusion pathway is always limited (less than **1 μm**).

Define the law that governs the pulmonary diffusive flux (J_{gas} mol/time) of gas

Fick's law of diffusion states that the flux of gas transferred across the alveolar–capillary barrier is directly related to the **solubility** (The **Bunsen solubility quotient** α) of the gas, the diffusion area of the pulmonary capillaries (A), the length of the diffusion pathway from the alveoli to the blood (L), and the driving pressure ($P_1 - P_2$):

$$J_{gas} = (D\,\alpha\,A\,1/L)\,(P_1 - P_2).$$

Although A is close to the size of half a tennis court, and the diffusion distance (L) is 0.5 – 1 μm, it is difficult to predict their size in different individuals. Therefore, Krogh developed **the individual lung diffusion capacity (D_L)** defined as the flux of gas transferred per pressure unit through the lung barrier of a certain person. Krogh incorporated solubility (Bunsens α), molecular weight (mol.weight), A, and L in her **lung diffusion constant (D_L). D_L is equal to a constant, K, multiplied with the solubility (α), and divided by the square root of the mol.weight** (see also Chapter 32). **Thus $D_L = K\,\alpha/\sqrt{\text{mol.weight}}$,** or for oxygen: $D_{LO_2} = K\,0.022/\sqrt{32}$.

Hereby she eliminated all the unknown variables, and for oxygen the above equation is simplified to:

$$J_{gas} = \mathring{V}_{O_2} = \Delta P_{O_2} \times D_{LO_2}.$$

Diffusion is rapid, because diffusion is a process with an extremely **small distance**

capacity. In normal lungs there are **transbarrier pressure gradients** in the correct directions for diffusion of both O_2 and CO_2. $\mathbf{D_L}$ is measured by measuring the carbon monoxide (CO) uptake and the driving pressure. Values for healthy persons are calculated later in this chapter.

Is the pulmonary exchange of gases diffusion or perfusion limited?

A gas such as **CO** that does not equilibrate across the **lung barrier**, such that its pressure gradient is maintained while the blood is still in transit through a pulmonary capillary, is purely **diffusion limited**, and is not dependent on the bloodflow ***per se*** (i.e. perfusion).

Gases, such as hydrogen, nitrogen, nitrous oxide, tritium, ^{133}Xe, and anaesthetic gases, that equilibrate rapidly across the **alveolo–capillary barrier** are **perfusion limited**. Soon after the blood has entered the pulmonary capillary, the **partial pressure for the gas in the blood becomes equal to that in the alveoli**. No additional diffusion of this gas can occur during the remaining transit of the blood in the pulmonary capillary.

The pulmonary transfer of O_2 and CO_2 is **perfusion limited** over a wide range of activity levels. Even though CO_2 is 24-times more soluble in water and diffuses 20 times faster through water than does O_2, the two gases have essentially the same pulmonary equilibration times (which is 0.25 s when blood has passed one-third of the way through the capillary). The equilibrium time is due to the small pressure gradient for CO_2 and the time needed for conversion of bicarbonate and carbamino-compounds to dissolved CO_2.

A perfect matching between alveolar air and capillary blood as described above does not occur throughout the whole lung. **Even in healthy persons alveolar ventilation/perfusion inequalities disturb the ideal exchange.**

What is the most frequent cause of low P_{aO_2} in patients with cardiopulmonary disease?

Alveolar ventilation/perfusion ratio ($\mathring{V}_A/\mathring{Q}$) inequality.

How much can $\mathring{V}_A/\mathring{Q}$ vary in the normal lung?

Normally, ventilation ($\mathring{V}_A$) and perfusion ($\mathring{Q}$) are matched and $\mathring{V}_A/\mathring{Q}$ is **between 0.8 to 1.2 with normal alveolar and blood gas tensions**. In the normal upright lung the $\mathring{V}_A/\mathring{Q}$ ratio is approximately 0.6 at the base and about 3 at the apex region. The **pulmonary bloodflow decreases** linearly from the base to the apex of the lung. Likewise, ventilation of the lung is also greatest in the basal part. Thus the regional ventilation–perfusion ratio ($\mathring{V}_A/\mathring{Q}$) varies from **zero at the basal region to infinity at the apical region of the upright lung**. At the base regional $\mathring{V}_A$ approaches zero, and at the top of the lung regional $\mathring{Q}$ approaches zero. Every regional deviation from the average total $\mathring{V}_A/\mathring{Q}$ in healthy subjects, will result in alveolo-arterial gas tension differences (Fig. 40-3).

What are the physiological factors causing mismatch of $\mathring{V}_A/\mathring{Q}$?

Gravity. Alveolar ventilation and bloodflow decrease from the base to the apex in the upright lung. The relative decrease in bloodflow exceeds that for alveolar ventilation. Consequently, there is a continuum of $\mathring{V}_A/\mathring{Q}$ ratios from the base to the apex. Lung regions at the base with low $\mathring{V}_A/\mathring{Q}$ have low P_{AO_2} and high P_{ACO_2}, relative to normal mean values. Upper lung regions with **high $\mathring{V}_A/\mathring{Q}$** have relatively **high P_{AO_2} and low P_{ACO_2}.**

Anatomic venous-to-arterial shunts. Normally, up to 5% of the venous return passes directly into the systemic arterial circulation. This shunt-blood includes nutrient bloodflow coming from the upper airways and collected by the bronchial veins. Also the coronary venous blood that drains directly into the left ventricle through the Thebesian veins is shunt-blood.

How does $\mathring{V}_A/\mathring{Q}$ inequality affect the blood gases?

Healthy and ill people have **three types** of alveoli: (1) Normal alveoli that are **both ventilated and perfused; (2) alveoli with ventilation but without blood supply**; and (3) **alveoli with blood perfusion but no ventilation.** The last two types of alveoli do not exchange gases with the blood.

1. This represents the normal situation in which $\mathring{V}_A$ and $\mathring{Q}$ are matched ($\mathring{V}_A/\mathring{Q} = 1$), P_{aCO_2} is 40 Torr or 5.3 kPa, and P_{aO_2} is 108 Torr or 14.4 kPa (Fig. 40-3).
2. In alveoli with wasted ventilation (i.e. alveolar dead space), because there is no bloodflow, the **$\mathring{V}_A/\mathring{Q}$ ratio approach infinity**, and alveolar gas tensions approach those found in **normal inspired air.** Regions of infinite $\mathring{V}_A/\mathring{Q}$ can occur anywhere in subjects with pulmonary vascular occlusive disease (Fig. 40-3).
3. In alveoli with perfusion but no ventilation, the $\mathring{V}_A/\mathring{Q}$ ratio is zero, and blood gas tensions approach those found in normal venous blood. Such regions have a **functional venous-to-arterial shunt** (Fig. 40-3).

Can we compensate for $\mathring{V}_A/\mathring{Q}$ mismatch?

Low P_{AO_2} in poorly ventilated alveoli, causes **arteriolar constriction,** which redistributes bloodflow to well-ventilated alveoli.

Low P_{ACO_2} in alveolar regions with a **high $\mathring{V}_A/\mathring{Q}$ ratio,** constricts the small airways leading to these alveoli. Their reduced ventilation results in redistribution of gas to alveoli with better bloodflow.

Both of these compensatory mechanisms create a **better match between regional ventilation and perfusion.**

Why is CO the gas chosen for measuring lung diffusion capacity (D_L)?

D_L is defined as the volume of gas diffusing through the lung barrier of a given person per min and per pressure unit. **$D_L = \mathring{V}/\Delta P$.**

Since the counter pressure of CO in the blood is virtually zero, a simple measure of **P_{ACO} provides us with the pressure gradient.** The **standard affinity of the haemoglobin–CO reaction is very large and 250 times greater than that of O_2.** The **standard affinity** is equal to the **reciprocal value of $P_{0.5}$.** The $P_{0.5}$ for haemoglobin–CO is just a fraction of 1 Torr, and the haemoglobin–CO dissociation curve is almost equal to the ordinate of Fig. 35-1 (hidden in the line).

D_L consists of a **barrier-factor** and a **haemoglobin-factor** (which reflects the binding rate of oxygen to haemoglobin). Haemoglobin permits blood to absorb 65-fold as much O_2 as the content in plasma at normal P_{aO_2}.

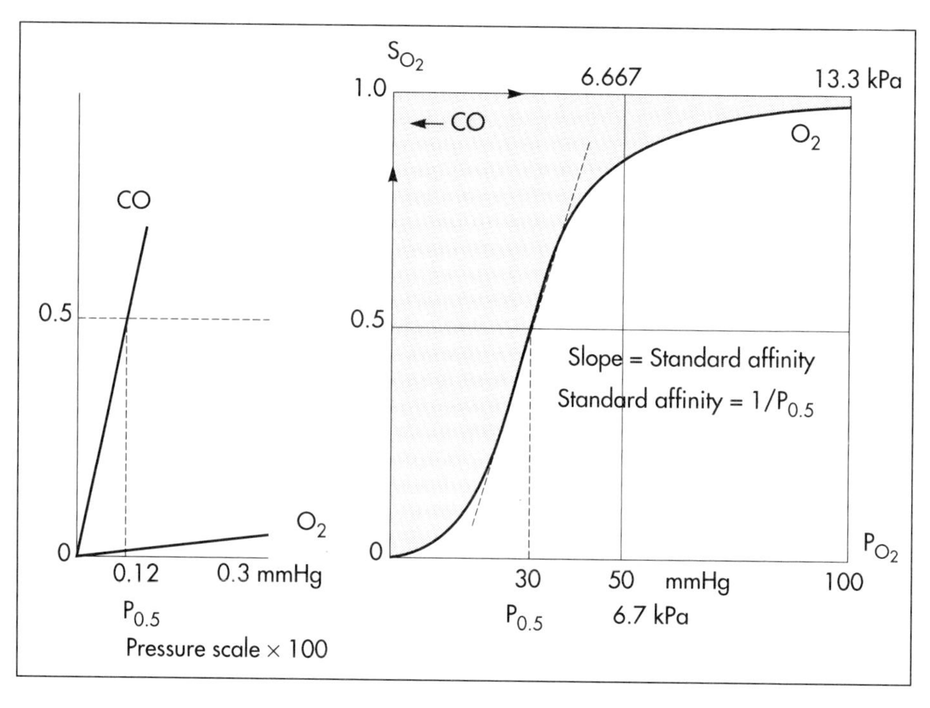

Fig. 35-1: Dissociation curves for oxy- and CO-haemoglobin.

How is D_{LO_2} and D_{LCO_2} calculated, when D_{LCO} is known?

The lung diffusion capacity for a certain gas is directly proportional to its solubility coefficient (α) and inversely proportional to the square root of its molecular weight:

$$D_{LO_2} = K \times 0.022/\sqrt{32};\ D_{LCO_2} = K \times 0.51/\sqrt{44};\ \text{and } D_{LCO} = K \times 0.018/\sqrt{28}.$$

$$D_{LO_2}/D_{LCO} = [K \times 0.022/\sqrt{32}]/(K \times 0.018/\sqrt{28}) = 1.14.$$

$$D_{LCO_2}/D_{LCO} = [K \times 0.51/\sqrt{44}]/(K \times 0.018/\sqrt{28}) = 22.6.$$

A mean value for **D_{LCO} for healthy persons at rest is 3.1 ml STPD s^{-1} kPa^{-1}** (25 ml CO min^{-1} $mmHg^{-1}$). This mean value implies a mean D_{LO_2} of **3.6 (29) of these units,** and a mean **D_{LCO_2} of 70 ml s^{-1} kPa^{-1} (565 ml min^{-1} $Torr^{-1}$) at rest.** During exercise the **average D_{LO_2} increases to 9 ml STPD s^{-1} kPa^{-1}** (72 ml STPD min^{-1} $Torr^{-1}$).

How does CO affect the O_2 transport?

CO competes with O_2 for binding sites on haemoglobin, and thus exposure to CO **reduces the O_2 binding to haemoglobin.** Persons with a normal blood [haemoglobin] breathing traces of CO block a large part of all binding sites by CO. The CO binding causes a leftward shift of the oxy-CO-haemoglobin dissociation curve. All the binding sites that are bound to CO, do not respond to falling P_{aO_2}. The **remaining O_2 molecules on the CO–haemoglobin molecule are much more avidly bound and unload slower than normal.**

By what means does D_{LO_2} and $\mathring{V}_{O_2}$ increase during exercise?

$$D_{LO_2} = \mathring{V}_{O_2} / \Delta P_{O_2}.$$

Accordingly:

$$\mathring{V}_{O_2} = \Delta P_{O_2} D_{LO_2}.$$

The increase in ΔP_{O_2} from 1.6 kPa at rest to 9.4 kPa (12 to 70 Torr) at maximum work is caused by a rise in P_{AO_2} towards 17.3 kPa (130 Torr), whereas $P_{\bar{v}O_2}$ falls towards zero.

D_{LO_2} also increases, because the **number of open lung capillaries is increased**, the surface area increases and the **barrier-thickness is reduced.**

In addition, O_2 transport is boosted further by the **rise in $\mathring{Q}$** from 5 to 30–40 l of blood min^{-1}.

What limits the $\mathring{V}_{O_2}max$ for an exercising person?

The **mass balance principle** provides an overview:

$$\mathring{V}_{O_2}\mathbf{max} = \mathring{Q}\mathbf{max}(C_{aO_2} - C_{\bar{v}O_2}).$$

Both the **maximum cardiac output** and the **maximum arterio-venous O_2 content difference** are limiting factors. Healthy persons have redundant ventilation and diffusion capacity in their lungs imposing no limitation.

Oedema or interstitial pulmonary fibrosis leads to thickening of the alveolar–capillary barrier, which will impede O_2 exchange. The reason is that the pulmonary vascular volume is reduced (reduced capillary transit time), and thus the diffusion equilibrium point is moved towards the end of the capillary. If patients with lung diseases try to exercise, this problem is further aggravated by the still more reduced capillary transit time. Thus exercise would impose a significant **diffusion limitation** on O_2 transfer.

Further reading

Cherniack, N.S., M.D. Altose and S.G. Kelsen (1983) The respiratory system. In: Berne, R.M. and M.N. Levy (Eds) *Physiology*. Sect. VI. C.V. Mosby, St Louis.
Nunn, J.F. (1987) *Applied Respiratory Physiology*, 3rd edn. Butterworths, London.
West, J.B. (1989) *Respiratory Physiology*, 4th edn. Williams & Wilkins, Baltimore.

35. Case History

A male, age 23 years and weight 70 kg, is breathing atmospheric air with traces of CO at 1 atmosphere. The man is at rest, and has an arteriovenous oxygen content difference of 50 ml l^{-1}. An arterial blood sample obtained after equilibrium between alveolar air and pulmonary blood is analysed with the following results: P_{aO_2} 13.3 kPa (100 Torr), C_{aO_2} 170 ml STPD l^{-1}, C_{aCO} (the concentration of CO in the blood) 28.3 ml STPD l^{-1}, and the [haemoglobin] 9.18 mmol l^{-1} (148 g l^{-1}). The standard affinity between haemoglobin and CO is 260 times greater than the standard affinity between haemoglobin and oxygen. The binding capacity for oxygen and CO is 1.34 ml STPD g^{-1} of hemoglobin.

1. *Define the concept standard affinity and P_{50}.*
2. *Calculate the dry CO-fraction in the alveolar air (F_{ACO}).*
3. *Is the oxygen supply to the tissues at the venous capillaries better for this CO-poisoned person than for a similar person without CO-poisoning, but with severe anaemia (haemoglobin 6.6 mmol l^{-1})?*

Try to solve the problems before looking up the answers in Chapter 74.

CHAPTER 36. PULMONARY BLOOD FLOW

Pulmonary vascular resistance (PVR) is minimal compared to that of the systemic circulation. The pulmonary vascular system is a low pressure system, with a bloodflow sensitive to gravity and to P_{AO_2}.

Is the pulmonary circulation fit to meet metabolic demands?

This part of the circulatory system is **basically a low-pressure, low-resistance, highly compliant vessel system,** which is **meant to accommodate the entire $\mathring{Q}$ during gas exchange** with the alveolar air – and not meant to meet special metabolic demands as in the case of the systemic circulation.

What are the pressures in the pulmonary vascular system?

The pressure in the right ventricle is **3.3 kPa systolic and −0.133 kPa diastolic (25/−1 Torr)** in a healthy, supine person at rest. The pressure in the pulmonary artery is about **3.3 kPa systolic and 1 kPa diastolic (25/8 Torr),** with a mean of 1.7 kPa (13 Torr or mmHg). The blood flow of the pulmonary capillaries pulsates and its mean pressure is **below 1 kPa.** The pressure in the left atrium is 0.7 kPa (5 Torr). This value implies a pressure drop across the pulmonary circulation of (1.7 – 0.7) =1 kPa or 8 Torr. This blood pressure gradient (ΔP) or **driving pressure** is **less than 1/10 of the systemic driving pressure.**

The walls of the pulmonary vessels are thin, hence their pressure must fall at each inspiration, because the intrapulmonic pressure falls.

Change of posture from supine to erect position will reduce the pressure toward zero in the apical vessels, whereas it increases the pressure in the basal vessels due to gravity.

When the **driving pressure** in the apical blood vessels approaches zero, the blood flow will also approach zero. Other than its implication for gas exchange, this phenomenon limits the supply of nutrients, so that lung disorders often occur in the apical regions.

Is the pulmonary vascular resistance equal to the systemic?

No. The **PVR** is the ratio between the pressure gradient and the bloodflow. A **peripheral resistance unit (PRU)** is the number of Torr ml^{-1} blood s^{-1}. The basic equation is: **PVR (PRU) = ΔP (Torr)/flow (ml s^{-1}).** At rest, the pulmonary driving pressure is 8 Torr and the bloodflow is 5 l min^{-1} (83 ml s^{-1}). The ratio is 8/80 = **1/10 PRU,** which shows that normal PVR is only 10% of the systemic resistance at rest (TPVR = 1 PRU). Calculated in kPa the **PVR is 1/80 kPa s ml^{-1}.** Such low values for PVR are only found in the lungs of healthy, non-smokers.

The **PVR** remains low in healthy persons, even when $\mathring{Q}$ increases to 30 l min^{-1}, because of the large storage capacity of pulmonary vessels. Stretch receptors, found in the left atrium and in the walls of the inlet veins, are believed to be stimulated by distention. Such a distention inhibits liberation of **vasopressin (antidiuretic hormone, ADH)** from the **posterior pituitary** (Chapter 64). When the vessels become distended by blood, the urine volume increases and the ECV decreases due to the low ADH level.

What factors can change pulmonary vascular resistance?

Changes in **PVR** are achieved mainly by **passive** factors, but also by **active modification**.

The larger arteries and veins are located outside the alveoli (extra-alveolar); they are tethered to the elastic lung parenchyma, and are exposed to the intrathoracic or intrapleural pressure. The pulmonary capillaries lie between the alveoli (intra-alveolar vessels) and are exposed to intrapulmonic or alveolar pressure.

1. **Alveolar volume.** The intra-alveolar vessels are wide open at low alveolar volumes, so that their PVR must be minimal. With increasing alveolar distention these vessels are compressed. This increases their **intra-alveolar PVR**. However, at low alveolar (lung) volumes, the extra-alveolar vessels are small because of the small transmural vascular pressure gradient, and their PVR is high. With increasing lung distention, the intrathoracic pressure becomes more subatmospheric. This elevates the transmural vascular gradient and is coupled with the radial traction on these vessels by the surrounding lung parenchyma as it expands. Thus, the **extra-alveolar PVR decreases.**

   The greatest cross-sectional area exists in the **many intra-alveolar vessels, hence increasing PVR in these vessels offsets decreased extra-alveolar PVR**. Thus, total pulmonary vascular resistance is increased at higher alveolar volumes when intra-alveolar PVR is high. PVR is minimal at functional residual capacity (FRC), where there is air enough to open the extra-alveolar vessels with minimal closure of the intra-alveolar vessels.
2. **Pulmonary artery pressure.** A healthy person at rest (FRC) has approximately **half of the pulmonary capillaries open**, but with increasing arterial pressure, the previously closed capillaries open (recruitment). As the arterial pressure continues to rise, the capillaries become distended. The net effect is a rise in the total cross-sectional area of the lung capillaries, leading to decreased PVR.
3. **Left atrial pressure.** Patients with **high left atrial pressure** have distended capillaries due to the venous back pressure. As a result of the reduced driving pressure their PVR is decreased further.
4. **Gravity.** The lung bloodflow per unit lung volume is greatest at the base of the lung and decreases towards the apex. Gravity creates a gradient of vascular pressures from the top to the bottom of the lungs. The intravascular pressure is much lower at the apex than at the base of the lung, unlike the **intrapulmonic, alveolar pressure which is essentially constant throughout the lung**. At the top of the lung all vascular pressures can approach zero (with the intrapulmonic, alveolar pressure as reference). No blood would flow through the apical region, and if it is still ventilated, it would be considered **alveolar dead space**.

Are the pulmonary vessels controlled actively?

The pulmonary blood vessels are sparsely innervated by both sympathetic and parasympathetic fibres. **Sympathetic stimulation** constricts the pulmonary vessels, whereas **parasympathetic stimulation** dilates them. **Vasoconstrictive agents** include: arachidonic acid, catecholamines, leucotrienes, thromboxane A, prostaglandin F, angiotensin-II and serotonin. The **vasodilatators** are ACh, bradykinin, nitric oxide (NO) and prostacyclin.

A **decrease in** P_{AO_2} in an occluded region of the lung produces hypoxic vasoconstriction of the vessels in that region. The reduced P_{AO_2} causes constriction of the precapillary muscular arteries leading to the hypoxic region. The hypoxic effect is **not nerve-mediated**. This reaction shifts blood away from poorly ventilated alveoli to better-ventilated ones. NO seem to dilate the vessels of the well-ventilated segments of the lung. Perfusion is hereby matched with ventilation.

Describe the pathogenesis of lung congestion and lung oedema

Oedema is particularly serious in the lungs because it widens the diffusion distance between the alveolar air and the erythrocytes. There is not enough time for oxygen to travel from the air to the individual erythrocyte. Thus, the blood leaving the lungs is only partially oxygenated. Both the vital capacity (VC) and the compliance is reduced.

1. **Increased pressure.** Patients with **left cardiac failure** (acute myocardial infarct, chronic myocardial failure, mitral stenosis, aortic stenosis, hypertension) can drown in their own plasma transudate. The increased venous back pressure distends all pulmonary vessels (lung congestion), and as soon as the pulmonary capillary pressure is higher than the **colloid osmotic pressure** (normally 3.3 kPa or 25 Torr), there is a filtration of plasma water into the pulmonary interstitial tissues and into the alveoli. The pulmonary vascular pressure rises in the supine position causing attacks of lung oedema to occur at night.
2. **Increased capillary permeability.** Lung oedema can be caused by capillary damage with chemical warfare agents, toxins, pneumonia, etc.
3. **Reduced plasma [protein]** increases net filtration at the arteriole end of the lung capillary and reduces net reabsorption of filtered fluid at the venule end.

Describe the treatment of lung oedema and its mechanisms

Primarily, it is important to find the cause, such as left cardiac failure, and correct the disorder.

Patients with **chronic cardiac failure** have reduced contractility, which improves by continuous treatment with digoxin.

Patients with **lung oedema** must sit up erect in bed, and calm down. This improves $\mathring{Q}$ and the effective filtration pressure is reduced.

Breathing of **air enriched with oxygen** reduces hypoxia and dilates the lung vessels. The filtration pressure is reduced.

Effective diuretics **increase the excretion of Na^+** and thus of water via the kidneys. The loss of fluid also implies oedemal fluid.

Positive pressure breathing is thought to minimize the difference between the central and the peripheral venous pressure, so the venous filling and thus $\mathring{Q}$ is reduced. The blockade of lung capillary blood flow in the overpressure-phase, and the fear of the patient (increases $\mathring{Q}$), does not make this treatment the best choice. The effect is probably similar to the earlier application of blood-letting tourniquet to reduce the pressure gradient from the left to the right atrium.

What part of the upright lung distends most during inspiration from FRC to TLC?

Milic-Emili has developed the elegant **onion skin illustration** of this problem (Fig. 36-1). The first 25% of the lower abscissae is the residual volume (RV), and this axis shows the total lung capacity (TLC) up till 100% TLC at maximal inspiration. The upper abscissae shows the VC from zero to 100%. The ordinate is the **regional ventilation in % of the regional TLC**. The regional TLC is any given lung region totally filled with air by a maximal inspiration (Fig. 36-1).

The slope of the **onion skin** line is constant, thus the fraction of tidal volume (TV) reaching each lung region must be constant during the whole inspiration (Fig. 36-1). The slope is larger in the lower than in the apical lung region, because the lower alveoli are compressed most by the intrapleural, intrathoracic pressure. Accordingly, they can distend most during inspiration. The upper alveoli are always more expanded than the lower due to the **pull** of the gravity, and they follow the **first in last out law**. During expiration to RV the upper alveoli are the last to empty (Fig. 36-1). During inspiration from RV, the lower alveoli are closed up to FRC (**closing volume** and **closing capacity** – see the horizontal curve in Fig. 36-1). Around FRC the lower alveoli open.

At the start of the inspiration from FRC the lower alveoli are the smallest, so any inspiration will always distend the lower alveoli most.

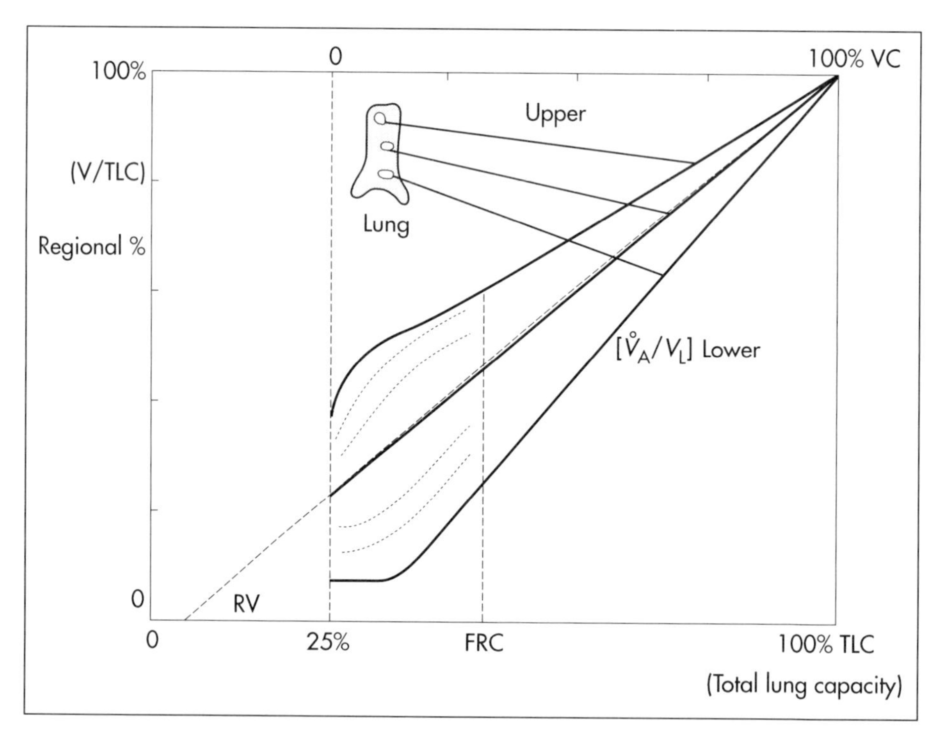

Fig. 36-1: The relative, regional ventilation (ordinate) depending upon total ventilation from RV to TLC (modified from Milic-Emili *et al.*, 1966).

The upper alveoli are always expanded by gravity. **At TLC all alveoli are assumed to be maximally distended** (Fig. 36-1). The alveoli and small airways are distended more and more from basis to apex of the lung. Accordingly, their compliance must decrease more and more. The intrathoracic or intrapleural pressure also decreases toward the apex.

What are the inborn errors of any human lung?

Both healthy and abnormal human lungs have: (1) mismatched regional $\mathring{V}_A/\mathring{Q}$; (2) venous-to-arterial shunting of blood (no ventilation); (3) alveoli with ventilation but without perfusion (**alveolar dead space**); and (4) low lung diffusion.

Further reading

Änggård, E. (1994) Nitric oxide: mediator, murderer, and medicine. *Lancet* **343**: 1199–1206.

Milic-Emili, J. *et al.* (1966) Regional distribution of inspired gas in the lung. *J. Appl. Physiol.* **21**: 749.

Nunn, J.F. (1987) Applied respiratory physiology, 3rd edn. Butterworths, London.

West, J.B. (1989) *Respiratory Physiology*, 4th edn. Williams & Wilkins, Baltimore.

36. Multiple Choice Questions

Each of the following five statements concerning pulmonary circulation have True/False options:

A. The pulmonary vascular pressure and the pulmonary vascular resistance (PVR) is only 1/10 of that of the systemic circulation.
B. The PVR is highest in intra-alveolar vessels at high lung volumes.
C. The PVR increases when pulmonary arterial pressure increases.
D. The pulmonary circulation is dependent on gravity but the pulmonary ventilation is not.
E. The P_{AO_2} has a direct effect on pulmonary circulation.

Try to solve the problems before looking up the answers in Chapter 74.

CHAPTER 37.
NEURAL CONTROL OF RESPIRATION

The control of respiration involves three components: **I.** The respiratory centres and other central neurons, **II.** Effectors, and **III.** Sensors.

I. CENTRAL NEURONS

Results of brainstem transections performed on animals over the past century indicate that the **central respiratory controller (RC)** is located in the brainstem. Such transection results also imply the presence of a **pneumotaxic centre** in the rostral pons, and an **apneustic centre** in the caudal pons (Fig. 37-1).

Respiratory movements are rhythmic and unconscious. What does this imply?

Respiratory rhythmogenesis is likely due to a **central pattern generator** located in these subcortical centres (Fig. 37-1). Respiratory rhythm can persist after removal of the entire brain above the brainstem (i.e. in a decerebrated animal).

Transections made at the midpons, with the vagi intact, cause slowing and deepening of respiration. When the vagi are also sectioned, midpontine transections produce **prolonged inspirations (I-spasms)** separated by short expirations (i.e. apneustic breathing).

Transection between medulla and the spinal cord causes respiratory arrest. What does this imply?

Respiratory arrest is termed **apnoea**. The respiratory centres are located above the point of transection. That is the pons and medulla (Fig. 37-1).

A **pneumotaxic centre is located in the rostral pons** (the **parabrachialis medialis** and the **Kolliker–Fuse nuclear complexes**). **Pneumo-taxis** means **air-order**. Impulses from this centre or from the vagi inhibit an **apneustic centre**, holding its inspiratory drive in check (Fig. 37-1).

Apneustic breathing is unmasked by vagal and upper pontine sections. Accordingly, normal activity in an **apneustic centre**, located caudally in the **pontine reticular formation**, is periodically inhibited by vagal and by pneumotaxic signals.

Removal of the **pontine apneustic centre** by transection between the pons and medulla produces a gasping, irregular pattern called **ataxic ventilation** (Fig. 37-1). This breathing pattern is still rhythmic, and it demonstrates that the neurons of the medulla themselves have a **spontaneous rhythmicity**. Thus, the role of the pontine centres is to make the discharges of the medullary neurons smooth and regular.

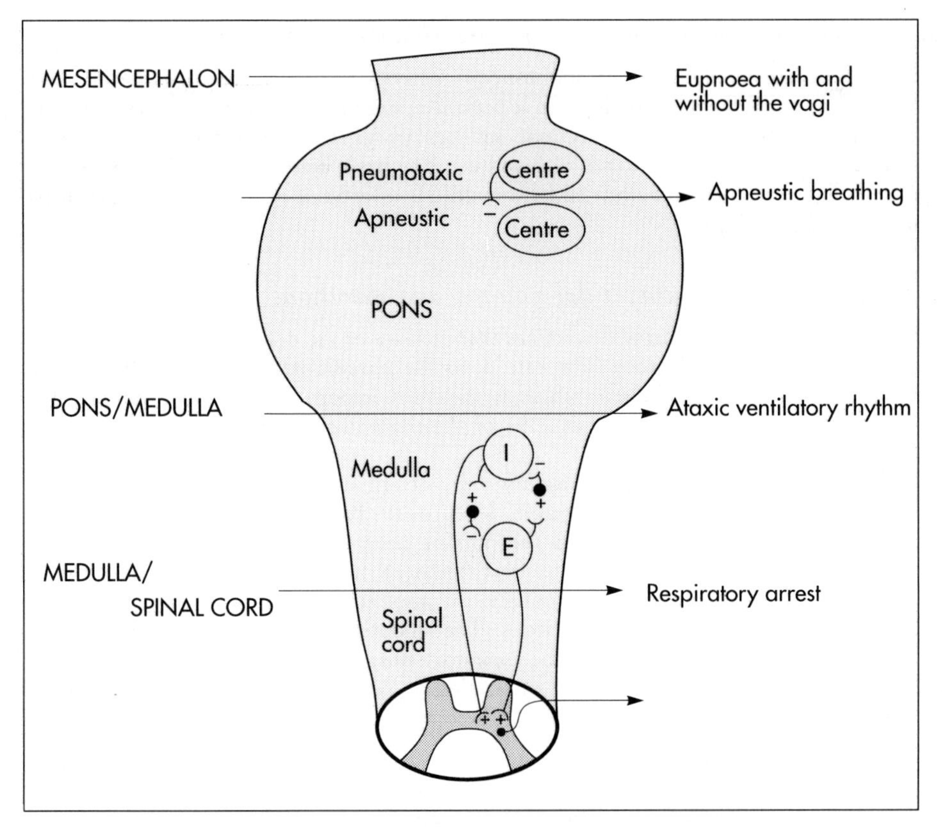

Fig. 37-1: Results of brainstem transections.

What is the respiratory controller (RC)?

RC is a collective term for a diffuse network of at least two types of neurons located in the **medullary reticular formation**: the **inspiratory I-neurons,** which fire just before and during inspiration, and the **expiratory E-neurons,** which fire just before and during expiration (Fig. 37-1). During inspiration, E-neurons are inhibited by hyperpolarization from the I-neurons (Fig. 37-1). During expiration, I-neurons are actively inhibited by hyperpolarization from the E-neurons in a feedback system (Fig. 37-1).

I- and E-neurons are localized in at least two medullary groups, the dorsal and the ventral respiratory groups.

The dorsal group contains mainly I-neurons, and they are located in the **ventrolateral nucleus of the solitary tract**. The axonal projections of these neurons terminate in the **cervical** and **thoracic anterior spinal motor neurons** of the phrenic and intercostal nerves.

The ventral group consists of the **Bøtzinger complex** close to nucleus retrofacialis, the **nucleus ambiguus, para-ambigualis and retro-ambigualis**. These diffusely located neurons include both I- and E-neurons, and their E-neurons project to expiratory **intercostal motor neurons.**

How can we explain the involuntary breathing rhythm?

The **I-neurons** show bursts of **spontaneous activity** (approximately 12–16 times per minute) separated by quiet periods, in a pattern mimicking the breathing at rest. The E-neurons, on the other hand, are not self-excitatory. The activity of the I-neurons correlates with the rate and depth of breathing. The **basic activity of the I-neurons,** like that of all pacemakers, is modulated by a multitude of external signals from the periphery and from areas of the CNS.

What determines inspiratory tidal volume and duration?

The **central inspiratory drive** determines the intensity of the desire to inspire. This drive is measured as the **inspiratory flow rate** or as the **phrenic nerve activity**. The phrenic nerve innervates the diaphragm, which is the most important inspiratory muscle. The diaphragm is responsible for about 75% of the inspiratory volume, and the intercostal muscles for the rest.

The duration of the central inspiratory drive determines the inspiratory time (i.e. the active phase of **medullary pacemaker I-neurons**). Both vagal stimulation by lung inflation above 1 l and input from the pontine neurons activate the **medullary I-off switch neurons** causing **inspiratory arrest (apnoea).** Inspiratory airflow equals driving pressure divided by resistance. Accordingly, the inspiratory flow rate is a function of the **pressure-generating respiratory muscles**. The **smooth muscles of the upper airways** determine upper airway resistance. The expiratory phase lasts as long as the **medullary I-off switch neurons** are active, and inspiration is not resumed. Expiration is passive, except at very high levels of ventilation, where expiratory muscles contract rhythmically to augment expiration.

What is the main, chronic stimulus of the RC?

The main stimulus is **CO_2 through the medullary chemoreceptors**, located close to the RC.

II. EFFECTORS

Do we have two efferent systems from respiratory neurons?

The **descending brain impulses** that drive the motor neurons, consist of those arising from voluntary and involuntary cerebral sources. These motor neurons in cranial and spinal nerves lead to the **respiratory muscles** and to the **smooth muscles of the upper airways**. The **corticospinal tract** transfers voluntary signals directly from the cortex to the motor neurons, or pass the same information through the pontomedullary centres before descending to the motor neurons via the reticulospinal tract. The **involuntary descending impulses** arise from the I-neurons in the **solitary tract nucleus** (via the phrenic nerve to the diaphragm), the mixed neurons in the **nucleus para- and retro-ambigualis** (the intercostal nerves to the intercostal muscles), and the **nucleus ambiguus** (the cranial nerves to the airways including smooth muscles).

Is the gamma-loop important to quiet breathing?

The intrafusal muscle fibres can shorten relatively more than the extrafusal (**autogenic facilitation**). Hence, a **simultaneous γ- and α-efferent discharge** releases the **gamma-loop servo system**, causing the respiratory muscles to shorten with exactly the necessary power. The respiratory muscles (except the diaphragm) have **gamma-loop servo**. Actually, this is a sensory feedback to ensure homeostasis, and we must look at the sensors of the control system.

III. Sensors

What are the external influences of RC?

1. **Higher brain centres** (cortex, hypothalamus, and diencephalon). Cortical, voluntary breath holding is possible until the **breaking point (i.e. the point where apnoea is disrupted)**, which is disturbed by bilateral blockade of the vagi and the glossopharyngeal nerves. Similar responses are released by stimulation of the diencephalon. A rise in hypothalamic temperature releases frequent breathing (**tachypnoea**) via the respiratory centres. During sleep and anaesthesia, breathing provides for metabolic needs primarily via the automatic homeostatic control system described above. During wakefulness, however, the breathing system subserves both homeostatic and voluntary, behavioural (non-homeostatic) needs. The behavioural needs include sucking, swallowing, speech, singing, laughing, crying, defecation, breath holding, hyperventilation, and coughing. Such voluntary acts affect P_{aCO_2}, P_{aO_2}, and $[H^+]$.
2. **Bronchopulmonary stretch receptors** are located in the smooth muscles of the trachea, larger bronchi and also in the lung parenchyma in the alveolar ductules and sacs. The activity of these **smooth muscle receptors** increase markedly with airway distention, and the activity ebbs slowly with time, hence they are called **slowly-adapting pulmonary receptors**. If the **tidal volume exceeds 1 l**, these receptors initiate signals that **inhibit the inspiratory drive** via myelinated vagal fibres, reinforcing the actions of the pontine centres and protecting the lungs from overexpansion. In humans this **Hering–Breuer reflex** plays no part in regulating ventilation during quiet breathing at rest, but the reflex is active during exercise.
3. **Rapidly adapting irritant receptors** are probably **free or modified vagal nerve endings** in the epithelium of the airways. These receptors are stimulated by irritants (smoke, allergens) and by inflammatory mediators such as prostaglandins and histamine. **Rapidly adapting irritant receptors** mediate the protective and sometimes pathological responses of **cough and bronchospasm**. The efferent limb of this reflex is the motor fibres in the vagus, which initiates bronchospasm.
4. **Juxta-pulmonary capillary receptors are terminals of non-myelinated vagal fibres (J-receptors)**. Distention of the interstitial space, as seen in a variety of cardiopulmonary disorders (microemboli, pulmonary oedema, pneumonia, fibrosis, atelectasis, and irritants), elicits increased ventilation (hyperpnoea), tachypnoea, bradycardia (slow heart rate), and low arterial pressure via vagal reflexes. **Atelectasis** means alveolar collapse.
5. **Peripheral arterial chemoreceptors** are found in the carotid bodies of humans (the aortic bodies account for only a small effect). Rapid changes in $\mathring{V}_E$ are registered by

the carotid chemoreceptors which transmit impulses to CNS via cranial nerves IX and X. The glomus cell is sensitive not only for **low P_{aO_2} (hypoxia), but also for increasing P_{aCO_2} (hypercapnia), $[K^+]$ and $[H^+]$.**

6. **Central, medullary chemoreceptors** are essential and particularly important to steady state ventilation (see Chapter 38).
7. **Thermoreceptors,** stimulated by a **rise in core temperature (so-called heat receptors),** elicit tachypnoea and tachycardia (frequent heart rate) via hypothalamus. Other thermoreceptors termed **cold receptors** are stimulated by a fall in shell temperature. Stimulation of these receptors elicits bradycardia via the regulating **hypothalamic temperature area. This so-called Survival Reflex (or "diving bradycardia")** protects the **brain, the heart and the lung** in emergency situations involving breath holding, not only in water, but also on land.
8. **Arterial baroreceptors** in the carotid sinus are **stretch receptors,** stimulated by increased transmural arterial pressure. Stimulation decreases arterial pressure, and inhibits heart rate and ventilation as in the **survival reflex.**
9. **Receptors in working muscles** stimulate ventilation via type III and IV afferents to the respiratory centres in the medulla. These receptors are involved in the exercise hyperpnoea.
10. **Protective, vagal reflexes** are related to vomiting, hiccup and swallowing. These reflexes protect us from inhaling vomit and thus being choked.

Further reading

Cherniak, N.S. (1984) Sleep apnea and its causes. *J. Clin. Invest.* **73**: 1501.

Euler, C.V. (1986) Brain stem mechanisms for generation and control of breathing pattern. In: Cherniak, N.S. and J.G. Widdicombe (Eds) *Handbook of Physiology, The Respiratory System,* Vol. II. Oxford University Press, New York.

37. Multiple Choice Questions

37.1. Each of the following five answers have True/False options:
Stem statement: The hyperventilatory response to hypoxia and hypoxaemia is mediated by the:

A. Bronchopulmonary mechanoreceptors.
B. Chemoreceptors in the carotid bodies.
C. Central chemoreceptors.
D. Irritant airway receptors.
E. Arterial baroreceptors.

37.2. Each of the following five statements have True/False options:

A. The primary muscle of inspiration is the diaphragm.
B. The air flow resistance is highest in the small terminal bronchioles.
C. The intrathoracic pressure is less than atmospheric pressure.
D. The VC is defined as the maximum volume of air that can be inhaled following a maximum expiration.
E. The lung compliance is reduced in patients with emphysema.

Try to solve the problems before looking up the answers in Chapter 74.

CHAPTER 38.
CHEMICAL CONTROL OF RESPIRATION

The respiratory system exerts its homeostasis by both **peripheral arterial** and by **central medullary chemoreceptors**. Both types of receptors are sensitive to changes in the $[H^+]$ around them. Such changes imply changes in the intracellular $[H^+]$, and changes in the ionic composition (Na^+, K^+, Ca^{2+}). Both types of receptors are activated by increases in P_{CO_2} independent of $[H^+]$. Hypoxia stimulates the **peripheral arterial chemoreceptors**, but has a **central depressant effect** on both chemoreceptors and regulatory neurons.

Where are the peripheral arterial chemoreceptors located?

The carotid and aortic bodies (glomera carotici et aortici) are small organs located in the tissue between the internal and external carotid arteries and at the arch of the aorta, respectively. The bodies contain **chemosensitive glomus cells**, fixed in a plasma-like tissue fluid surrounded by lots of sinusoidal capillaries which have an extremely large flow (2000 FU) of arterialized blood.

Glomus cells (type I) are sensitive to **decreased P_{aO_2}** (hypoxaemia mediating hyperventilation), but also to **increased P_{aCO_2}** (hypercapnia) and **increased $[H^+]$** (acidaemia), **and $[K^+]$** (hyperkalaemia) in the arterial blood (Fig. 38-1). The carotid bodies are predominant in humans.

These bodies are innervated by the carotid sinus nerve, which is a branch of the glossopharyngeal nerve (IX).

The carotid chemoreceptors, short latency and response periods are closer to the heart and lung, and stimulate ventilation reflectorily. The carotid bodies are therefore responsible for the immediate reactions to changes in P_{aO_2}, P_{aCO_2}, $[H^+]$, and $[K^+]$.

How do the carotid bodies work?

Stimulation of the carotid bodies by **hypoxaemia, hypercapnia, and acidaemia** increases the impulse frequency of the carotid sinus nerve and the glossopharyngeal nerve to the medullary respiratory centres (RC).

Increased activity from RC to the respiratory striated muscles and the upper airway smooth muscles causes increased ventilation. This has a homeostatic effect on the initial stimuli – partially relieving **hypoxaemia, hypercapnia and acidaemia**. This is a **negative feedback loop**.

What makes the carotid bodies important at high altitude?

The P_{aO_2} must fall below approximately 7.3 kPa (55 Torr) before the glomus cells are sufficiently stimulated to increase ventilation (Fig. 38-1 right). Patients with a P_{aO_2} below this threshold are **hypoxic** and suffer from **hypotonic hypoxia**. At that threshold the hypoxic stimulation outweighs the inhibitory effect of the fall in P_{aCO_2} due to hyperventilation. Compensatory hyperventilation **maintains adequate oxygenation** in the face of acute decreases of P_{IO_2} from the normal 20 kPa (150 Torr), until P_{aO_2}

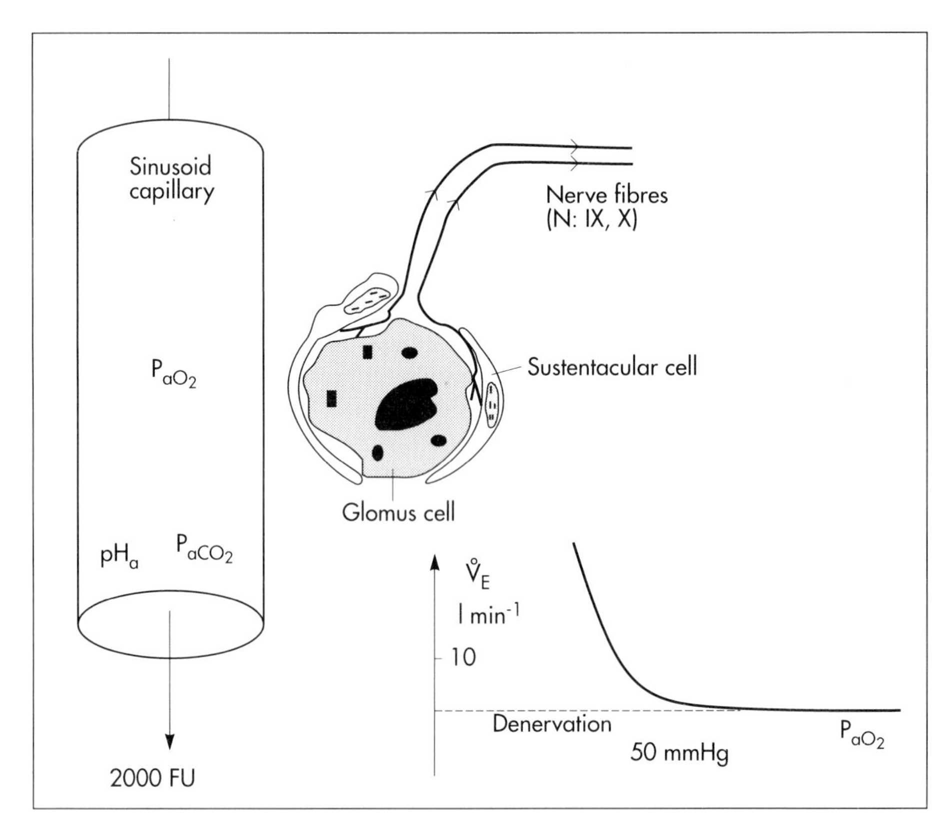

Fig. 38-1: Glomus cell with arterialized blood supply. The IXth cranial nerve is from the carotid bodies and the Xth (vagus) nerve from the aortic bodies. Ventilation starts to increase below a P_{aO_2} of 7.3 kPa (55 Torr).

decreases below 4 kPa (30 Torr), where consciousness is lost. There is an inverse or hyperbolic relationship between $\dot{V}_A$ and P_{aO_2}. At the peak of Mt Everest the P_{aO_2} is just below 4 kPa in trained mountain climbers, and they can only walk with great difficulty.

Do patients with CO poisoning and severe anaemia hyperventilate?

These patients have reduced C_{aO_2} in spite of normal P_{aO_2}. Since P_{aO_2} represents the stimulus to the glomus cells of the carotid bodies, and not the oxygen concentration of the arterial blood, such conditions are seldomly associated with compensatory hyperventilation. A person with CO poisoning can die **without increasing his or her breathing** and without having felt shortness of breath (**dyspnoea**). Under certain conditions CO acts as metabolic poison at the carotid bodies and increases ventilation to some extent.

Describe the medullary chemoreceptors

Persons without peripheral arterial chemoreceptors can still increase their ventilation in response to increases in P_{aCO_2} or $[H^+]$. This suggests that there exists other chemoreceptors to these two stimuli.

Three superficial (subpial) areas have been defined on the ventrolateral surface of the medulla, designated L (Loeschke), M (Mitchell) and S (Schläfke) after the scientists that located the **medullary chemoreceptors**. Areas L and M are believed to be chemosensitive, and nerve fibres from L and M converge on area S, where they pass deeper into the medulla to reach the medullary RC. The medullary chemoreceptors are located 100–200 μm below the ventrolateral surface. These receptors communicate with RC neurons located deeper in the medulla and which are the regulatory centres. The main stimulus to the chemoreceptors is the $[H^+]$ of the **brain ECF**, which is in close proximity to the cerebrospinal fluid (CSF) that bathes these receptors.

How can the brain ECF keep a $[H^+]$ in equilibrium with CSF $[H^+]$, and also a P_{CO_2} close to P_{aCO_2}?

The thin **pia layer** between CSF and the brain ECF is **highly permeable to CO_2**, and even to ions, whereas the **blood–brain barrier resists ionic diffusion** (Fig. 38- 2). It takes several minutes for changes in blood $[H^+]$ to be reflected in the brain ECF.

The blood–brain barrier is formed by the tight junctions between the endothelial cells of the cerebral capillaries (Fig. 38-2). This is why brain ECF has a $[H^+]$ that follows changes in CSF, and a P_{CO_2} that follows changes in P_{aCO_2} (Fig. 38-2).

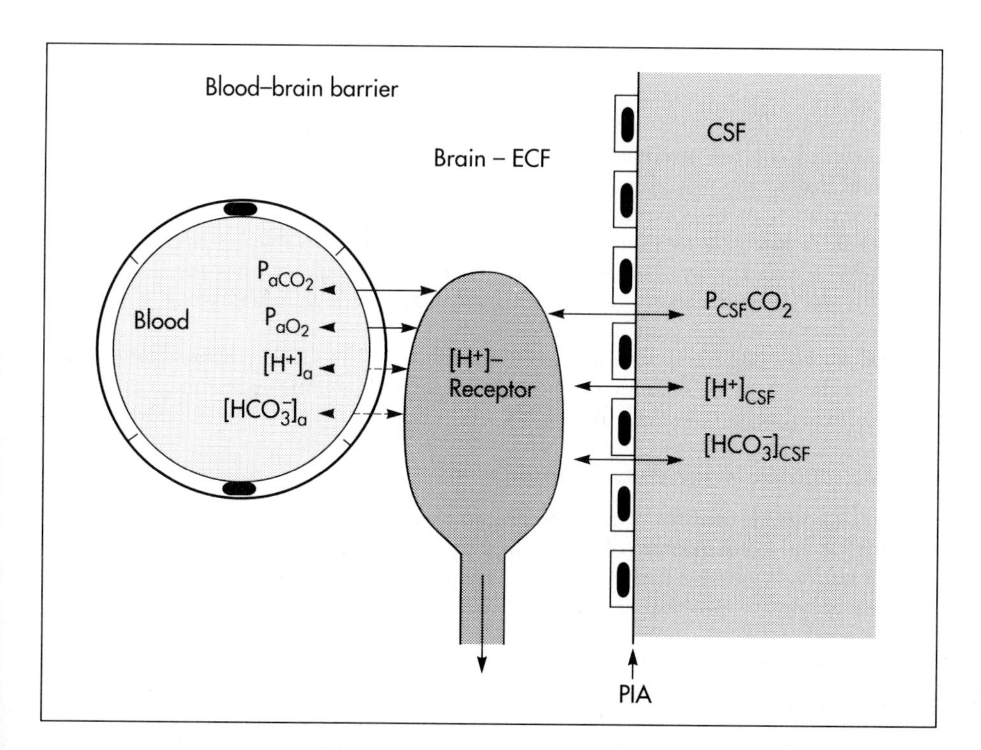

Fig. 38-2: The blood–brain barrier and the medullary chemoreceptors.

In **metabolic acid–base disorders**, the steady-state changes in brain ECF $[H^+]$ are much smaller than the changes in blood $[H^+]$, because the blood–brain barrier has **transport proteins** that can move ions in and out of ECF. The blood–brain barrier can be passed by control molecules in area striata, which has widely fenestrated capillaries and no blood–brain barrier. The **medullary chemoreceptors** are stimulated by **increased brain ECF $[H^+]$**, and the information reaches the **RC neurons**.

A compensatory hyperventilation is initiated. Breath-holding and hypoventilation leads to **CO_2 accumulation** with acute increase of P_{aCO_2}.

Since CO_2 is lipid soluble it diffuses rapidly across the blood–brain barrier and increases brain ECF $[H^+]$ by hydration to carbonic acid followed by dissociation. Not only this H^+ but also P_{CO_2} itself stimulate the **medullary chemoreceptors**. The gain of the medullary chemoreceptors to changes in $[H^+]$ exceeds that of the carotid chemoreceptors. Most of the steady-state ventilatory rise is mediated by the medullary receptors, whereas fast and transient changes are first detected by the **carotid chemoreceptors**.

Describe the CO_2 response curves obtained during hypoxia, acidosis and exercise

The quantitative relationship between the ventilation and P_{ACO_2} is called the **CO_2 response curve** (Fig. 38-3). The **slope** of such a curve shows the sensitivity or the **gain** of the RC. At low P_{AO_2} there is no hyperventilation due to a rise in P_{ACO_2} (see the hockey stick curve in **Fig. 38-3A**). This is because the **threshold** of the RC has not been reached.

Panel A: The A-curve to the right shows the rise in ventilation by hypercapnia at normal P_{AO_2}. The left curve shows the response at P_{AO_2} 5 kPa (37 Torr). The slope is steeper which implies an increased sensitivity or gain of the chemosensitive feedback loop during hypoxia. The hypoxia also lowers the threshold for the CO_2 stimulus.

The combined effect is **multiplicative**. The shift to the left of the curve is called **reduced threshold** due to the new stimulus which is hypoxia (Fig. 38-3A). The steeper slope of the hockey stick curve is typical for **high altitude acclimatization**.

Panel B: A low pH_a as seen in chronic, metabolic acidosis releases hyperventilation, so the P_{ACO_2} falls further. This is a reduction of the threshold but not the sensitivity to P_{aCO_2} of the chemoreceptors (see also Chapter 34). A high blood $[H^+]$ is the new stimulus for the acidotic patients of **Fig. 38-3B.**

Panel C: Exercise has a similar effect (**Fig. 38-3C**). The CO_2 response curve is also shifted to the left, again without a change in the slope. Here, the extra stimuli are signals from the working muscles and from CNS.

What is adaptation or acclimatization?

These concepts are used for acute and chronic or **long-lasting changes of physiological systems as a consequence of long lasting or repeated stress.** An example of acclimatization is observed in inhabitants exposed to **chronic hypoxia** at high altitudes. Another example is **physical training.**

Describe the acid–base changes at high altitude adaptation

Acute adaptation

A person just arriving at high altitude experiences immediate hyperventilation. This high $\mathring{V}_E$ persists for the rest of his or her life, if they become a permanent high altitude

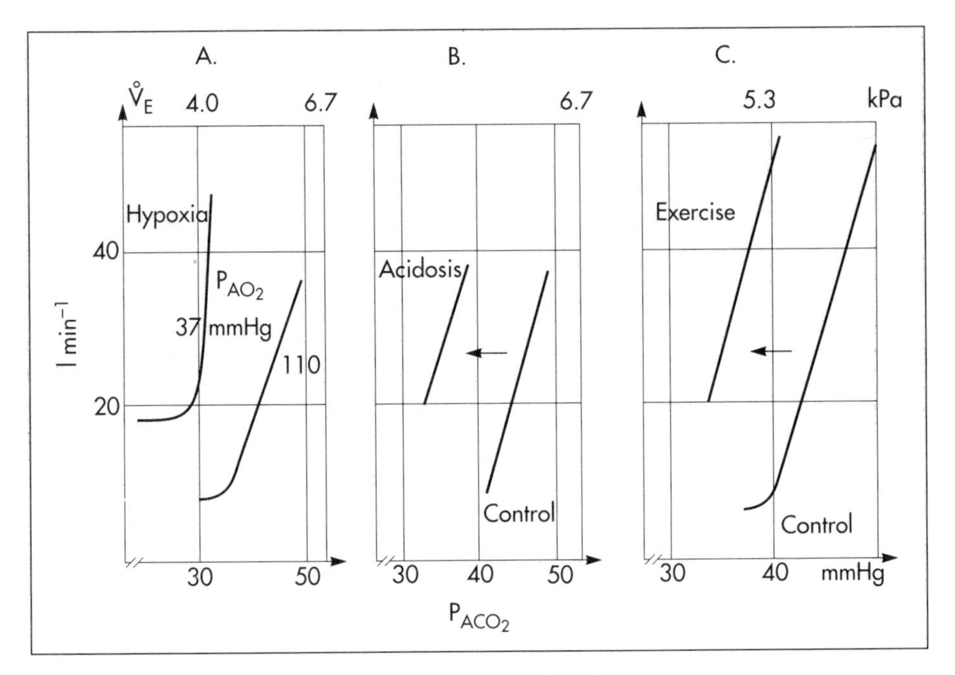

Fig. 38-3: CO_2 response curves: A, results from persons exposed to a combination of hypoxia and hypercapnia; B, the change from a normal control curve to the response of acidotic patients; C, the change from control to steady state exercise.

resident. The rise in ventilation is mainly caused by the low P_{aO_2} stimulating the carotid chemoreceptors. Ventilatory adaptation is an effect upon the peripheral chemoreceptors. The ventilatory response implies excess clearance of CO_2 with reduced P_{aCO_2} and **acute respiratory alkalosis**, according to the alveolar ventilation equation.

The P_{IO_2} in the moist tracheal air will fall with **falling F_{IO_2}** or with falling **atmospheric pressure** (P_B) at altitude (Fig. 38-4). $\mathring{V}_E$ starts to rise at P_{IO_2} values below 13.3 kPa (100 Torr), which corresponds to a P_{aO_2} of 7.5 kPa, and the rise is hyperbolic (Fig. 38-4). The rise in $\mathring{V}_E$ in itself reduces the chronic P_{aCO_2} stimulus (Fig. 38-4). The data confirm the presence of an **acute respiratory alkalosis** with a high pH (Fig. 38-4).

Chronic changes

When the low P_{aCO_2} has persisted for 5 days, adaptive mechanisms in the renal tubule cells are optimized. The renal compensation consists of **reduced tubular H^+ secretion** and thus **reduced tubular bicarbonate reabsorption**. This then elicits a fall in arterial and CSF [bicarbonate] and a rise of the low [H^+]. In the 10 days until adaptation is fully accomplished, the arterial [bicarbonate] is reduced in proportion to the fall in P_{aCO_2}. Thus pH_a is only mildly elevated, and the final condition is called a **compensated respiratory alkalosis** (i.e. increased pH_a, reduced P_{aCO_2}, reduced [bicarbonate] and negative base excess).

The **CSF contains fewer buffers and lower buffer concentrations than the blood.**

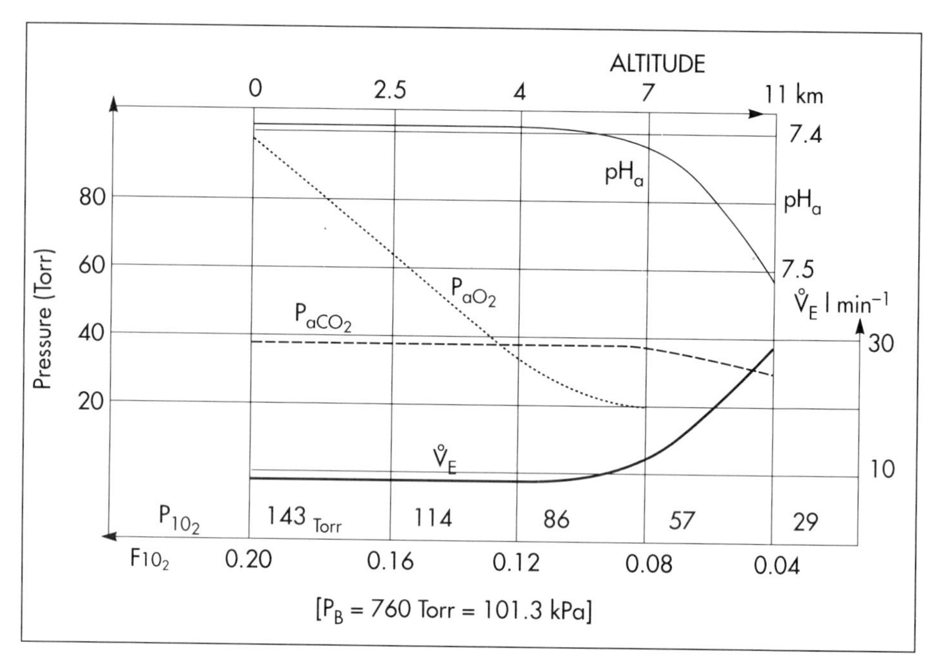

Fig. 38-4: High altitude (pH thin curve, P_{aO_2} punctured curve, P_{aCO_2} stippled curve, $\mathring{V}_E$ solid curve).

High altitude acclimatization shifts the CO_2 response curve to the left (i.e. reduced threshold), and the slope becomes steeper illustrating a higher sensitivity or gain (Fig. 38-3A).

Describe the circulatory adaptations to altitude

The acclimatized person has similarities to the well-trained athlete with high oxygen capacity (high haematocrit, haemoglobin and erythrocyte count – also similar to chronic polycytaemia). The adaptation is associated with a rise in the circulatory oxygen transport, because both $\mathring{Q}$ and the oxygen capacity of the blood are increased.

The acclimatized person has a large blood and plasma volume, growth of the pulmonary arterial wall and the right ventricular muscle mass in response to hypoxic pulmonary vasoconstriction. There is constantly a high sympathetic tone which decreases the renal bloodflow (RBF). Eventually the hypoxic vasoconstriction leads to pulmonary hypertension, and **right ventricular failure (cor pulmonale)** can develop as in mountain sickness.

Stimulation of the α-adrenergic receptors of the pulmonary vessels causes pulmonary vasodilatation and is used in the treatment of mountain sickness.

Is it possible to calculate P_{ACO_2} at a given altitude?

During climbing to high altitude, the falling P_B implies a fall also in: $P_{IO_2} = (P_B - 6.3)F_{IO_2}$ and thus in P_{aO_2}. When P_{aO_2} is below 7.3 kPa (55 Torr), the ventilation must increase more and more, whereby P_{aCO_2} is reduced. The **alveolar gas equation** can be solved for:

$$(P_{IO_2} - \boldsymbol{P_{AO_2}}) = \boldsymbol{P_{ACO_2}} \cdot [\boldsymbol{F_{IO_2}} + (1 - \boldsymbol{F_{IO_2}})/\boldsymbol{R}].$$

At $R = 1$ and $F_{IO_2} = 0.21$ the value within the brackets [] is one, and we have the **simplified alveolar gas equation**. Accordingly:

$$(13.3 - 7.3) = \mathbf{6\ kPa\ or\ 45\ Torr}\ \text{in}\ P_{ACO_2}.$$

This theoretical calculation of P_{ACO_2} is only mathematically correct. The argument is clearly wrong when P_{AO_2} is below 7.3 kPa, because at this level the hyperventilation will diminish P_{ACO_2}. The P_{ACO_2} must fall exactly in proportion to the fall in barometric pressure, if the metabolic rate and F_{ACO_2} $[= P_{ACO_2}/(P_B - 6.3)]$ is assumed constant.

Ventilatory acclimatization elicits a long lasting rise in BTPS-ventilation (measured at altitude), inversely proportional to the fall in P_B, and to the fall in P_{CO_2} and $[H^+]$ in blood and body fluids. A doubling of the BTPS-ventilation at 0.5 atmosphere implies that the ventilation measured in STPD-units is unchanged. According to the **alveolar ventilation equation,** F_{ACO_2} is equal to the **ratio between $\mathring{V}_{CO_2}$ and $\dot{V}_A$** both measured at STPD. Since both of these volumes are unchanged after total adaptation, it follows that also F_{ACO_2} must be unchanged. Hence, **P_{ACO_2} must fall proportional to the fall in P_B.**

What is de-acclimatization?

Following return to sea level a **de-acclimatization** takes place over the next 3 weeks with falling ventilation and heart rate, falling blood pressure and pulmonary vascular resistance.

Adaptation to high altitude and to physical training is **lost within a few weeks.**

Further reading

Honda, Y. (1992) Respiratory and circulatory activities in carotid body-resected humans. *J. Appl. Physiol.* **73** (1): 1–8.

Levitzky, M.G. (1991) *Pulmonary Physiology*, 3rd edn. McGraw-Hill, New York.

Loeschcke, H.H. (1983) Central chemoreceptors. In: Pallot, D.J. (Ed.) *Control of Respiration*, pp. 41–77. Croom Helm, London.

Pokorski, M. and S. Lahiri (1983) Relative peripheral and central chemosensory responses to metabolic alkalosis. *Am. J. Physiol.* **245**, R873–R880.

West, J.B. (1991) *Respiratory Physiology*, 4th edn. Williams & Wilkins, Baltimore.

38. Case History

A healthy female, with an anatomic V_D of 0.12 l and a respiratory frequency (f) of 14, is flying in an open-cockpit airplane at 2000 m, where her P_{IO_2} is 16 kPa (120 Torr) and her $\mathring{V}_A$ is increased to 5.6 l STPD min^{-1} (from 4.2 at the ground). She has had a mixed diet meal before take-off, and her R value remains at 0.8. Her P_{AO_2} is restored to 13.3 kPa (100 Torr) by the rise in ventilation.

1. *Calculate her P_{ACO_2} by the simplified alveolar air equation.*
2. *Calculate her $\dot{V}_E$.*

Try to solve the problems before looking up the answers in Chapter 74.

CHAPTER 39. METABOLISM AND EXERCISE

The final product of metabolism is **carbon dioxide**. The **control of breathing during exercise** is not completely understood. However, the **carbon dioxide molecule** is most likely the controlled variable, perhaps as P_{aCO_2}.

Is the oxygen uptake limited by ventilation at extreme exercise?

Increasing work rate (e.g. 15 W each min) leads to a marked increase in pulmonary ventilation ($\mathring{V}_E$) without any ceiling being reached even at **maximal oxygen uptake** (**$\mathring{V}_{O_2}$max** in Fig. 39-1). The steeper rise in $\mathring{V}_E$ is shown by its deviation relative to the dashed line towards the right (Fig. 39-1). Light exercise often increases $\mathring{V}_E$ by an increased tidal volume (V_T). With increasing work rate also the respiratory frequency

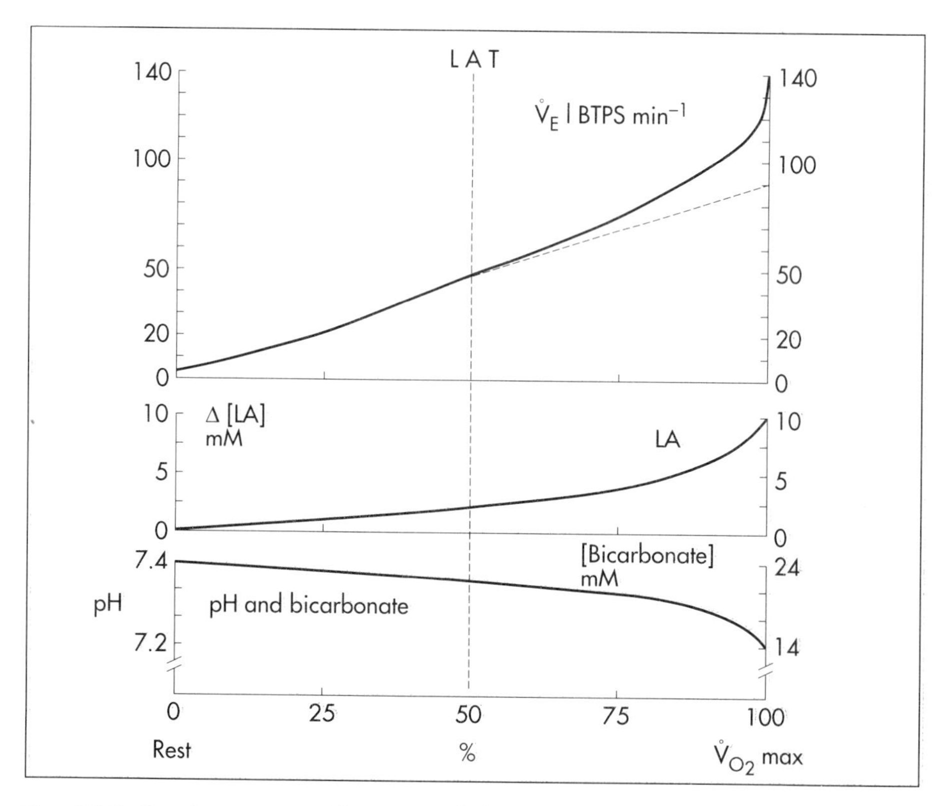

Fig. 39-1: Respiratory variables, arterial [lactate] & [bicarbonate], and pH at rest and during an incremental work test on a cycle ergometer.

must rise from 10 to 50 respirations per minute. The tidal volume can increase to half the value of VC (6 l), which corresponds to a $\dot{V}_E$ of 150 l min^{-1}. At exhaustion the $\dot{V}_E$ is much greater than at the point where $\dot{V}_{O_2}$max is already reached. At $\dot{V}_{O_2}$max many individuals can increase $\dot{V}_E$ further voluntarily.

Thus **ventilation is not the limiting factor in healthy persons.**

What happens to the alveolar and arterial blood gases during exercise?

P_{AO_2} and P_{ACO_2} are essentially maintained during most work rates. At maximal work rate the P_{AO_2} increases and P_{ACO_2} decreases 5–10%. This fact illustrates effective gas exchange or adequate $\dot{V}_E$ during non-exhausting exercise.

What is the lactic acid threshold?

The **lactic acid threshold (LAT)** is the oxygen uptake at which a substantial rise in blood [lactate] is found. (**LAT** is indicated by a vertical broken line in Fig. 39-1.) Just after the LAT is passed, the $\dot{V}_E$ increases proportional to the increase in $\dot{V}_{CO_2}$ (i.e. so-called **normo-capnic buffering**). Accordingly, $\dot{V}_E$ increases linearly with $\dot{V}_{CO_2}$.

Beyond the range of isocapnic buffering, $\dot{V}_E$ rises out of proportion with $\dot{V}_{CO_2}$, causing P_{aCO_2} to fall (hyperventilation). The $\dot{V}_{CO_2}$ and $\dot{V}_E$ will increase faster than $\dot{V}_{O_2}$, because bicarbonate reacts with the lactic acid produced, so CO_2 is liberated and added to the metabolic CO_2 production. Note the rise in plasma [lactate] of 10 mmol l^{-1}, which is equal to the fall in plasma [bicarbonate] from 24 to 14 mmol l^{-1} (Fig. 39-1).

Exercise levels above the **maximal aerobic capacity ($\dot{V}_{O_2}$max)** is called **supra-maximal work.** Here the anoxia leading to a **metabolic or lactic acidosis** contributes with a **large ventilatory drive,** as shown in the steep component of $\dot{V}_E$ (Fig. 39-1).

Describe the relationship between $\dot{V}_E$ and $\dot{V}_{O_2}$ in trained and untrained individuals

Results from an untrained person with a $\dot{V}_{O_2}$max of 2.4 l STPD min^{-1} (LAT 1.3 l min^{-1}), and from a top athlete with 6 l min^{-1} (LAT 3.6 l min^{-1}) are shown in Fig. 39-2. Several studies have shown $\dot{V}_{O_2}$ to remain at $\dot{V}_{O_2}$max despite increasing work rates, $\dot{V}_{CO_2}$ increasing too. These curves also illustrate that $\dot{V}_E$ is not the limiting factor for $\dot{V}_{O_2}$max.

How can we explain the steep rise in ventilation at exhaustion?

If the athlete is suddenly breathing oxygen instead of atmospheric air, while working at a high level (5–6 l oxygen min^{-1}), a drastic fall in $\dot{V}_E$ will occur within 30 s. This is not a chemoreceptor response, since there is no stimulus. The oxygen breathing reduces the blood [lactate], but not within 30 s. Oxygen breathing abruptly increases the rate of diffusion from haemoglobin to the muscle mitochondria. This is a possible key to the explanation of the steep rise in ventilation at exhaustive exercise. In conclusion, the steep rise in ventilation and the limit to $\dot{V}_{O_2}$max is due to the **reduced rate at which oxygen can move by diffusion** from haemoglobin to the muscle mitochondria during exhaustive exercise with metabolic acidosis.

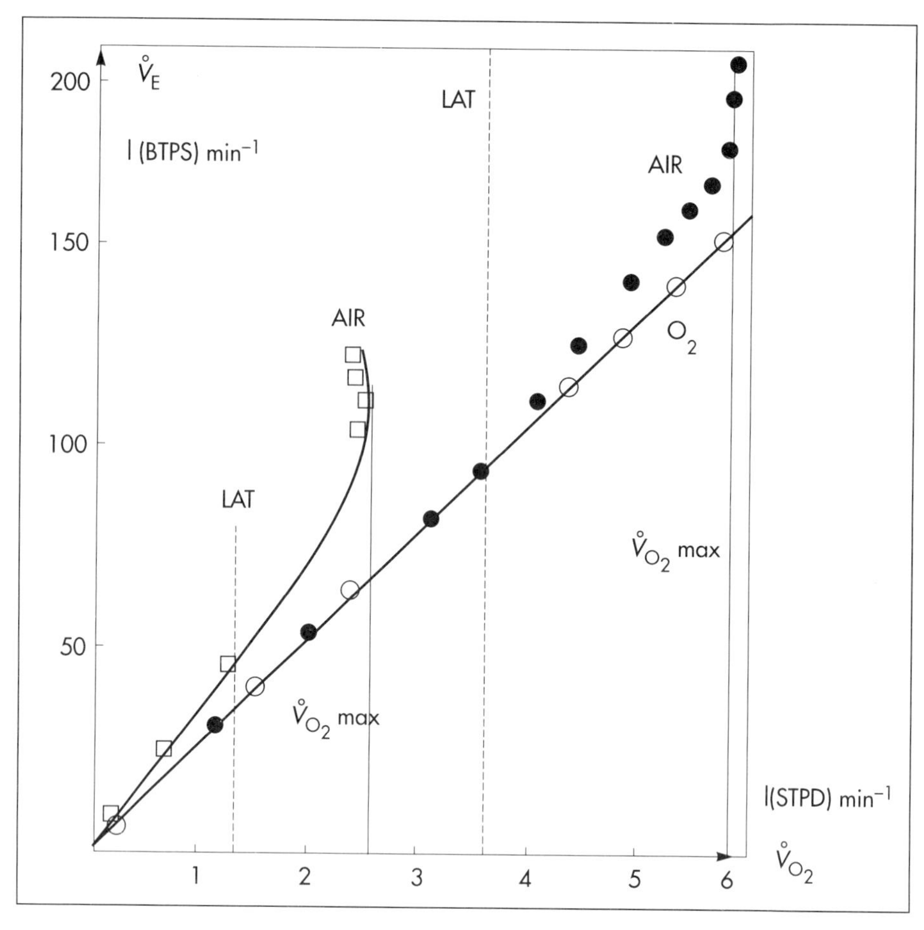

Fig. 39-2: Ventilation and oxygen uptake in an untrained person (□) and in a top athlete, with a $\mathring{V}_{O_2}$max of 6 l min^{-1} breathing air (●) or oxygen (○).

Is lactate always produced during exercise?

Lactate is produced even at light exercise, but only minimal amounts are liberated to the blood. Trained subjects do not increase blood [lactate] until $\mathring{V}_{O_2}$ is quite high. Untrained subjects produce lactic acid at lower levels of exercise, and at any $\mathring{V}_{O_2}$, they have higher $\mathring{V}_E$ and heart rate than the trained. The LAT in untrained persons is often about 50% of $\mathring{V}_{O_2}$max, whereas in athletes it approaches 80%. Patients with heart disease increase their blood [lactate] with minimal activity.

Calculate maximum values of $\mathring{Q}$ and D_{LO_2} during exercise

The law of mass balance states the following: $\mathring{V}_{O_2} = \mathring{Q} \cdot (C_{aO_2} - C_{\bar{v}O_2})$.

Typically, at rest the $\mathring{V}_{O_2}$ is often 250 ml STPD min^{-1}, and $\mathring{Q}$ is 5 l blood min^{-1}; the **arterio-venous oxygen content difference** is thus 50 ml STPD l^{-1} of blood.

In well-trained athletes (body weight 70 kg) a $\mathring{V}_{O_2}max$ of 6 l STPD min^{-1} is often found, corresponding to 86 ml STPD O_2 kg^{-1} min^{-1} (the **fitness number**). As the maximal **arterio-venous oxygen content difference** is 150 ml STPD l^{-1} of blood, it follows that such athletes can attain a maximal $\mathring{Q}$ of 40 l of blood min^{-1}. At maximal work the **lung diffusion capacity for oxygen (D_{LO_2})** rises to 9 ml STPD s^{-1} kPa^{-1} (from 3.6 at rest, see Chapter 35). Pulmonary bloodflow equals $\mathring{Q}$, so to accommodate the increase during exercise, lung capillaries must open up. Thus, unlike at rest, the apical parts of the lungs become well perfused, improving $\mathring{V}_A/\mathring{Q}$, and increasing the area available for gas transport.

The 24-fold increase of oxygen uptake in the lungs is also made possible by an increased oxygen tension gradient: $\mathring{V}_{O_2} = (D_{LO_2} \cdot \Delta P_{O_2})$. The ΔP_{O_2} is equal to P_{aO_2} minus $P_{\bar{v}O_2}$. The substantial rise in ΔP_{O_2} is caused by a fall in $P_{\bar{v}O_2}$ due to more complete oxygen extraction in exercising muscles, and to the increase in P_{aO_2} because $\mathring{V}_A$ increases out of proportion to $\mathring{V}_{O_2}$ above the LAT.

How is oxygen delivery to the mitochondria increased during strenuous exercise?

The **oxyhaemoglobin dissociation curve is moved progressively to the right** as exercise intensity increases due to the rise in 2,3-DPG and to the rise in temperature.

Above the LAT, when oxidative metabolism is maximal, extra mechanical output is financed by anaerobic energy generation. The end product is lactic acid and 2,3-DPG. The **lactic acidosis** and 2,3-DPG causes a **further shift** to the right of the oxyhaemoglobin dissociation curve easing oxygen delivery to the mitochondria. Lactate and adenosine also dilate muscle vessels and increase the number of open capillaries, thus decreasing the diffusion distance for oxygen from capillary blood to the mitochondria.

Strenuous exercise is also associated with a rise in plasma concentration of catecholamines ([catecholamines]), and a rise in core temperature approaching 41°C. The sensitivity of most receptors is increased in an overheated body. Increased activity of the **arterial chemoreceptors** causes hyperventilation in situations where P_{aO_2} is dangerously low, and the person approaches exhaustion and collapse.

Is it possible to estimate $\mathring{Q}$ from $\mathring{V}_{O_2}$ during work?

The following relation is valid for exercising healthy males (work rates up to 70% of their $\mathring{V}_{O_2}max$):

$$\mathring{Q}\ \mathrm{l\ min^{-1}} = 3.07 + 6.01 \cdot \mathring{V}_{O_2}.$$

This calculation of $\mathring{Q}$ allows for estimation of the rarely available $C_{\bar{v}CO_2}$ from Fick's principle:

$$C_{\bar{v}CO_2} - C_{aCO_2} = \mathring{V}_{CO_2}/\mathring{Q} \text{ or } C_{\bar{v}CO_2} = C_{aCO_2} + \mathring{V}_{CO_2}/\mathring{Q}.$$

What is the basis for muscular fatigue?

1. **Neuromuscular fatigue** is probably caused by progressive depletion of ACh stores during prolonged, high-frequency muscular activity.
2. **Isolated muscular fatigue** is due to depletion of ATP stores, whereby the actin–myosin filaments form a fixed binding and develop rigour or cramps.

Is it possible to explain exercise hyperpnoea and the rise in $\dot{Q}$?

The proportional increase in $\dot{V}_E$ and $\dot{Q}$ with increasing $\dot{V}_{O_2}$ suggests a common control system. The integrator consists of **sensory and motor cortical areas**, and the brainstem neighbour-centres for respiratory and cardiovascular control. The link between the respiratory and the circulatory control system is probably established in the neural network of the brainstem centres.

The **nucleus of the tractus solitarius** is the site of central projection of both chemoreceptors and baroreceptors. Such a **single factor concept** is supported by recent data, where time constants for heart rate, $\dot{V}_E$, $\dot{V}_{O_2}$, and $\dot{V}_{CO_2}$, were much longer in elderly than in young women, with a strong link between the relative slowing of the four components. The respiratory and the cardiovascular system are linked together during most forms of dynamic exercise (Fig. 39-3). Since the two functions are not linked in all forms of dynamic exercise, they must also be able to **operate differentially**. There is a sharp rise in $\dot{V}_E$ within the first breath at the onset of exercise, and $\dot{Q}$ also increases abruptly (Fig. 39-3). Both variables increase progressively over minutes until a steady state is reached. At the off-set of exercise, $\dot{V}_E$ and $\dot{Q}$ falls instantly (Fig. 39-3).

The cardiopulmonary adjustments to exercise comprise an integration of I. neural and II. humoral factors.

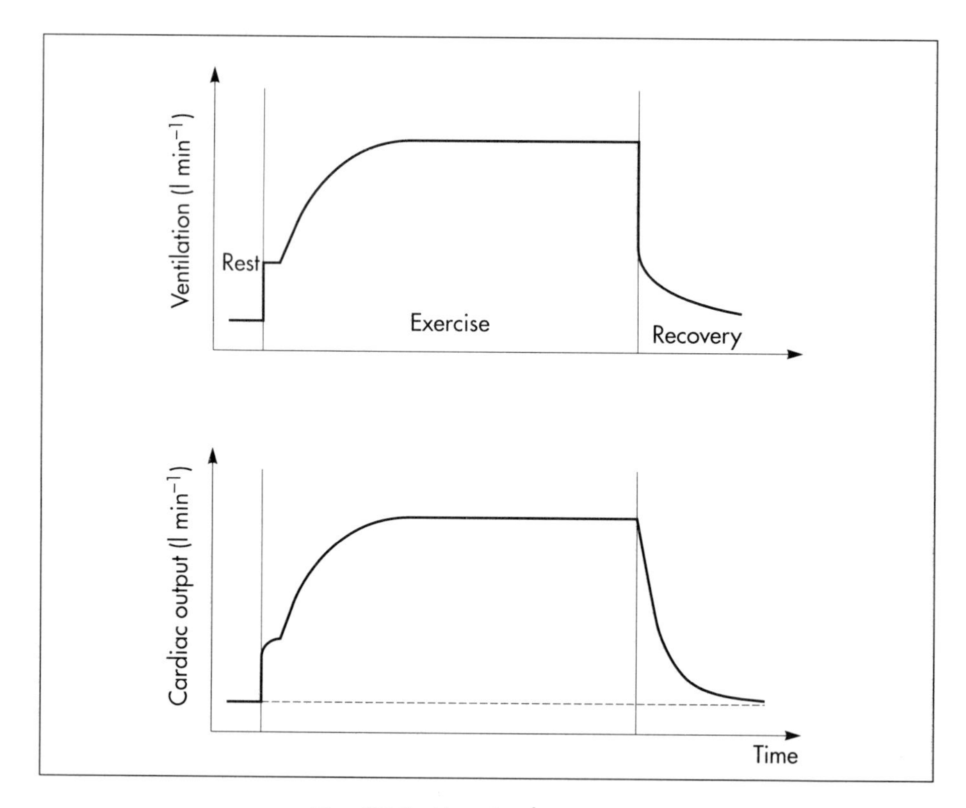

Fig. 39-3: Exercise hyperpnoea.

I. The neural factors consist of: (1) signals from the brain; (2) reflexes originating in the contracting muscles; (3) the baroreceptor reflex; and (4) the peripheral chemoreceptors.

1. Signals from the brain to the active muscles pass the **reticular activating system (RAS)** in the reticular formation of the medulla, which includes the respiratory controller (**RC**) neurons and cardiovascular centres. This is called **irradiation** from the motor cortex to the RC, and proposed as an explanation of the **exercise hyperpnoea**. The mesencephalon and hypothalamus are also involved in the irradiation hypothesis called **central command**. Cortical activation of the sympathetic nervous system accelerates the heart, increases myocardial contractility, dilates the muscular arterioles and contracts other vascular beds such as the splanchnic region. Speculative mechanisms such as irradiation or central command are so-called **feedforward hypotheses.**
2. Afferent signals from proprioceptors in the active muscles through thin myelinized and unmyelinized fibres in the spinal nerves (type III and small unmyelinated type IV) to RC is the best documented **feedback hypothesis**.
3. Baroreceptor reflexes, originating in the **carotid and aortic sinuses**, are tonically active and regulate blood pressure. An increase in arterial pressure stretches the baroreceptors and reflexly inhibits the vasomotor centre in the medulla causing vasodilatation and bradycardia. A decrease in arterial blood pressure stimulates the vasomotor centre and causes vasoconstriction and tachycardia.
4. Peripheral chemoreceptors are located in the **carotid and aortic bodies**. The pH, P_{aO_2}, and P_{aCO_2} are normal during moderate steady state exercise. However, during transitions from rest to exercise and during severe exercise the peripheral chemoreceptors are stimulated. Stimulation of peripheral chemoreceptors increases the rate and depth of respiration and causes vasoconstriction.

II. The humoral factors that influence skeletal muscle bloodflow, $\mathring{Q}$, and ventilation are metabolic vasodilatators and hormones. Neural and chemical control mechanisms oppose each other. During muscular activity the local vasodilatators supervene. The local vasodilatators have not been identified. Ischaemic mitochondria in fast oxidative muscle fibres release many vasodilatators such as adenosine, AMP, ADP and NO.

However, it is possible to block many of the neural and humoral factors without disturbing the proportional exercise hyperpnoea and the rise in $\mathring{Q}$. These experiences suggest that the human body has a redundancy of overlapping control systems. The **redundancy-hypothesis**, with neural factors dominating at the start of work and peripheral feedback control during steady-state, is a logical compromise.

Further reading

Cunningham, D.J.C., P.A. Robbins and C.B. Wolff (1986) Integration of respiratory responses to changes in alveolar partial pressures of CO_2 and O_2 and in arterial pH. In: Cherniak, N.S. and Widdicombe, J.G. (Eds) *Handbook of Physiology*, Section 3. *The Respiratory System*, pp. 475–528. *Am. Physiol. Soc.*, Bethesda, MD.

Cunningham, D.A., J.E. Himann, D.H. Patterson, and J.R. Dickinson (1993) Gas exchange dynamics with sinusoidal work in young and elderly women. *Resp. Physiol.* **91** (1): 43–56.

Wagner, P.D. (1992) Gas exchange and peripheral diffusion limitation. *Med. & Sci. in Sports & Exercise* **24** (1): 54–8.

39. Case History

A female, 20 years of age, with a body weight of 62 kg, is exercising on a bicycle ergometer during steady state. Her cardiac output ($\dot{Q}$) is measured to 25 l min^{-1} by the mass balance principle with CO_2 as indicator, and her arteriovenous O_2 content difference is measured to 170 ml STPD l^{-1}.

1. *Calculate her oxygen-uptake per minute.*
2. *What assumption must be made in order to calculate her fitness?*

Try to solve the problems before looking up the answers in Chapter 74.

CHAPTER 40. RESPIRATORY INSUFFICIENCY

Hypoxia is a condition with insufficient oxygen supply or oxygen utilization in tissues.

Acute hypoxia with low P_{aO_2} stimulates the carotid bodies. This triggers a rise in ventilation (primary hyperventilation). The hyperventilation reduces P_{aCO_2} and $[H^+]$, which limits the initial rise in ventilation, because it decreases the carotid body and central chemoreceptor stimuli.

Chronic hypoxia increases breathing in another way. The primary hyperventilation leads to an acute respiratory alkalosis. This disorder is partially compensated by renal excretion of bicarbonate. Hereby, the $[H^+]$ returns toward normal. The low [bicarbonate] in the ECF, including brain interstitial fluid, is partially replaced by lactate from the hypoxic brain. The carbon dioxide response curve is shifted to the left and much steeper than normal (Fig. 38-3A).

What is hypoxia and hypercapnia?

Hypoxia denotes **oxygen deficiency in tissues** due to insufficient access to oxygen (low P_{aO_2}) or insufficient oxygen utility (normal P_{aO_2}). **Hypotonic hypoxia** is characterized by a P_{aO_2} less than 7.3 kPa (55 Torr). Below this threshold the ventilation starts to increase by carotid body activity.

Describe types of hypoxia

Hypoxia with low P_{aO_2}

The most frequent cause of this hypoxia is **uneven matching of ventilation and pulmonary bloodflow** (i.e. alveolar ventilation/perfusion ratio $[\mathring{V}_A/\mathring{Q}]$ inequality). To optimize the efficiency of gas exchange and achieve **normal blood gas levels,** the alveoli must be both adequately ventilated and perfused. Abnormal mismatching leads to **abnormal $\mathring{V}_A/\mathring{Q}$ ratios and to subsequent hypoxaemia and hypercapnia** (see below).

Persons with **venous-to-arterial shunts** in the heart and in the vessels, and persons with diffusion impairment all have low P_{aO_2} values. Hypoxia also hits persons breathing air with reduced P_{IO_2} (reduced P_B or reduced F_{IO_2} – see Chapter 38).

Hypoxia with normal P_{aO_2}

Ischaemic or stagnant hypoxia is caused by reduced bloodflow either locally or generally.

Stenosis of the coronary arteries leads to **cardiac cramps or angina pectoris.** Arteriosclerosis and stenosis of the leg arteries causes **intermittent walking or claudicatio intermittens.** The poor circulation in shock conditions is a **generalized hypoxic ischaemia.**

Anaemic hypoxia with reduced oxygen capacity is a common phenomenon (e.g. **anaemia and cyanosis; methemoglobinaemia; cherry red cyanosis in CO-haemoglobinaemia**).

Histotoxic hypoxia is caused by insufficient capacity for oxygen utility. This condition is present in cyanide poisoning with blockade of mitochondrial oxidation, and in exhaustive (supramaximal) work, where the O_2 utility is excessive.

Describe symptoms and signs of acute and chronic hypoxia and hypercapnia

1. **Acute hypoxia. In the CNS:** headache, nausea, **black out,** cramps and unconsciousness. **In the peripheral tissues: cyanosis** of the skin and the mucous membranes. The **threshold for cyanosis** is the presence of **more than 50 g reduced haemoglobin per litre of average capillary blood.**
   In principle, cyanosis is caused by **either reduced arterial saturation** or **increased arteriovenous oxygen difference.**
2. **Chronic hypoxia:** nausea, vomiting, lack of appetite, increased ventilation, increased erythropoiesis, increased brain bloodflow, right ventricular failure or cor pulmonale (**lung–heart**) and mountain sickness.
3. **Acute hypercapnia** (CO_2-poisoning): the patient is flushed, nervous and horrified of death, and has increasing dyspnoea. The death-horror and hallucinations are the last feelings before he or she loses consciousness and respiration is arrested. The blood gases show increased P_{aCO_2} and reduced pH (**acute respiratory acidosis**) with normal standard [bicarbonate] and base excess (Chapter 34).

Is it risky to treat chronic hypercapnic and hypoxic patients with 100% oxygen?

Hypoxia is dangerous because its effects are irreversible, while hypercapnia is reversible and unpleasant. The oxygen treatment increases P_{aO_2}, which is vital, so oxygen therapy should be administered instantly to hypercapnic, hypoxic patients. A few patients may have adverse effects, when the hypoxic drive for the peripheral chemoreceptors is eliminated. This is probably the last drive for the patient's ventilation. The ventilation will fall, which elicits a substantial rise in P_{aCO_2}. Carbon dioxide is a strong anaesthetic substance, and RC is depressed leading to respiratory arrest.

What is hypocapnia?

Hypocapnia or hyperventilation is a disorder with abnormally reduced P_{aCO_2}. The hyperventilation reduces P_{aCO_2} and produces an **acute respiratory alkalosis**, characterized by **increased pH, and normal** or unchanged [Standard Bicarbonate]/Base Excess (see Chapters 34 and 60). Changes in **[Standard Bicarbonate]/Base Excess** are effected by renal mechanisms which take days to develop.

How can two functional classes of lung disorders be distinguished?

1. **Obstructive lung disorders** are characterized by **low expiratory air flow** (low FEV_1 in Fig. 40-1). The FEV_1 is above 80% of the forced vital capacity in healthy persons.
   The obstructive patient with reduced lung recoil is forced to live with expanded lungs and a barrel-formed thorax with an abnormally high residual volume (RV in Fig. 40-1). The obstructive disorder is acute and episodic as in **bronchial asthma** or chronic as in **chronic bronchitis with emphysema.** Widespread degeneration of

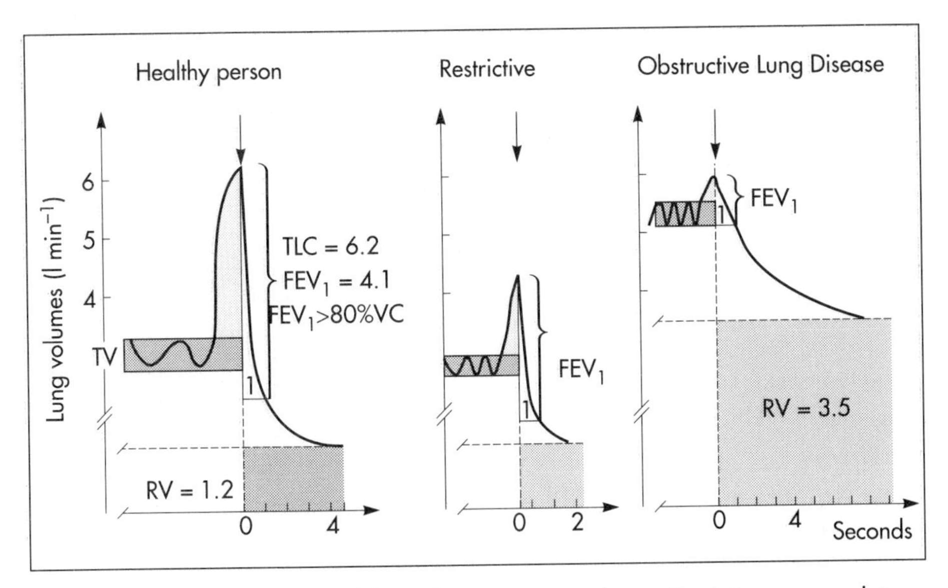

Fig. 40-1: Vital capacity and forced expiratory volume (1 s) is measured in a healthy person and in patients with restrictive and obstructive lung disease, respectively.

septal tissue with lung capillaries reduces the area of diffusion, and the lung diffusion is minimal in **emphysematous patients.**

2. **Restrictive lung disorders** are characterized by **low lung volumes** (Fig. 40-1). Both the IRV and ERV are small. FEV_1 is normal relative to the forced vital capacity, but small in absolute terms. Examples are lung fibrosis, lung congestion, and lung oedema.

What is the consequence of abnormal blood gases for lung patients?

Abnormal blood gas values are indicators of the severity of the disorder. The first phase is that characterized by **normal blood gases** at rest – often called **ventilatory insufficiency.**

The gas exchange of the chronically ill patient is reduced over the years, and **abnormal arterial blood gas tensions at rest** develop. This late stage of lung disease is often called **terminal respiratory insufficiency**, as the patient may die at any moment.

What is the advantage of oxygen enriched air treatment?

The advantage can be shown by an example. A patient with asthma is hospitalized with a **P_{aO_2} of 5.5 kPa or 41 Torr and a S_{aO_2} of 0.75** (Fig. 40-2). Oxygen-enriched air is valuable to such a patient. **Oxygen-enriched air** is administered accurately with a simple plastic mask using the **Venturi principle.**

A small increase in the oxygen concentration of atmospheric air from 21 to 24% leads to a rise in P_{IO_2} (3% of 95 kPa is 2.9 kPa; 3% of 713 Torr is 21.4 Torr). The major part

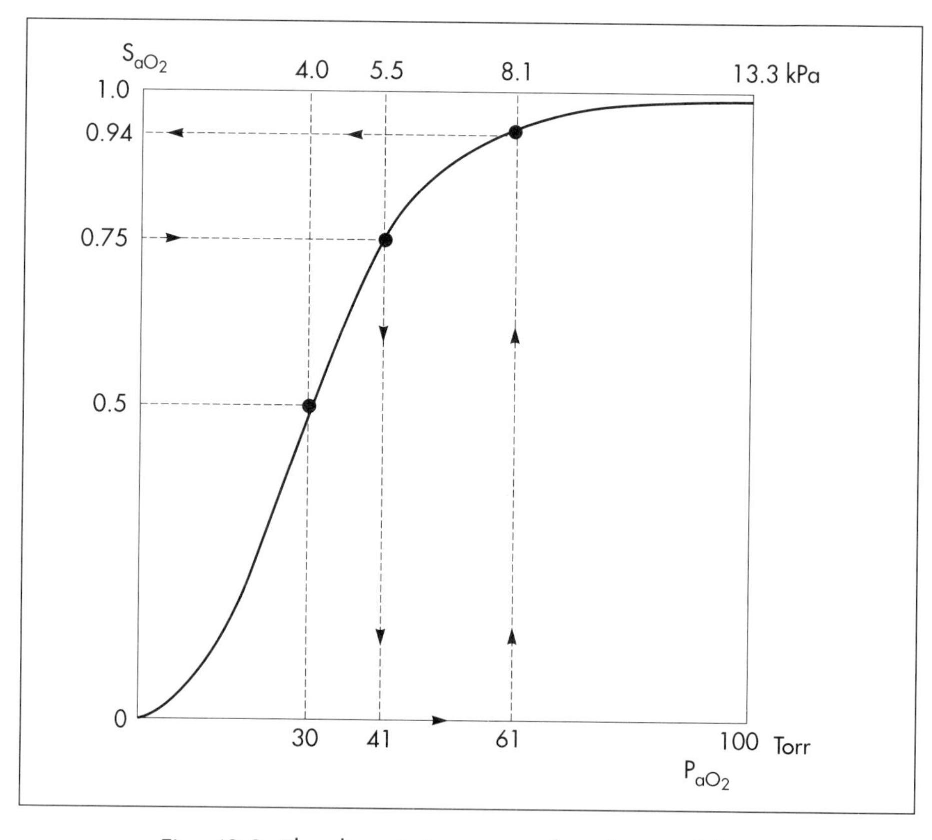

Fig. 40-2: The dissociation curve of oxyhaemoglobin.

of this rise reaches the arterial blood (2.6 kPa or 20 Torr) and this rise in P_{aO_2} from 41 to 61 Torr is often enough to save the patient, because S_{aO_2} **increases to 0.94** (Fig.40-2). This depends upon a normal haemoglobin concentration and a normal $\mathring{Q}$, and the P_{aO_2} is of course of utmost importance.

What is the spectrum of possible $\mathring{V}_A/\mathring{Q}$ ratios in the lung?

Referring to Fig. 40-3:

A. is the **extreme mismatch of venous to arterial shunting of blood**, where $\mathring{V}_A$ is zero and $\mathring{Q}$ is normal ($\mathring{V}_A/\mathring{Q} = 0$). The blood gas tensions approach those of venous blood ($P_{\bar{v}O_2} = 45$ and $P_{\bar{v}CO_2} = 46$ Torr in Fig. 40-3, equal to 6 and 6.1 kPa).

B. represents the **normal situation in which $\mathring{V}_A$ and $\mathring{Q}$ are matched** ($\mathring{V}_A/\mathring{Q} = 1$), P_{aCO_2} is 40 Torr or 5.3 kPa, and P_{aO_2} is 108 Torr or 14.4 kPa.

C. shows a lung unit changed to an **alveolar dead space** by a lung embolus. The $\mathring{V}_A$ is normal, but there is no $\mathring{Q}$, so $\mathring{V}_A/\mathring{Q}$ approach infinity. In such a lung unit, alveolar gas pressures approach the levels in inspired air (Fig. 40-3).

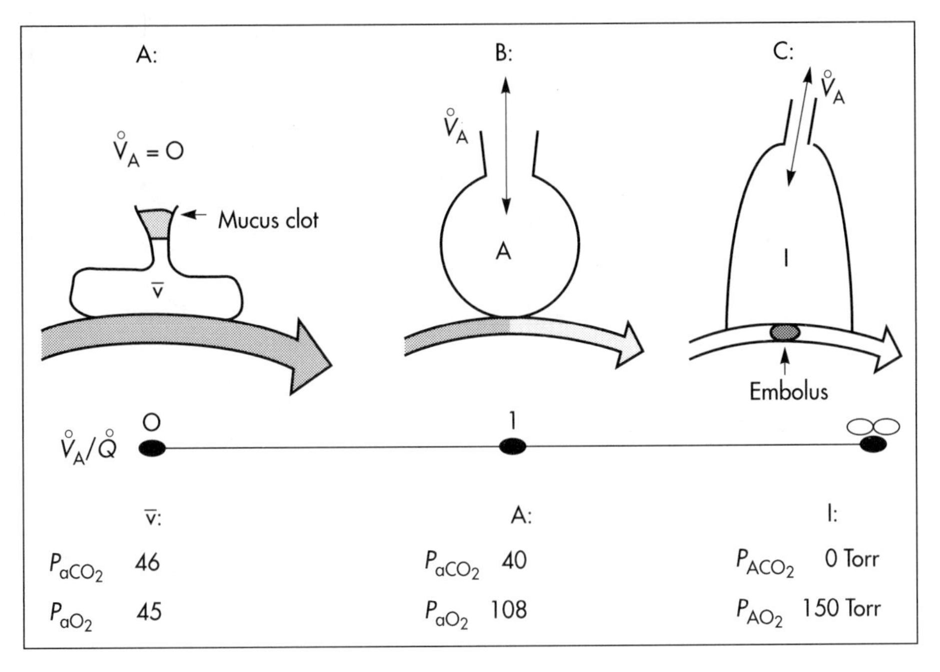

Fig. 40-3: Three alveoli representing $\dot{V}_A/\dot{Q}$ ratios from zero to infinity.

What are the pathological factors causing mismatch of $\dot{V}_A/\dot{Q}$?

Uneven distribution of tidal volume can eventuate from **uneven resistance to airflow** within the lung (bronchoconstriction, collapse and compression of airways) or from **uneven regional lung compliance** (insufficient surfactant, loss of elastic recoil as in destruction of alveolar tissue, and increase of elastic recoil as in connective tissue scarring or fibrosis with stiff lungs). **Hypoperfusion** can be caused by **compression of pulmonary vessels, obliteration of vessels** by fibrosis, or **blockage by emboli or thrombosis.**

Functional shunts arise with any consolidation of alveolar regions that continue to have bloodflow (pneumonia, oedema, haemorrhage, cell necrosis, lack of surfactant).

Is it possible to estimate the size of a veno-arterial shunt simply by arterial blood gases?

In a **$P_{O_2} - P_{CO_2}$ diagram** (Fig. 40-4) a $\dot{V}_A/\dot{Q}$ relationship, representing **all $\dot{V}_A/\dot{Q}$ ratios from zero to infinity**, can be drawn. Such a curve connects the points for $\dot{V}_A/\dot{Q} = 0$ ($P_{\bar{v}CO_2}$ and $P_{\bar{v}O_2}$) and for $\dot{V}_A/\dot{Q}$ = infinity (P_{IO_2} and P_{ICO_2} in Fig. 40-4). The curve is constructed as the set of intersections between the straight lines representing varying ***R* values** (respiratory exchange ratio, see below), and the curved blood–*R* lines. The shape of the **blood–*R* lines** is dictated by the oxyhaemoglobin dissociation curve, the CO_2–bicarbonate dissociation curves, the Bohr- and the Haldane-shifts. The blood–*R* lines fan out from the venous point (*v*), whereas the gas–*R* lines fan from the *I* point.

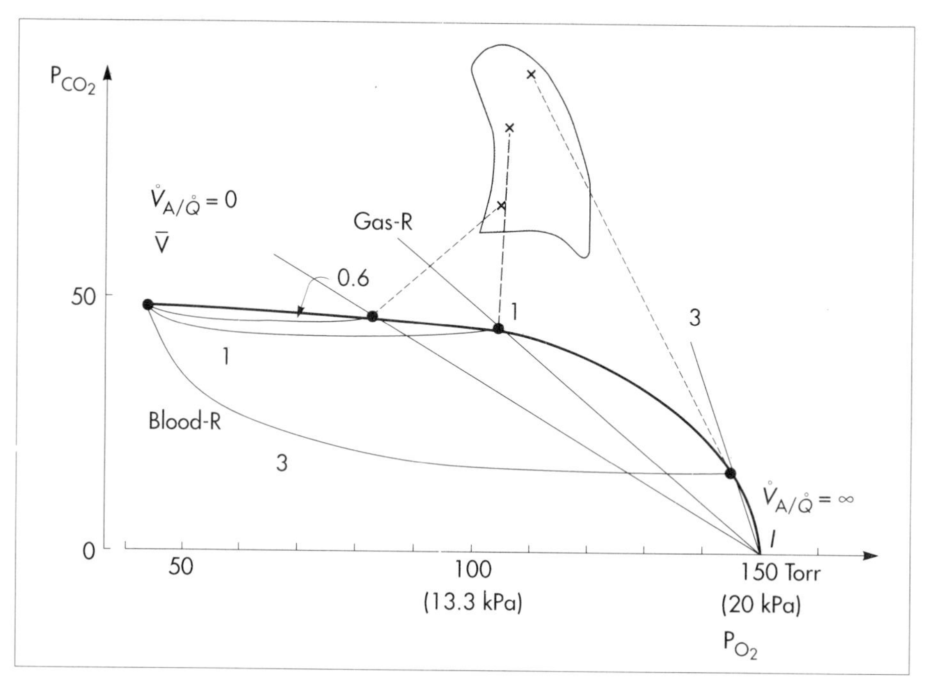

Fig. 40-4: The $\dot{V}_A/\dot{Q}$ curve and R values.

Each point of intersection represents an alveolar gas composition with which blood can equilibrate. Three gas–R lines show that the **lower lung regions are relatively underventilated** ($\dot{V}_A/\dot{Q}$ is below 1, and R is 0.6), the **middle lung regions are well matched** ($\dot{V}_A/\dot{Q}$ is 1, and R is also 1), and the **upper lung regions are overventilated** ($\dot{V}_A/\dot{Q}$ is above 1 and approaching infinity just as R). Underventilated and overperfused alveoli have increased P_{AN_2} and thus increased P_{aN_2}, whereas overventilated alveoli reduce their P_{ACO_2} almost as much as they increase P_{AO_2} (Fig. 40-4). Hereby, P_{aN_2} can become greater than P_{AN_2}, and a precisely measured difference used as a measure of mismatch.

Each point in a P_{O_2}/P_{CO_2} diagram represents a **gas mixture with which blood can be equilibrated.** If the alveolar gas exchange with the ratio 1, the blood must do the same, and the **venoarterial CO_2 content difference** divided by the arteriovenal O_2 content difference must also be 1.

Just as the gas–R lines deviate from the inspiratory point, I, also the blood–R curves deviate from the venous point, $\bar{v}$ (Fig. 40-4). Curves of **similar values cross each other in the $\dot{V}_A/\dot{Q}$ curve with all solutions from zero to infinity.**

Arterial blood gas tensions can deviate from the ideal alveolar point all the way to the venous point, where the venous shunt is 100% and no blood is oxygenated (Fig. 40-4).

Further reading

Rahn, H. and W. O. Fenn (1962) *A Graphical Analysis of the Respiratory Gas Exchange.* Am. Physiol. Soc., Washington, DC.

West, J.B. (1989) *Respiratory Physiology.* 4th edn. Williams & Wilkins, Baltimore.

40. Multiple Choice Questions

A 30-year-old female is anaemic. Her haemoglobin concentration is only 65 g l^{-1}. The patient has given birth to five children.

Each of the following statements concerning her condition have True/False options:

40.1.:

A. A normal P_{aO_2}.
B. A normal S_{aO_2}.
C. A rightward shift of the oxyhaemoglobin dissociation curve.
D. A smaller oxygen capacity than normal.
E. A $P_{\bar{v}O_2}$ below normal.

40.2.:

A. Her working capacity is reduced.
B. Her serum iron concentration is normal.
C. Her $\dot{V}_A$ is normal.
D. Her $\dot{Q}$ is increased.
E. Her max$\dot{V}_{O_2}$ is reduced.

Try to solve the problems before looking up the answers in Chapter 74.

CHAPTER 41. DIVING AND FLYING

I. FLYING AND SPACE

High-altitude flight encounters disorders due to:

1. Changes in pressure (cf. barotrauma, decompression sickness).
2. Forces of acceleration and weightlessness.
3. Fall in P_{IO_2} (i.e. oxygen deficiency – see also altitude, Chapter 38).
4. Fall in ambient temperature.
5. Increased radiation.

Commercial airplanes have pressurized cabins, partially offsetting the effects of the low P_B at high altitude. The cabins are pressurized to a P_B of 80 kPa, corresponding to an altitude of 2000 m. The main problem is hypobarotrauma or squeeze during take-off and landing.

Space flight requires pressure suits or cabins ensuring adequate oxygen and pressure to protect against hypoxia and decompression sickness. Recycling techniques have been developed for the reuse of O_2. The spacecraft must carry along enough CO_2 absorbent to prevent CO_2 intoxication.

What are the dangers of space radiation?

A **lethal radiation dose** is rapidly reached when flying around the Earth in the **Van Allen radiation belts**, consisting of **high-energy level protons and electrons**. Therefore, flying takes place essentially below the inner belt (500 km). Take-off and landing of long distance space flights take place as close to the magnetic poles (minimum radiation energy) as possible.

What are the reactions to gravity?

Forces of acceleration are measured in **Gravity units (G)** based on the gravity of the Earth. These forces cause pooling of blood in the lower extremities, a critical drop in arterial blood pressure with **orthostatic hypotension or collapse** due to reduced venous return, but also cause nausea, vomiting, and spatial disorientation.

What are the acceleratory forces on the astronaut at take-off?

The start acceleration of a spacecraft is approximately linear, and the astronaut is exposed to a tremendous acceleration – often close to 10 times the gravity of the earth (10 G) at the first stage of a three-stage blast-off.

The body of the astronaut is located transverse to the axis of acceleration, and the astronaut must carry an anti-G suit in order to prevent pooling of blood in the legs (Chapter 26).

What disorders are caused by weightlessness?

Gravity acts on both the astronaut and the spacecraft in space, so astronauts are floating inside the satellite. **Weightlessness** produces diverse reactions caused by the absence of gravitational forces. The space pilot experiences **space sickness** (nausea and vomiting), falling total blood volume, muscular atrophy, Ca^{2+} loss from the bones and, after living in outer space for weeks, **adaptative difficulties** to life on Earth upon return. These difficulties are low working capacity and a tendency to faint (Chapter 26). The bone loss continues long after the return to Earth, because stimulation of bone formation requires physical activity in a gravity field.

What forces are involved in a parachute jump?

The acceleratory force of gravity (G = 9.8 or approximately 10 m s^{-2}) has the same size at a given latitude. Thus in an environment without air molecules and air resistance, all objects must move with the same acceleration when leaving the same aircraft or satellite. This is regardless of the size and character of the objects.

Assuming that a parachuter is **only exposed to gravity** he or she will fall towards the Earth with the acceleration 10 m s^{-2}. Following 10 s the velocity is approximately (G t) = (10 m $s^{-2} \times$ 10 s) or 100 m s^{-1}.

In real life there is no such thing as a free fall from an aircraft, because the air resistance soon reduces the velocity and exactly outbalances the acceleration following 10-12 s. Here the parachuter reaches the terminal velocity determined by the relation between G and air resistance.

As the parachute opens, the parachuter experiences an **opening shock load** of approximately 600 kg. The area of the parachute reduces the terminal velocity to 1/9. The force of impact at landing is $(1/9)^2 = 1/81$ of the landing force without parachute. This is equal to a jump from a height just above 2 m.

II. DIVING

Diving also encounters the above described problems, where the cause is change of pressure, temperature and body position in space.

Circulatory aspects of diving and flying are described in Chapter 26.

What is special about breath-hold diving?

Diving with no equipment at all is dangerous for several reasons.

1. **Hyperventilation** prior to the dive can cause **acute respiratory alkalosis** with dizziness and convulsions even before the dive.
2. **Oxygen deficiency** with **black out/grey out**. Prior hyperventilation with only three deep respirations increases P_{AO_2} to 17.3 kPa (130 Torr), whereas P_{ACO_2} is only 2.6 kPa (20 Torr). Cerebral P_{O_2} can fall below 4 kPa or 30 Torr (zone of unconsciousness) before a sufficient P_{aCO_2} is built up to cause breathlessness. Hypoxia in itself is insufficient to trigger inspiration. Drowning is likely, if this occurs underwater.
3. **Cutaneo-visceral reflexes** can cause malignant arrhythmia or a slow heart rate

termed **diving bradycardia** with vasovagal syncope in cold water. Heart rates around 30 beats min^{-1} have been observed in small children, who have been victims of cold water near-drowning for periods up to 40 min. Such heart rates are similar to those found during hypothermic surgery. These arrhythmias are triggered from cold receptors in the skin.

How do we take up and eliminate inactive gases?

By diffusion. During diving the ambient pressure will rise by **1 atmosphere for every 10 m of depth**.

Inactive gases (i.e. N_2, H_2, He, Ne, Ar, etc.) increase their partial pressure when present in the breathing medium, and the blood of the lung capillaries is in immediate equilibrium with the partial pressure in the alveolar air. Atmospheric air is the commonly used breathing gas for diving. The **solubility coefficient for N_2 in blood** is low, but five times larger in fat tissue. Thus a large $\mathring{Q}$ is necessary to transport substantial amounts of N_2 to the tissues. The modest perfusion rate of fat tissues ensures long duration of saturation dives, where the nitrogen tension of fat tissue reaches equilibrium with P_{aN_2}.

The human body is a complex of tissue types, each with an individual, exponential uptake rate or half-life for saturation (Fig. 41-1). Often a model with **half-life tissues** of 5-, 10-, 20-, 40-, 80-, 120-, 240-, 480-min (of half-life) is used.

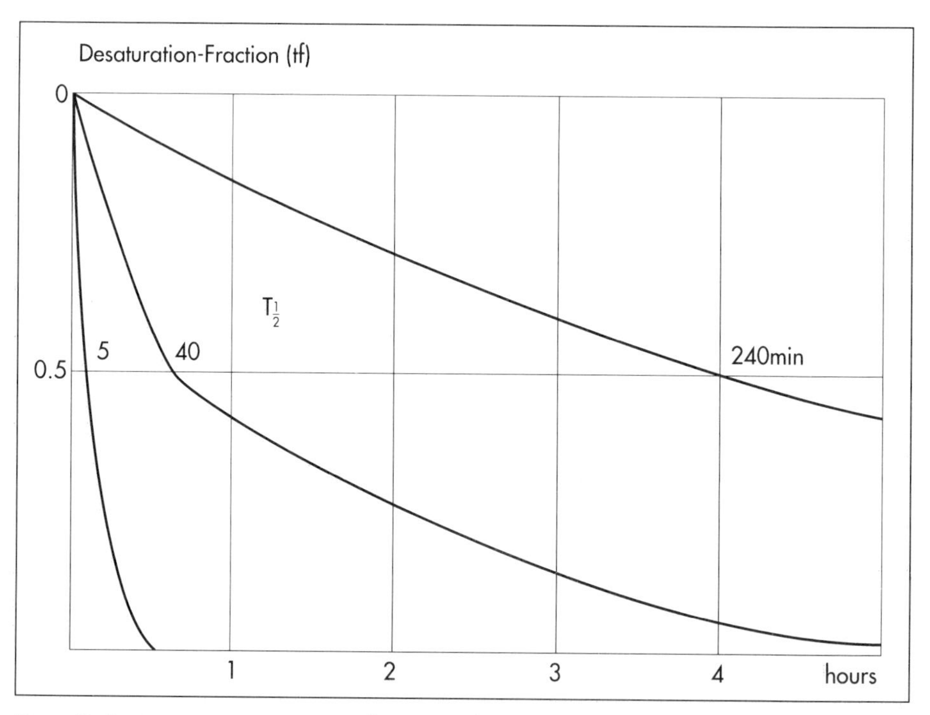

Fig. 41-1: Desaturation curves showing elimination of nitrogen from tissues with different half-life.

What is decompression sickness?

If decompression is too rapid **inert gases stored in the tissues form bubbles in the blood and tissues**, in the same way as bubbles are formed when a bottle of soda water is opened.

Symptoms and signs of decompression sickness are the following. Pains in the joints (**bends**) and in the muscles. A most alarming symptom is thoracic pains (**chokes**), which are caused by life-threatening air bubbles plugging the pulmonary capillaries. CNS symptoms and signs comprise dizziness, paralysis, collapse and unconsciousness.

How do we avoid decompression sickness? How is it treated?

No decompression diving limits (cf. decompression tables) are limits of depth and time spent at the depth. Use of the tables sometimes allows ascent and emergence slowly and systematically without decompression stops. Any saturation dive to less than 10 m allows ascent without decompression stops (Haldane's rule). Fast ascent during flight seldom causes problems at altitudes below 10 km. At altitudes above 20 km (where P_B <6.5 kPa or 47 Torr) the blood of a space traveller with normal body temperature will boil, if the pressure within his or her suit is lost. This is because the partial pressure of saturated water vapour at body temperature is 6.5 kPa or 47 Torr.

Stage decompression according to tables, with a rate below 18 m min^{-1} between stages, allows most people to ascent without decompression sickness (bends, caisson disease).

Rational treatment of bends requires immediate **recompression** in a tank, where sufficient pressure can be established to eliminate the bubbles causing the disease. The nitrogen gradient from the diver's body to the air in the decompression tank is preferably increased for rapid removal of nitrogen from the body. This is done by increasing the P_{O_2} of the alveolar air.

What is saturation diving?

Diving at great depth (80–300 m) makes it necessary to live in large compression tanks for longer periods with the body **saturated** with inert gases.

Extremely deep dives are performed with He instead of N_2 as the inert gas. The solubility coefficient for He is lower than that of N_2, He is far less toxic than N_2, and He has a much lower density than N_2. The low density reduces the breathing resistance, whereby it is possible to work at great depth.

He is used together with a small O_2 fraction, in order to avoid acute O_2 poisoning (see below). At 200 m of depth only 1% O_2 is necessary in the **He–O_2** mixture (so-called **heli-ox**).

What is inert gas narcosis?

When breathing compressed air, the first symptoms of this **rapture of the depth** appear at 40 m of depth, with an euphoric behaviour similar to alcohol intoxication. The intoxication increases in intensity with depth (Martinis law) with anxiety, lack of judgement, lack of concentration, and lack of muscular co-ordination. Above 10 atm. absolute pressure a typical **narcotic condition** (anaesthesia and unconsciousness) develops. The diver has reduced nerve conduction velocity and α-waves in EEG.

The inactive gases are lipid-soluble and thus diffuse easily into fatty tissues, plasma membranes, and intracellular structures, where they bind to **active sites** or receptors. **Ar** has a larger narcotic effect than N_2 (larger **lipid solubility**, larger **energy content** or van der Waal-forces).

The limit for compressed air diving should be 50 m.

What is hyperbaric oxygen therapy?

Hyperbaric oxygenation therapy is used for disorders with either local or global oxygen deficiency, and for general diseases. Several medical centres can treat patients at high oxygen pressures in pressure tanks. Oxygen is given at partial pressures just below 2 atmospheres (see acute O_2 poisoning).

Treatment of **gas gangrene, leprosy, air embolism, CO poisoning, tetanus and myocardial infarction** is successful. Accumulation of oxygen free radicals seem able to destruct the anaerobic microorganisms causing clostridial gangrene, tetanus, and leprosy infections.

What is acute and chronic oxygen toxicity?

Acute toxicity

Oxygen free radicals such as the superoxide O_2^- and hydrogen peroxide are produced from the dissolved oxygen in tissue fluids. The production equals the enzymatic removal as long as tissue P_{O_2} is normal. Above 2 atmospheres absolute pressure of 100% oxygen the enzymatic removal fails. Use of more than 2 atmospheres of O_2 partial pressure for more than 30 min leads to an **acute, cerebral intoxication** with fasciculations of the mimic face muscles, nausea, vertigo, hiccup, coma and universal cramps.

The retinal cells of the eye are actually brain cells. **Acute O_2 intoxication** causes retinal vasoconstriction, vascular proliferation and retinolysis (i.e. **retrolental fibroplasia**).

Chronic toxicity

A patient breathing **80% O_2 at 1 atmosphere absolute for more than 12 h** frequently develops **pulmonary hypertension** with retrosternal pain, coughing and pulmonary oedema with bleeding. Such a patient loses surfactant and develops atelectases. Several patients have died of hypoxia in spite of the high P_{O_2}. The exposure of lung tissue to high P_{O_2} is direct and total, without the protection of the haemoglobin buffer system.

What is squeeze?

Squeeze or **hypobarotrauma** is a consequence of the **Boyle–Mariotte law**, the product of pressure and volume is constant at constant temperature. The hypobarotrauma is caused by a negative pressure difference across the wall of a non-collapsible air-space in the body (middle ear, sinuses, compressed lungs, etc.). A diver who has caught a cold will descend with sinus openings closed, and develop pain with depth as the pressure in his or her occluded sinuses becomes more negative compared to the surroundings.

Is it possible to dive deeper than 30 m on breath-holding?

When a healthy person, with TLC 6 l and RV 1.5 l, **breath-hold dive** to 30 m or 4 atmospheres of pressure, his or her TLC is compressed by a quarter to 1.5 l, which is equal to his or her RV at the surface. This depth is assumed to be the maximal diving depth. In such a calculation is implied that the RV at the surface is equal to the RV at the bottom. However, this is not true.

The **world record is beyond 105 m**, which is unbelievable when compared to the maximal diving depth calculated above. The following two phenomena explain this world record. **RV decreases** with increasing diving depth, because the diaphragm is pushed upwards like the piston in a syringe, and **blood is pushed into the pulmonary circulation**. At some further depth, the capillaries will rupture and blood/oedemal fluid reach the alveoli (**alveolar squeeze**). This is a **hypobarotrauma** or **barotrauma of descent**, and it has been observed in some of the record holders.

How does air embolism develop?

This is a **hyperbarotrauma or barotrauma of ascent**. When a diver breathes compressed air at depth, there is a risk of developing **air embolism**, where alveolar air enters the blood through damaged vessel walls. An anxious diver may not expire during ascent and so retain air in the lungs. The **lung volume increases** during ascent according to the **Boyle–Mariotte law**, the thin walled thoracic vessels are compressed and the **alveoli dilate until they burst**. The **air dissects its way** into the skin (subcutaneous emphysema) and into the pleural cavity (pneumothorax). The terminal stage of this condition involves large amounts of air into the lung veins (air embolism), which causes death within minutes due to blocked circulation to brain and heart.

Escapes from submarines at depths of 100 m are successful, when the person expires during ascent, and thus avoids air embolism.

What is the key problem in drowning?

In more than 90% of drownings and near-drownings the **lungs are flooded with water.**

Fresh water is hypotonic and rapidly absorbed, diluting plasma to become hypotonic and causing bursting of the red cells (**haemolysis**). The victim often dies within 3 min.

Sea water is hypertonic and draws fluid from the blood plasma into the lung alveoli and interstitial fluid, so that plasma volume decreases. This causes **haemoconcentration** and shock. The victim often dies within 6 min.

The **rational therapy**, if early enough, is immediate resuscitation and treatment of the respiratory and circulatory disorders (haemolysis or haemoconcentration).

How are light and sound distorted in water?

The light intensity decreases rapidly with depth, and 100 m below sea level, it is permanently dark.

The **cornea–air interface** becomes a **cornea–water interface**, when in water without goggles. The refractive index for water and for the cornea itself is much the same (1.33), eliminating most of the refractive power in air. A skin diver without goggles is **hypermetrope** under water.

Sound propagates more rapidly in soft tissue and water than in air (1540 m s^{-1} instead of 340 m s^{-1}). Therefore **underwater sound sources** appear nearer than they actually are. Because of the shortened delay between the two ears, localization of sound sources becomes extremely difficult. The voice of a diver speaking in a helium–oxygen atmosphere resembles the voice of Disney's Donald Duck.

Further reading

Edmonds, C., C. Lowry and J. Pennefather (1981) *Diving and Subaquatic Medicine*, 2nd edn. The Diving Medical Center, Biomed. Marine Serv., Seaforth, Australia.

Hayward, J.S., C. Hay, B.R. Matthews, C.H. Overwell and D.D. Radford (1984) Temperature effect on the human dive response in relation to cold water near-drowning. *J. Physiol.* **56**: 202–6.

Paulev, P.-E., M. Pokorski, Y. Honda, B. Ahn, A. Masuda, T. Kobayashi, Y. Nishibayashi, Y. Sakakibara, M. Tanaka and W. Nakamura (1990) Facial cold receptors and the survival reflex 'diving bradycardia' in man. *Jpn. J. Physiol.* **40**: 701–12.

41. Case History

A male farmer, 18 years of age, is stabbed through the left foot while working as a stableman. He develops a fulminant infection with Clostridium tetani, *which only grows under anaerobic conditions and produces a potent toxin causing permanent depolarization of the motor endplate. He is brought to hospital in a moribund condition, which is not improved by antitoxin and antibiotics. The farmer is transferred to a pressure chamber, where hyperbaric oxygen therapy is provided at 3 atmospheres pressure for 20 min. The haemoglobin concentration of the patient is 170 g l^{-1}. P_B is 1 atmosphere (760 Torr), P_{ACO_2} is 5.3 kPa (40 Torr), and the alveolar P_{water} is 6.5 kPa (47 Torr). The solubility coefficient (α) for oxygen in the blood is 0.022 ml STPD ml^{-1} $Torr^{-1}$. One Torr equals 133.3 Pascal.*

1. *Calculate the concentration of physically dissolved O_2 in the blood leaving the lung capillaries of the patient.*
2. *Calculate the concentration of chemically dissolved O_2 in the blood.*
3. *Does excessive oxygenation lead to gas transport problems?*

Try to solve the problems before looking up the answers in Chapter 74.

SECTION V.
Metabolism and Gastrointestinal Function

Starvation on one side and obesity resulting in atherosclerosis on the other are major problems in different parts of the world. Related topics are treated in this section. Solving the hunger problem with genetic manipulation and the introduction of ruminant enzymes such as **cellulase** are not attractive propositions. Problems related to luxury food consumption are easier to handle by the medical profession, and may contribute to the solution of hunger problems.

CHAPTER 42.
ENERGY EXCHANGE AND EXERCISE

It is generally believed that nutrients are necessary in order to produce energy in the human body. However, **this is impossible.** The **first law of thermodynamics** states that energy can neither be created nor destroyed, but can **only be transferred** from one form or one place to another.

Life is the maintenance of an infinite row of non-equilibrium reactions in such a way that appear to be in a stationary condition, **a steady state.** Real life is **chaos,** a steady state only being maintained as long as we derive chemical energy from food. Only part of the dietary energy is available for ATP formation in humans; cellulose, for example, passes through the digestive tract without being absorbed. The absorbable chemical energy passes through the intestinal mucosa, and in the body is transformed to energy-rich phosphate bindings in ATP (**Gibbs energy,** ΔG).

ATP is broken down to ADP during muscular contractions. Muscular contractions stimulate the oxidation of fatty acids and carbohydrates in the muscle cells which liberate more energy for rephosphorylation of ADP to ATP. The energy is used for the maintenance of chemical syntheses, electrochemical potentials and for the net transport of substances across membranes.

The **Gibbs energy** is the free chemical energy available for exercise and life in general. However, 75% is lost as heat energy, and the **mechanical efficiency** of exercise is therefore only 25%. The ratio between **external work** (W') and the total energy used during work ($-\Delta U$) is called the **mechanical efficiency**. In this case $\Delta U = \Delta G$. The **mechanical efficiency** is always less than 1 and often only 0.25 as stated above. The energy which is not transferred to external work is released as heat energy ($-Q$) or is accumulated in the body as heat. At the onset of exercise 50% of the total energy from hydrolysis of ATP is converted into mechanical energy in the myofibrils. The remaining 50% is dissipated as **initial heat**. However, as shown above the mechanical efficiency is only 25%. This is because energy recapturing recovery processes (oxidative regeneration of ATP, etc.) occur outside the myofibrils. Hereby, half of the energy is dissipated as so-called **recovery heat**.

Heat energy is **low prize energy**. In contrast to ATP energy, it is not available for work in the body. The sum of heat energy generated and work performed is constant:

$$1.\ \textbf{Law of thermodynamics}: Q + W' = \Delta G.$$

When no work is performed W' is zero, and all body reactions are reflected by the liberated heat energy ($-Q$), which is equal to the decrease in Gibbs energy ($-\Delta G$).

When the pressure–volume work is zero, we have a special energy concept: the **heat content** or **enthalpy**, H, which sums up all energy. The sum of liberated heat energy ($-Q$) and liberated work ($-W'$) is thus equal to the fall in enthalpy:

$$(-\Delta H) = (-Q) + (-W').$$

The decrease in enthalpy of the human body ($-\Delta H$) is equal to the fall in potential, chemical energy stored in the body.

Entropy is the tendency of atoms, molecules and their energies to spread in a maximum space. The **Gibbs energy** (G) is the difference between enthalpy (H) and entropy (S) when multiplied with the absolute temperature (T):

$$G = H - TS.$$

G determines if a certain reaction occurs, since G is minimum at equilibrium. According to the formula, entropy is important at high temperatures, and energy is most important at low temperatures.

The **decrease in Gibbs energy** covers **almost the total energy**, except for the **pressure–volume work**. Since oxygen consumption is almost equal to the carbon dioxide output, the pulmonary volume change is negligible and this work is negligible.

What is the definition of metabolism?

The **metabolism** of a person is defined as the **sum of all chemical reactions in which energy is made available and consumed in the body** in a given time period. The bindings between hydrogen and carbon in nutrients are a source of energy for animals. Such substances are changed into **metabolic end-products** (eliminated as bilirubin, urobilin, urea, uric acid, creatinine, etc.) and to **metabolic intermediary products** (i.e. products that participate in other chemical reactions). The **net metabolism** is the **sum stochiometry of the single net reactions in the body**.

The decrease in enthalpy in a given time period ($-\Delta H$ **per min**) is the **Metabolic Rate (MR)**.

What are the most important reactions in net metabolism?

The **oxidation of fuel (carbohydrates, glycerol, fatty acids)** to CO_2 and water is the primary pathway for generation of energy and subsequent heat energy liberation. Protein can also serve as an important energy source during prolonged exercise, but it must first be broken down to amino acids, which are then partially oxidized (to CO_2, H_2O, NH_4^+ etc.). The daily production of **metabolic water** is 350 g and of urea 30 g.

Diabetes mellitus and hunger (**hunger diabetes**) are conditions where fatty acids produce ketone bodies.

During forceful exercise, energy is obtained primarily from **non-oxidative sources (glycolysis)**. There is, therefore, a net formation of lactic acid from glycogen. Following anaerobic exercise the **lactate elimination** accounts for an extra O_2 consumption called **oxygen debt (Fig. 42-2).**

Oxidation of alcohol contributes to metabolism. The energetic value of alcohol is 30 kJ g^{-1}. An adult person of **70 kg body weight** can combust **7 g of alcohol per hour**. The chemical energy liberated is $(7 \cdot 30) =$ **210 kJ h^{-1}** or 70% of his or her resting MR (300 kJ h^{-1} or 83 W).

Most of the chemical reactions in our body are **degradative or catabolic** – they break a molecule down to smaller units. These reactions are often also **exothermic** (heat releasing) and **exergonic** (the content of Gibbs energy decreases during these reactions). **The synthetic or anabolic reactions** (e.g. the formation of protein from amino acids) are obviously coupled to these degradative reactions. Synthetic reactions are most often also endothermic and endergonic.

Describe the metabolism of alcohol

Alcohol diffuses easily in the human body. Twenty per cent of the intake is readily absorbed in the stomach. The **absorption is fast** and is stimulated by CO_2 in wine such as champagne.

Alcohol distributes in the total water of the body within 1 h. The fraction of the body weight which is the distribution pool for alcohol is called ***r*** by the Swedish scientist Widmark (**mean-*r*** for females is 0.55 kg per kg body weight and is 0.68 kg per kg body weight for males).

The most important **elimination of alcohol is by oxidation**. The rate of oxidation is constant ($\beta = 0.0025‰\ \text{min}^{-1}$) and is independent of the blood [alcohol]. The **blood [alcohol]** is measured in the unit **g kg**$^{-1}$ (‰). The absolute amount of **alcohol eliminated per minute** is thus: ($\beta \cdot \boldsymbol{r} \cdot$ **body weight**).

The constant rate is due to the primary, partial oxidation to acetate via acetaldehyde in the liver by alcohol dehydrogenase: $C_2H_5OH + O_2 \xrightleftharpoons{\text{Alcohol dehydrogenase}} CH_3COOH + H_2O$. Acetate is broken down in nearly all tissues. The total oxidation of alcohol: $C_2H_5OH + 3\ O_2 \leftrightarrow 2\ CO_2 + 3\ H_2O$ implies an RQ of 2/3. A healthy person with a MR of 1 mol O_2 h^{-1} can partially oxidize almost 1/6 mol of alcohol per hour, by using almost 1/6 of his or her MR in the liver (1 mol = 46 g alcohol; 46/6 or about 7 g alcohol h^{-1}).

If this person receives an alcohol infusion of 7 g h^{-1} and has a normal hepatic bloodflow of 90 l h^{-1} (1.5·60 min), his or her maximal alcohol elimination rate corresponds to a blood [alcohol] of (7/90) = 0.08 g l^{-1}. This is a blood [alcohol]

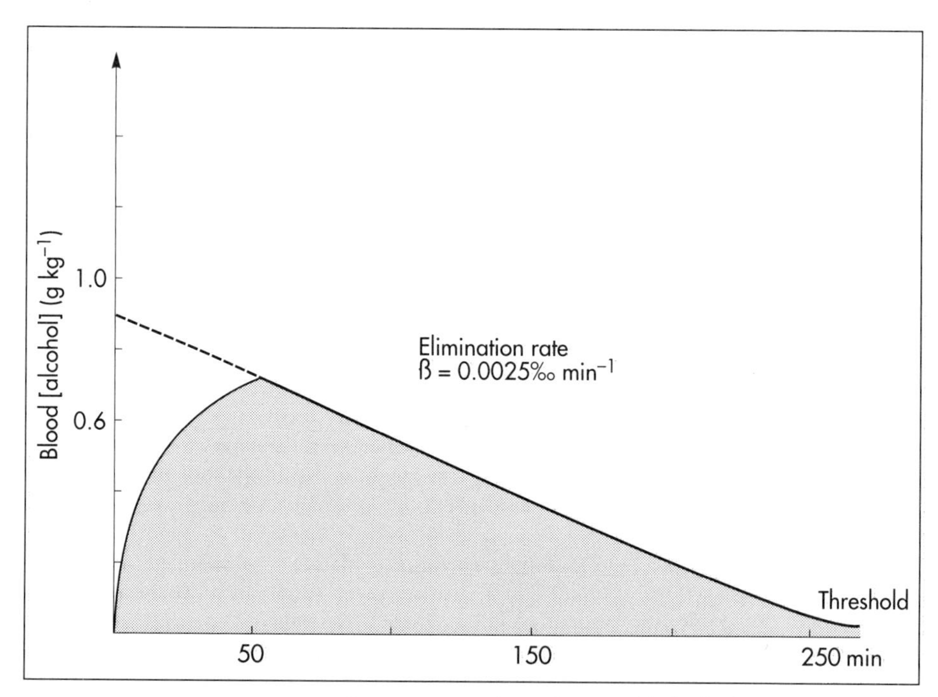

Fig. 42-1: Absorption and oxidation of alcohol.

threshold below which the oxidation rate decreases (Fig. 42-1). There are two other enzymes, apart from **hepatic alcohol dehydrogenase**, that can oxidize alcohol. These are catalase and MEOS (microsomal ethanol oxidation system). A small amount of alcohol dehydrogenase is found in the gastric mucosa.

What happens during alcohol intoxication?

The sequence of events in **acute alcohol intoxication** proceeds with an increasing sense of warmth, flushing of the face, dilated pupils, dizziness and euphoria. There is a general sense of well-being with unjustified optimism and the feeling of increased strength and energy. The subject shows boisterous behaviour with increased psychomotor activity, which is clumsy, and social inhibitions are dissolved.

With increasing intoxication the symptoms and signs of CNS depression become apparent. The subject may become drowsy, argumentative, angry or weepy, and eventually they may vomit and complain of diplopia (double vision). Later an examination reveals areflexia, loss of muscular tension, loss of sphincter control, rapid heart rate and respiratory frequency, falling arterial pressure and MAP leading to shock. The subject develops hypothermia (Chapter 44) and increasing stupor, anaesthesia, coma or death.

The intoxication depresses the myocardium and dilates the peripheral vessels. This is why the MAP falls together with cardiac performance.

Is excretion of alcohol important?

The excretion of alcohol molecules takes place through (1) **expiratory air**; and (2) **urine and sweat.**

1. **Ventilation**. A resting athlete with a blood [alcohol] of 1 g kg^{-1} has a small alcohol partial fraction in blood and alveolar air (1/2000). With an alveolar ventilation of 5 l BTPS min^{-1} at rest, this person excretes $(1 \cdot 5 \cdot 1/2000) = 0.0025$ g min^{-1} or **0.15 g h^{-1}** via expiratory air.
2. **Urine and sweat**. The concentration of alcohol in plasma water, sweat and urine is 20% (1.2) higher than the blood [alcohol]. A resting person with a **blood [alcohol] of 1 g kg^{-1}**, a **diuresis of 1 ml min^{-1} or 0.06 l h^{-1}**, excretes alcohol by renal ultrafiltration at a rate of only $(1.2 \times 0.06) =$ **0.072 g h^{-1}**. If the person also has a sweat loss of 0.1 l h^{-1}, he or she further excretes $(1.2 \times 0.1) =$ **0.12 g h^{-1}**. The total excretion at rest is **0.342 g h^{-1}, which is negligible.**

   However, during 1 h of exercise in a warm climate (with 1 g kg^{-1}), when $\mathring{V}_A$ is 80 l min^{-1}, and when water loss is 4 l h^{-1} (sweat and evaporation), his or her total alcohol excretion flux is: $(1 \times 80 \times 60 \times 1/2000) + (1.2 \times 4) =$ **7.2 g h^{-1}**. This **total excretion is larger** than the **amount broken down by maximal oxidation: 7 g h^{-1}.**

Is the oxidation rate of alcohol variable?

The rate (β) increases with **increasing temperature**, with **increased metabolism** (thyroid hormones, dinitrophenol), and **decreases under the influence of enzyme inhibitors**. Fructose increases β. The sum [NAD + $NADH_2$] is constant. Hepatic alcohol oxidation causes [$NADH_2$] to rise, so that **[NAD] becomes the limiting factor** in oxidation.

What is the respiratory quotient (RQ) and what is R?

RQ is the **hypothetical, metabolic ratio** between carbon dioxide output ($\mathring{V}_{CO_2}$) and oxygen consumption ($\mathring{V}_{O_2}$) of all the cells of the body. RQ is an indicator of the type of foodstuff metabolized.

***R* is the measurable ventilatory ratio ($\mathring{V}_{CO_2}/\mathring{V}_{O_2}$)** for the person quantified by gas exchange equipment.

Respiratory steady state is a condition where ***R* equals RQ**, and the gas stores of the body are unchanged.

How does RQ reflect the type of foodstuff oxidized?

Compared to the oxidation of carbohydrates (RQ= 1), fat oxidation has a distinctly low RQ (0.7), and protein is oxidized with a RQ of 0.8.

Carbohydrates are rich in oxygen compared to the minimum in fats. Overfeeding with carbohydrates results in a partial conversion to fats. The corresponding release of oxygen diminishes $\mathring{V}_{O_2}$, and RQ becomes **greater than 1**.

The diminished glucose metabolism during fasting and in diabetics, lowers the RQ towards 0.7, because of the increased conversion rate of fat.

What happens to the R value during hyperventilation?

Hyperventilation decreases the amount of exchangeable CO_2 in the large body stores, without altering oxygen uptake. The tissues and blood cannot store additional oxygen. As a consequence the $\mathring{V}_{CO_2}/\mathring{V}_A = F_{ACO_2}$ is reduced. This implies a fall in P_{ACO_2} and in P_{aCO_2}. *R* is distinctly increased during hyperventilation often up to 2–3.

Hypoventilation reduces *R* towards zero at apnoea.

Metabolic acidosis is characterized by low pH and base deficit (negative base excess) in the ECF. **Metabolic alkalosis** is characterized by a high pH and a positive base excess. **Metabolic acidosis** is compensated by hyperventilation implying a rise in *R*, and metabolic alkalosis is compensated by hypoventilation with a fall in *R*.

What happens to R during exercise?

***R* does not change** when a person on a mixed diet (RQ = 0.83), or when a person on a high fat diet (RQ = 0.7) exercises moderately, because the fat combustion dominates.

***R* will fall**, however, when a person on **carbohydrate-rich diet** (RQ = 0.96 – 1) works for hours.

Strenuously heavy exercise implies a substantial, initial rise in *R* ($R > 3$), because the lactate liberated will release CO_2, which is then eliminated in the lungs in much larger volumes than oxygen is taken up.

Calculate RQ for the oxidation of glycogen and glucose

$$\text{Glycogen: } (C_6H_{10}O_5)n + 6n\ O_2 = 6n\ CO_2 + 5n\ H_2O, \text{ that is } \mathbf{RQ} = 1.$$

$$\text{Glucose: } C_6H_{12}O_6 + 6\ O_2 = 6\ CO_2 + 6\ H_2O, \text{ that is } \mathbf{RQ} = 1.$$

The enthalpy released per mole of glucose is 2826 kJ. One mole of glucose has a mass of 180 g, and 6 mole of oxygen have a volume of $(6 \cdot 22.4) = 134.3$ l STPD. The **enthalpy** per gram of glucose is thus 2826/180 = 15.7 kJ g^{-1}, and the **energy equivalent**, which expresses the energy with respect to the oxygen consumed, is 2826/134.3 = 21 kJ l^{-1} STPD.

Calculate the size of human oxygen and CO_2 stores

The **oxygen stores of the lungs** are $(ERV + RV)F_{AO_2} = (2700 \times 0.15) = 405$ ml STPD or **18 mmol**, and of the **blood** (2 l of arterial and 3 l of venous blood): $(2C_{aO_2}) + (3C_{\bar{v}O_2}) = (400 + 350) = 850$ ml STPD or **38 mmol**. We also have small O_2 stores (**7 mmol**) in myoglobin and in the tissues, a total of $(18 + 38 + 7) =$ **63 mmol.** Hypoxic brain damage can occur after 5 min of apnoea and after 5 s of cardiac arrest (**black out and grey out**).

The **CO_2 stores of the lungs** are $(2700\ F_{ACO_2}) = (2700 \times 0.056) = 151$ ml STPD, and those of the blood are $(2C_{aCO_2} + 3C_{\bar{v}CO_2}) = (1000 + 1650) = 2650$ ml STPD or **118 mmol**. The ECV of an adult person contains $(13 \cdot 24) =$ **312 mmol** bicarbonate. Intracellularly there is a $(26 \cdot 10) =$ **260 mmol** mobile bicarbonate pool, and there are large amounts of CO_2 fixed in bones. These exchangeable CO_2 stores comprise $(118 + 312 + 260) =$ **690 mmol** in total. The size is subject to changes by alterations of **ventilation or acid–base status.**

What is meant by nitrogen balance?

The dietary protein-nitrogen is equal to the nitrogen excretion in the urine when the person is in nitrogen balance. Protein retention during growth, training, protein-rich diet, pregnancy and convalescence is called **positive nitrogen balance** (not urea accumulation in uraemia). Protein loss during inactivity, bedrest, fever, blood loss, burns and lesions is called **negative nitrogen balance**. In any event of negative nitrogen balance, nutritional support is necessary. The aim of the diet formulation regimen is to achieve a positive nitrogen balance. This condition can usually be obtained by giving up to 5 g of nitrogen in excess of output. A useful formula for calculation of the nitrogen loss is given below:

$$\text{Nitrogen loss (g per 24h)} = \text{Urinary urea (mmol per 24h)} \times 28(\text{g/mol}) + 2(\text{g/24h})$$

where the 2 represents non-urinary nitrogen excretion.

What is the mechanical efficiency during exercise?

The **net mechanical efficiency (E_{net})** is the **ratio of external work rate ($N \cdot m\ s^{-1} = J\ s^{-1}$) to chemical energy expenditure ($J\ s^{-1}$) during work**. E_{net} is 20–25% in isolated muscles and also in humans during aerobic cycling. Its size increases with the amount of training, because the untrained individual does not use the muscles effectively. Leg work has the largest E_{net}, since arm work necessitates fixation of the shoulder belt. The work rate is measurable with a cycle-ergometer. A measurable blocking force is applied to a wheel with a given radius (r) and with a given rotation-frequency (measured in revolutions per minute, RPM). The **work rate** or power (force · velocity) is now determined, because the force is known (N) and the distance per second is: $(2 \cdot \pi \cdot r\ \text{RPM})/60$. The **work rate is thus measured in J s^{-1} or W.**

What is the basis of oxygen deficit?

The **O_2 deficit** is defined as the **difference in O_2 volume of an ideal, maximal O_2 uptake and an actual uptake** (see Fig. 42-2). The lacking O_2 volume is the deficit.

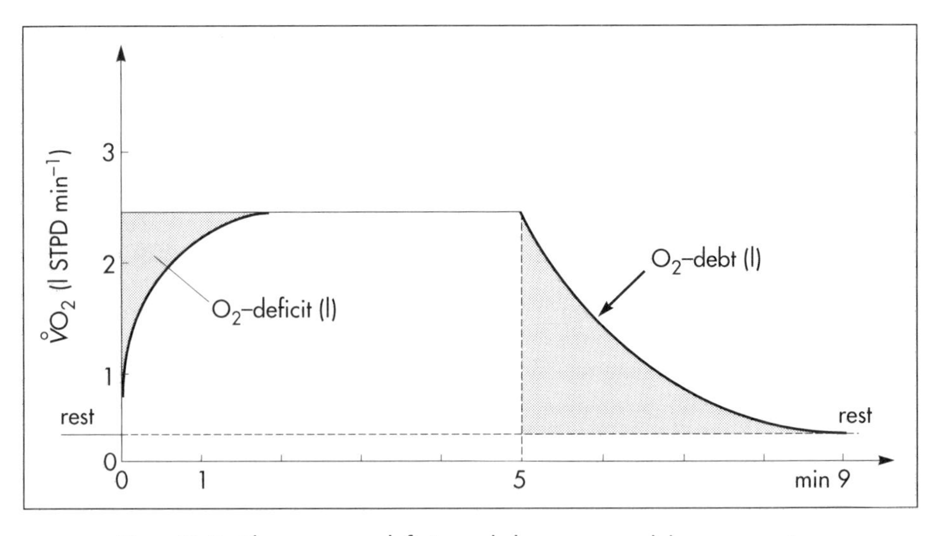

Fig. 42-2: The oxygen deficit and the oxygen debt at exercise.

The **energy demand increases instantaneously** at the start of the 5-min working period, but the actual O_2 uptake via the lungs lags behind for 2 min. The oxygen demand deficit is provided for by the O_2 stores (oxymyoglobin) and by anaerobic energy.

What is the basis of oxygen debt?

Oxygen debt is defined as the **extra volume of O_2 that must be taken into the body after exercise to restore all the energetic systems to their normal state** (Fig. 42-2). The **non-lactic O_2 debt** following moderate work is characterized by normal **resting blood [lactate] around 1 mM**. The value is maximally 3 l, used for regeneration of the phosphocreatine and for refilling the O_2 stores. The **lactacide O_2 debt** following supramaximal work (100–400 m **dash**) can amount to 20 l and the **blood [lactate] can be as high as 20–30 mM**. This extra O_2 debt is used for oxidation of 75% of the lactate produced and for the formation of 25% of the lactate to glycogen in the liver. Restoration of phosphocreatine, etc. following activity, is a process referred to as **repayment of the O_2 debt**. However, it is very uneconomical, since the debt is often twice as high as the O_2 deficit.

From where is the energy for muscular contraction derived?

The predominant source of energy is **oxidation of fuel in the mitochondrion**. Here, high-energy compounds such as phosphocreatine and ATP are formed. Glucose is oxidized by NAD^+, so by glycolysis two pyruvate molecules are formed in the cytosol, transported to the mitochondrion, and transformed to a **co-enzyme-A derivative (acetyl-CoA)**, which then is involved in the **tricarboxylic acid (TCA)** cycle (Fig. 42-3). Provided a certain oxygen flux from the lungs to the mitochondria is present, the electron transport chain (the **glycero-phosphate shuttle**) will reoxidize ($NADH+H^+$) and $FADH_2$ to NAD^+ and FAD (Fig. 42-3).

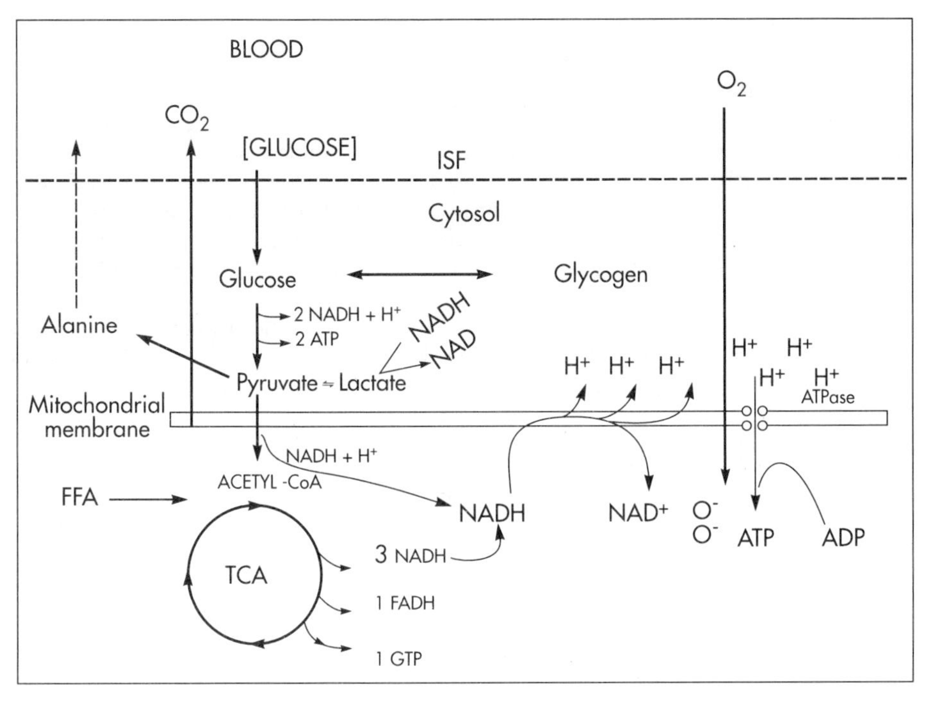

Fig. 42-3: Biochemical pathways for ATP production.

How many ATP units are produced per glucose molecule by oxidative phosphorylation?

In the glycolysis, one glucose molecule is converted to 2 molecules of pyruvate, with the other products being 2 ATP and (2 NADH + H^+) – see Fig. 42-3. Through the oxidation of pyruvate in the TCA-cycle, three (NADH + H^+), one $FADH_2$, and one GTP are formed (Fig. 42-3). If **complete oxidation** occurs in the glycerophosphate shuttle of the mitochondrion, 1 NADH equals 3 ATP, and $FADH_2$ equals 2 ATP. Since the NAD^+ reduced in the glycolysis is cytosolic, it usually equals 2 ATP only, depending on the shuttle used.

When pyruvate is transformed to acetyl-CoA, one molecule of ($NADH^+ + H^+$) is formed.

The total production by oxidative phosphorylation is 36 ATP per glucose molecule (six from the glycolysis, six from the transformation and 24 from the TCA cycle).

Oxidation of one glucose molecule implies the use of six oxygen molecules. Accordingly, the $P:O_2$ ratio is $36/6 = 6$, which is equal to a **P:O ratio of 36/12 = 3.** The free fatty acids (FFA) from the cytosol (intramuscular or extramuscular origin) are transformed to acetyl-CoA (Fig. 42-3).

The pyruvate production rises with increasing glycolysis rate, and pyruvate is the substrate for alanine production (Fig. 42-3). Alanine is liberated to the blood and its concentration increases linearly with [pyruvate] during rest and exercise.

During anaerobic conditions – an insufficient oxygen supply – (NADH+H^+) is reoxidized by the pyruvate–lactate reaction, and the glycolysis continues. The anaerobic ATP production does not block the aerobic ATP production, but functions as an **emergency supply**.

The largest rise in blood [lactate] takes place at work intensities above 50% of $\dot{V}_{O_2}$max. This is the so-called **anaerobic threshold** or **lactate threshold**. Lactic acid is a fixed acid – in contrast to the volatile H_2CO_3 – produced during exercise, and in a muscle cell with a pH of 7 such an acid is essentially totally dissociated (pK = 3.9). Since the proton associated with lactate production reacts immediately with bicarbonate within the cell, its CO_2 production must increase by 1 mol CO_2 for each mole of **bicarbonate buffering lactic acid**.

Lactate accumulates in the muscles and blood, if the glycolysis proceeds at a rate faster than pyruvate can be utilized by the mitochondria, or if (NADH + H^+) is not reoxidized rapidly enough. See also Chapter 6.

Are the carbohydrate stores of importance during exercise?

We possess **100 mmol of glucose (stored as glycogen) per kg of wet muscle weight**, or **3.5 mol** in the muscle tissue. Muscle tissue does not contain glucose-6-phosphatase. The 5 l of circulating blood only contains 5 mM or 25 mmol (5 g) of glucose as a total. During exercise the muscle uptake of glucose increases considerably, but the blood [glucose] does not fall. The blood [glucose] is kept normal by an increased flux of glucose from the liver (Fig. 42-4).

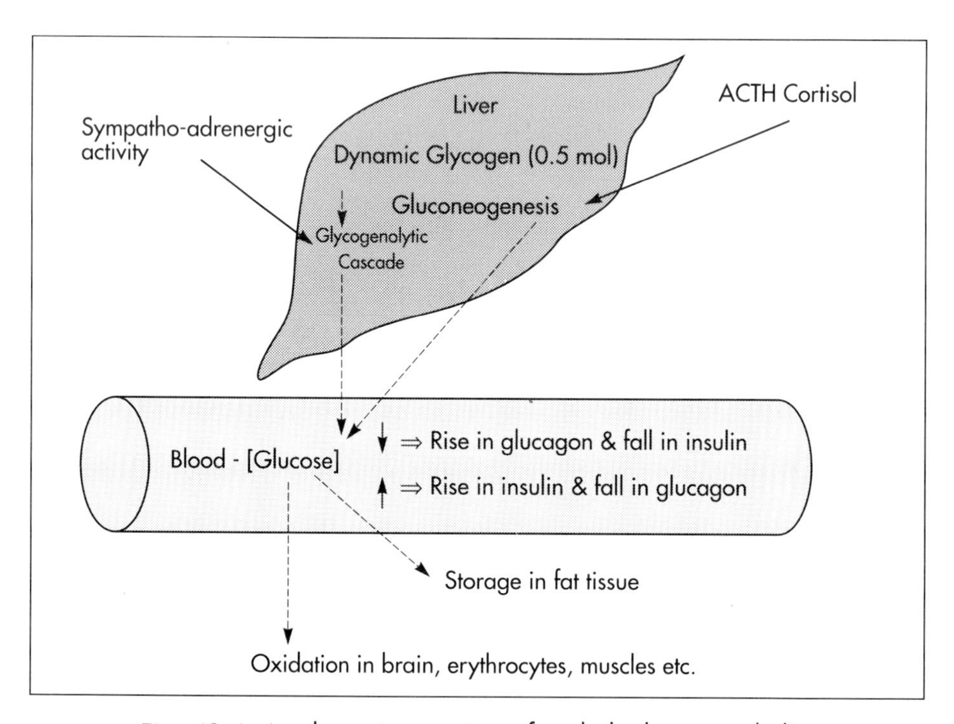

Fig. 42-4. A schematic overview of carbohydrate metabolism.

1. With increasing intensity and duration of exercise, the sympatho-adrenergic activity and the blood [catecholamines] increase. This is a strong stimulus to the **hepatic glucose production**. The liver contains 50–100 g of **dynamic glycogen**. This liver glycogen is easily broken down into glucose by **glycogenolysis** and released to the blood. Any fall in blood [glucose] during exercise will increase the blood [glucagon] and decrease [insulin] toward zero. Glucagon is bound to hepatocyte receptors, and via cAMP a **glycogenolytic cascade** is started (Fig. 42-4). Hereby the hepatocytes produce large amounts of glucose, sparing muscle glycogen and delaying the onset of fatigue. The lack of insulin inhibits the glucose transport across the cell membranes.
2. Glucose is also produced by **gluconeogenesis** in the liver from glycerol, lactate, pyruvate, and glucogenic amino acids. The gluconeogenesis is stimulated by pituitary ACTH and by cortisol from the adrenal cortex (Fig. 42-4).

   With prolonged exercise the blood [glucose] will fall at the end, when hepatic and muscle glycogen stores are depleted, and the **compensating gluconeogenesis** is also running out of energy sources.
3. Complete exhaustion is delayed considerably in trained athletes, because they utilize lipids, so the glycogen stores are spared by **oxidation of FFA**.

What is the main source of energy during exercise?

Skeletal muscles contain **lipid stores** (20 g triglycerides kg^{-1} wet weight or 700 g in a person with 35 kg muscles). A standard 70 kg man also contains **extramuscular fat stores** of triglycerides (15 kg).

Sympathetic activity and catecholamines **increase lipolysis** (i.e. hydrolysis of the stored adipose tissue to FFA and glycerol) via activation of **adenylcyclase**, increase in cAMP, phosphorylation and activation of the **hormone sensitive lipase**.

Increased blood [lactate] and **glucose intake** reduce lipolysis during exercise. The fat stores are the **ideal energy stores** of the body, because a large quantity of ATP is available per gram. This is due to the relatively low oxygen content of lipids – the point being that the necessary oxygen is inhaled at request.

Are amino acids metabolized during exercise?

At rest we have a slow turnover of muscle protein, but during exercise alanine is released in appreciable amounts by transamination of pyruvate in the muscle cells, and the blood [alanine] is doubled – without any important change in other amino acids. Alanine is produced via the **pyruvate–alanine cycle**, and the amino groups are from valine, leucine and isoleucine. The muscle alanine is transported by the blood to the liver, where its carbon skeleton is used in the **gluconeogenesis**. The blood [alanine] also stimulate the pancreatic islet-cells to **increased glucagon secretion**. Glucagon activates the **glycogenolytic cascade** (see above) in the liver cells, further stimulating glucose output from the liver. These are the two factors in the **alanine-liver** cycle of exercise.

What are the factors determining the oxygen diffusion gradient from blood to mitochondria?

The ventilatory, the cardiovascular and the metabolic systems are coupled, and dependent on the following factors.

The primary factor is the **size of P_{aO_2}**, but the **blood oxygen store** is of similar importance in keeping P_{aO_2} as high as possible. The blood oxygen store depends upon the **haemoglobin concentration, the haemoglobin–oxygen affinity as well as 2,3-DPG, temperature, and P_{aCO_2}.**

The **total oxygen flux** to a certain population of mitochondria also depends upon the bloodflow (i.e. $\dot{Q}$, muscle bloodflow, lung perfusion, etc.).

Further reading

Lamb, D.R. and R. Murray (Eds) (1988) *Perspectives in Exercise Science and Sports Medicine*, Vol. 1. *Prolonged Exercise*. Benchmark Press, Indianapolis.

Sims, E.A.H. and E. Danforth Jr (1987) Expenditure and storage of energy in man. *J. Clin. Invest.* **79**: 1019.

42. Case History

A male, 23 years of age, has a daily metabolic rate of 12 600 kJ (12.6 MJ) and he eats a mixed diet resulting in a RQ of 0.83. The enthalpy equivalent for oxygen is 0.46 kJ mmol^{-1}. His arterial pH is 7.30 and pK for ammonia is 9.3.

1. *Calculate the ratio between ammonia and NH_4^+ in his blood.*
2. *Calculate his carbon dioxide output in mol day^{-1}.*
3. *Calculate the amount of carbon dioxide eliminated per day in combination with ammonia, assuming 1 mol day^{-1} of ammonia to be involved in the urea production.*

Try to solve the problems before looking up the answers in Chapter 74.

CHAPTER 43. ENERGY BALANCE AND OBESITY

Describe indirect methods of calorimetry

Indirect measures of enthalpy (MR in kJ min^{-1}) are easily applicable both at rest and in an exercise setting. Expired air is collected in a Douglas bag (volumetric principle) for subsequent air analysis, and the **volume of oxygen consumed per min** is calculated ($\mathring{V}_{O_2}$). It is convenient also to determine the carbon dioxide production in the same period ($\mathring{V}_{CO_2}$), because the $\mathring{V}_{CO_2}/\mathring{V}_{O_2}$ ratio is the RQ.

A person on a mixed diet has a RQ of 0.83 and a heat energy yield of 20 kJ l^{-1} or 0.45 kJ $mmol^{-1}$ of O_2. The **estimated volume (l min^{-1}) of O_2 consumed** is multiplied by 20 to calculate MR in kJ min^{-1}. The **heat energy yield** vary with RQ and is found in a table (see Chapter 75). A metabolic ratemeter (a spirometer with CO_2 absorber) is practical for determination of $\mathring{V}_{O_2}$.

A more detailed calculation is performed as follows. The $\mathring{V}_{O_2}$ and $\mathring{V}_{CO_2}$ is measured volu- or gravimetrically together with determination of the nitrogen content in 24 hours' urine from the person examined. The urine nitrogen expresses the protein combustion, since protein contains 16% nitrogen. Subtraction of the gas volumes for protein combustion (see data in Chapter 75) from the total, results in residual volumes only related to the fat (F g min^{-1}) and carbohydrate (C g min^{-1}) combustion. Thus F and C can be calculated by solution of two equations with these two unknowns. By multiplication with the nutritive equivalents for O_2 and for CO_2 (mmol gas g^{-1} in Chapter 75) the mass balance states:

$$F \text{ and } C \text{ rel. } \mathring{V}_{O_2} \text{ mmol min}^{-1} = (37 \text{ mmol g}^{-1} \cdot C \text{ g min}^{-1}) + (91 \text{ mmol g}^{-1} \cdot F \text{ g min}^{-1})$$

$$F \text{ and } C \text{ rel. } \mathring{V}_{CO_2} = (37 \text{ mmol g}^{-1} \cdot C \text{ g min}^{-1}) + (64 \text{ mmol g}^{-1} \cdot F \text{ g min}^{-1})$$

Now the mass of protein, fat and carbohydrate combusted per minute is found, and a very precise indirect measure of MR in kJ min^{-1} is obtained by multiplication with their enthalpy equivalents (17, 39 and 17.5 kJ g^{-1}, respectively).

Disadvantages of indirect calorimetry are that it ignores the O_2 debt, and that the gas stores of the body and the nitrogen balance must be maintained.

Describe the direct calorimetric method

The total output of heat energy in the body is most precisely measured in a whole-body calorimeter. The **Atwater–Rosa–Benedict's human calorimeter** has been used to verify the first law of thermodynamics in humans. The heat energy delivered from the chamber is only equal to the MR, provided the external work is zero, and neither equipment nor the human body alters temperature.

What factors determine human metabolic rate (MR)?

The major single factor is **muscular activity**, which can increase MR with a factor of 20 even for hours in marathon running. Inactive persons can have a daily MR of 9600 kJ, whereas heavy occupational labour requires 20 000 kJ (20 MJ).

Dietary intake can increase MR by 20–30% (see specific dynamic activity, below).

Increased energy demand in heart and lung diseases, or in rapid growing cancer will increase MR importantly. Energy is also lost in other disease states such as proteinuria, glucosuria, ketonuria, diarrhoea, and exudation of plasma through lesions in the skin or in the mucosa. An **extra physiological energy loss** takes place during pregnancy and during nursing.

Deposition of heat energy in the body (**enthalpy accumulation** as in fever and hyperthermia) can increase MR.

MR is considered to be 10% lower in females than in males.

MR decreases with age and, of course, depends on the body size (height, weight, body surface area).

*What factors determine basal metabolic rate (*BMR*) in persons of the same race?*

No work is done under basal conditions, so that all energy is ultimately liberated in the body as heat energy. The liver and the resting skeletal muscles account for half of the basal metabolic rate (BMR).

Measurement of the BMR requires the subject to be awake in the morning, fasting and resting horizontally. The ambient temperature must be neutral, which is the temperature at which compensatory activities are minimal.

BMR is rarely used for diagnosis of thyroid disease, because radioimmunoassays for thyroid hormone analysis are specific and uncomplicated in use (Chapter 65).

Does the BMR *vary with size, age and sex?*

1. **The surface law** states that the **BMR per body surface area is much more uniform** than the BMR per kg of body weight in individuals of the same species but of different form and size. Among different animal species the large animals (elephants) have the smallest relative surface area (i.e. surface area per kg), so **elephants must have small BMR per surface area compared to mice.**

   Body surface area (BSA) is estimated with the approximation formula of the DuBois family: BSA (cm^2) = $Weight^{0.425}$ (kg) · $Height^{0.725}$ (cm) · 71.84. The BMR is 45 W m^{-2}, so a person with a body surface area of 1.8 m^2 has a BMR of 80 W. This is a daily BMR of ($60 \cdot 1440$ min $\cdot 80 \cdot 10^{-3}$) = 6912 kJ.
2. BMR **decreases with age** in both sexes.
3. The **female BMR values are 10% below the male** values throughout life.

Describe the specific dynamic activity (SDA) of the diet

Intake of meals as such increases **MR**. This is the **specific dynamic activity of the diet (SDA)** or **dietary thermogenesis**. SDA is less than 10% of the intake energy for carbohydrates and for fat, but 30% for proteins.

A glucose-loaded person forms glycogen and fatty acids out of glucose within an hour, even before the glucose can be oxidized. Accordingly, the SDA caused by glucose can be due to an obligate formation of glycogen and fatty acids. The thermogenic response to carbohydrate seems to include a muscular component activated by adrenaline via β_2-**receptors and a non-myogenic component activated by NA via** β_1-receptors.

Proteins have no SDA in hepatectomized animals, so the SDA of proteins must be caused by hepatic intermediary processes. These **intermediary processes include formation of urea from NH_4^+, breakdown of amino acids, etc.**

In general, SDA can also be related to **mass action** due to increased supply of nutrients, and to temperature increase by the activity (increases all enzymatic processes).

Is fasting unhealthy?

Fasting is a **total stop of food intake**. After 12 h conditions are optimal to measure BMR or to analyse the chemical composition of blood (fasting blood values are predictable and easy to interpret). The 12 h are the methodological criterion for the correct minimum BMR, but continued fasting for days results in a much lower value (65% of BMR). Following the first 2 weeks, the normal body weight is reduced to 85%, whereas the resting MR is stable at 65% of BMR, which is constant to the end of the fasting period (either voluntary or by death).

Glygogen stores are broken down in a few days, since only small stores prevail (liver and muscles). Then **urine nitrogen increases** as a sign of renewed protein combustion (gluconeogenesis). In general the **fat combustion dominates**, until the fat stores are used. Healthy people contain 5–15 kg fat, but monstrous amounts have been recorded in a 540 kg male from the *Guinness Book of Records*.

Oxidation of fat stores – including the partial hepatic oxidation to ketonic bodies – implies development of ketoacidosis and a diabetic glucose tolerance test. Such a **hunger diabetes**, with ketonaemia and ketonuria as in diabetes, has been found in healthy individuals even after only 24 h of fasting or after extremely fatty meals.

Serious illnesses develop after a few weeks of fasting, because the cell structure proteins are broken down. The proteins of the cell nuclei produce uric acid, which accumulates in the heart (**cardiac disease**) and in the articulations (**uric acid arthritis or podagra**).

Describe the control of energy balance in humans

A person with a body weight of 70 kg contains 550 MJ of combustible enthalpy, and if allowed to eat naturally, at least 10 MJ is consumed every day. If the person is fasting for some days they will lose body weight and their MR will fall to 6.6 MJ daily, so a certain **input control** is hereby documented. The loss in body weight is rapidly compensated for when feeding is resumed. If enough food is available the person will automatically eat more (towards a doubling) with increasing work load (MJ day^{-1}), so a certain **output control** is also documented. The internal feedback signals operating in this output and input control are uncertain.

Signals from gastrointestinal centres inhibit the feeding centre through afferent nerves (Fig. 43-1). Chyme in the duodenum containing HCl and fatty acids liberates enterogastrones in the blood. The enterogastrone family consists of secretin, somatostatin, cholecystokinin (CCK) and gastric inhibitory peptide (GIP). Enterogastrones reduce gastric activity, stimulate the satiety centre (Fig. 43-1), and increase the production of bicarbonate-rich bile and pancreatic juice. A glucose-rich chyme in the duodenum liberates members of the incretin family into the blood. The incretin family consists of gut glucagone, glucagone-like peptide 1 & 2, and GIP.

Incretins produce a rapid rise in insulin secretion, which causes the energy stores to increase (Fig. 43-1). A hypothalamic **glucostat or satiety centre** is assumed to be sensitive to blood [glucose], blood [amino acids], FFAs and glycerol (Fig. 43-1). Also

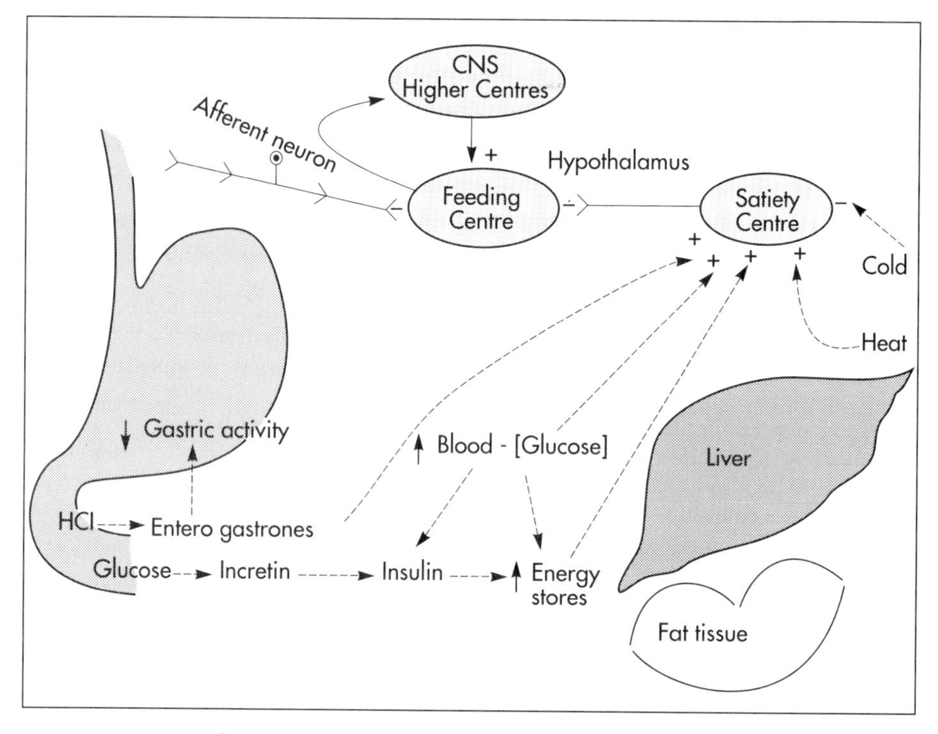

Fig. 43-1. A schematic overview of the regulation of food intake. The hypothalamic 'feeding centre' is located in the medial region, whereas the 'satiety centre' (also called a glucostat) is laterally located in the hypothalamus.

thermoregulatory signals from heat receptors, and signals from energy stores stimulate the satiety centre (Fig. 43-1). In workers a minimal work activity threshold must be passed in order to trigger the hypothalamic weight control, but above this threshold it is obvious that eating increases proportional with the work load, and the body weight is constant. If the work rate is extremely high the hypothalamic control is also broken, but now the dietary intake and the body weight cannot cope with the high combustion, and the body weight falls drastically. The cybernetics of appetite control are not only feedback factors. As in all human behaviour, cerebral feedforward factors can dominate. Cerebral feedforward factors are exercise habits, eating habits, **social inheritance**, and these can be of extreme importance to the individual.

The hypothalamus controls food intake and metabolism, mainly by **autonomic effects on the islets of Langerhans** (secreting insulin, glucagone, pancreatic polypeptide and gastrin), **hepatocytes and adipocytes**. Neuroendocrine–behavioural disturbances seem to be involved in abnormal eating patterns such as **anorexia nervosa, bulimia and obesity**. We seem to regulate our appetite by a **combination of negative feedback and essential feedforward factors**.

What is the physiological basis for obesity?

The **ideal weight** is the weight associated with the highest statistical life expectancy. The **Broca index** is a popular and easy method of determining the recommended weight. For a male the Broca index is the number of kg which equals his height in cm minus 100, and for a female 110 is subtracted. **Obesity** is defined by WHO as an **actual body weight exceeding the ideal weight by more than 20%** (if not explained by an above-average muscle and bone mass). Established obesity is therapy-resistant.

Hyperplastic fatness (too many adipose cells) is often found in babies from rich parts of the world.

Hypertrophic obesity (too large adipose cells) is related to exercise habits and to psychological factors, such as habitual or social relationships. The major factor is **physical inactivity**. Persons who exercise can increase their MR by a factor of 10–20 several hours a day. The second choice for obese persons is to reduce the dietary intake of nutrients. However, even a reduction to half the usual amount of food, an heroic and probably futile project, would be inefficient when compared to an increased combustion factor of 10–20.

Giving oral glucose to obese people triggers a **reduced thermic response and an impaired activation of the sympathetic system**. After weight loss the thermic effect of glucose is still reduced as is the arterial [NA] response. A defective sympathetic nervous system may be the cause of obesity. Obese people have a small SDA, because they avoid physical activity with a high metabolic rate in skeletal muscle and in adipocytes. They thus have a low MR for their daily activities. Obesity is the fate of people **dominated by their parasympathetic activities and minimizing the use of the sympathetic nervous system.** Thus, obesity is a psychological condition in many people who have chosen a physically inactive life style.

Extremely rarely, obesity is caused by hormonal, metabolic diseases (insulinoms, hypercorticism, diseases of the thyroid gland, hypothalamic lesions, etc.). For persons with insulinoms, the **high food intake** (hyperphagia) can be a question of life and death.

Is obesity a risk factor?

A **risk factor** is an epidemiological term for conditions statistically correlated with shortened life expectancy. Obesity paves the way for diabetes mellitus, high blood pressure, myocardial infarction and stroke; obesity is clearly a risk factor.

Further reading

Astrup, A., T. Andersen, N.J. Christensen, J. Bülow, J. Madsen, L. Breum and F. Quaade (1990) Impaired glucose-induced thermogenesis and arterial norepinephrine response persist after weight reduction in obese humans. *Am. J. Clin. Nutrition.* **51** (3): 331–7.

Bjørntorp, P. and B.N. Brodoff (Eds) (1992) *Obesity.* J.P.Lippincott, Philadelphia.

Lamb, D.R. and R. Murray (Eds) (1988) *Perspectives in Exercise Science and Sports Medicine*. Vol. 1. *Prolonged Exercise*. Benchmark Press, Indianapolis.

Leibowitch, S.F. (1992) Neurochemical–neuroendocrine systems in the brain controlling macronutrient intake and metabolism. *Trends in Neurosciences* **15** (12): 491–7.

Sims, E.A.H. and E. Danforth Jr (1987) Expenditure and storage of energy in man. *J. Clin. Invest.* **79**: 1019.

43. Case History

A 70 kg male, 22 years of age, is in a room, where the temperature is 20°C and P_B is 101.3 kPa. P_{AO_2} is 18.45 kPa, and P_{AN_2} is 74 kPa. The $\dot{V}_{CO_2}$ is 600 ml STPD min^{-1}. The urinary nitrogen excretion is 10 mg min^{-1}. The pressure of water vapour in the alveoli is 6.2 kPa, and F_{IO_2} is 0.2093.

1. *Calculate $\dot{V}_A$ and F_{ACO_2}.*
2. *Estimate P_{aCO_2}, and provide reasoning for a possible hyperventilation in this condition.*
3. *Calculate $\dot{V}_{O_2}$ for this person.*
4. *Calculate the pulmonary exchange quotient (R).*
5. *Is the pulmonary exchange quotient different from the RQ?*

Try to solve the problems before looking up the answers in Chapter 74.

CHAPTER 44. THERMOREGULATION

What is the enthalpy of the human body?

The **heat content** (*H*, **enthalpy**) of the body is reflected by its temperature. The mean body temperature is 37°C, but there are considerable variations between the core and the shell. The **skin** is the **main heat exchanger** of the body. The skin or shell temperature is determined both by the temperature and humidity of the environment and by the needs of the body to exchange energy.

What is the core- and shell-temperature of the body?

The **core temperature** is the temperature in the **deeper parts of the body** and in the proximal extremity portions.

The **shell temperature** is measured on the **total skin surface and deep portions of the hands and feets**. The shell temperature is several degrees lower than the temperature in the core. The shell temperature and the size of the shell varies with the environment. Venous blood draining active muscles and the liver is likely to be warmer than pulmonary venous blood, since this has undergone evaporative cooling in the alveoli. The **heat capacity of water** and of the **human body** is 4.19 and 3.35–3.47 kJ kg^{-1} $°C^{-1}$, respectively.

What is meant by mean body temperature?

The **mean body temperature** is calculated with the following equation:

$$T_{body} = (0.7 \cdot T_{core} + 0.3 \cdot T_{shell}),$$

where the **mean shell temperature** is estimated from a series of representative skin temperatures. The obvious assumption in this equation is that 70% of the body weight is **core**, and the balance is **shell**, but the size of the shell varies with the environmental temperature.

What is heat energy balance and thermic steady state?

Heat energy balance is a condition, where a resting person deposits no heat energy, so that the **metabolic rate is equal to the heat dissipation.**

Thermic steady state is a condition in which a resting person is in **heat energy balance as well as having no internal energy flux** in the body.

A patient with high fever can be in **thermic steady state,** if both core and shell temperature are constant, and no internal energy flux occurs for a period.

What characterizes warm-blooded and cold-blooded animals?

Warm-blooded animals (**homeotherms**), such as humans, can change their metabolism in order to keep their heat production equal to the heat dissipation. Such animals have

temperature regulation and thereby maintain a **rather constant core temperature**. However, the human core temperature falls during the oestrogen phase of the menstrual cycle and during sleep.

Cold-blooded animals (**poikilotherms**) have no temperature control, so that their core and shell temperature varies with the environment. Reptiles, premature and newborn babies are cold-blooded. The babies have a minimal thermoregulatory capacity. However, their capacity for heat production is 5–10 times as great per unit weight as that of adults. Adults have a **homeothermic core** and a **poikilothermic superficial shell**. Adults exposed to general anaesthesia lose thermoregulation.

How is heat transferred?

1. **Radiation** describes a transfer of energy between objects in the form of an electromagnetic process (photons). This includes ultraviolet, visible (sun light) and infrared or warm radiation.

   Radiative heat transfer can be calculated for a naked person:

   $$Q_R = 0.5A(T_{skin} - T_{obj}),$$

   where 0.5 is kJ min^{-1} m^{-2} K^{-1}, A is the area of the human body (radiating or receiving radiation), and T_{obj} is the temperature of the object exchanging energy.
2. **Conduction** describes a **transfer of kinetic energy by contact between skin and metals or other immobile objects**. Sitting on a cold stone is a typical example of conduction loss.
3. **Convection** is defined as the **transfer of kinetic energy by contact between the skin and moving air or water**.

Diving illustrates the importance of conduction and convection in heat energy transfer. The **dry diving suit** excludes water from contact with the skin and traps low-conductance air in protective clothing worn inside the watertight suit. The **wet suit** traps water next to the skin but prevents its circulation. The water is warmed through contact with the skin, and the low energy conductance of the wet suit, with its many air pockets, minimizes the rate of heat energy loss from the trapped air and water to the surrounding water. Air is a poor heat conductor. High pressures compress these air pockets and reduce the insulation properties of wet suits.

How does the body respond to cold?

Vasoconstriction lowers skin temperature, and thereby reduces the **conductive–convective energy loss** that is determined by the temperature gradient from the skin to the environment. Cutaneous vasoconstriction leads more of the venous blood back to the core through the deep veins and the commitant veins. The veins are located around the arteries with warm blood, so that the venous blood receives part of the energy from the arterial blood – the so-called **counter current**. The vasoconstriction is so effective, that the bloodflow in the fingers can fall to below 1% of the flow at normal temperature. The cooling of the shell is immediate, and the size of the shell increases. The resistance vessels of the hands will be open periodically to nourish the tissues.

Shivering is a **reflex myogenic response** to cold with **asynchronous** muscle contraction. No outside work is obtained except shivering, so all energy is liberated as metabolic heat energy. Heat production is also increased by thyroid gland activity and by release of catecholamines from the adrenal medulla.

External work, such as running, is helpful in maintaining body temperature when freezing. Freezing increases the motivation for **warm-up exercises** and illustrates the voluntary, cortical or feedforward influence on temperature homeostasis. People adapt to prolonged exposure to cold by increasing their BMR up to 50% higher than normal. This **metabolic adaptation** is found in Innuits and other people continuously subject to cold.

How does the body respond to heat?

1. **Sweat secretion**. Three million sweat glands produce sweat at a rate of up to **2 l h^{-1}** or more during extreme conditions. Such high sweat rates lead to circulatory failure and shock. Sweat resembles a dilute ultrafiltrate of plasma. Healthy humans cannot maintain their body temperature, if the environmental air reaches body temperature and the air is saturated with water vapour. Primary sweat is secreted as an iso-osmotic fluid into the sweat duct, and subsequent NaCl reabsorption results in the final hypo-osmotic sweat. **Thermal sweating** is abolished by atropine, proving that the postganglionic fibres are cholinergic. Cholinergic drugs (eg neostigmine, edrophonium, acetylcholine, etc) provoke sweating just as adrenergic agonists (eg adrenaline, terbutaline, noradrenaline, phenylephrine, etc) do. **Evaporation** of water on the body surface eliminates 2436 J g^{-1} (or kJ kg^{-1}). Evaporation of large volume rates of sweat ($\mathring{V}_{sweat}$) implies a substantial loss of energy (Q_E J min^{-1}) according to the equation:

$$Q_E\ (\mathrm{J\ min^{-1}}) = \mathbf{2436}\ (\mathrm{J\ g^{-1}}) \cdot \mathring{V}_{sweat}\ (\mathrm{g\ min^{-1}}).$$

2. **Condensation** of water on the skin, gains energy which is stored in the body.
3. **Vasodilatation** in warm environments permits increased skin bloodflow as $\mathring{Q}$ also begins to rise. The **arterio-venous anastomoses** are dilated, and the bloodflow can rise up to 100 FU in hands and feets. The **shell bloodflow** determines how much heat energy can be carried from the core to be dissipated on the surface. The heat energy is transferred from core to shell by **convective transport in the blood**. A substantial part of the heat energy is dissipated through the superficial veins working as **cooling ribs**. A piece of steak has the same composition as human skin but of course no blood flow and no sweat evaporation. Thus the steak will be cooked at an air temperature that humans can survive. A person can stay in a room with dry air at 128°C for up to 10 min during which time the steak is partially cooked.

What is emotional sweating?

This is a **paradoxical response** in contrast to the **thermal sweating** of thermoregulation. Emotional stress elicits **vasoconstriction** in the hands and feets combined with **profuse sweat secretion** on the palmar and plantar skin surfaces.

How does metabolic rate vary with environmental temperature?

The energy transfer by radiation (Q_R), by convection and conduction (Q_C), and by evaporation (Q_E) is measured in kJ h^{-1}. **Heat energy balance** is maintained:

$$\mathbf{MR} = Q_R + Q_C + Q_E,$$

so only at the highest temperatures, is energy available of deposition (Fig. 44-1). The dressed person is in **thermic steady state** and MR is almost constant in the **thermoneutral zone** between 20 and 30°C (Fig. 44-1). The **law of metabolic reduction** reflects the tendency for heat production to match the rate of heat loss. The thermoneutral zone, where minimal compensatory activity is required, is separated in the **lower vasomotoric** and the **upper sudomotoric control zone.** In the lower, comfortable zone (20–26°C) the **total heat dissipation** is maintained equal to MR by cutaneous, vasomotoric alterations. The evaporation loss is termed insensible perspiration; this is low and rather constant at this MR (Fig. 44-1), which is almost BMR.

In the **upper sudomotoric zone** above 26°C the evaporation loss rises, as does sweat secretion and bloodflow through the skin. At 35°C total energy dissipation occurs via evaporation (Fig. 44-1).

When the environmental temperature falls the MR increases – first by increasing muscle tone and then by shivering. The **chemical or metabolic temperature control** is the region from 20°C and below (Fig. 44-1), where shivering, decreased bloodflow through skin and non-myogenic heat production take place. Here, metabolism controls the core temperature by increasing MR with falling temperature in the environment. Above 20°C, the **physical temperature control** takes over, as an autonomic capacity for alterations in heat dissipation. In this thermoneutral zone the body temperature is kept constant almost without either heat-producing mechanisms or sweat secretion. The **thermal comfort point** for lightly clothed, seated persons is about 26°C when the humidity is 50%, 28°C for nude persons, and about 36°C sitting in water to the neck.

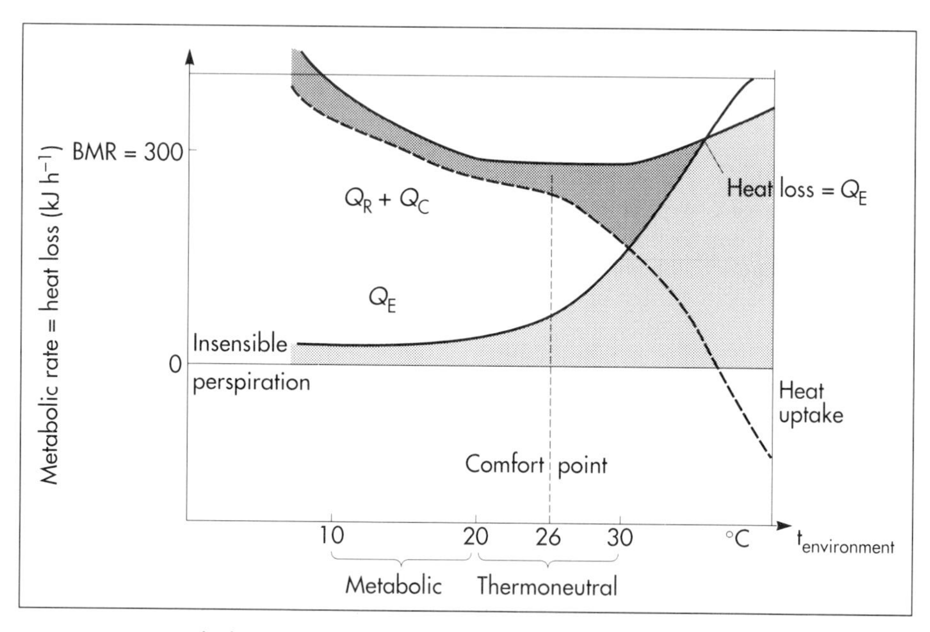

Fig. 44-1: Metabolic rate (MR) and environmental temperature in a fasting, dressed human at rest.

Do we have a temperature control?

Humans have a rather **constant core temperature**. This implies that **control** is exerted.

Information about the external temperature is provided by **peripheral thermosensors,** which are located in the skin, abdominal organs, and muscles. Internal or blood temperature is monitored by **central thermosensors** in the medulla and the preoptic hypothalamus.

A rise in **hypothalamic temperature** causes vasodilatation in the skin and reduces muscular tone. Then thermic sweat is observed, and after some time reduced activity of the adrenal cortex and of the thyroid gland is also observed.

A **fall in hypothalamic temperature** releases cutaneous vasoconstriction together with increased muscular tone and shivering.

What is a dynamic gain and a set-point control system?

A **dynamic gain system** responds **continuously to feedback signals**, regardless of the size of the core temperature. With rising tissue temperature, the neural activity of heat sensors increases linearly, whereas the activity of cold sensors decreases (Fig. 44-2 **A**).

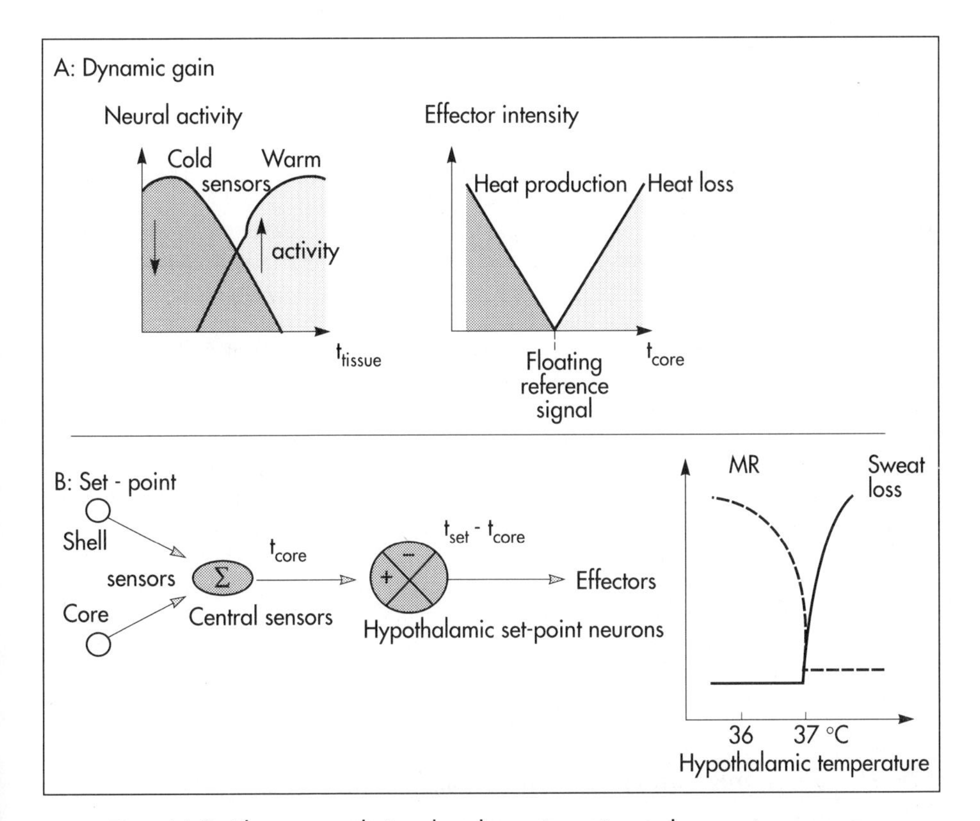

Fig. 44-2: Thermoregulation by dynamic gain and set-point systems.

This combined sensory input increases the core temperature and thus increases the activity of **heat loss effectors**, while inhibiting **heat production effectors** (Fig. 44-2 **A**). This determines the **reference signal**.

The dynamic gain system has a floating **reference signal** moving with the continuous heat loss and heat production (Fig. 44-2 **A**).

A **set-point system** does not respond to a rise in temperature before a certain set-point is reached. The set-point is the core temperature at which neither heat loss mechanisms nor heat production mechanisms are active.

When a thermic disorder reaches a certain **set-point** in the hypothalamus, signals pass to the effectors. The desired core temperature (t_{set} = **set-point temperature**) is compared to the actual value (t_{core} in Fig. 44-2 **B**). The caudal hypothalamus works as a thermostat. The hypothalamus uses the rest of the body to stabilize its own temperature. Error signals – a deviation from t_{set} – evoke responses that tend to restore core and hypothalamic temperature toward the set-point. When the core and hypothalamic temperature rises above the desired set-point such as 37°C, effectors are turned on, and the compensatory heat energy loss is almost linear (Fig. 44-2 **B**). These compensatory mechanisms (vasodilatation, sweat, reduced muscle tone) do not turn off until the temperature drops to the set-point (i.e. an all-or-none response).

When the core and hypothalamic temperature is below the set-point, the compensatory mechanisms (vasoconstriction and shivering) are relatively inactive. The hypothalamic set-point is elevated in fever by **pyrogens from microorganisms** and the rise in metabolism is mainly accomplished by shivering.

Describe the human thermocontrol system

Human temperature control exhibits both **dynamic gain** and **set-point** characteristics. **The control system** implies widespread cutaneous and deep sensors. Their afferents converge towards the hypothalamic integrator and **thermostat**. The hypothalamus also contains thermosensors in the preoptic region, and inhibitory neurons perform **crossing inhibition** (Fig. 44-3). The **central heat drive** from the **preoptic hypothalamus** is maintained (Fig. 44-3). The stability of the core temperature is increased by the **large heat capacity of the body mass**, and by the deep thermosensors, which are dominant.

Shivering is released from cutaneus cold sensors firing maximally at 20°C. These cold sensors are silent above 33°C. Deep cold sensors in the preoptic hypothalamus are activated by the cold shell (Fig. 44-3). This increases heat production by shivering. The preoptic thermostat simultaneously reduces heat loss by crossing inhibition (Fig. 44-3). **Sweat secretion** is released by preoptic warm sensors as soon as their temperature is 37°C or above (t_{set}). Cutaneous cold sensors inhibit sweat secretion at shell temperatures below 33°C, since they are silent above this temperature. In conclusion, preoptic warm sensors show set-point characteristics below the set point, and preoptic cold sensors show set-point characteristics above the set-point. Apart from that, preoptic sensors show **dynamic gain**: with rising tissue temperature, the neural activity of heat sensors increases linearly, whereas the signal frequency of cold sensors increases with falling temperature (Fig. 44-2 **A**). **Cutaneous sympathetic vasodilatation** is probably also released by preoptic warm sensors above set-point (Fig. 44-3). A fall in skin temperature below 33°C will reduce skin bloodflow by crossing inhibition (Fig. 44-3).

Describe the thermoregulatory effector systems

The **sympathetic** and the **somatomotor nervous system** participate in thermoregulation (Fig. 44-4).

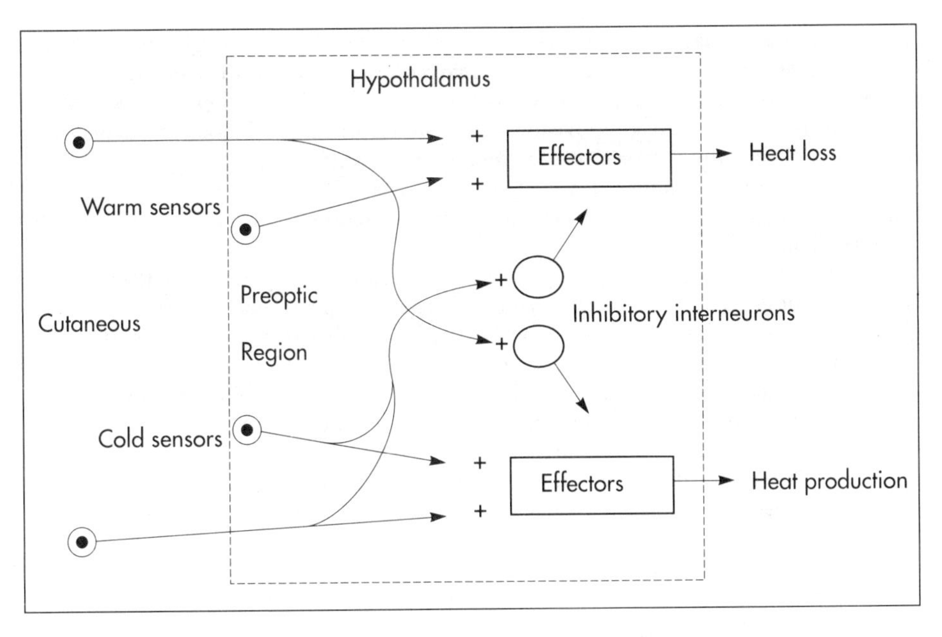

Fig. 44-3. The hypothalamic thermostat and its connections.

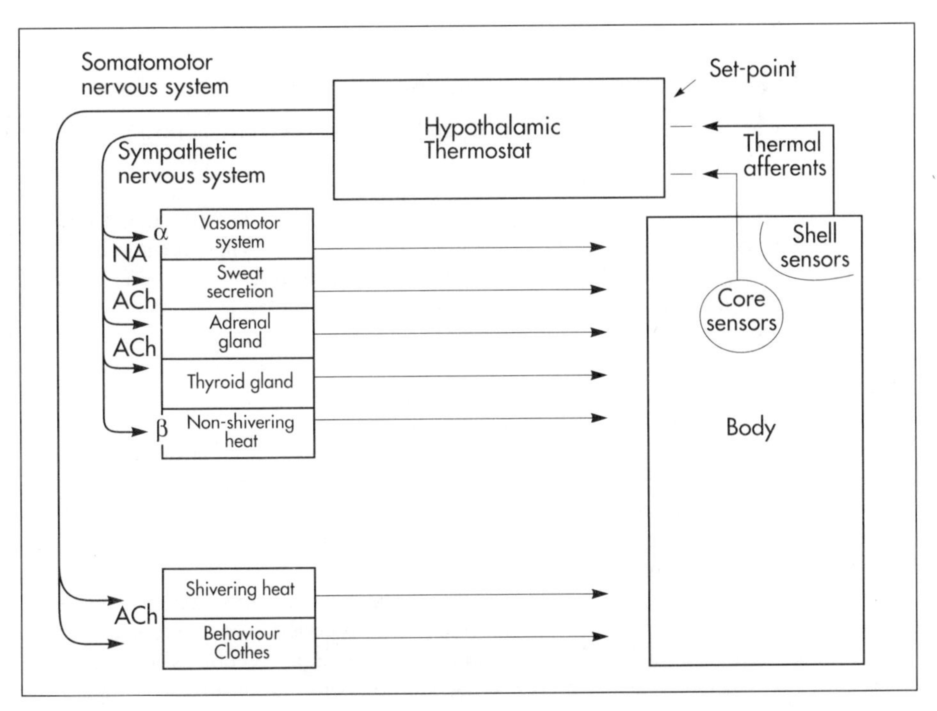

Fig. 44-4. The thermoregulatory feedback system.

Noradrenergic sympathetic neurons control the bloodflow through fingers, hands, ears, lips and nose. **Arterioles and arteriovenous anastomoses** constrict (thermic insulation) following an increase in sympathetic tone, and dilate following a decrease in tone. When arterioles and arteriovenous anastomoses open, the bloodflow is markedly increased and thus the **convective heat loss** is increased.

Sweat secretion is controlled by cholinergic sympathetic fibres. The vasodilatator **bradykinin** is secreted in sweat. Profuse sweat secretion is always accompanied by vasodilatation.

Sympathetic activation releases **thyroid hormones** from the thyroid gland and **catecholamines** from the adrenal medulla. These hormones liberate fatty acids and glucose for combustion. A reduced sympathetic tone also reduces the activity of the adrenal and the thyroid gland.

The thermogenic response to cold also involves a **non-myogenic or non-shivering component** probably in adipocytes. Non-shivering heat production is controlled by the sympathetic nervous system via adrenergic β-receptors. The NA released at the nerve terminals close to the adipocytes, stimulates the liberation of FFAs and their subsequent oxidation. **Non-myogenic heat transfer** includes a small contribution from the brown fat of babies.

Shivering is induced by way of the motor system. The **central shivering pathway** passes from the hypothalamus to the motor nuclei. Shivering is abolished by blockade of the neuromuscular endplate with curare.

Thermoregulatory behaviour such as fanning and adding or removing clothing is effective in changing the thermal insulation. Several layers of clothing with trapped air is a good insulator.

What is the difference between fever and hyperthermia?

Fever occurs when the **core temperature of the body** is raised above normal steady state levels. The body reacts as if it is too cold. Fever implies shivering combined with vasoconstriction, headache, dedolation pains, and general discomfort. Fever is the result of one of two phenomena: the set-point may be reset to a higher level, or the efficacy of the temperature control system may be impaired. Fever implies **hyperthermia**; however, many cases of hyperthermia do not constitute fever. Fever results from the action of **endogenous pyrogens** on the **hypothalamic heat control centre** (they increase the set point via prostaglandins). Exogenous pyrogens from microbes cause these **endogenous polypeptides** to be released from the defence cells of the body (i.e. the reticuloendothelial system, RES). Antipyretic drugs inhibit cyclo-oxygenase activity, hereby interfering with the synthesis of prostaglandins and thromboxanes.

Physiological hyperthermia is an increase in body temperature caused by **extreme heat stress** such that the heat loss capacity is exceeded. During hyperthermia the heat loss effectors are strained to the utmost. The high body temperatures of exercise activate cooling mechanisms such as sweat loss, which strive to return the core temperature to its normal level. **Hyperthermia** can increase the core temperature to more than 41°C. At this temperature **brain damage** with **delirium** and **universal cramps** occurs. Irreversible protein denaturation occurs above 44°C. **Malignant hyperthermia** is seen in persons allergic to halogen-substituted ethane anaesthesia. The allergic reaction is a result of an opening of Ca^{2+} channels in the muscle cells. The following influx of Ca^{2+} implies the generation of enormous amounts of heat, which is life threatening.

What is heat exhaustion and heat syncope? Heat stroke?

Heat stress, such as life in the tropics and **long-distance running,** is a threat to the water–salt balance and the circulation. Profuse sweat secretion leads to dehydration, reduced bloodflow, decreased sweat production, extreme vasodilatation, falling blood pressure, rising heart rate (baroreceptor response), increasing body temperature, brain confusion (**heat exhaustion**) and unconsciousness (**heat syncope**). The last two effects are caused by low brain bloodflow. **Heat stroke** (or in the sun, **sun stroke**) is actually heat syncope that occurs suddenly. The person suddenly falls into coma. The brain damage blocks the hypothalamic control, and sweat secretion ceases. The condition is terminal (**heat death**), except in cases where the patient is immediately cooled down.

What is hypothermia?

A hypothermic person loses consciousness when the temperature falls below 32°C. **Below 30°C a spontaneous cure** to a normal state **is impossible**, because heat-generating biochemical processes (shivering and non-myogenic heat liberation) are abolished or occur at a very low rate. The lower limit is difficult to define, but **life-threatening ventricular arrhythmia** develops at 25°C – **deep hypothermia. Artificially induced hypothermia** is used in brain- and heart-surgery, where the usual thermocontrol is inactivated by general anaesthesia. The procedure is dangerous because of **ventricular fibrillation.**

Further reading

Astrup, A., L. Simonsen, J. Bülow, J. Madsen and N.J. Christensen (1989) Epinephrine mediates facultative carbohydrate-induced thermogenesis in human skeletal muscle. *Am. J. Physiol.* **257** (3): E340–345.

Benzinger, T. H. (1961) The human thermostat. *Sci. Am.* **204**: 134–47.

Wasserman, D.H. and A.D. Cherrington (1991) Hepatic fuel metabolism during muscular work: Role and regulation. *Am. J. Physiol.* **260**: E811.

44. Multiple Choice Questions

Each of the following statements have True/False options:

A. Humans can survive at a temperature that would cook a piece of steak, because the steak cannot dissipate core heat energy.

B. Heat conductance of air at high pressure exceeds that at low pressure, so more heat energy is lost from the diver's body by conduction through air inside a diving bell deep under water than at 1 atmosphere of pressure.

C. Temperature homeostasis is present when heat energy input equals output.

D. BMR is lower before rather than after a meal.

E. The neutral temperature defines the level, where the resting metabolic rate is minimum.

44. Case History A

A male with a body weight of 70 kg stops his malaria prophylaxis with primaquine when leaving the endemic area. Three weeks later, a sudden attack of fever increases his core temperature from 37 to 40°C within 30 min. The heat capacity of the human body is 3.47 kJ kg^{-1} °C^{-1}. The metabolic rate of the subject increased substantially during the rise in temperature. Following the cold stage with uncontrollable shivering, the patient develops a delirious condition with severe headache. Two hours later the patient develops a profuse sweating. Partial evaporation of the water in 32 ml sweat min^{-1} (25% evaporates) occurs from the body surface (eliminating 2436 J g^{-1}), during which body temperature drops.

1. *Calculate the extra heat energy stored in his body after 30 min.*
2. *Calculate the smallest possible metabolic rate in the 30-min period.*
3. *What causes the extra heat energy stored in the body?*
4. *Calculate the reduction in body temperature by ingestion of 1 l of ice water, when the fever is at its highest level.*
5. *Calculate the time it takes to lose the accumulated heat energy by evaporation.*

44. Case History B

A male sedentary person, weight 70 kg, has a daily food intake of 400 g carbohydrate (17.5 kJ g^{-1}), 100 g fat (39 kJ g^{-1}), and 100 g protein (17 kJ g^{-1}). There is a metabolic water formation of 32 mg kJ^{-1}. The man excretes 2100 ml of water (i.e. 1200 ml in the urine, 100 ml in faeces, and 800 ml through lungs and skin).

1. *Calculate the metabolic rate (in kJ day^{-1} or MJ day^{-1}).*
2. *The man is in water balance by a water intake of 1700 ml as a total. Explain this apparent imbalance.*

Try to solve the problems before looking up the answers in Chapter 74.

Chapter 45. Gastrointestinal Nervous System

Describe the autonomic control of the digestive system

The hypothalamus and the other limbic structures are included in the **central autonomic system** (Fig. 45-1). These structures co-operate in situations of survival character such as defence, feeding and drinking.

The digestive system is innervated with nerve fibres of both the sympathetic and parasympathetic divisions, although the **parasympathetic control is dominant** (Fig. 45-1). The parasympathetic system **increases** digestive activity (secretion and motility), and the sympathetic system has a net **inhibitory** effect.

The vagus nerve contains both efferent and afferent fibres.

The efferent fibres enhance digestive activities by stimulating local neurons of the **intrinsic, enteric nervous system** located in the gut wall (i.e. the myenteric and the submucosal plexus of Fig. 45-1). The smaller intrinsic neurons stimulate the smooth muscles and gland cells.

The mucosa and other wall structures contain mechano- and chemoreceptors with afferent fibres to the intramural plexuses. Many afferent, sensory fibres in the vagus nerve inform the **central autonomic system** about the condition of the gut and its content.

The generally inhibitory digestive effects of the **sympathetic nervous system** are caused indirectly, by vasoconstriction which reduces bloodflow in the digestive tract.

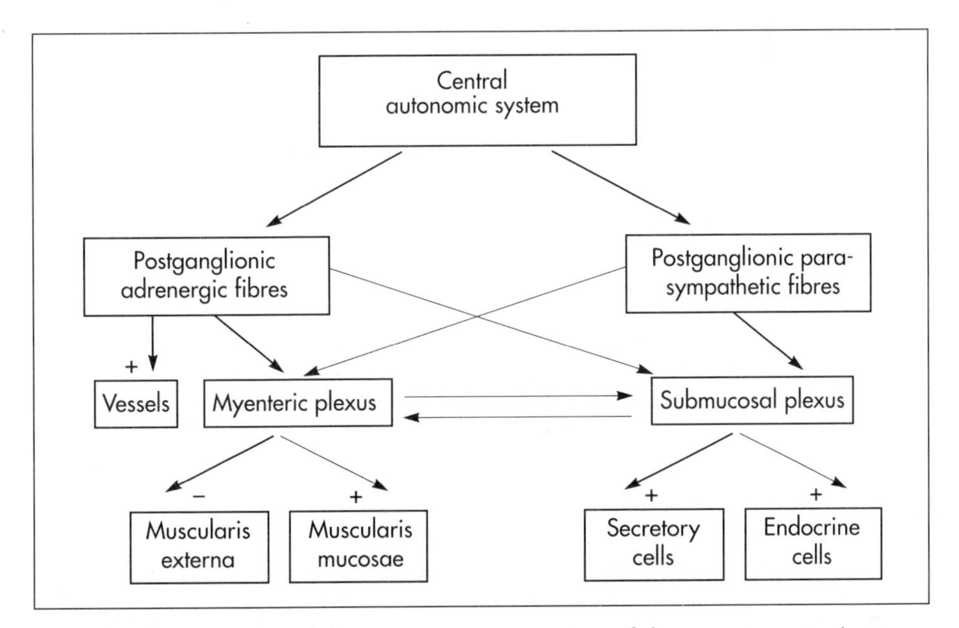

Fig. 45-1. An overview of the autonomic innervation of the gastrointestinal system.

What is the intrinsic, enteric nervous system?

This system consists of two sets of nerve plexi. The **submucosal (Meissner) plexus** mainly regulates the digestive glands, whereas the **myenteric (Auerbach) plexus**, located between the outer longitudinal and the middle, circular muscle layers, is primarily connected with gut motility. The plexi contain local sensory and motor neurons as well as interneurons. Motor neurons in the myenteric plexus release ACh and substance P. ACh contracts smooth muscle cells, when bound to muscarinic receptors. Inhibitory motor neurons release **vasoactive intestinal peptide (VIP)** and **nitric oxide (NO)**. These molecules relax smooth muscle cells (for the NO mechanism see Fig. 5-3). **Sensory neurons** are connected to chemoreceptors which detect different substances in the gut lumen, and to stretch receptors which respond to the tension in the gut wall caused by the food and chyme. The **short effector neurons** increase digestive gland secretion and induce smooth muscle contraction. The myenteric and submucosal plexi in the same region communicate with each other and with plexi further in the digestive system via **interneurons**. The large number of neuronal connections constitute the **intrinsic, enteric nervous system**, mediating **brain influence** on **digestive functions**.

What is the cephalic phase of regulation of the digestive system?

The thought of food can evoke salivary and some gastric secretion. This is a **conditioned reflex**. The smell and the taste of food substantially increase salivary and gastric juice secretion and trigger the secretion of pancreatic juice.

The main CNS centres regulating digestive functions are located in the brainstem, where the sensory taste fibres from gustatory, tactile and olfactory receptors terminate on the cell bodies of the **motor vagal and salivary nuclei**. The higher cortical and olfactory centres influence these **brainstem motor centres** and their **parasympathetic outflow**.

What is the gastric phase of regulation?

The food distends the stomach and stimulates **stretch receptors** and peptide sensitive chemoreceptors in the gastric mucosa. These receptors pass the new information to three targets: The **enteric plexi**, the **brainstem centres** and the **antral gastrin producing G cells**.

What is the intestinal phase of regulation?

Chyme entering the duodenum initiates the **intestinal phase** of nervous control. Gastric secretion and motility are at first increased to promote further digestion and emptying. This fills the small intestine with acidic and fatty chyme, which stimulates the intestinal cells to release the inhibitory hormones (**enterogastrones**, see Chapter 46) that decrease stomach activity in order to prolong emptying and allow time for intestinal digestion.

Further reading

Schultz, S.G., J.D. Wood and B.B. Rauner (Eds) (1989) In: *Handbook of Physiology*, Sect.6. "*The Gastrointestinal System.*" American Physiological Society, Bethesda, MD.

Yamada, T. (1991) *Textbook of Gastroenterology*, Vol. 1. J.B. Lippincott, Philadelphia.

45. Multiple Choice Questions

Each of the following statements have True/False options:

A. The receptive relaxation response of the stomach decreases gastro-oesophageal reflux.

B. The intrinsic innervation of the digestive, secretory epithelium responds to parasympathetic input with decreased secretion.

C. The sympathetic nerve fibres to the gut acts presynaptically to inhibit ACh release in the myenteric ganglia and activate α-receptors. Hereby, sphincter muscles are contracted, blood vessels are constricted, and secretion is inhibited.

D. Relaxation of the lower oesophageal sphincter is not caused by increased vagal inhibitory fibre discharge.

E. Oesophageal reflex activity is controlled by primary peristalsis that is co-ordinated by a swallowing centre in the solitary tract nucleus, vagal nuclei, and reticular formation. Local distention stimulates the secondary peristalsis.

45. Case History

A male with a body weight of 83 kg joins a cocktail party at 8 o'clock p.m. and drinks steadily until he leaves the party at 1 a.m. He drives his car and near home at 2 o'clock, he causes a traffic accident, where a 16-year-old girl is killed instantly. He is brought into custody by the police, and at 3 a.m. he delivers a blood sample containing 1.3 g of alcohol per l of blood plasma.

1. *Calculate his distribution volume for alcohol.*
2. *Calculate the estimated alcohol concentration in the plasma of the driver at the time of the accident (2 a.m.).*
3. *Calculate the amount of alcohol ingested by the driver.*

Try to solve the problems before looking up the answers in Chapter 74.

Chapter 46. Gastrointestinal Hormones

All gastrointestinal functions seem to be under **hormonal control**. Many of these organs also have **autonomous nerve fibres** and **internal nerve networks**.

Gastrointestinal hormones are **peptides**. The total mass of scattered mucosal cells producing peptides is greater than that of the hypothalamo-hypophyseal system. These peptides are released into the blood by products of digestion and by vagal nerve signals in response to feeding; their response can be mimicked by infusion of the peptide and blocked by infusion of a specific antagonist. Gastrin, cholecystokinin, secretin, and gastric inhibitory polypeptide (GIP) comply with these criteria.

Peptide hormone families are groups of hormones that exhibit **sequence homology**. This means that they possess a common amino acid sequence, such as the **gastrin family** which has **sequence homology** in their terminal penta-peptide.

What is the definition of hormones?

For the definition of hormones and hormone receptors, for chemical types of hormones, synthesis of peptides, elimination of peptides, and mechanism of action for peptides, see Chapter 63.

What is the function of peptides?

Peptide hormones have **autocrine and paracrine functions** in the gastrointestinal tract.

Peptide hormones function as **neurotransmitters for peptidergic neurons** – especially in the CNS. They are so-called **regulatory peptides**, which work in a **diffuse neuro-endocrine system** which is found in the walls of the gastrointestinal tract.

What happens when a meal reaches the intestine? What is incretin?

When a meal reaches the intestine, several peptides forming the **incretin family** are released during the **intestinal secretion phase**. Typical representatives of the incretin family are **gastric inhibitory peptide (GIP or glucose-dependent insulin-releasing peptide), glicentin (intestinal glucagon), and glucagon-like peptides (GLP-1 and -2)**. For further information, see Chapter 73.

What is the function of gastrin?

Gastrin is the **most potent stimulant of gastric acid secretion**, which is derived from **parietal or oxyntic cells** in the stomach (Fig. 49-2). When stimulating gastric acidity, gastrin relaxes the gastric muscles, thus retarding the passage of chyme into the duodenum.

Feeding induces the secretion of gastrin to the interstitial fluid and then to the blood. Neural signals pass through the vagal nerve to the **gastrin-secreting G cells of the gastric antrum and duodenum** (Fig. 46-1). The afferent input begins with the smell and

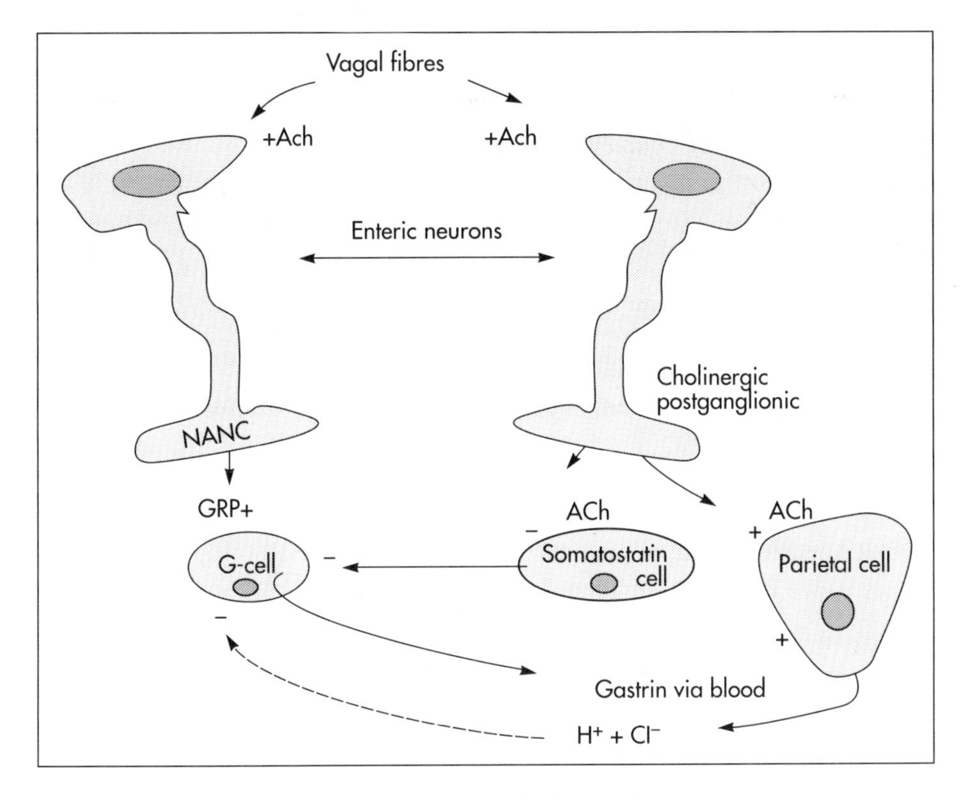

Fig. 46-1: Gastric HCl secretion.

taste of food, and is reinforced by vago-vagal reflexes elicited by oesophageal and gastric distention. **Digested protein** (polypeptides and amino acids) act directly on G cells.

Vagal, cholinergic preganglionic fibres transfer signals to the G cells via **non-adrenergic, non-cholinergic (NANC) postganglionic neurons** in the intrinsic system (Fig. 46-1). These postganglionic nerve fibres release **gastrin-releasing peptide (GRP)**. GRP stimulates the G cells to produce gastrin. The gastrin released reaches the parietal cells through the blood and causes HCl secretion (Fig. 46-1). An indirect vagal route to the G cells is via **postganglionic cholinergic enteric neurons** to **somatostatin cells** which are located close to the **G cells** (Fig. 46-1). When these enteric neurons release ACh, the response of the somatostatin cells is inhibition of somatostatin release. Somatostatin inhibits G cell secretion by paracrine action. The result of both vagal inputs to the G cells is gastrin release (Fig. 46-1). An elevated $[H^+]$ in the duodenum inhibits gastrin release Fig. 46-1). There are two major forms of gastrin in the plasma, **normal gastrin or G-17** and **big gastrin or G-34**. They are 17 and 34 amino acid polypeptides, respectively. Gastrin is produced by G cells of the gastric antrum and duodenum. The duodenal Brunner glands secrete half of the G-34.

Gastrin also imposes **tropic (growth-stimulating) actions** on the parietal cells, the mucosa of the small and large intestine and possibly the pancreas. Gastrin stimulates the **pepsin secretion** from peptic cells, and the **glucagon secretion** from the α-cells of the pancreatic islets.

Nitric oxide (NO) is a possible neurotransmitter between the preganglionic and the NANC postganglionic neurons.

What is the function of cholecystokinin (CCK)?

CCK, according to its function and structure, belongs to the **gastrin family**. However, CCK has a higher affinity for receptors stimulating gallbladder contraction and pancreatic enzyme secretion. Both gastrin and CCK release glucagon from the α-cells of the pancreatic islets.

CCK is cleaved from **prepro-CCK** in the **duodenum, upper jejunum** (I-cells) and in the **brain**. CCK molecules consist of a group of peptides. CCK-8, CCK-22 and CCK 33 are the dominant forms in the blood.

1. CCK stimulates gallbladder contraction and empties the gallbladder, as its name implies.
2. CCK stimulates pancreatic secretion of an enzyme-rich fluid. CCK has a maximal effect only in the presence of secretin (**potentiation**) and normal vagal influence. The most important stimulus for CCK liberation is amino acids and fatty acids, which reach the duodenal mucosa. Bile is ejected into the duodenum, where fat is emulsified to ease its absorption. CCK also acts as an **enterogastrone** – an intestinal hormone that inhibits gastric activity and emptying. This leaves more time for the bile to emulgate fat.
3. CCK also stimulates gut motility with its main location being at the upper small gut.

What are the hormones of the secretin–glucagon family? How does secretin work?

Secretin exhibits sequence homology with **pancreatic glucagon, vasoactive intestinal peptide (VIP), growth hormone-releasing hormone (GHRH) and GIP**. A family of five genes code for these five hormones.

Secretin is secreted by S cells in the mucosa of the upper small intestine, when acid chyme (pH below 4.5) arrives to the first part of the duodenum. Fatty acids from fat digestion also contribute to secretin release.

Secretin stimulates the secretion of bicarbonate and water by pancreatic duct cells, and of bicarbonate-rich aqueous bile. Secretin potentiates the action of CCK including an **enterogastrone effect**. Secretin antagonizes the gastrin action.

Describe two effects of gastric inhibitory polypeptide (GIP)

1. **Enterogastrone effect**. GIP is one of several enterogastrones secreted by the duodenum and upper jejunum, in order to **limit the amount of HCl here**. When gastric acid, fats and hyperosmolar solutions have entered and distended the duodenum, GIP and other enterogastrones (somatostatin, CCK and secretin), are released and suppress gastric acid secretion and motility of the stomach.
2. **Incretin effect**. GIP is also called **glucose-dependent insulin-releasing peptide**. GIP is liberated to the blood as gastric chyme enters the duodenum – and before the glucose of the chyme can be absorbed. GIP increases **insulin secretion** from the β-cells of the pancreatic islets much earlier and to a greater extent, than when the blood [glucose] is elevated by intravenous infusion (see **incretin effect** in Chapter 73).

What is somatostatin and glucagon?

Somatostatin or **growth hormone-inhibiting hormone (GHIH)** is a build-up by 14 amino acid moities, and produced from a prehormone of 28 amino acids. Further information on somatostatin and glucagon is given in Chapter 73.

What hormones are secreted by the pancreatic islet cells?

Glucagon is produced by the **α-cells**, insulin by the **β-cells**, and somatostatin and gastrin are produced by the **δ- and G-cells** of the pancreatic islets. Meals rich in protein and fats release **pancreatic polypeptide** (PP) from the PP cells (Fig. 72-2). Pancreatic polypeptide inhibits both the enzyme secretion from the pancreas and the emptying of bile into the small intestine. This leads to a **delay in the absorption of nutrients** including **glucose**.

What is the function of vasoactive intestinal polypeptide (VIP) and substance P?

VIP is built by 28 amino acid moities, and produced by brain neurons, vessel wall neurons and by digestive tract neurons.

VIP is generally vasodilatating, and the increased bloodflow **increases intestinal secretion; the exception is that VIP inhibits HCl secretion. VIP is also involved in penile vasodilatation during erection and in bronchiolar dilatation.**

Substance P is built up of 11 amino acid residues. It is produced in peptidergic neurons and glands of the digestive tract and in brain neurons. **Substance P stimulates gastrointestinal motility.**

What is the function of gastrin-releasing peptide (GRP) and villikrinin?

Vagal, cholinergic preganglionic fibres transfer signals to the **G cells** via **non-adrenergic, non-cholinergic (NANC)** postganglionic neurons (Fig. 46-1). These enteric neurons liberate **gastrin-releasing peptide (GRP)** to the gastrin producing G cells. The gastrin released reaches the parietal cells through the blood and increase the HCl secretion (Fig. 46-1). GRP thus releases gastrin and hereby stimulates the secretion of gastric acid. GRP consists of 27 amino acid moities and is also released from neurons in the brain.

Villikrinin from the duodenum and ileum stimulates the villi to contract rhythmically.

Further reading

Calver, A., J. Collier and P. Vallance (1993) Nitric oxide and cardiovascular control. *Exp. Physiol.* **78**: 303–26.

Del Valle, J., and T. Yamada (1990) The gut as an endocrine organ. *Ann. Rev. Med.* **41**: 447.

Furness, J.B. *et al.* (1992) Roles of peptides in the enteric nervous system. *Trends Neurosci.* **15**: 66.

Johnson, L.R. (Ed.) (1987) *Physiology of the Gastrointestinal Tract*, Vol.2, 2nd edn. Raven Press, New York.

46. Multiple Choice Questions

Each of the following statements have True/False options:

I.

A. Gastrin originates in the antral and duodenal mucosa, where it is released from G-cells.
B. Secretin is a hormone that is released from the duodenum in response to HCl.
C. Pancreozymin (CCK) contracts the sphincter of Oddi.
D. GIP stimulates insulin secretion.
E. GRP is involved in vagal gastrin secretion.

II.

A female, 67 years of age, has a 51-year history of pulmonary tuberculosis. Surgery has included removal of more than one lung and one kidney. She has been cured for tuberculosis 15 years previous to the present status.

The patient is walking like a drunk, and she is now complaining of nausea, vertigo, and fatigue. She only wants to rest without much food and fluid. The doctor prescribed the patient to drink two glasses of tap water at two daily meals beyond her usual intake, and the result was a dramatic relief within hours of all symptoms and signs.

Each of the following five diagnoses have True/False options:

A. Pulmonary tuberculosis.
B. Brain tumor.
C. Dehydration.
D. Renal insufficiency.
E. Hypoxia.

Try to solve the problems before looking up the answers in Chapter 74.

46. Essay

Describe the gastrointestinal peptide hormones and their relationship to the nervous system.

CHAPTER 47. GASTROINTESTINAL MOTILITY

The process of chewing or **mastication** requires co-ordination of the chewing muscles, the cheeks, the palate and the tongue. Chewing is normally a reflex action. The forces involved in grinding and cutting the food are enormous, and sufficient to fragment cellulose membranes. Finally, the food is mixed with **saliva** and formed into a **bolus**. The bolus is pushed back into the pharynx when the tongue is pressed against the **hard palate**.

The gastrointestinal tract moves ingested materials and secretions from the mouth to the anus. These movements, as well as non-propulsive contractions, are called **motility**.

What characterize visceral smooth muscle cells?

See the 10 characteristics in Chapter 7.

Describe swallowing and its control

Swallowing (**deglutition**) begins as a **voluntary process** by which the tongue pushes a portion of the food back against the soft palate. Elevation of the soft palate closes the nasopharynx, and the food enters the pharynx, the larynx is elevated closing the epiglottis and respiration stops. The upper pharyngeal constrictor contracts, initiating **sequential contractions** of the other pharyngeal constrictors. These contraction waves are **involuntary** and push the food towards the oesophagus. Peristalsis in the oesophagus is started as the pharyngeal wave passes through the upper oesophageal sphincter. When the propulsive wave reaches the **lower oesophageal sphincter (LES)**, the relaxed muscle wall preceding the bolus momentarily relaxes the LES, and the food passes the cardia to enter the stomach. The upper third of the oesophagus is composed of striated muscle, the middle third contains mixed smooth and striated muscle, and the lower third contains only smooth muscle.

Swallowing is controlled by brainstem neurons. They form a **swallowing centre**. The vagus nerve contains both **somatic motor neurons** (originate in the nucleus ambiguus) that form motor endplates on striated muscle fibres, and **visceral, preganglionic motor neurons** (from the dorsal motor nucleus to the myenteric plexus). Vagal stimulation relaxes the upper sphincter and LES.

Sympathetic stimulation contracts the LES, mediated by NA acting on α-receptors. When a swallow is initiated in the pharynx, or when the lower oesophagus is distended by a bolus, it will relax the LES by **reflexes in inhibitory vagal fibres** joining the **enteric nervous system**. VIP and NO act as transmitters.

What is achalasia?

Achalasia is a condition where the **LES** cannot relax completely during swallowing. Vomiting and weight loss are major symptoms. There is no **receptive relaxation**, because the myenteric plexus does not work.

Gastro-oesophageal reflux with oesophagitis is caused by incomplete closure of the LES. Gastric contents with acid reaction then reflux into the oesophagus causing inflammation, erosion and bleeding.

Describe gastric motility

In the stomach digestion continues (**salivary amylase**) and the stomach regulates emptying of its content into the duodenum. The fundus has a **high compliance**, so food can accumulate without much increase in pressure. This **receptive relaxation** is mediated by vagal fibres releasing VIP to inhibitory neurons of the myenteric plexus. The body of the stomach (corpus ventriculi) mixes and grinds the food with gastric juice – also by **retropulsion** – and propels the content toward the antrum and pyloric region for regulated emptying. The distal stomach reduces solids to a fluid consistently composed of particles less than 2 mm. Here is a **forceful peristalsis**, so the pyloric sphincter opens and the chyme is ejected into the duodenum.

Along the greater curvature of the stomach is a region of rapid spontaneous depolarization, which is called the **gastric pacemaker** establishing the maximum rate of gastric contractions. The gastric smooth muscle wall generates two types of electrical activity. **Slow waves (basic electrical rhythm)** are slow depolarizations occurring at a frequency of 3–6 min^{-1}. The slow waves change the resting membrane potential from −70 to −55 mV (15 mV). The slow depolarizations of the membrane opens potential sensitive Ca^{2+}-channels, causing a Ca^{2+}-influx to the smooth muscle. **Spike potentials** are **periodic fast waves of depolarization** that most often follow a slow wave, and then always initate gastric contractions (elicited by a rise in cytosolic [Ca^{2+}]). Spike potentials are caused by opening of more Ca^{2+} channels, and these potentials are long lasting, since the **Ca^{2+} channels** open slowly and longer than the Na^{+} channels. Spike potentials are elicited by vagal signals, by ACh, by myenteric signals and by gastrin. Catecholamines (Ad and NA) relax smooth muscle by hyperpolarization through **α-receptors.** Relaxation occurs when intracellular Ca^{2+} is returned to the ECF and to the ER.

Describe the motility of the small intestine

The small intestine is about 8 m long and commonly divided into three segments: the duodenum, jejunum and ileum. The intestinal contents must be moved in a manner that brings them into contact with the mucosa, and propels the contents along this tubular organ. The slow waves are controlled by several **pacemaker regions** in the small intestine. The **pacemaker rate is highest in the duodenum** (about 12 min^{-1}), and decreases down to 8 waves min^{-1} in the terminal ileum.

During the fed state, segmentation serves to mix chyme with enzyme-containing digestive fluid, and brings the mixture into contact with the mucosal surface. Segmentation divides the small intestinal content into many segments by localized circular smooth muscle contractions.

Propulsive motility is accomplished by **peristalsis**. Peristalsis is a propagating contraction of successive sections of circular smooth muscle preceded by a dilatation. The dilated intestinal wall is drawn over its content in this reflex mechanism which has been called the **law of the gut**. Peristaltic contractions usually travel along a small length of the small intestine, except for the **peristaltic rush** related to the **migrating motor complex**. The migrating motor complex cleanses the digestive tract of non-absorbable substances, and provides an effective emptying of the tract all the way.

The ileocaecal sphincter prevents retrograde flow of colonic matter. The sphincter regulates emptying of ileum 5 h after a meal. The emptying of ileum is stimulated by **gastrin**, possibly via the **gastroileal reflex**, but the emptying is inhibited by a distended colon. The ileocaecal sphincter is normally passed by **1 l of faecal matters per day**.

Describe the mechanism of vomiting

Vomiting or emesis is initiated by the feeling of nausea, and an array of sympathetic and parasympathetic responses. **Sympathetic responses** include sweating, pallor, increased respiration and heart rate and dilatation of pupils.

Parasympathetic responses include profuse salivation, pronounced motility of the oesophagus, stomach, and duodenum, and relaxation of the lower and upper oesophageal sphincters. Duodenal contents can be forced into the stomach by antiperistalsis. During the expulsion of gastric contents, the person **takes a deep breath**, the **pylorus is closed,** the **glottis is closed** so respiration stops, and the stomach is **squeezed between the diaphragm and the abdominal muscles**, causing rapid emptying. Vomiting is co-ordinated by the **vomiting centre in the medulla**. Vomiting is stimulated in certain areas of the brain (hypothalamus) and the cerebellum through sensory stimuli or injury. Vomiting is also provoked by certain labyrinthine signals, and from the **chemoreceptive trigger zone** located on the floor of the 4th ventricle close to **area postrema**.

Is vomiting dangerous during deep anaesthesia?

During deep anaesthesia the vomiting and swallowing mechanisms are **paralysed**. Any patient must abstain from food and water for at least 6 h before deep anaesthesia is administered. Otherwise, the patient may vomit into the pharynx, and suck his or her own vomit into the trachea. Over the years, many patients have choked to death due to this mechanism. Such an event is a clear case of malpractice for the anaesthesiologist. The survivors develop **aspiration pneumonia.**

The swallowing mechanism is also paralysed by injury of the 5th, 9th, or 10th cranial nerve, by **poliomyelitis**, by **myasthenia gravis** and by **botulism**.

How is the pH of the ECF regulated following substantial vomiting?

An acute loss of H^+ from the ECF creates a **metabolic alkalosis** (high pH with high base excess and standard bicarbonate). Such a condition is life threatening, and effective control systems are present to bring back the pH of the ECF to the normal value 7.40. Buffer systems exist – not only those in the ECF, but also in the cells. However, the only way to equilibrate the body systems to normal is to use the control of the lungs and the kidney.

1. **The ventilation is reduced** through the lack of stimulation of the carotid chemoreceptors. The superficial ventilation prevents CO_2 from being eliminated. The rise in P_{aCO_2} eventually stimulates both the peripheral and the central chemoreceptors, so the respiratory frequency increases sufficiently to cover the demand for O_2.
2. The renal **ultrafiltration of bicarbonate** increases proportionally to the rise in plasma and ECF [bicarbonate]. The three mechanisms by which the renal tubules normally secrete H^+, are also used to avoid H^+-secretion.
   (a) Reduced H^+-secretion implies a fall in both the proximal and the distal tubular

reabsorption of bicarbonate. The most important of these processes is the reduction of the **distal H^+-secretion**. The lack of H^+ inhibits the **luminal Na^+/H^+ antiport,** so the Na^+ reabsorption is minimal, and the increased flux of Na^+ in the tubular fluid drags out water and results in an **osmotic diuresis** with a high content of bicarbonate and thus a high pH. The fall of $[H^+]$ in plasma also reduces plasma $[K^+]$, because these ions are exchanged over the cell membranes. The cells deliver H^+ to the ECF in exchange of K^+. In order to avoid hypokalaemia, the kidneys must try to reduce K^+-excretion by **reduction of the aldosterone secretion** from the adrenal cortex.

(b) The NH_4^+ production in the kidneys falls, because the glutamine source from the liver falls. Hereby, elimination of H^+ is avoided, and more urea must be eliminated to get rid of the nitrogen.

(c) The phosphate buffer in the blood is forced to the basic side, and **more secondary phosphate** is ultrafiltered and excreted. Thus loss of H^+ is avoided.

Describe colonic motility

Colonic transit is **measured in days**. Mixing occurs in the **ascending colon**, because peristalsis is followed by antiperistalsis. **Slow waves of contraction** move the content in the oral direction to delay propulsion and increase reabsorption of water and electrolytes. **Segmentation** is mixing the content by regular segments called **haustrae. Prominent haustration** along the length of the colon is characteristic for the X-ray image of the normal colon. The colon must provide an optimal environment for bacterial growth. **Peristaltic rushes** in the colon occur several times per day. They often start in the transverse colon as a **tight ring**, continuing as a **long contraction wave.** Peristaltic rushes are promoted by **gastro-colic** and **duodeno-colic reflexes assisted by gastrin and by CCK**.

Describe the mechanism of defecation

Defecation is an act involving colon, rectum, anal sphincters and many striated muscles (diaphragm, abdominal and pelvic muscles). **Defecation** is a temporal release of anal continence brought about by a reflex. The rectum is usually empty, and it has a rich sensory supply. Distention of the rectosigmoid region with fecal matter releases awareness of the urge to defecate, an **intrinsic defecation reflex**, and a **strong, spinal reflex**.

The **smooth internal anal sphincter muscle** maintains a tonic contraction during continence, due to its sympathetic fibres from the **lumbar medulla** (through hypogastric nerves and the inferior mesenteric ganglion) and **parasympathetic sacral fibres**. The muscle relaxes due to its parasympathetic fibres in the pelvic splanchnic nerves (S_2–S_4). The strong spinal reflex produces relaxation of the smooth muscles of the internal anal sphincter and contraction of the striated muscles of the external anal sphincter (pudendal nerves) inhibiting the reflex and causing **receptive relaxation**. This is the last decision – before defecation.

The **levator ani muscle** contributes to the closure of anus, because contractions increase the angle between the rectum and the anus.

Destruction of the lower sacral medulla (the **defecation center**), destroys the spinal reflex and thus the normal defecation. Higher spinal lesions destroy the voluntary control. An acceptable status is obtainable by mechanical release of the reflex once daily following a meal.

Describe five disorders of colonic motility

1. **IRRITABLE BOWEL SYNDROME (COLON IRRITABILE)** is a condition with painful spasms causing constipation alternating with mucous diarrhoea. The condition is related to stress and sedentary life style, and is relieved by daily exercise.
2. **DIVERTICULAR DISEASE WITH DIVERTICULITIS** is a condition with multiple small hernia of mucosa through the muscle wall of the colon, caused by increased intraluminal pressure.
3. **HIRSCHSPRUNG'S DISEASE OR AGANGLIONOSIS** is caused by congenital absence of the myenteric plexus at the region where the colon passes into rectum. The cause is mutation of a gene localized on chromosome 10. The aganglionic segment is permanently contracted, and the colon distends above forming the **megacolon**. Large amounts of fecal matters accumulate, because peristalsis and mass movements are impossible.
4. **CONSTIPATION** is often caused by **irregular defecation habits**, and irrational use of laxatives. Such habits suppress the natural reflexes.
5. **COLON CANCER** is related to slow passage of fecal material with **cancerogenous substances** through the colon. **Carcinogens** are chemicals whose byproducts bind to DNA and damage it.

Sedentary persons have a high frequency (morbidity) of **constipation** and a high mortality of **colon cancer**, but not of **rectal cancer**. The colon cancer is clearly related to an **inactive life style**, and regular exercise reduces morbidity and mortality.

Further reading

Del Valle, J. and T. Yamada (1990) The gut as an endocrine organ. *Ann. Rev. Med.* **41**: 447.

Ederly, P. *et al.* (1994) Mutations of the RET proto-oncogene in Hirschsprung's disease. *Nature* **367**: 378–80.

Johnson, L.R. (Ed.) (1986) *Physiology of the Gastrointestinal Tract*, Vol. 2, 2nd edn. Raven Press, New York.

Read, N.W. and J.M. Thims (1986) Defecation and pathophysiology of constipation. *Clin. Gastroenterology* **15**: 937–65.

47. Multiple Choice Questions

Each of the following five statements have True/False options:

A. The basic electrical rhythm is an electrical event which always causes contractions in the digestive system.
B. The basic electrical rhythm determines the maximal rate of peristaltic contractions.
C. Slow waves in the colon cannot result in antiperistalsis.
D. The major role of the human colon is to reabsorb water and electrolytes.
E. The only entirely voluntary motor process of the motility patterns in the digestive tract is chewing.

47. Case History

A female diabetic, with a body weight of 60 kg, develops dedolations and high fever, with a rise in body temperature from 36.9 to 41.4°C within 2 h. She is in a comatous stage with arterial pH of 7.09, P_{aCO_2} of 3.5 kPa (26 Torr) and a base deficit of (BE: −20 mM).

1. *Calculate the extra heat energy stored in her body.*
2. *Characterize her acid–base status.*
3. *How does she compensate this condition?*
   Later she develops a large sweat secretion, and 20% of its water content evaporates.
4. *Calculate the sweat volume necessary to eliminate the extra heat energy stored.*
5. *How does the ionic concentrations of her pancreatic juice vary with increasing secretion rate?*

Try to solve the problems before looking up the answers in Chapter 74.

CHAPTER 48. SALIVARY SECRETION

What is saliva and what is its importance?

Saliva is a mixture of secretions from three pairs of glands. The **parotid** is the largest and serous (watery saliva), the **sublingual** is mucous (viscous, containing mucin), and the **submandibular salivary gland** is built of mucous acini surrounded by serous **half moons**. The primary saliva is produced in the acini, but **secondary processes** in the salivary ducts (secretion and reabsorption) are involved in the final saliva production. Salivary glands have a high bloodflow. Up to 1 l of saliva is produced per day. The maximal secretion rate is one ml of saliva per gram salivary tissue per min (i.e. 60 times that of pancreas).

Saliva is a watery solution of electrolytes and organic substances. Salivary **mucin (a glycoprotein) and water** lubricate food, solubilize particles, and salivary enzymes initiate digestion. α-Amylase or **ptyalin** cleaves α 1–4 glycoside bindings in starch. **Salivary buffers** maintain the pH-optimum (6.8) of amylase during the first period in the stomach. The saliva dilutes injurious agents.

Saliva cleans the mouth and pharynx (prevents caries), and eases swallowing.

Salivary lysozyme lyses bacterial cell walls. The salivary **epidermal growth factor** promotes the healing of wounds. Animals instinctively lick their wounds. Saliva contains immunodefensive secretory globulin A **(IgA)**, amino acids, urea, and blood-type antigens in secreting persons. Saliva may inactivate human immunodeficiency virus (HIV). The most common infection of the salivary glands is **acute parotitis** caused by the mumps virus.

The virus causing infectious mononucleosis is probably transferred with saliva by 'deep kissing'. **Infectious mononucleosis** is a disease characterized by lymphadenopathy, lymphocytosis and a duration longer than an ordinary tonsillitis. The condition is dangerous, because spontaneous rupture of the spleen occurs.

How is the primary saliva produced?

The **primary salivary secretion** in the acini resembles an **ultrafiltrate of plasma**.

1. Neural or humoral stimulation of cholinergic, muscarinic receptors (ACh) on the basolateral membrane of acinar cells leads to a rise in intracellular [Ca^{2+}].
2. This rise triggers **luminal Cl^-- and basolateral K^+-channels**. Hereby, K^+ is transferred to ISF and Cl^- to the acinar lumen in a balanced relationship (Fig. 48-1). Therefore, Cl^- flows easily down its electrochemical potential gradient into the lumen of the acinus, and K^+ down its electrochemical potential gradient to the ISF through activated **channels**. These ion flows create a negative electric field in the lumen.
3. The initial fall in intracellular [K^+] increases the driving force of the electroneutral **Na^+-K^+-$2Cl^-$ co-transporter** to transport Cl^- into the cell together with Na^+ and K^+ (Fig. 48-1). Thus the electrochemical potential of Cl^- and K^+ is greater in the cell, than in the ISF and in the saliva.

4. The negative field provides an electric force that drives a **passive Na^+ flux** into the acinar lumen through leaky **tight junctions** (Fig. 48-1). The NaCl flux into the lumen is followed by osmotic water transport (through **leaky tight junctions** and transcellularly through water channels in the cell membranes).
5. The basolateral membranes of acinar cells contain a **Na^+-K^+-pump** that provides the energy for the primary salivary secretion (Fig. 48-1). The rise in intracellular [Na^+] from (3) activates the **Na^+-K^+-pump**, whereby [Na^+] is kept almost constant. Ouabain inhibits salivary secretion, because it blocks the pump.

Is salivation a conditioned reflex?

Salivation is controlled by **unconditioned reflexes** (taste-, olfactory- and mechanoreceptors) as well as **conditioned reflexes** (the thought of food). These signals reach the **brain stem salivary centres**, which activate the parasympathetic nerves to the salivary glands.

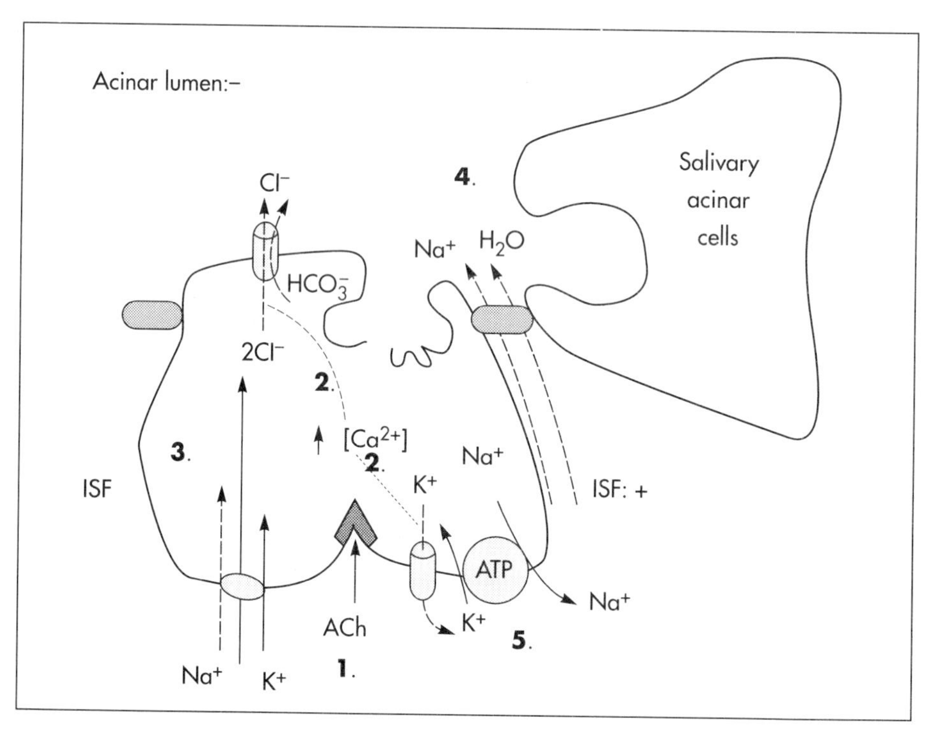

Fig. 48-1: Ion and water transport by salivary acinar cells. The transcellular Cl^- transport is coupled to the paracellular Na^+ transport. The net result is an isosmotic NaCl transport produced by a secondary active Cl^- secretion. Solid and dashed arrows indicate active and passive transport, respectively. Circles are carrier molecules, whereas tubes symbolize transport channels.

How can we control salivary secretion?

Salivary secretion is controlled by the autonomic nervous system, and minimally influenced by hormones.

Parasympathetic, cholinergic fibres, originating in the **salivary nuclei of the brainstem**, synapse with postganglionic neurons close to the secretory cells. These neurons transmit signals to the **cholinergic, muscarinic receptors** (ACh in Fig. 48-2). Parasympathetic activity can release maximal salivary secretion and bloodflow resulting in an amylase-rich saliva with mucins (glycoproteins). **Atropine** blocks the **muscarinic, cholinergic receptors** (e.g. during anaesthesia where the mouth becomes dry). The rise in bloodflow is atropine resistant and caused by the vasodilatating VIP, which is released from peptidergic nerve terminals that also contain ACh. **β_1-Adrenergic agonists and VIP** elevate cAMP in the acinar cells (Fig. 48-2), an effect potentiating the secretory effect of ACh. The vascular smooth muscle relaxation by VIP is probably also mediated via cAMP.

Sympathetic nerve signals, and circulating catecholamines via β-adrenergic receptors, inhibit the bloodflow and the secretion of serous saliva (β_1-receptors in Fig. 48-2). A small, transient, mucous secretion with a high [K^+] and [bicarbonate], and a low [Na^+] is produced, because of the low secretion rate. NA stimulates both α_1-adrenergic and β_1-adrenergic receptors. Binding of NA or β-adrenergic agonists elevates intracellular cAMP,

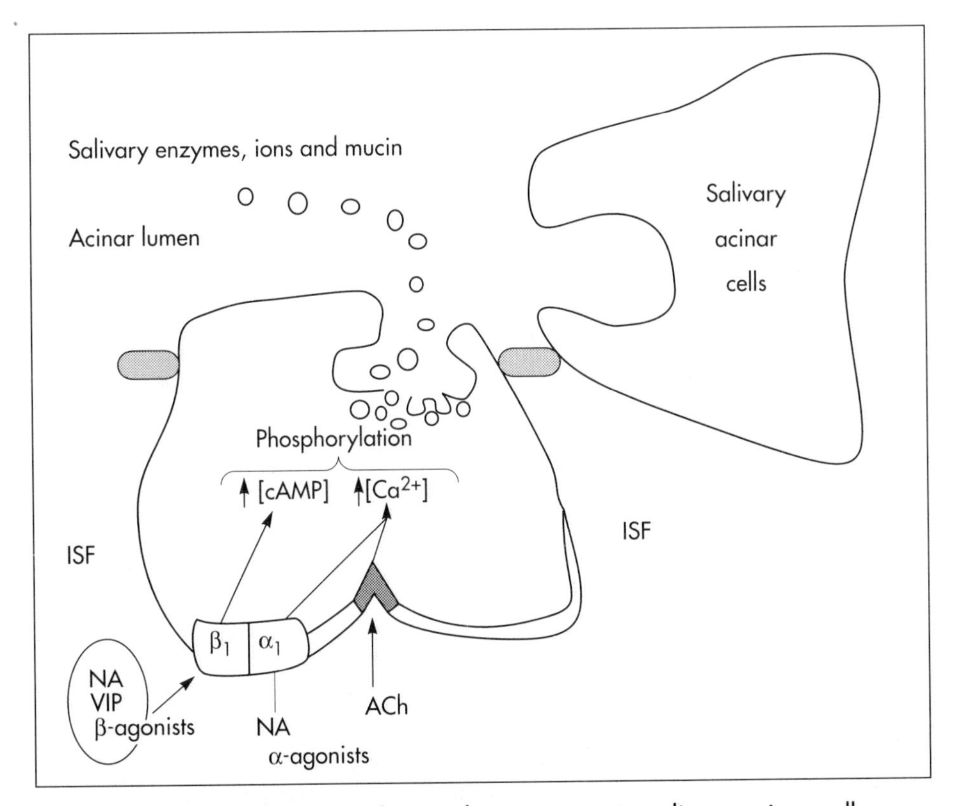

Fig. 48-2: Receptors and second messengers in salivary acinar cells.

which correlates with a small increase in **primary salivary secretion** (Fig. 48-2). This explains why the mouth becomes dry during events, where the sympathetic system dominates (anxiety, excitement, etc.).

What happens in the salivary ducts?

The salivary ducts are almost water-tight. Therefore, the final salivary flow is dependent upon the primary salivary secretion rate in the acini.

The duct systems, in particular the small striated ducts with a substantial O_2 consumption **reabsorb large amounts of Na^+ and Cl^-**, whereas **bicarbonate and K^+** are secreted. Saliva becomes more and more hypotonic at low secretion rates, because the **Na^+ and Cl^- reabsorption dominate.**

1. The **reabsorption of Na^+ and the secretion of K^+** are processes stimulated by the mineralo-corticoid, **aldosterone**. Aldosterone (Aldo in Fig. 48-3) stimulates Na^+-influx through the **luminal Na^+-H^+-exchanger** (Fig. 48-3). Na^+ enters the cell in exchange with H^+. The resulting intracellular rise in $[Na^+]$ activates the basolateral **Na^+-K^+-pump**. Thus, Na^+ is reabsorbed transcellularly from the salivary duct (Fig. 48-3). The pump maintains the electrochemical potential gradients of Na^+ and K^+.
2. The Cl^- follows passively, and is partly exchanged with bicarbonate along the duct system through a **luminal Cl^--bicarbonate exchanger** (Fig. 48-3). The secretion of bicarbonate is so great that its concentration in the final saliva exceeds that in plasma.

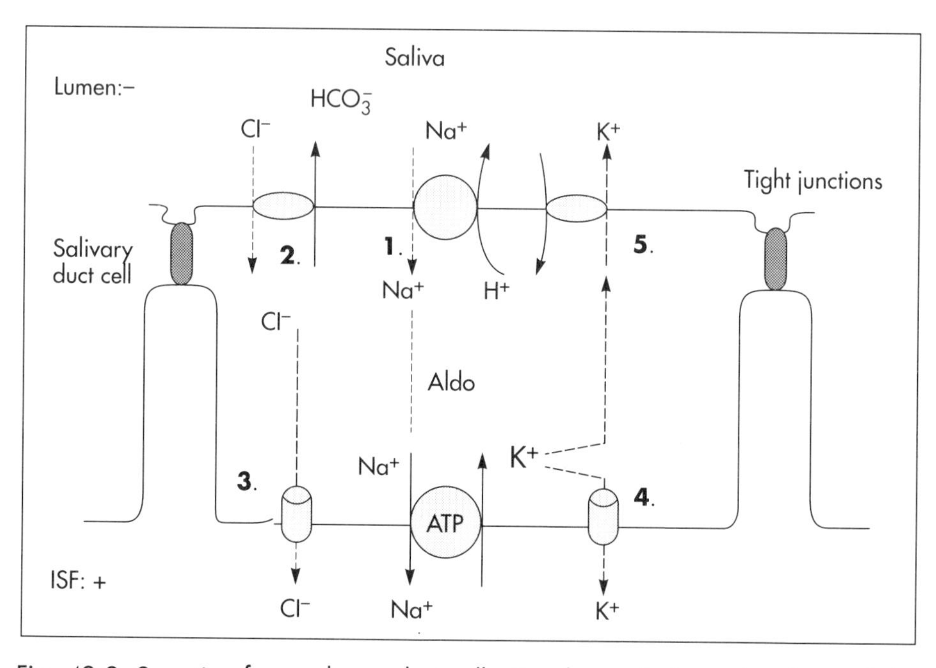

Fig. 48-3. Secretion from salivary duct cells. For the symbols used in this and the following chapters see legend of Fig. 48-1.

3. At the basolateral membrane Cl^- leaves the cell via an **electrogenic Cl^- channel,** while Na^+ is pumped out.
4. K^+, taken up by the Na^+-K^+-pump, leaves the cell through **K^+-channels** in the basolateral membrane, recycling K^+ to balance the Cl^- efflux (Fig. 48-3).
5. Some of the K^+ leaves the cell by luminal **H^+-K^+-exchange.** At low secretion rates the **H^+-K^+-exchanger** (antiport) in the luminal membrane transfers sufficient K^+ for the $[K^+]$ in the final saliva to exceed the concentration in plasma. The net result is K^+-secretion from blood to the duct lumen.

The final **salivary $[Na^+]$ and $[Cl^-]$ increase with increasing salivary secretion rate,** because the high flow provides less time for reabsorption in the duct system. **Bicarbonate** may be secreted even without Cl^--reabsorption. At low salivary secretion rates the final saliva becomes hypotonic down toward **half of the osmolarity of plasma.**

The aldosterone effects described above (increased Na^+ reabsorption and increased K^+ secretion) are similar to those in the distal, renal tubules and in the sweat glands.

Is dry mouth a disease?

Patients with a rare autoimmune disorder (**the Sjögren syndrome**) suffer from dry mouth (**xerostomia**), dry eyes (**xerophthalmia**) and rheumatoid arthritis.

In patients **lacking functional salivary glands** xerostomia, infections of the buccal mucosa, and dental caries are prevalent.

In most cases of xerostomia the condition is therapy resistant and unexplained. Some cases are caused by dehydration or by antidepressants.

Cystic fibrosis or mucoviscidosis also alters lung and salivary secretions – see Chapter 50.

Further reading

Petersen, O.H. (1986) Calcium-activated potassium channels and fluid secretion by exocrine glands. *J. Physiol. (London)* **448**: 1–51.

Schultz, S.G., J.D. Wood and B.B. Rauner (Eds) (1989) In: *Handbook of Physiology*, Sect. 6: *The Gastrointestinal System*. Am. Physiol. Soc., Bethesda, MD.

48. Multiple Choice Question

Each of the following five statements have True/False options:

A. Hot and acidic liquids are buffered by saliva in the mouth, and the salivary epidermal growth factor promotes the healing of wounds.
B. The parotid secretion is watery and serves to solubilize food, so it can be tasted.
C. Salivary buffers maintain the activity of amylase during the first period in the stomach.
D. Saliva has bacteriocide effects due to lysozymes.
E. AIDS is transferred via saliva.

Try to solve the problems before looking up the answers in Chapter 74.

Chapter 49. Gastric Secretion

Where is the gastric juice produced?

The stomach is divided into three main regions: the fundus, corpus and pyloric antrum. The gastric mucosa is highly invaginated and is mainly composed of gastric glands, with **mucous neck cells**, parietal cells in the fundus and corpus secreting HCl, and **peptic (chief) cells** secreting pepsinogen. The parietal cells also secrete the peptide **intrinsic factor**, which is necessary for absorption of vitamin B_{12} (Fig. 52-1). G cells in the mucosa produce the hormone **gastrin**. The gastric secretions include hydrochloric acid (HCl), pepsin and mucus. Mucus contains **mucin** (glycoproteins) and salts.

The gastric juice is hyperosmotic (325 mOsmol l^{-1}), contains 10 mM of K^+ and is low in Na^+ at moderate and high secretion rates; the $[H^+]$ is 170 mM and the $[Cl^-]$ is 180 mM. Gastric juice has an approximate pH of 1, forming a million-fold gradient of H^+ across the gastric mucosa to the blood. The HCl activates pepsinogen, maintains the optimal pH for pepsin activity and denaturates proteins and microbes.

How is pepsin secreted?

The **peptic cells**, located in the base of the gastric gland, produce **pepsinogen**. Pepsinogen is stored in granules of the peptic cell. Pepsinogen secretion is stimulated by cholinergic, muscarinic substances and by β-adrenergic agents, but peptic cells have no histamine receptors. Exocytosis releases pepsinogen into the gastric juice, where it is cleaved into **pepsin**, if HCl is present. Pepsin is the major hydrolytic enzyme in the stomach, but it is only active in the **acidic gastric juice**.

Describe the two component theory for the HCl secretion

Adult humans produce up to 2 l of gastric juice daily. The gastric juice is produced from two different sources: The **parietal cell juice** with 170 mM [HCl], 10 mM $[K^+]$, and a low $[Na^+]$.

A juice with an ionic composition similar to that of plasma is produced from other cells – the **non-parietal juice**. Each of the two secretion products have almost a constant composition.

Increased secretion of gastric juice means increased secretion of **parietal cell juice**. This explains why the [HCl] increases more and more, whereas $[Na^+]$ falls with increasing secretion rate.

How is the gastric secretion controlled?

The gastric secretion related to a meal occurs in three phases (cephalic, gastric and intestinal).

1. **The cephalic phase** is elicited even before food arrives to the stomach. The thought, smell, sight, or taste of food signals to the brain (i.e. the limbic system including the

hypothalamus) with an intensity dependent upon the appetite. Efferent signals pass from the dorsal motor nuclei of the vagi to the stomach. ACh is released from the short postganglionic vagal fibres and directly stimulates parietal cells to secrete HCl. Chief cells are stimulated to secrete pepsinogen and G cells to release gastrin. The cephalic phase accounts for less than 20% of the total meal related gastric secretion.

2. **The gastric phase** is brought about when food enters the stomach. Distension of the body and the antrum of the stomach stimulates stretch receptors. They provide afferent signals for both long, central vagovagal reflex loops as well as local, enteric reflexes. Signals in these fibres reach cholinergic, muscarinic receptors on the basolateral membrane of the parietal cells. The external vagal fibres work together with **intrinsic, peptidergic neurons** containing vasoactive intestinal peptide (VIP) and gastrin-releasing peptide (GRP). **Histamine** from the granules of mast cells, is a major, local, paracrine stimulator of HCl secretion. **Somatostatin** (GHIH, the multipotent inhibitor) is produced in cells close to the parietal cells and inhibits HCl secretion by paracrine action, and by reduced gastrin secretion. VIP controls the bloodflow of the gastric mucosa, and GRP releases **gastrin** from the G cells. Of the **hormonal regulators**, gastrin is the most important (Fig. 49-2).

   Vagal reflexes explains why distension of the body of the stomach can release gastrin from the antral mucosa. Most of the daily gastric secretion of 1.5 l is accounted for by the gastric phase.

   Acetylsalicylic acid and other non-steroid anti-inflammatory drugs (NSAIDs) deplete the gastric mucosa for prostaglandins which leads to mucosal damage. Strong alcoholic beverages also damage the gastric mucosal barrier and stimulate acid secretion. Caffein stimulates gastric acid secretion.

3. **The intestinal phase** is elicited by duodenal and jejunal mechanisms that both stimulate and inhibit gastric acid secretion. Acid chyme reaching the duodenum with peptides and amino acids, releases gastrin from duodenal G cells, which causes secretion of small amounts of gastric secretion including HCl. Normally, the inhibitory intestinal mechanisms dominate, because the pH of the chyme is low. Acid chyme in the duodenum (including the duodenal bulb) causes release of secretin (from S cells) and bulbogastrone. These hormones inhibit gastrin-stimulated acid secretion. Acid chyme containing long-chain fatty acids and peptides stimulates the duodenal mucosa to release two inhibitory hormones: cholecystokinin (CCK) and gastric inhibitory peptide (GIP). These hormones inhibit gastrin release from the G cells, and they inhibit acid secretion by the parietal cells. All these **entero-gastric inhibitory factors** are called **enterogastrones** (i.e. CCK, somatostatin, GIP, neurotensin, VIP and secretin).

Explain the mechanism of gastric HCl secretion

Stimulation of the parietal cells with ACh, histamine and gastrin has two consequences for their content of second messengers. The cellular [Ca^{2+}] and [cAMP] is elevated (see later in this chapter).

1. These second messengers activate **luminal Cl^-- and K^+-channels**. Cl^- and K^+ pass into the lumen, whereby their cellular concentrations decrease. The luminal [K^+] activates the **K^+-H^+-pump**. In addition, more pumps are inserted into the luminal membrane from cellular tubulo-vesicles.
2. The fall in cellular [Cl^-], and a rise – see below – in cellular [bicarbonate], stimulates the **basolateral Cl^--bicarbonate exchanger**, whereby the cellular [bicarbonate] is

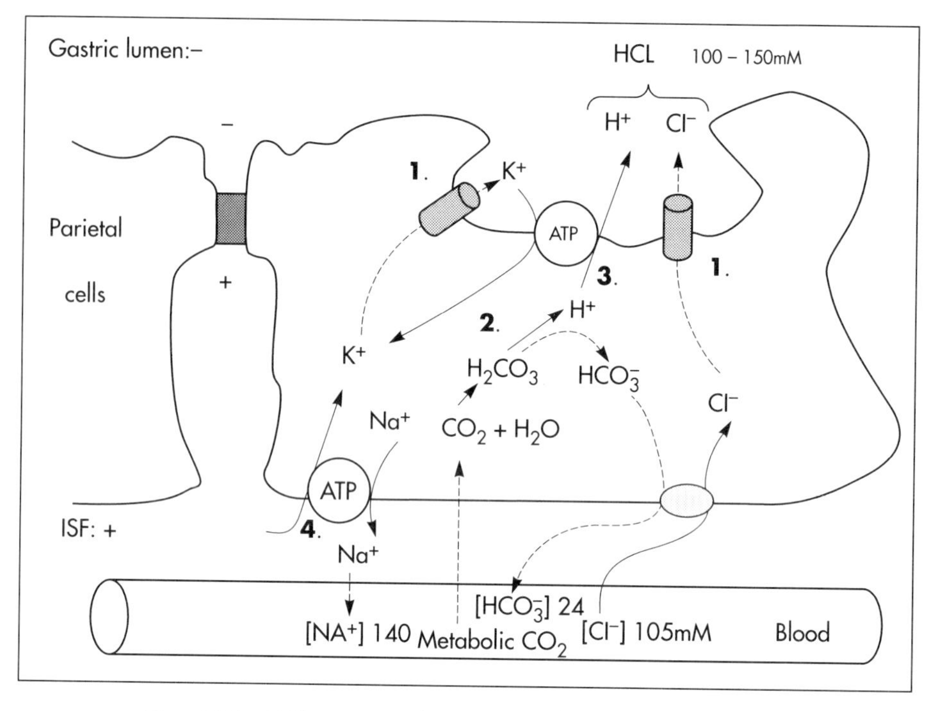

Fig. 49-1: HCl secretion from parietal cell in the stomach.

reduced. The fall in cellular [H^+] and [bicarbonate] stimulates formation of H^+ and bicarbonate, under the influence of **carbonic anhydrase**. The H^+ and bicarbonate are derived from metabolic carbon dioxide from the blood. Bicarbonate diffuses from the ISF into the blood. Every time the gastric juice receives one **H^+**, the blood will receive one **HCO_3^-**. This explains why the pH of the gastric venous blood increases after a meal – the **alkaline tide**.

3. Cellular [H^+] is a substrate for the luminal gastric proton pump (the **K^+-H^+-pump**), already activated by K^+. The net result is H^+-secretion to the lumen in a balanced relationship to Cl^--secretion (Fig. 49-1). The surface of the gastric mucosa is always **electrically negative** with respect to the serosa. H^+ moves against a large electrochemical gradient into the gastric lumen. The intracellular [H^+] of the parietal cells is 10^{-7} mol l^{-1}, so with a [H^+] of 10^{-1} mol l^{-1} in the gastric juice, a million-fold concentration gradient is present across the luminal membrane. Accordingly, energy is required for the transport of both ions. The HCl secretion requires ATP.
4. The cellular concentration of **cations** is maintained by the **basolateral Na^+-K^+-pump**.

The gastric proton pump has sequence homology with the Na^+-K^+-pump and other members of the ion-transporting ATPase family. The parietal cells contain **more mitochondrial mass per volume** unit than any other cells in the body, indicating a rich oxidative metabolism.

Describe the receptors of the parietal cell

Histamine, ACh and gastrin stimulate acid secretion. We have two types of histamine receptors in the human body: H_1-receptors (blocked by diphenhydramine) and H_2-receptors. Only H_2-receptors are located on the parietal cells.

1. The **H_2 receptors make histamine a potent stimulant of HCl secretion** (Fig. 49-2). When histamine is bound to the H_2-receptor it activates **adenylcyclase,** an enzyme generating cAMP from ATP. This increase in intracellular [cAMP] is specific for histamine (Fig. 49-2). The cAMP binds to and activates **cAMP-dependent protein kinase** (consisting of a regulatory and an active catalytic subunit). The cAMP binding releases the active catalytic subunit which phosphorylates a variety of target proteins.

   H_2-receptor antagonists (cimetidine and ranitidine) prevent histamine from binding to the H_2-receptors of the basolateral membrane of the parietal cells, which reduces acid secretion. Synthetic analogues of prostaglandin E can inhibit both the cAMP and the Ca^{2+} release mechanisms, thus promoting ulcer healing (see later).
2. ACh is released by vagal stimulation which leads to a stimulation of acid secretion (Fig. 49-2). This secretion is inhibited by atropine. Thus the parietal cells contain **muscarinic, cholinergic receptors (M_3).**

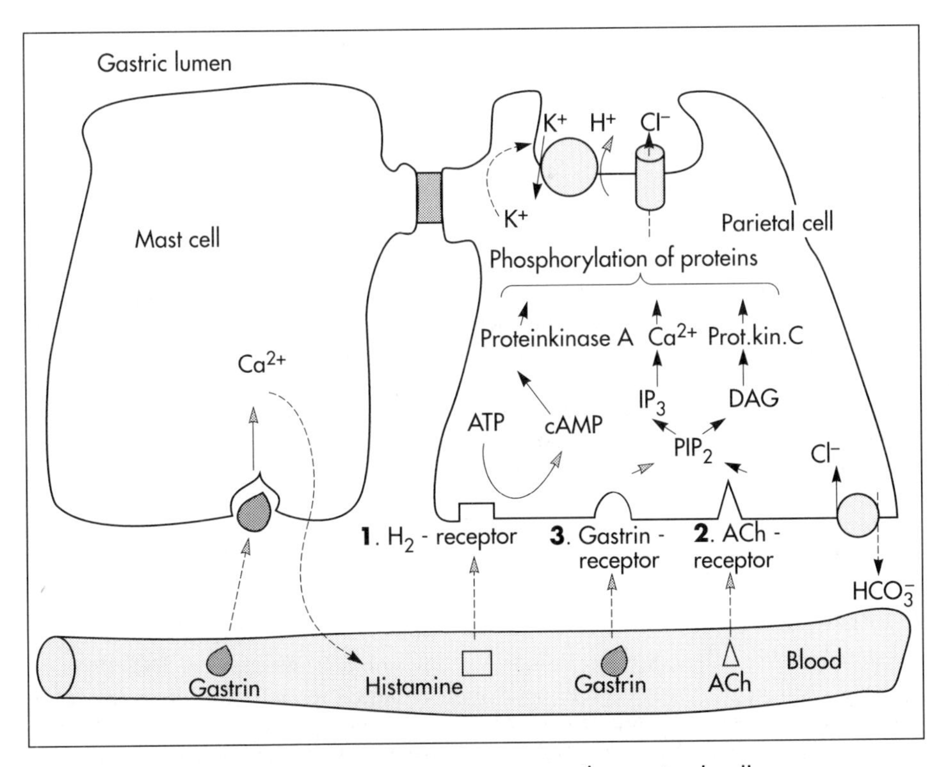

Fig. 49-2: Secretory receptors on the parietal cell.

3. Gastrin is the most potent stimulant of acid secretion in humans. Gastrin receptors were previously supposed not to be present on human parietal cells. Gastrin from G cells were thought to release **histamine** from mast cells in the gastric glands (Fig. 49-2). This is probably not the case. A direct gastrin effect on human gastrin receptors occurs, and an additional indirect effect via histamine increases the HCl secretion markedly (H_2 receptors in Fig. 49-2). However, the three-receptor hypothesis is still under debate.

   Gastrin and ACh release **inositol-(1,4,5)-trisphosphate** (IP_3), which is produced with diacylglycerol (DAG) by a **membrane phospholipase** (Fig. 49-2). The target system for IP_3 is a **Ca^{2+}-channel protein** located in the ER. Ca^{2+} is released from the reticulum, and Ca^{2+} also enters the cell through the basolateral membrane.

   Combined stimulation of all three receptors results in maximal gastric secretion (potentiation).

What are gastric and duodenal ulcers? Are they dangerous?

The normal stomach produces enough **mucus and alkaline juice** to protect the gastric and duodenal mucosa against HCl. The mucine molecules swell and form a **non-stirred layer** covering the mucosa. In the duodenum the pancreatic bicarbonate creates a pH of 7.5 at the luminal membrane of the mucosa.

Ulcers are caused by **excess acid secretion** caused by ***Helicobacter pylori*** infection. This infection also destroys the protective system. Hence the protective system of the mucosa can be regenerated by antibiotic treatment of the *Helicobacter pylori* infection.

Bleeding from ulcers can be **fatal**. Upper gastrointestinal tract bleeding implies a significant loss of blood into the lumen of the foregut. Such a bleeding is demonstrated by haematemesis and melaena. **Haematemesis** is defined as vomiting of whole blood or blood clots. **Melaena** is defined as passage of dark tarry stools (coal-black, shiny, sticky, and foul smelling).

How do gastric antisecretory agents work?

The H_2-receptor antagonists, cimetidine and ranitidine, prevent histamine from binding to the **H_2-receptors** on the basolateral membrane of the parietal cells. These drugs are widely used for reducing HCl secretion when treating duodenal ulcers.

The antisecretory agent, omeprazole, inhibits the **gastric proton pump** (Fig. 49-1). Omeprazole and similar antagonists to the gastric proton pump are especially effective in treatment of **persistent HCl-secretion caused by gastrin-secreting tumours of the pancreas (the Zollinger–Ellison syndrome)**. This is a special condition associated with duodenal ulcers, whereby there is excess HCl secretion due to overgrowth of parietal cells.

Prostaglandin E analogues (eg. nisoprostol) promote ulcer healing, due to their antisecretory and protective properties.

Further reading

Handbook of Physiology. Sect. 6. Vol. 3. Am. Physiol. Soc. Washington, DC, 1989.

Mezey, E. and M. Palkovits (1992) Localization of targets for anti-ulcer drugs in cells of the immune system. *Science* **258**: 1662–5.

Rabon, E.C. and M.A. Reuben (1990) The mechanism and structure of the gastric H^+-K^+-ATPase. *Ann. Rev. Physiol.* **52**: 321.

49. Multiple Choice Questions

Each of the following five statements have True/False options:

A. Gastric acid secretion is stimulated by ACh, gastrin and histamine.
B. H_2-blockers bind to histamine receptors at the basolateral membrane.
C. The parietal cells increase their O_2 consumption, acid secretion, intracellular [cAMP] and [Ca^{2+}], when stimulated by histamine.
D. The H^+-K^+-ATPase is responsible for gastric acid secretion.
E. Gastrin and ACh does not release IP_3.

49. Case History

In a healthy 22-year-old female at rest, the total diffusion pathway from the gastric lumen to her blood plasma, with a pH of 7.40, is considered to be one barrier. The intracellular [H^+] of the parietal cells is 10^{-7} mM, and the [H^+] is 10^{-1} mM in the gastric juice. The active transport of H^+ against a million-fold concentration gradient requires 85 kJ mol^{-1} of H^+. Glucose is assumed to deliver free energy to this active process with 15.5 kJ per gram of glucose oxidized (glucose MW=180). The enthalpy equivalent of oxygen is 20 kJ l^{-1} STPD of oxygen (on a mixed diet). During a histamine test the total mass of parietal cells secrete 35 mmol H^+ h^{-1}.

1. *Define the equilibrium potential for a given ion over a membrane (Chapter 2).*
2. *Calculate the equilibrium potential for H^+ over the total barrier.*
3. *Calculate the equilibrium potential for H^+ over the luminal membrane.*
4. *How much glucose must be oxidized in order to transport 1 mol of H^+ from the cytosol of the parietal cell into the gastric juice?*
5. *Calculate the oxygen uptake necessary to oxidize sufficient amounts of glucose for the transport of 35 mmol h^{-1}.*
6. *Assume a likely total oxygen uptake for the person, and compare this uptake to the result from 5.*

Try to solve the problems before looking up the answers in Chapter 74.

Chapter 50. Pancreatic Exocrine Secretion

Pan kreas is Greek and means **all meat**. This is the classical mixed gland with both endocrine and exocrine elements. The exocrine pancreas is an **abdominal salivary gland.** The endocrine pancreas is described in Chapter 72.

Secretions from the zymogen containing acinar cells collect in the acinar duct and travel through a network of **converging ducts** to the **main pancreatic duct**, which run into the common bile duct entering the duodenum at the duodenal papilla (Vateri), where the sphincter of Oddi is located.

The **exocrine glandular tissue** consists of **acinar cells** producing a primary secretion, with an ionic composition similar to that of plasma, and **duct cells** forming the secondary secretion by modification of the primary secretion.

Where is the primary pancreatic juice produced?

The organic components, secreted by acinar cells, are the major enzymes necessary for digestion of dietary nutrients. The acinar cells also secrete mucus and ions.

1. Upon stimulation of the duodenal mucosa with acid chyme containing peptides and long chain fatty acids, **CCK from I-cells** is released to the blood, whereby it can reach the pancreatic acinar cells. They carry specific receptors which bind the **gastrin-family** (gastrin and CCK competing for the same receptor) as well as the neurotransmitter, ACh.

   Receptor–ligand binding activates IP_3, thus elevating $[Ca^{2+}]$ in the cells (Fig. 50-1). Ca^{2+} triggers exocytosis of the enzymes from the zymogen granules, utilizing either a **Ca^{2+}-calmodulin complex** or a **Ca^{2+}-phosphatidyl serine-dependent protein-kinase C**. The secretin family (includes VIP) potentiates the action of CCK.
2. The rise in $[Ca^{2+}]$ opens a **luminal Cl^-- and a basolateral K^+-channel** (Fig. 50-1), whereby these ions are leaving the cell in a balanced relationship and produce a negative electric field in the acinar lumen. A small amount of bicarbonate also leaks through the anion channel.
3. The fall in intracellular $[Cl^-]$ and $[K^+]$ activates a **basolateral Na^+-K^+-2 Cl^--co-exchanger** through which NaCl enters the cell from the ISF (Fig. 50-1).
4. The negative electric field in the acinar lumen provides a force that drives a passive Na^+- and water transport into the acinar lumen through leaky tight junctions as an isotonic solution.
5. The secretory energy is from the **basolateral Na^+-K^+-pump**, which maintains the intracellular ion composition.

What happens in the pancreatic ducts?

As the primary juice leaves the acini and proceeds down the pancreatic ducts, it is supplied isotonically with **water and electrolytes (mainly bicarbonate salts)** from the duct cells (Fig. 50-2).

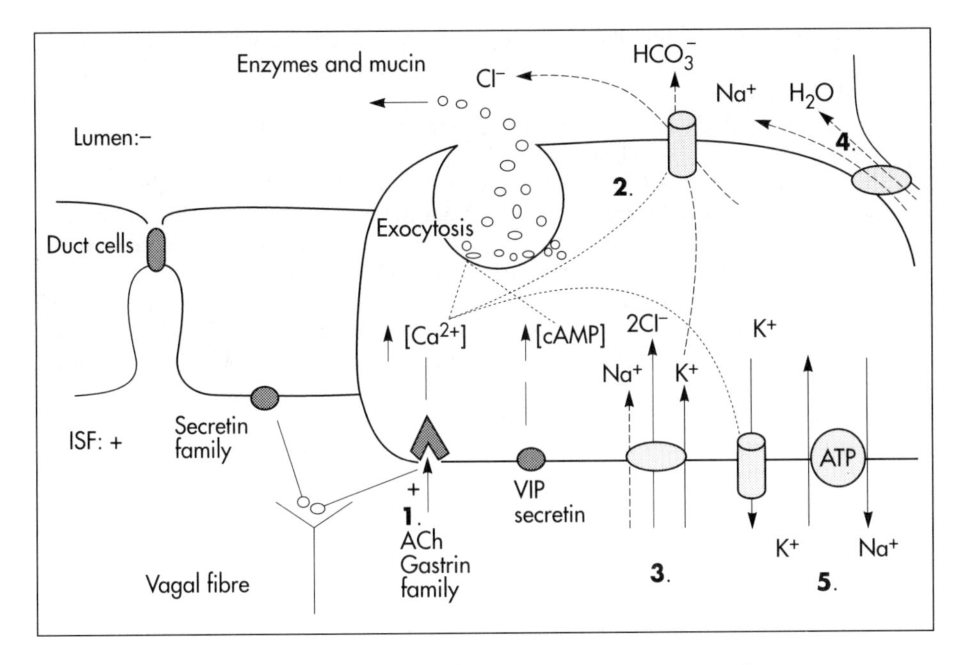

Fig. 50-1: Secretion from pancreatic acinar cells.

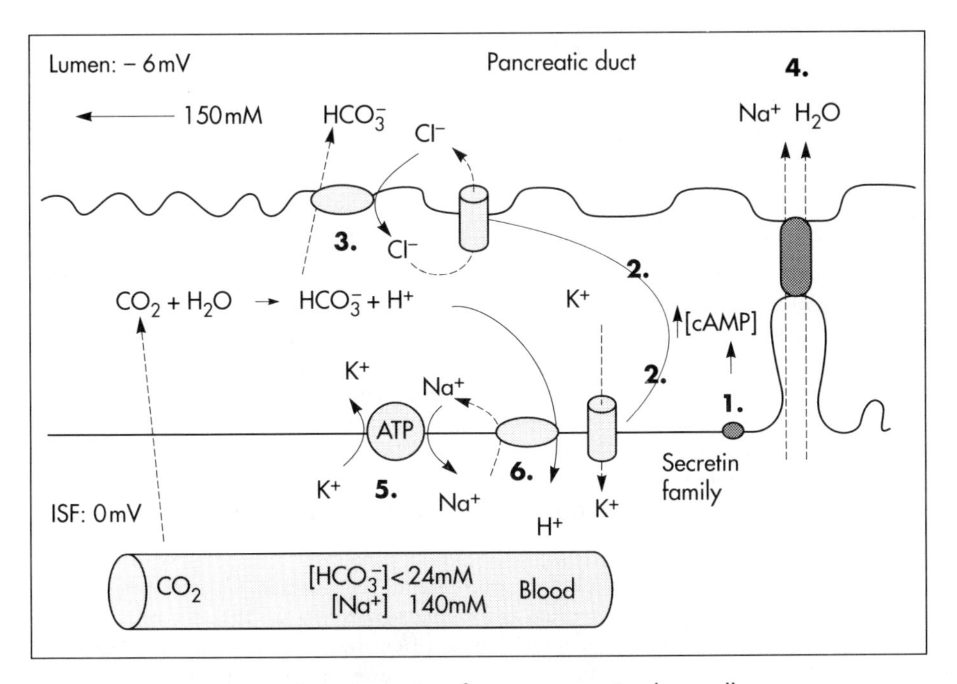

Fig. 50-2: Secretion from pancreatic duct cells.

1. Upon stimulation of the duodenal mucosa with acid chyme, **secretin** is secreted to the blood and transported to the pancreatic duct cells. This induces an important rise in cellular [cAMP].
2. The rise in [cAMP] activates **luminal Cl^-- and basolateral K^+-channels**, so these ions leave the cell in a balanced relationship.
3. This triggers a **luminal Cl^-/HCO_3^--exchanger** through which the cell eliminate the bicarbonate produced by carbon dioxide from the blood. Cellular carboanhydrase is essential for the bicarbonate production. A certain luminal [Cl^-] is necessary for recycling (Fig. 50-2). The net result is a secretion of bicarbonate.
4. This net secretion of bicarbonate induces a lumen negative transepithelial potential difference (−6 mV), which constitutes the driving force for the paracellular transport of Na^+ and K^+. The net secretion of salt drags water transepithelially in isosmotic proportion.
5. The fall in cellular pH upon secretion of bicarbonate activates a basolateral **Na^+/H^+-exchanger**, whereby the cells eliminate H^+ to the blood.
6. The secretory energy is from a **basolateral Na^+-K^+-pump**. The whole system is analogous to the formation of saliva.

What is the composition of pancreatic juice?

The pancreas (weight 100 g) of adult humans is capable of elaborating approximately 1.5 l of pancreatic juice per day, and its pH increases with increasing secretion rate. The maximal secretion rate is 1 ml g^{-1} of tissue each hour (i.e. 60 times less than that of the salivary glands).

The juice is a **clear fluid, isosmolar with plasma**. The basic reaction is due to bicarbonate, and the [bicarbonate] can approach the [H^+] in gastric juice (150 mM).

With increasing secretion rate, the [bicarbonate] in the final pancreatic juice increases at the expense of [Cl^-], whereas the [Na^+] and [K^+] remain relatively constant (Fig. 50-3). Pancreatic juice (pH 8) thus buffers the **extremely acid gastric juice** and protects the duodenal mucosa against **erosion**. Buffering of gastric juice also optimizes the activity of pancreatic digestive enzymes in the duodenum.

How is the exocrine pancreatic secretion controlled after a meal?

The pancreatic secretion is regulated by two intestinal hormone families: The **secretin (secretin and VIP)** and the **gastrin family (gastrin and CCK)**, as well as by the autonomic nervous system.

Signals in **cholinergic, vagal fibres** stimulate both pancreatic secretions via Ach-receptors (Fig. 50-1), whereas noradrenergic, sympathetic stimuli inhibit secretion via α-receptors. The secretion is also stimulated by signals in **peptidergic nerve fibres**. The free radical gas **nitric oxide** (NO) stimulates the exocrine pancreatic secretion, and simultaneously inhibits the non-adrenergic, non-cholinergic intestinal activity.

Related to the meal there are **three phases** of pancreatic secretion (cephalic, gastric and intestinal).

1. **The cephalic phase** is elicited before food reaches the stomach. Olfactory signals (via the limbic system), as well as visual and tactile signals (via the thalamic relay station) are processed in the brain, and vagal signals reach the antral mucosa. Here gastrin is released from G cells. Gastrin induces the secretion of a low volume of pancreatic juice with a high enzyme content.

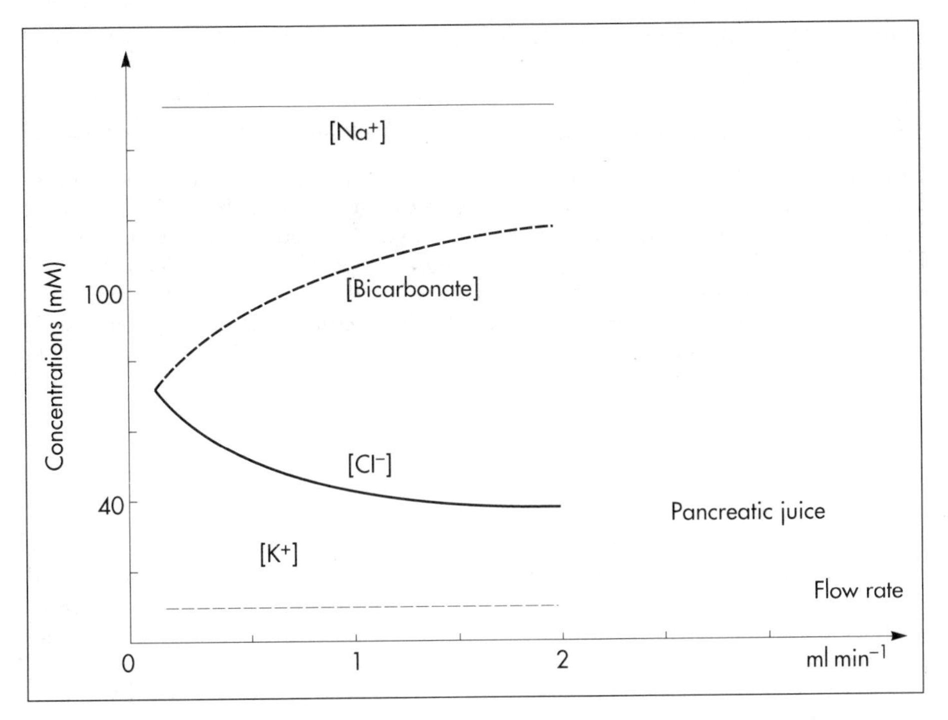

Fig. 50-3. Concentrations of ions (mmol l^{-1} or mM) in pancreatic juice as functions of the secretory flow rate.

2. **The gastric phase** is elicited by the presence of food in the stomach. Gastric distension and peptides reaching the antral mucosa trigger the release of more gastrin from the G cells. Hereby, the secretion of a small volume of pancreatic juice rich in enzymes, is continued.
3. **The intestinal phase** is elicited by duodenal and jejunal mechanisms. When chyme enters the duodenum both secretin and CCK are released for different reasons. **Secretin** is secreted by **S cells in the mucosa** of the upper small intestine, when acid chyme (pH below 4.5) arrives to the first part of the duodenum. This is an appropriate arrangement, because **secretin** stimulates both the secretion of **bicarbonate and water** by pancreatic duct cells (Fig. 50-2), and of **bicarbonate-rich bile** by small biliary ductules. Secretin inhibits gastric secretion. Secretin inhibits both the gastrin release by the antral G cells, and the gastrin effect on the parietal cells.

CCK from the **duodenal I cells** stimulates gallbladder contraction as its name implies, and stimulates pancreatic acinar secretion of an enzyme rich fluid (Fig. 50-1). The most important stimulus for CCK liberation is when an acid chyme with amino acids, peptides and long chain fatty acids reach the duodenal mucosa. This is essential, because CCK contracts the gallbladder and stimulates the pancreatic secretion of an enzyme rich juice. Bile is ejected into the duodenum, where fat is emulgated to ease absorption. CCK also acts as an **enterogastrone** – an intestinal hormone that inhibits gastric activity and emptying. This leaves more time for the bile to emulgate fat and for the digestible enzymes to work.

Describe a pancreatic enzyme which is secreted in an already active form

Pancreatic α-amylase does not pose any danger to pancreatic tissue. Pancreatic α-amylase – like the salivary α-amylase – cleaves the large dietary carbohydrate molecules at the internal 1,4-glycosidic bonds, but cannot hydrolyse terminal 1,4- bonds or 1,6-bonds. The end-products are oligo- and disaccharides like maltose (two glucose), maltriose (three glucose) and branched oligosaccharides known as α-limit dextrins. Other enzymes such as maltase and lactase, secreted by the intestinal mucosa, digest these end-products into **monosaccharides** (glucose, fructose and galactose). The carbohydrate absorption is shown in Fig. 53-1.

Describe the effects of pancreatic proteases

The protein digestion is continued in the duodenum and jejunum, where the protein breakdown products are attacked by the **proteolytic enzymes of the pancreas** (trypsin, trypsinogen, chymo-trypsinogen, pro-carboxy-peptidase, and pro-elastase). The pancreatic proteases are secreted as inactive proenzymes, and they are crucially important. The proenzymes are normally not activated before they arrive in the intestinal lumen. Trypsin catalyses its own activation (**autocatalysis**), and also activates chymotrypsinogen and the pro-carboxypeptidases in the **trypsin cascade**.

Duodenal enterokinase cleaves trypsinogen to trypsin, and hereby activates the **trypsin cascade**. When the chyme is pushed into the duodenum, the pancreatic juice neutralizes the chyme and the pepsin activity is stopped. The peptides and amino acids are absorbed as shown in Fig. 53-2.

Normally the trypsin cascade is inhibited by **trypsin inhibitor** from the pancreas, but cases of acute pancreatitis cannot inhibit the trypsin cascade, so **autodigestion** occurs.

How are lipids attacked?

Enzymes for the breakdown of fats are pancreatic lipase, phospholipase A, and lecithinase (see Chapter 53).

Pancreatic lipase and co-lipase cleave triglycerides into free glycerol and fatty acids or to monoglycerides (MG) and fatty acids (Fig. 53-3). Free glycerol is readily absorbed. The lipolytic activity requires the emulsifying action of bile salts in order to solubilize triglycerides in water. Once liberated fatty acids and monoglycerides participate into **bile salt micelle formation**. Micelles are passed through the unstirred water layer of the intestinal lumen to reach the absorbing mucosa (Fig. 53-3).

Is it possible to survive without a pancreas?

Pancreatic failure is caused by **pancreatitis** (inflammation not seriously affecting the islets), **blockage of the pancreatic duct, pancreatic carcinomas** and **surgical removal** of the pancreatic head. Loss of pancreatic juice means lack of pancreatic lipase, pancreatic amylase, trypsin, chymotrypsin, carboxypolypeptidase and elastase. Lack of these enzymes means that half of the fat entering the small intestine passes unabsorbed to the faeces, and one-third of the starches and proteins. A copious, fatty faeces is found (called **steatorrhoea**).

The major metabolic disorder caused by loss of pancreatic endocrine secretion is diabetes mellitus.

Thus the answer is yes; removal of pancreas is compatible with **survival as long as both the exocrine and endocrine vital substances are supplied artificially**. However, the five-year survival rate is miserably low.

What is pancreatitis?

Pancreatitis is an inflammatory disease with interstitial oedema and acute necrosis of the pancreas. Infection does not seem to account for many cases, but the incidence of pancreatitis in alcoholics is high. The normal balance between trypsin and **trypsin inhibitor** is destroyed. In both acute and chronic pancreatitis, premature activation of the pancreatic enzymes trypsin, phospholipase A and elastase, causes the **pancreas to digest itself.**

What is pancreatic cystic fibrosis?

This is a recessive genetic defect caused by dysfunction of exocrine glands. Fully manifested cases suffer from defect mucous secretion (**mucoviscidosis**) with chronic respiratory disease, cystic pancreatic fibrosis with pancreatic insufficiency, and abnormally high [NaCl] in sweat. An important contributor to this life-threatening condition is a genetic defect in the β-adrenergic-gated **Cl^--channels** of the glands in the airways, the pancreas, and in the sweat glands, thus the glands produce a viscoid secretion that causes cystic dilatation of the duct systems.

Further reading

Grotmol, T. *et al.* (1990) Secretin dependent HCO_3^- from pancreas and liver. *J. Intern. Med.* **27–8** (suppl. 1): 47.

Hug, M., C. Pahl and I. Novak (1994) Effect of ATP, carbachol and other agonists on intracellular calcium activity and membrane voltage of pancreatic ducts. *Pflügers Arch.* **426**: 412–18.

Johnson, L.R. (Ed.) (1991) *Gastrointestinal Physiology*, 4th edn. Mosby-Year Book, St Louis.

50. Multiple Choice Questions

Each of the following five statements have True/False options:

A. With increasing rate of pancreatic secretion its $[Cl^-]$ will increase.

B. Carbonic anhydrase is an important enzyme for the secretion of pancreatic bicarbonate.

C. Duodenal acidification stimulates the pancreatic secretion.

D. Duodenal chyme with a pH of 7 inhibits pancreatic bicarbonate secretion.

E. Duodenal enterokinase cleaves trypsinogen to trypsin.

Try to solve the problems before looking up the answers in Chapter 74.

Chapter 51. Fluid and Electrolyte Absorption

What is the net daily fluid intake?

A healthy person on a mixed diet (12 600 kJ day^{-1}) receives about 2 l of fluid per day (350 ml metabolic water, drinks 1 l, and 750 ml with the food). Balance is maintained as long as the **water loss is the same** (urine 1200 ml, faeces 100 ml and 800 ml through the skin and lungs).

What is the total daily fluid volume entering the small intestine through its walls and the pylorus?

The **ingested and metabolic water volume** is approximately 2 l. The **digestive secretions are 7–8 l daily** (1 l of saliva, 2 l of gastric juice, 3 l of pancreatic and intestinal juices, and 1 l of bile). This yields a **total fluid volume of 9–10 l entering the small intestine daily**. Normally, the small intestine absorbs 8 l, and 1–1.5 l passes the ileocaecal valve. Most of this water is absorbed in the colon of healthy people, and only 100 ml is found in the daily faeces. The colonic absorptive capacity (i.e. the **colonic salvage**) is much higher – around 4500 ml of fluid.

How are small ions and water absorbed?

The intestinal content is isosmolar with plasma, and the water is absorbed from the lumen to the blood by **passive osmosis**. The membranes of the intestinal mucosal cells and even the **tight junctions** are highly permeable to water. Hereby, **active transport** of Na^+ and Cl^- from the lumen to the small interstitial space builds up a forceful osmotic gradient, drawing water the same way by a passive process. In the small interstitial space water creates a hydrostatic overpressure. Since the capillary and lymph endothelial membranes are no barrier for Na^+, Cl^- and water, a **bulk flow** of fluid from the interstitial space passes into the blood- and lymph vessels. The intestinal mucosa possesses elevations called **villi**, and pitted areas called **crypts**. The villous cells have a typical brush border responsible for **net absorption of ions and water**, whereas the crypt cells contain secretory mechanisms causing **net secretion**.

The villous cells absorb Na^+ through the luminal brush border membrane by three mechanisms: (1) an inward **diffusion gradient through a Na^+-channel**; (2) a **Na^+-H^+-exchange**; and (3) a **Na^+-solute coupled co-transport** (the solute being glucose, galactose, bile salts, water soluble vitamins and amino acids).

1. The $[Na^+]$ is kept low (14 mM) in the cell, whereas $[Na^+]$ is 140 mM in the intestinal lumen. This **concentration gradient** works together with an **electrical gradient,** since the cytosol of the cell is −40 mV with the intestinal content as a reference (Fig. 51-1). Thus Na^+ can easily pass the luminal brush border membrane passively. The intestinal mucosa has ion permeable **tight junctions** – it is **leaky**. This paracellular transport is so great that the net absorption of Na^+ and Cl^- through the cells only amounts to 10% of the total transport through the mucosa.

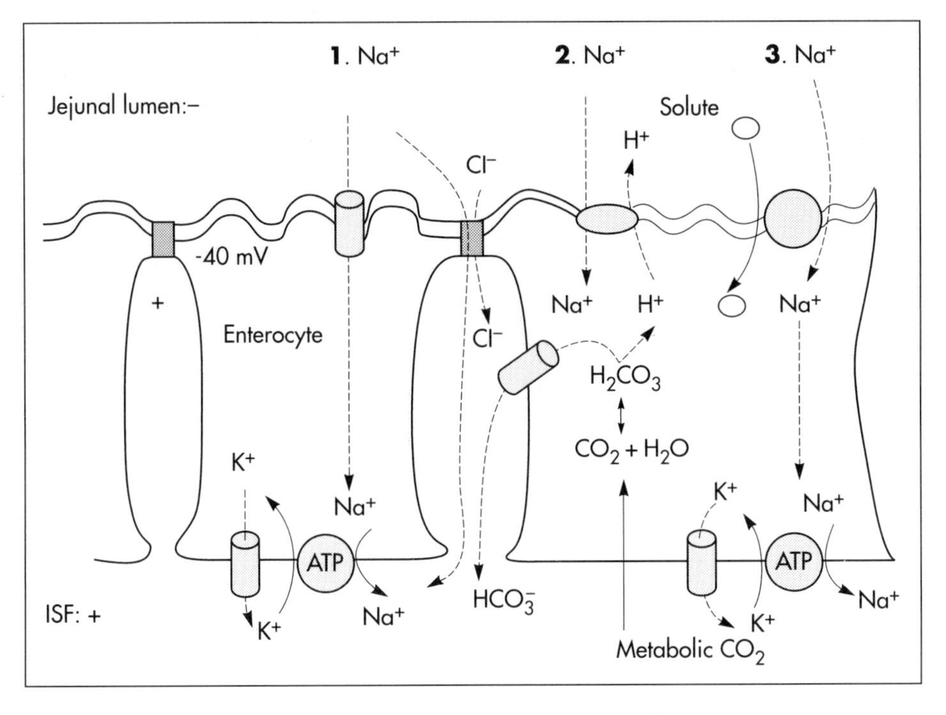

Fig. 51-1: Ion transport processes in jejunal enterocyte.

2. The transport of Na^+ into the enterocyte (Fig. 51-1) is through a **co-exchange protein (Na^+/H^+)**. Part of the energy released by Na^+ moving down its gradient is used to extrude H^+ into the intestinal lumen (Fig. 51-1). Here H^+ reacts with bicarbonate from bile and pancreatic juice to produce CO_2 and water, thus reducing the pH of the intestinal fluid.
3. **Na^+-solute coupled co-transport** (see Fig. 53-1). The basolateral membrane of the enterocyte contains a **Na^+-K^+-pump**, which maintains the inward directed Na^+-gradient (Fig. 51-1). The pump is energized by the hydrolysis of ATP, which provides the driving force for Na^+ entry. Thus by an active process Na^+ is pumped out in the small interstitial space and K^+ is pumped into the cell. The basolateral membrane also contains **many K^+-channel proteins**, so K^+ will leak back to the interstitial space almost as soon as it has entered the cell. The K^+ is absorbed by diffusion – a daily net total of 80 mmol.

How do crypt cells become net Cl^- secretors?

The Cl^- gradient, with an elevated intracellular $[Cl^-]$, is maintained by a **Na^+-K^+-2 Cl^- co-transporter** located on the basolateral membrane (Fig. 51-2). This transporter drags Cl^- from the ISF.

The transporter system uses the **electrochemical Na^+ gradient** to transport K^+ and Cl^- into the cell (Fig. 51-2). The crypt cells hereby can secrete Cl^- through the luminal membrane via an electrogenic channel. The Cl^- secretion produces a net luminal

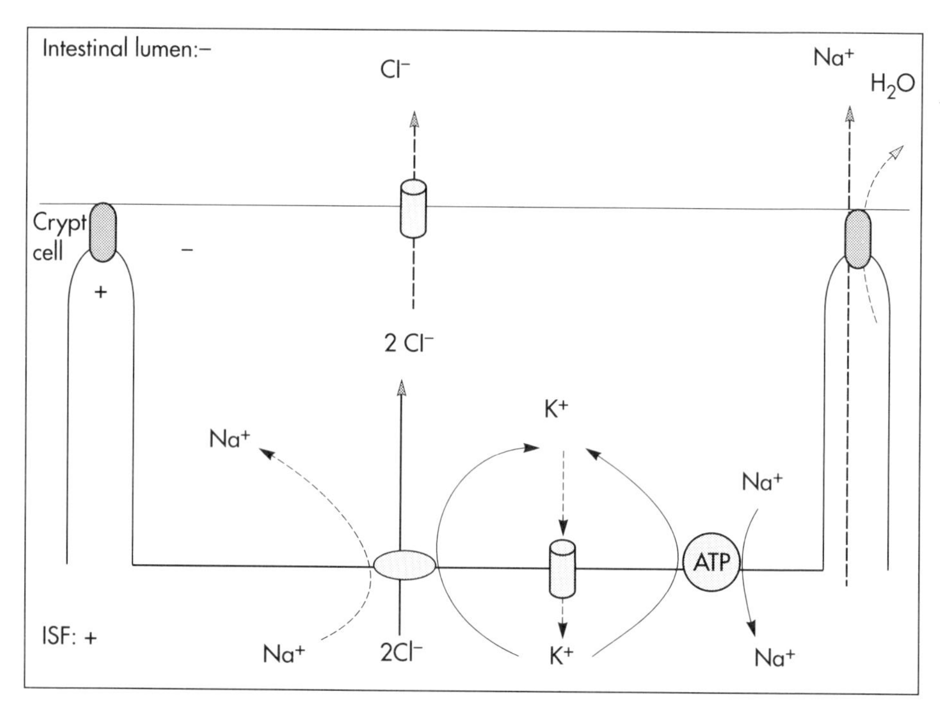

Fig. 51-2: Net Cl^--secretion by crypt cells of the small intestine.

electronegativity, which drags Na^+ across the tight junctions resulting in net secretion (Fig. 51-2). Water (about 2 l daily) is secreted by **passive osmosis** (Fig. 51-2).

A dramatic rise in Cl^- and water secretion – caused by gut inflammation with cholera – can lead to **secretory diarrhoea**.

How are electrolytes and fluid absorbed in the colon?

Fluid absorption in the colon is determined by the absorption of NaCl. The Na^+ transport involves (1) **electrogenic Na^+ transfer via Na^+ channels**; and (2) **Na^+-co-exchange** as in the small intestine (Fig. 51-1). Both transport processes are driven by the Na^+ gradient maintained by the basolateral **Na^+-K^+-pump**. (The Na^+-solute coupled co-transporter is not present in the human colon). The **colonic Na^+-K^+-pump** is more sensitive to **aldosterone** than that in the small intestine. Aldosterone is a steroid hormone. Steroids bind directly to cytosolic receptors and do not need second messengers. The colonic Na^+-K^+-pump activity accumulates K^+ in the enterocyte, and this gradient drives the **K^+ secretion** across the luminal K^+ channel. The Cl^- absorption is accomplished by diffusion along a Cl^--gradient, and by a **luminal Cl^--bicarbonate exchanger** producing **bicarbonate secretion**. We have a **bicarbonate–chloride shift** just as in the red cells. Since electrolyte absorption exceeds secretion, there is a net water absorption in the healthy colon (1–1.5 l daily and with a colonic salvage capacity of 4500 ml).

What are the mechanisms of diarrhoea?

Nutrient malabsorption of the small intestine increases the fluid volume delivered to the colon and can provide an osmotic effect in the colon with diarrhoea. Up to 4600 ml of fluid normally passes the ileocaecal valve without causing diarrhoea. In conditions such as **cholera,** the excess fluid from the ileum exceeds the **colonic salvage,** leading to life-threatening diarrhoea. The cholera toxin can enhance the **Cl^--secretion** drastically and cause **secretory diarrhoea** with large quantities of Cl^- and water.

In **inflammatory diseases** of the colon, the colonic salvage capacity is markedly reduced, resulting in **colonic diarrhoea.**

Describe the absorption of Ca^{2+}

The dietary content of Ca^{2+} is 1000 mg per day of which only 400 mg are absorbed in the intestine by an active, saturable process in the brush borders and transported across the cell by a cytosolic **Ca^{2+}-binding protein** for delivery to the blood. This transport protein is made in the mucosal cells under stimulation by **activated vitamin D** (1,25-dihydroxy-chole-calciferol), which binds to **specific nuclear receptors.**

PTH (parathyroid hormone) enhances the renal production of **activated vitamin D,** and thus stimulates intestinal absorption of Ca^{2+} indirectly.

The daily secretion of Ca^{2+} is 150 mg into digestive fluids, so the **net uptake is 250 mg of Ca^{2+} per day** in persons with a normal transport protein capacity. The **daily renal excretion is 200 mg and 50 mg are excreted via the skin.** PTH not only stimulates intestinal absorption of Ca^{2+} via vitamin D, it also stimulates renal reabsorption and bone resorption. The main effect is to **increase plasma [Ca^{2+}].** Vitamin D stimulates intestinal absorption of Ca^{2+} and bone resorption. Lack of vitamin D leads to insufficient bone formation, because the osteoid matrix does not calcify. This disease is called **rickets (latin: rachitis) in children and osteomalacia in adults.**

Describe the iron absorption

1. Ascorbate in the food reduces Fe^{3+} to Fe^{2+}, and forms a soluble complex with iron, thereby effectively promoting the iron absorption. We normally ingest about 20 mg iron daily, and less than 1 mg is absorbed in healthy adults, because iron forms insoluble salts and complexes in the gastrointestinal secretions.
2. **Iron** is transported from the lumen of the upper jejunum, across the mucosa, and into the plasma by an iron-binding protein called **gut transferrin.**
3. Receptor proteins in the brush border membrane bind the **transferrin–iron complex,** and the complex is taken up into the cell by **receptor-mediated endocytosis** (Fig. 51-3).
4. When blood containing products are ingested, proteolytic enzymes release the haeme groups from the haemoglobin in the intestinal lumen. Haeme is absorbed by facilitated transport (Fig. 51-3). Approximately 20% of the haeme iron ingested is absorbed.
5. When intracellular iron is available in excess, it is bound to **apoferritin,** an ubiquituos iron-binding protein, and stored within the mucosal cells as **ferritin** (Fig. 51-3). The synthesis of apoferritin is stimulated by iron. This translational mechanism protects against excessive absorption.
6. Iron exists in one of two states in the cytosol: **the ferrous state** (Fe^{2+}) or the ferric

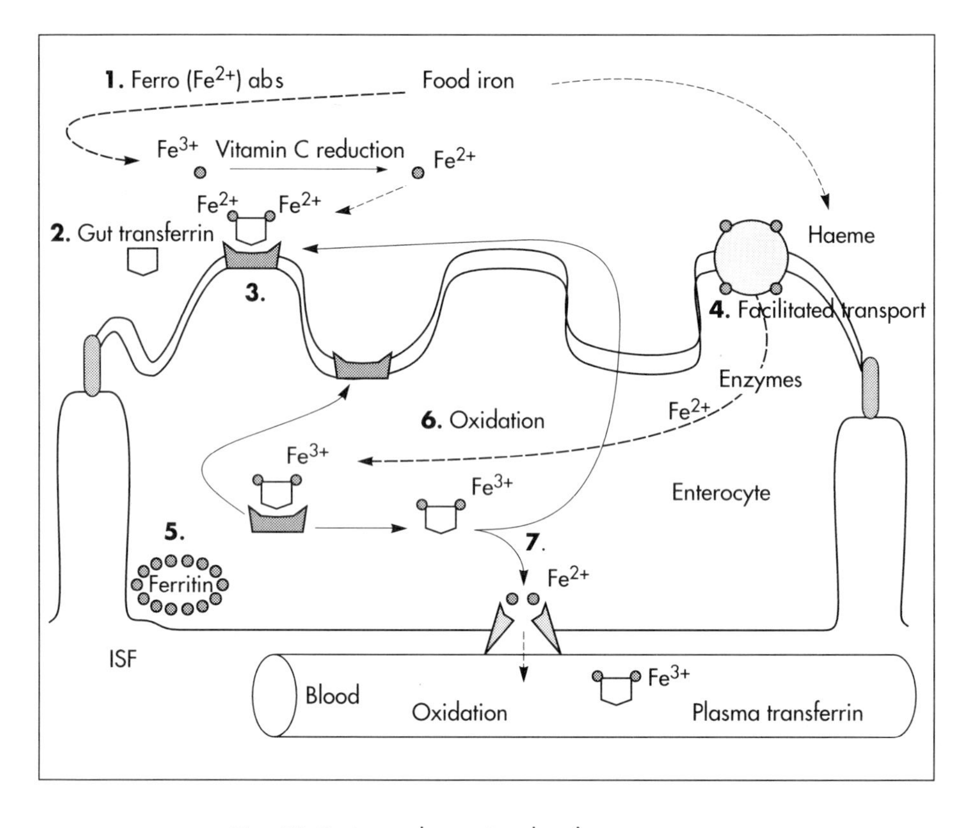

Fig. 51-3: Iron absorption by the enterocyte.

state (Fe^{3+}). The Fe^{2+} ions, after absorption into the mucosal cell, are oxidized to Fe^{3+} (Fig. 51-3).

7. At the basolateral membrane the Fe^{3+} are reduced to Fe^{2+} and pass from the interstitial space to the blood. Here Fe^{2+} are again oxidized to Fe^{3+} and bind to **plasma transferrin** (Fig. 51-3). Normally, **serum-iron is 12–36 μmol l^{-1}**, which is about one-third of the plasma **total iron-binding capacity** in adults. This means that one-third of the **circulating plasma transferrin** is saturated with iron.

Is iron of biological importance?

Iron is an important part of the haeme group. The **ability to transport O_2** depends on the presence of haeme within the haemoglobin molecule. Haeme gives the red cell its characteristic red colour. Only **haemoglobin with iron in the ferrous state binds O_2,** whereas the dark red methaemoglobin, with the iron in ferric state, cannot bind O_2. Iron plays an important role in the **cytochrome systems of all cells**, and in the myoglobin of muscle cells. Ferritin is further saturated with iron to form **haemosiderin** in the liver and elsewhere, when abnormal amounts are ingested over months.

Further reading

Field, M. and R.A. Frizzel (Eds) (1991) *Handbook of Physiology* Sect. 6, Vol. 4, Am. Physiol. Soc., Bethesda, MD.

Johnson, L.R. (Ed.) (1991) *Gastrointestinal Physiology*, 4th edn. Mosby-Year Book, St Louis.

Sullivan, S.K. and M. Field (1991) Ion transport across mammalian small intestine. In: Field, M. & Frizzell, R.A. (Eds) *Handbook of Physiology*, Sect. 6, Vol. 4, Am. Physiol. Soc., Bethesda, MD.

51. Multiple Choice Questions

Each of the following five statements have True/False options:

A. An elevated concentration of cAMP in the intestinal mucosal cells inhibits Na^+ absorption.
B. The intestinal Na^+ absorption is parallel to the Cl^- absorption.
C. Increased intracellular $[Ca^{2+}]$ increases intestinal Na^+ absorption.
D. The basolateral Na^+-K^+-pump (ATPase) maintains an essential electrochemical gradient with a high intracellular $[Na^+]$.
E. The intestinal Na^+ absorption is secondary to the water transport across the mucosal cells.

51. Case History

A male patient with caecal cancer still maintains his colon function. Approximately 1.5 l of intestinal fluid passes the ileocaecal valve in 24 h, and only 150 ml is found in the daily faeces. The intestinal fluid has a $[Na^+]$ and $[K^+]$ of 120 and 4 mM, respectively. The 75% water in the faecal volume of 150 ml contains 20 mM Na^+ and 5 mM K^+.

1. *Calculate the water absorption in the colon and rectum.*
2. *Calculate the net Na^+-and K^+-absorption in the colon.*
3. *Calculate the loss of Na^+ and K^+ with faeces.*
4. *Removal of the caecum and the ascending colon with the tumour necessitates an ileostomy. Calculate the loss of Na^+ and K^+ with an unchanged fluid flux through the terminal ileum.*
5. *Are dietary measures important for the ileostomy patient?*

Try to solve the problems before looking up the answers in Chapter 74.

Chapter 52.
The Liver and Bile in Digestion

What is the importance of the liver?

1. ***Metabolism* in hepatocytes.** The liver is responsible for the **key elements of intermediary metabolism**, regulating the metabolism of carbohydrates, lipids, and proteins essential for life: carbohydrate to fat, glycogenesis (formation of glycogen from glucose), ketogenesis (beta-oxidation), urea-genesis (from ammonia and carbon dioxide), and alcohol to acetate. Hepatocytes synthesize cholesterol, which is essential for cortical and sex hormones. Hepatocytes synthesize very low density lipoproteins (VLDL) which are a major source of lipids for use by the cells of the body. Essential proteins in the blood are synthesized in the hepatocytes: major plasma proteins (albumin, globulin, fibrinogen), other coagulation factors, angio-tensinogen, trypsin-inhibitor, etc.
2. ***Glucose exchanger* for the hypothalamic glucostat.** The carbohydrate store liberates glucose from hepatocytes to blood during prolonged fasts, and the liver is the main organ in control of blood [glucose]. Hepatic glucose production by **glycogenolysis (glycogen to glucose)** and **gluconeogenesis** is life saving, when blood [glucose] is low.
3. ***Secretion*.** Secretory function from hepatocytes to bile capillaries. The liver produces **bile**, which facilitates fat digestion and absorption, as well as cholesterol excretion.
4. ***Excretion*.** The liver is an important **excretory organ**, because the hepatocytes excrete bile pigments, and deactivate hormones, toxins and drugs by hydroxylation, proteolysis, and hydrogenation. Many drugs are coupled to glucuronic acid, sulphate, acetate or glycine. The Kupffer macrophages in the liver sinusoids eliminate microbes.
5. ***Storage*.** The liver is an important store of **vitamins (A, B_{12}, D and K)**, of **iron** (ferritin), and of **coagulation factors**. The store of vitamin B_{12} is normally sufficient for several years (see case history 52).
6. ***Lipid transfer*.** The **liver lymph** covers 50% of the total lymph produced, although the normal liver is only 1.5 kg of the total body weight. Chylomicrons packed with lipids are transferred to the systemic circulation via the liver lymph.
7. ***Kinetics of alcohol oxidation*.** The primary oxidation of alcohol takes place in the liver. This is why the elimination rate for alcohol is independent of its concentration above the threshold (Chapter 42).
8. ***Hepatic blood store*.** Normally, the pressure in the hepatic veins is zero. A small rise in central venous pressure results in **hepatic congestion**, and even a pressure rise of 5 mmHg (0.7 kPa) produces **hepatic stasis** with **ascites**. Up to 400 ml of blood can be stored in the liver. Hepatic stasis and other types of liver insufficiency lead to jaundice (icterus) and fatty stools (steatorrhoea).

How does the liver produce bile?

The **basic hepatic unit** is the **liver lobule**. The liver cells (hepatocytes) are arranged into walls of cells, which are separated by highly porous capillaries called **venous sinusoids**.

The portal vein brings blood from the intestine, and the hepatic artery brings arterialized blood from the heart. The **venous sinusoids** are lined with fenestrated endothelial cells and with specialized, reticuloendothelial Kupffer cells. The large O_2 and nutrient demand is extracted from this pool of mixed blood by the hepatocytes; then the blood drains into the **central veins** and leaves the liver through the **hepatic vein**. The hepatocytes form bile and secrete it into **bile canaliculi**, which converge to form a **ductal system**, where bile flows in the opposite direction of the blood. Hereby, cleared blood passes 'new' bile. The many bile ducts converge to form the **hepatic duct**. Secretin stimulates the secretion of bicarbonate from the bile ductal system.

What is bile?

Bile has a golden colour, a pH of 8, and it is nearly isotonic with blood plasma. Salts (NaCl and bicarbonate) exist in the bile at concentrations similar to those of plasma, but more Ca^{2+} is bound in bile (to bile acids) than in plasma. There is an isosmotic NaCl and Na-bicarbonate absorption across the gallbladder epithelium. We normally produce **0.5 l of hepatic bile per day with bile salts and 1.5 g of bile pigments**. The normal gallbladder can concentrate the hepatic bile by a factor of 5. The bile salts (cholic acid and deoxycholic acid) are made from cholesterol, which is also abundant in bile. The formation of **mixed micelles**, containing cholesterol, phospholipid and bile salts, provides solubilized concentrations of both phospholipid and cholesterol far exceeding their normal solubility in water. Red blood cells are continuously being degraded, and the haeme released is taken up from the blood by the hepatocytes to produce bilirubin. Bilirubin is conjugated to glucuronic acid by a transferase in the liver cell to form the golden yellow **bilirubin mono- and diglucuronide**. These conjugates are much more water-soluble than bilirubin, and is thus easily excreted in the bile of the bile capillaries. The liver contains a **large store of vitamin B_{12}**. Only 0.1% of the store is lost daily in the bile, because most of its content is reabsorbed. Even if absorption totally ceases the hepatic vitamin B_{12} store lasts for **5–6 years**. In the absence of **vitamin B_{12}** the maturition of erythrocytes is retarded, and **pernicious anaemia** results.

How is bile acid absorbed by terminal ileum?

Most bile salts are rapidly reabsorbed by the terminal ileum to the portal blood (the **enterohepatic bile salt circuit**). The total bile acid pool in the body is about 3 g, and this pool can be recycled up to 12 times per day.

By contrast only a small fraction of the bile pigments are reabsorbed by the intestine (the **enterohepatic bile pigment circuit**).

1. The terminal ileum of humans absorbs conjugated bile salts efficiently by an **active Na^+-dependent co-transporter (symport)**, that is similar to the **glucose/Na^+** and **amino acid/Na^+ co-transporter** in the duodenum–jejunum (Fig. 52-1).
2. Bile acids also cross the brush border by diffusion in unconjugated form.
3. In the cytosol the bile acids are probably bound to macromolecules, and they traverse the basolateral membrane by **facilitated or active transport**.
4. The absorbed bile salts reach the liver, where they are conjugated and reprocessed, and the hepatocytes clear the portal blood from bile acids in a single passage. The reabsorbed bile acids are essential stimuli for the liberation of bile with new and reprocessed bile acids, but when entering the liver they **inhibit** the synthesis of new bile acids. Normally, the daily use of bile acids is covered by 85% reabsorbed molecules and 15% newly synthesized bile acids.

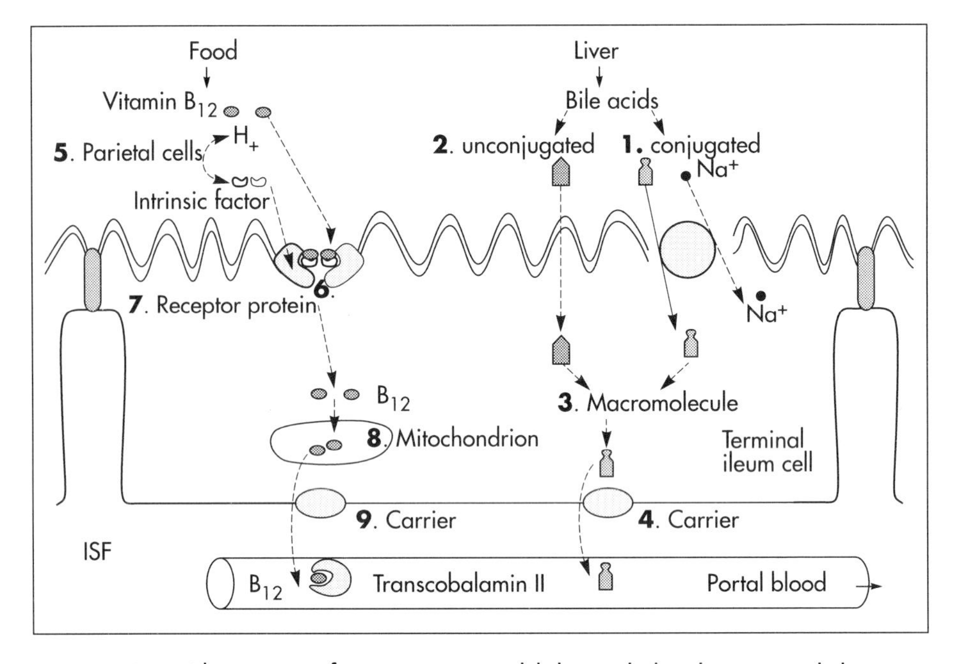

Fig. 52-1: Absorption of vitamin B_{12} and bile acids by the terminal ileum.

How is vitamin B_{12} absorbed?

1. **Intrinsic factor** is a cobalamin-binding protein that is secreted by the parietal cells parallel to the rate of gastric acid secretion. **Intrinsic factor–cobalamin complexes** are resistant to pancreatic proteases (Fig. 52-1: 5).
2. These enzymes separate vitamin B_{12} (cobalamin) from the **gastric R proteins**, so that two cobalamin molecules are tranferred to a dimer form by intrinsic factor in the upper ileum (Fig. 52-1: 6).
3. In the terminal ileum the brush border membranes contain a **receptor protein** that recognizes and binds the **intrinsic factor–vitamin B_{12} dimer** (Fig. 52-1: 7). The free vitamin B_{12} enters the enterocyte by active transport, and the intrinsic factor stays in the lumen.
4. In the cell the vitamin is delayed for several hours due to conversion in the mitochondria (Fig. 52-1: 8).
5. Vitamin B_{12} exits from the cell by facilitated or active transport, and appears in the portal blood bound to the globulin, **transcobalamin II** (Fig. 52-1: 9). The hepatocytes clear the portal blood for vitamin B_{12} by **receptor-mediated endocytosis.**

What is terminal ileitis?

Crohn's disease (regional ileitis) is a transmural inflammatory bowel disease often with fistules. Crohn's disease is mainly localized to the **terminal ileum**. These patients suffer from malabsorption of bile salts and vitamin B_{12} (Fig. 52-1). The Crohn patients often look emaciated, because they excrete large amounts of bile salts with faeces together with fat, proteins and water.

What are the causes of pernicious anaemia?

Pernicious anaemia is caused by atrophy of the gastric mucosa, resulting in **insufficient synthesis of intrinsic factor**. The patients cannot secrete intrinsic factor, hydrochloric acid and pepsin.

Pernicious anaemia has three forms: (1) most patients have an **autoimmune disorder**, with plasma antibodies against their own parietal cells; (2) rarely, newborne babies suffer from **congenital intrinsic factor deficiency**, but pepsin and acid secretion are normal; and (3) vitamin B_{12} absorption is deficient because of a defect in the **intrinsic factor–B_{12} receptors** in the terminal ileum.

What regulates the gallbladder function?

CCK is released by the duodenal mucosa in response to contact with fat and essential amino acids. CCK reaches the gallbladder wall via the blood, and it causes contractions of the gall bladder and relaxation of the sphincter of Oddi. **Gastrin** has a small CCK effect, and **VIP/ACh** inhibits gallbladder contractions.

What are the functions of bile salts?

Cholate and desoxycholate are **fat-solubizing agents**. They have **fat-soluble hydrocarbon rings** that enable them to mix with fats and **several charged groups** that enable them to mix with water. Large fat droplets in the duodenal chyme become dispersed, forming smaller fat particles – a process called **emulsification**. The bile salts contribute to emulsification.

These smaller fat particles are efficiently digested by the **water-soluble pancreatic lipases**, forming glycerides and fatty acids in special fatty aggregates called **micelles** (Fig. 53-3).

The micelle contents are readily absorbed by the enterocytes. The co-lipase helps the lipase to eliminate the inhibitory bile salts from the surface, so that the **lipase is fixed to the lipids**.

The pancreatic lipase cleaves the ester linkage of tri-acyl-glycerol at the 1- and 3-position, releasing two fatty acids and 2-monoglyceride (2 MG), or occasionally a free glycerol molecule (Fig. 53-3). Free glycerol is readily absorbed. A protein – **fatty acid binding protein** (FABP; MW 12 000) – is present in the cytosol of the enterocytes. **FABP binds fatty acids to re-esterify the fatty acids and to protect the cell from adverse effects of cytotoxic fatty acids.** Once the fatty acids are formed, the fatty acids and 2 MG participate in the emulsification process, but the fatty acids still require bile salts for complete water solubility. Micelles are passed through the aqueous bowel lumen to reach the absorbing mucosa (Fig. 53-3). **Lipids do not adequately form micelles**, if there is no bile present.

What causes the disorder, gallstones?

There are two types of gallstones – those composed of cholesterol and those composed of bile pigment. Cholesterol stones only develop when bile is supersaturated (ie, bile that has an excess of cholesterol relative to bile salts and phospholipids).

Excessive removal of water in the gallbladder can be pathogenic. Enlarged gallstones can obstruct the common bile duct thus causing bile with bilirubin to flow back into the liver and leak into the blood plasma (**jaundice or icterus**).

What is clinical jaundice?

The criterion for jaundice is a hyperbilirubinaemia exceeding 15 μmol l^{-1} of plasma, where the yellow colour is evident in the skin and sclerae.

What is acute biliary tract disease?

These disorders are diagnosed by **biliary pain** and confirmed by ultrasonographic or computer tomographic evidence of a distended/inflammed gallbladder. The disorders are subdivided into **inflammatory** (e.g. acute cholecyctitis) or **obstructive** (e.g. stone in the common bile duct).

Further reading

Johnson, L.R. (Ed.) (1987) *Physiology of the Gastrointestinal Tract*. New York, Raven Press.

La Russo, N.F. (1984) Proteins in bile: How they get there and what they do. *Am. J. Physiol.* **247**: 6199.

Sellinger, M. and J.L. Boyer (1990) Physiology of bile secretion and cholestasis. In: H. Popper and F. Schaffner (Eds) *Progress in Liver Disease*, Vol IX, pp. 237–60. W.B.Saunders, Philadelphia.

52. Multiple Choice Questions

Each of the following five statements have True/False options:

A: Bile acids are essential for solubilizing cholesterol and phospholipids by formation of micelle aggregates.

B: Bilirubin binds to cytoplasmic proteins within the hepatocyte.

C: The primary bile acids are deconjugated and dehydroxylated to form the secondary bile acids.

D: Intrinsic factor–cobalamin complexes are inactivated by pancreatic proteases.

E: Cholate and desoxycholate have water-soluble hydrocarbon rings.

52. Case History

A female, age 42 years, is admitted to hospital due to fatigue. She describes a serious gastrointestinal infection of which she was cured some years ago. Suspicion of vitamin B_{12} deficiency leads to the following findings: a seriously low vitamin B_{12} concentration and plasma antibodies against her own parietal cells in the gastric mucosa.

Assume that the absorption of vitamin B_{12} totally ceased at the time where her parietal cells were destructed by autoimmune disease. Assume further that she had a normal liver store of 5 mg vitamin B_{12}, and that she has lost 1‰ daily of the hepatic store in the bile.

1. *Calculate the half-time period necessary to reduce the hepatic vitamin B_{12} store by 50%.*
2. *Calculate the number of years it takes to empty the hepatic vitamin B_{12} store down to 0.5 mg (manifest pernicious anaemia).*

Try to solve the problems before looking up the answers in Chapter 74.

CHAPTER 53. INTESTINAL DIGESTION AND ABSORPTION

Almost all of the dietary nutrients, water and electrolytes that enter the upper small intestine are absorbed. The small intestine, with its epithelial folds, villi, and microvilli, has an internal surface area of 200 m^2 or about the **size of a tennis court**.

I. CARBOHYDRATES

What is the most important energy-giving component of the diet?

Carbohydrates are the most important energy-containing components of the diet. The energetic value of most carbohydrates is 17.5 kJ g^{-1}, so that a daily diet of 400 g carbohydrates covers 7000 kJ, which is 56% of the usable energy in a diet of 12 500 kJ day^{-1}. The formation of **metabolic water** on a mixed diet is **0.032 g J^{-1}**.

The common sources of digestible carbohydrates are **starches (amylose), table sugar, fruits and milk**. Plant and animal starch (amylopectin and glycogen) are branched molecules of glucose monomers. Indigestible carbohydrates are present in vegetables, fruits and grains (cellulose, hemicellulose, pectin) and in legumes (raffinose). Indigestible carbohydrates are also referred to as **dietary fibres**. Digestion of starches to simple hexoses occurs in two phases: the **luminal phase** begins in the mouth with the action of **salivary amylase (ptyalin)**, but most of this phase occurs in the upper small intestine as **pancreatic α-amylase** reaches the chyme. The starch polymer is reduced to maltose, maltriose and α-limit dextrins (Fig. 53-1). The three substrates are pushed through the intestine and are now ready for the **brush border phase**. Some of the substrate molecules get into contact with the brush borders of the absorbing mucosal cell via the **unstirred water layer**. Enterocytes carry disaccharidases and trisaccharidases (oligosaccharidases) on their surface that cleave these substrates to glucose (Fig. 53-1).

Milk sugar (lactose) and cane sugar (sucrose) only require a **brush border phase** of digestion, since they are **disaccharides** (Fig. 53-1). Sucrose is reduced to **glucose and fructose**, and **lactose to glucose and galactose** by the action of disaccharidases (sucrase and lactase).

How is carbohydrate absorbed?

Glucose in the intestinal lumen is absorbed by active glucose transport (Fig. 53-1).

1. The mechanism of active glucose transport is a carrier-mediated, **Na^+-glucose co-transport**. As the luminal [glucose] falls below the fasting blood [glucose], active glucose transport becomes essential and sequesters all remaining luminal glucose into the blood. Glucose and Na^+ bind to apical membrane **transport proteins** (a glucose-transporter, GLUT). The two substances are deposited in the cytoplasm, because of conformational changes in GLUT, whereby the **affinity of GLUT for glucose-Na^+**

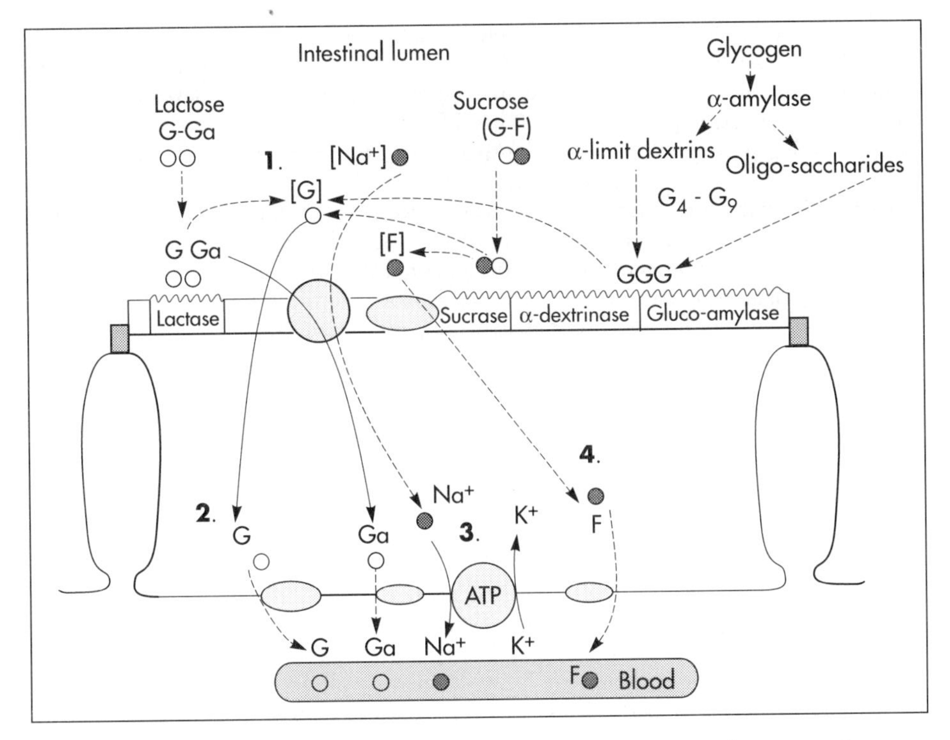

Fig. 53-1: Absorption of carbohydrates by the enterocyte. G, glucose; F, fructose; GA, galactose.

changes from high to low. Glucose accumulates inside the cell to a level that exceeds blood [glucose].

2. Glucose therefore diffuses down its concentration gradient, through a specific uniport carrier in the basolateral membrane, out into the interstitial space and into the blood (Fig. 53-1). The **basolateral uniport carrier for glucose** (GLUT 2) does not depend upon Na^+. Galactose is also actively transported by the luminal glucose carrier system, and is a competitive inhibitor of glucose transport. Phlorhizin blocks the glucose absorption, when its **glucose moiety** binds to the transporter instead of glucose.
3. Cytoplasmic Na^+ is actively pumped out through the **basolateral membrane by the Na^+-K^+-pump**. The low intracellular [Na^+] creates the Na^+ gradient and energizes the transport of hexoses over the luminal enterocyte membrane.
4. **Fructose** has no effect on the absorption of glucose and galactose. Fructose is not actively transported by the enterocytes, but is absorbed by a **carrier-mediated, facilitated diffusion system**, where energy is not required (Fig. 53-1).

What is the most common chronic disease in humans?

The most common chronic disorder in humans is **lactose malabsorption or hypolactasia** (lactose-induced diarrhoea or lactose intolerance), which is due to a **genetical deficiency**

of lactase in the brush-border of the duodeno-jejunal enterocytes. More than 50% of all adults in the world are lactose intolerant. Infants with the rare **congenital lactase intolerance** are borne without lactase in their brush-border. They develop diarrhoea, when they are breast fed. This can result in a life-threatening dehydration. The amount of lactose entering the colon determines the size of the **osmotic diarrhoea**. Milk made from soya beans and fructose is well tolerated.

Sucrase–isomaltase deficiency is an autosomal recessive genetic condition with sucrose intolerance. This disorder is found in 10% of Innuits, which is not surprising, since they must have lived for thousands of years on a natural diet without sucrose.

Glucose–galactose malabsorption is a rare genetic disorder caused by a defect in the brush border system for glucose and galactose absorption (GLUT). Fructose is well tolerated.

II. Proteins

How much dietary protein does an adult need per day?

The typical Western diet contains **100 g of protein**, which is equivalent to an energy input of 1700 kJ day^{-1}, although an adult needs only less than 1 g kg^{-1} of body weight. This luxury consumption is an inappropriate use of global resources. Moreover, a high protein intake implies a long-term risk of **uric acid accumulation** from purine degradation (**hyperuricaemia**). Meats, fish, eggs, and dairy products are high in proteins and expensive. Vegetable proteins are not as expensive as animal proteins.

Residents of areas with carbohydrate-dominated nutrition and protein hunger develop diseases of **protein deficiency**, such as **Kwashiorkor**, characterized by hypoproteinaemia with thin limbs and a tremendous oedema of the abdominal cavity (ascites).

Intact proteins and large peptides can be absorbed by humans to an extent that is sufficient to trigger an **immunological or allergic response**.

What is hyperuricaemia?

Hyperuricaemia is a condition with an abnormally high concentration of uric acid ([urate]) in the blood plasma and ECF (above **420 μmol l^{-1}**). Hyperuricaemia is asymptomatic for long periods. In patients with **gout (uric arthritis)** the hyperuricaemia becomes clinically significant through recurrent attacks of acute arthritis.

Uric arthritis is genetic or acquired, with accumulation of urate and other crystals in articulations (podagra), kidneys (urate nephropathy), ureter (kidney stones) and connective tissues. Most forms of metabolic gout are a result of overproduction of uric acid caused by accelerated purine synthesis from amino acids, formate and CO_2, whereas dietary purines play a minor role. **Xanthine oxidase** oxidizes hypoxanthine to xanthine and xanthine to uric acid. During purine degradation large quantities of NH_4^+ are liberated. The acidosis leads to crystallization of urate. Hyperuricaemia is treated with **allupurinol**, a purine analogue which is oxidized to alloxanthine by xanthine oxidase. Alloxanthine inhibits xanthine oxidase for hours, so a daily dose of allupurinol reduces plasma [urate] effectively, together with a rise in the more soluble hypoxanthine and xanthine. See Chapters 57 for renal handling of urate, and 62 for the relation between urate and uraemia.

What is gluten enteropathy?

Gluten enteropathy (coeliac disease, non-tropical sprue) is probably an immunological disease in the intestinal mucosa. Gluten is present in wheat and rye. In allergic persons **gluten** causes an immunological reaction with desquamation of the luminal part of the intestinal mucosa, in particular most of the microvilli. The marked fall in area available for absorption causes malabsorption. In severe cases the malabsorption involves fats causing steatorrhoea, Ca^{2+} causing osteomalacia, vitamin K causing bleeding disturbances, vitamin B_{12} causing pernicious anaemia, and folic acid causing folic acid deficiency.

Non-tropical sprue is cured by removal of wheat and rye flour from the diet. **Tropical sprue** is found in tropical areas, and probably caused by gastrointestinal infection, although the bacterial diagnosis is seldom confirmed. Tropical sprue is often curable with antibacterial agents.

How are dietary proteins digested?

Digestion of dietary proteins begins in the stomach, with the action of the gastric enzyme **pepsin** (pH optimum is 1), which cleaves proteins to proteoses, peptones and polypeptides. **Pepsin is produced from pepsinogens in the presence of HCl.** Pepsinogen is secreted by the **gastric chief cells**. The digestion is continued in the intestine by proteolytic enzymes of the pancreas. **Enteropeptidase** converts trypsinogen to trypsin. Trypsin acts **autocatalytically** to activate trypsinogen, and also converts chymo-trypsinogen, pro-carboxy-peptidases A/B, and pro-elastase to their **active form**. When the chyme is pushed into the duodenum, the pancreatic juice neutralizes the chyme and the activity of pepsin is stopped. The proteolysis in the small intestine plays the major role, because the digestion and absorption of dietary protein is not impaired by total absence of pepsin.

Cytosolic peptidases from the enterocytes and **brush border peptidases** from the brush borders of the villous cells then cleave the small peptides into **single amino acids** (enteropeptidase, amino-polypeptidase and dipeptidases). The end products of protein digestion by pancreatic proteases and brush border peptidases are di- and tripeptides and amino acids. The cytosolic peptidases are abundant and particularly active against di- and tripeptides.

How are amino acids absorbed in the small intestine?

Hydrolytic digestive products such as tripeptides, dipeptides and amino acids can be absorbed intact across the intestinal mucosa and into the blood. Two transport routes are dominant:

1. A **peptide transporter**, with high affinity for di- and tripeptides, absorbs the small peptides (Fig. 53-2). The system is stereospecific and prefers peptides of physiological L-amino acids. This peptide transport across the brush border membrane is an H^+ coupled **secondary active process** indirectly powered by the electrochemical potential difference of Na^+ across the membrane via H^+/Na^+ exchange. The total amount of each amino acid that enters the enterocytes in the form of small peptides is greater than the amount that enters as single amino acids.
2. The absorption of **single amino acids** from the intestinal lumen is an active process which involves a **Na^+-dependent, carrier-mediated co-transport system** similar to that for glucose (Fig. 53-2). **Active transport** is in this case also characterized by

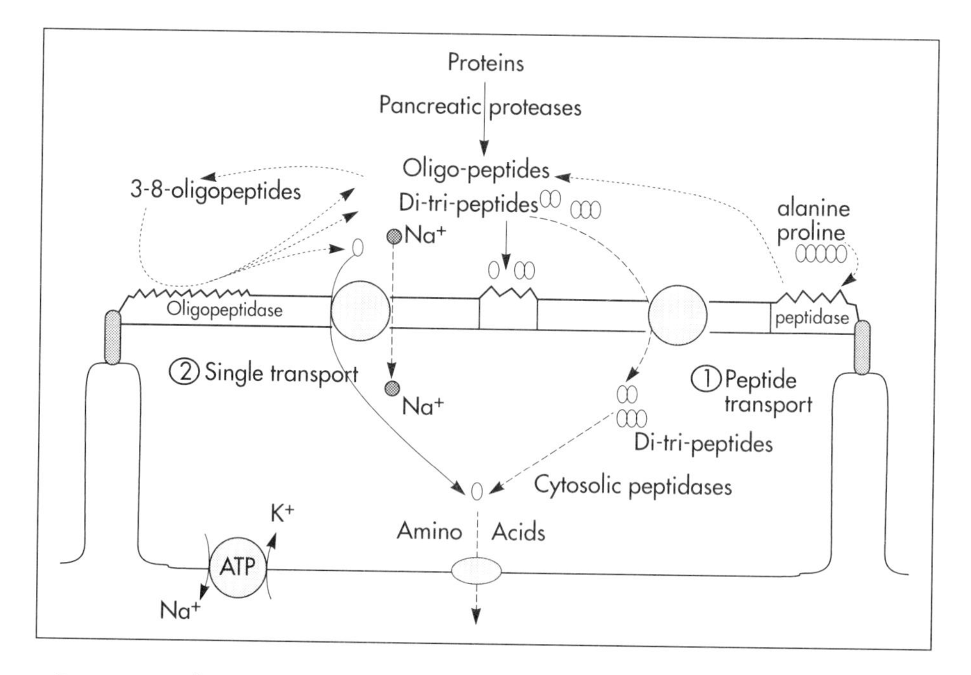

Fig. 53-2: Absorption of peptides and single amino acids by the enterocyte.

competitive inhibition, saturation kinetics, Na^+ dependency, and expenditure of metabolic energy.

Selective carrier systems appear to be present for certain groups of amino acids: neutral, acidic, imino and basic groups. The **neutral brush border (NBB) system** transports most of the neutral amino acids. The **imino acid system** handles proline and hydroxyproline.

Basic amino acids and phenylalanine, are absorbed primarily through **facilitated diffusion** from the gut lumen to the blood.

The basolateral membrane is more permeable to amino acids than is the brush border membrane. Therefore diffusion is more important for the basolateral transport, especially for amino acids with hydrophobic side chains.

The amino acids are carried in the blood to the **liver via the portal vein. Half of the amino acids absorbed in the intestine** are from the diet, the remaining half are from digestive secretions and from desquamated mucosal cells. Only 1 % of the dietary protein is excreted in the faeces, the remaining faecal protein is derived from microorganisms and desquamated cells.

Is the amino acid transport similar in renal and in intestinal cells?

The reabsorption of amino acids (and glucose) in the renal tubules shares many similarities to the active absorption mechanism in the intestine.

A rare genetic disease involves **defective intestinal absorption** of neutral amino acids and a similar **defective renal reabsorption**. This condition is called **Hartnups disease,** which is caused by defects in the NBB transport system of the brush border coated epithelial cells of the jejunum and the proximal renal tubules.

Also in the genetic disease **cystinuria** there is defective tubular reabsorption and jejunal absorption of cystine and basic amino acids lysine, ornithine and arginine.

III. Lipids

What fraction of the dietary energy ought to be lipids?

The typical Western diet contains **100 g of lipids (3900 kJ) per day**. Most of the dietary lipids consumed are triglycerides (only 2–4% is made up of phospholipids, cholesterol, cholesterol esters, etc.). Lipids would comprise just above 30% (i.e. 100 g = 3900 kJ) of a standard diet of 12 500 kJ day^{-1}. An optimal diet should contain **only 20% lipids**, such as the lipids of fish oil and olive oil.

Absorption of excess lipids results in accumulation (obesity). The consequences of **long-term obesity** are described in relation to **diabetes mellitus** in Chapter 73. **Essential dietary fatty acids** are polyunsaturated and cannot be synthesized in the body (linoleic acid, linolenic acid and arachidonic acid).

How are lipids digested?

Dietary triglycerides are broken down into simpler molecules, to facilite absorption. A small fraction of the triglyceride is digested in the mouth and stomach by **salivary, lingual lipase**.

Most dietary triglycerides (TG) are digested in the small intestine. However, two problems must be solved before digestion can occur. **TG are insoluble in water**, and the chyme in the intestine is an emulsion of large fat particles in water. **All the lipase proteins by contrast are water soluble.** It follows that triglycerides must be solubilized in the aqueous phase before they can be digested. The lipolytic activity requires the **emulsifying action of bile salts in order to solubilize TG in water**. Pancreatic lipase binds to the surface of the small emulsion particles.

How are bile micelles and mixed micelles built?

Simple bile micelles are aggregates of **bile salt monomers** that form spherical structures with a diameter of 5 nm, and the micelles have a negative charge. Following a meal, bile micelles are formed above a certain concentration of bile salts, called the **critical micellar concentration**. The **lipophilic, hydrophobic, apolar end** of the bile acids faces inward creating a hydrophobic core (Fig. 53-3). The **hydrophilic polar end of the bile salts** (hydroxyl-, carboxyl- and amino- groups) points outward, so that they are mixed with the **polar water molecules**. The simple lipids must pass a diffusion barrier – an **unstirred water layer**, which is the water layer immediately adjacent to the mucosa, where the intestinal flow rate is essentially zero (Fig. 53-3). This water layer contains the water soluble lipases and cholesterol esterases. A large concentration of **bile salts** helps the **lipid-laden micelles** to get access to the absorbing surface.

Lipids then diffuse easily out of the **lipophilic micellar core** and into the **lipid layer of the apical membrane** of the mucosal cell (Fig. 53-3).

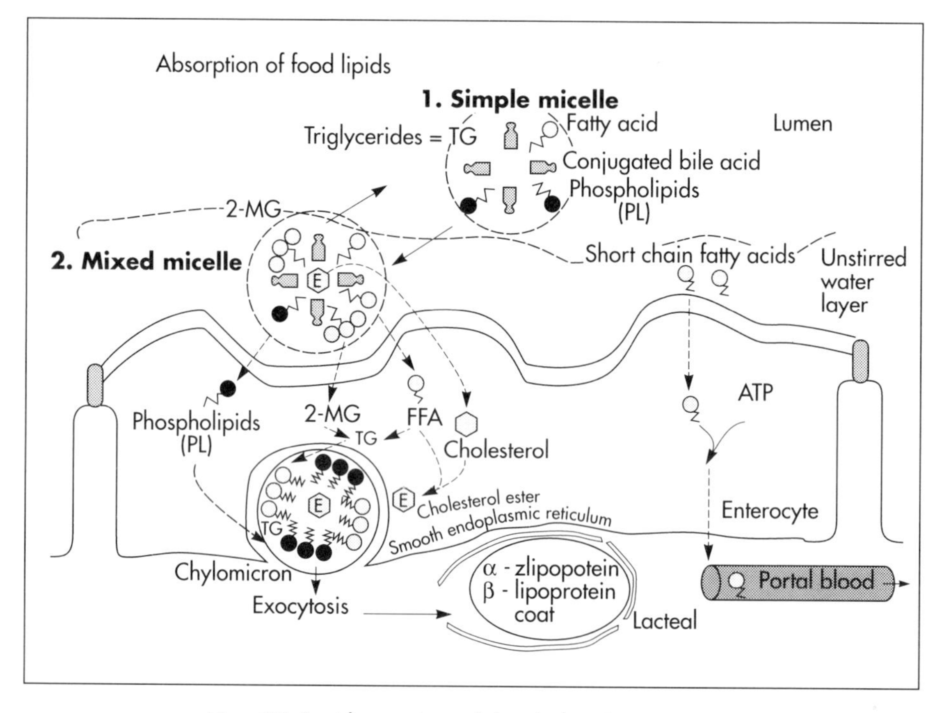

Fig. 53-3: Absorption of lipids by the enterocyte.

Mixed micelles. Simple lipid molecules (cholesterol, phospholipids, fatty acids, 2-MG, fat-soluble vitamins and lysolecithin) diffuse into the lipophilic core of the **simple bile micelles** and form a **mixed micelle** (Fig. 53-3). A solution of micelles is water-clear and stable.

The **mixed micelles** carry the major part of all the lipids that are absorbed by the intestinal microvilli. When the lipids of the mixed micelle have diffused into the enterocyte, the **empty bile micelle** is recycled by emulsifying more hydrolysed lipids (Fig. 53-3). **Neither bile salt micelles nor bile salt molecules diffuse into the enterocyte** (Fig. 53-3).

Are small amounts of lipid absorbed otherwise?

The **fatty acids with a short chain** (up to 12 C atoms) are more hydrophilic than the rest. They can diffuse **directly to the portal blood as fatty acids**. Once fatty acids enter the enterocyte, they are primarily activated to **acetyl co-enzyme A** by a process that requires ATP and **acetyl co-enzyme A synthetase**. Acetyl co-enzyme A enters one of two pathways: the 2-MG and the α-glycerol phosphate pathways. Both bring about the resynthesis of TG in the enterocyte.

What is the chylomicron mechanism?

In the enterocyte the **lipids are reformed to TG, cholesterol, phospholipids, etc.** (Fig. 53-3). The reformed TG, cholesterol, phospholipids, fatty acids, esters and fat-soluble vitamins reach the ER, where they are packed in another **lipid-carrying particle: the chylomicron.**

The centre of the chylomicron is a cholesterol ester (E in Fig. 53-3). Chylomicrons are packed into vesicles in the Golgi-system. These vesicles reach the basolateral membrane, and their contents pass through this membrane by **exocytosis** (Fig. 53-3). Thus the chylomicrons reach the lymphatic channel of the villus (the **central lacteal**). The **lymph delivers the chylomicrons to the blood** through the thoracic duct. **Plasma is milky (lipaemic) following a fatty meal.**

What is the fat content of normal faeces? What is fat malabsorption?

All of the dietary lipid is normally absorbed in the intestine. **Faecal fat derives from bacterial lipids and lipids of desquamated mucosal cells.** Disorders such as gallstones, pancreatitis, Crohn's disease, and liver disease can lead to **fat malabsorption (steatorrhoea or fat-diarrhoea).**

What makes lipid absorption special?

Lipids are mainly absorbed through the enterocyte and **transported by the lymph,** which reaches the blood via the **thoracic duct.** Lipids thus reach the liver through the hepatic artery, with the exception of **short-chain fatty acids that enter the portal blood directly.** Other nutrients are absorbed directly to the blood and reach the liver through the portal vein.

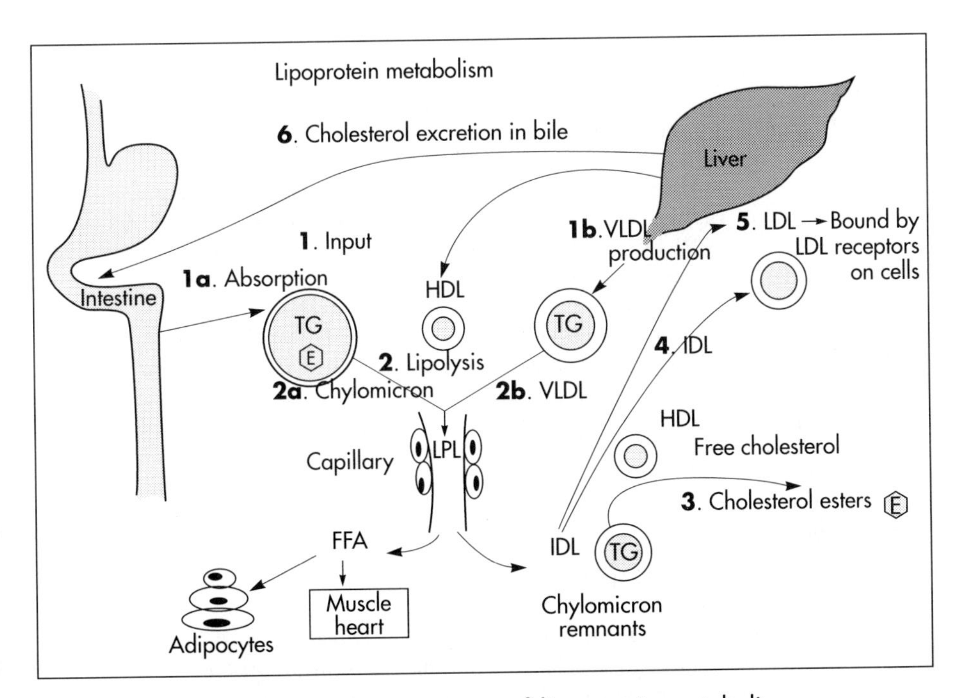

Fig. 53-4. An overview of lipoprotein metabolism.

Describe the lipoprotein metabolism in humans

TG and cholesterol circulate in the blood as complex lipoprotein particles. **The lipoprotein particle** is built up by a non-polar core containing TG and cholesterol esters. Its polar shell consists of phospholipid, apoproteins and cholesterol (Fig. 53-4). TG (1.2 mM) and total-cholesterol (5 mM) form the major components of fasting plasma lipids, supplied with small amounts of FFA.

1. INPUT

1.a. **Absorption.** Chylomicrons are formed from dietary fat after a meal and absorbed from the intestine into the blood. They have a half-life of 5 min.

1.b. **Very low density lipoproteins (VLDL),** which contain mainly TG, are synthesized by the liver and liberated from the liver in the postabsorptive phase. Hepatic synthesis of cholesterol varies inversely with the dietary intake.

2. LIPOLYSIS

2.a. **Chylomicrons** are hydrolysed by the enzyme **lipoprotein-lipase (LPL)** on the endothelial cell surfaces in the capillaries (Fig. 53-4). The FFA liberated by hydrolysis are absorbed by the cells for resynthesis and stored as intracellular TG (Fig. 53-4).

2.b. **VLDL** are hydrolysed by lipolysis much the same way as chylomicrons. Thus they lose TG and form the small **intermediate density lipoproteins (IDL).**

3. **Cholesterol esters.** The remains of the chylomicrons (chylomicron remnants) co-operate with IDL and **high density lipoproteins (HDL)** to form cholesterol esters (Fig. 53-4). Cholesterol esters are then exchanged for TG in VLDL and chylomicrons by the **cholesterol ester-transfer protein,** whereby HDL_3 changes to the less dense HDL_2. HDL transfer lipids between organs and between other lipid particles. HDL facilitates the flux of TG back to the liver, and the flux of cholesterol/TG to peripheral tissues (Fig. 53-4).

4. **IDL transfer.** IDL particles are partly taken up by the liver and partly converted to **low density lipoprotein (LDL)** in the blood and liver.

5. **LDL** is the largest cholesterol fraction in blood plasma, and has a half-life of 24 h. LDL transport cholesterol to the peripheral tissues (Fig. 53-4). LDL are here bound to the cells by **LDL receptors** (Fig. 53-4). **The genetic LDL receptor deficiency** elevates the ratio of LDL to HDL in blood plasma, and a ratio greater than 4 is a high risk factor for **cardiovascular disease.**

   During fasting conditions, HDL concentrations in the blood plasma are generally increased in females, by oestrogens, by exercise, and by moderate alcohol intake. Similarly, fasting HDL concentrations are reduced (and LDL increased) in males, by androgens, by smoking, by obesity, and by an inactive sedentary life-style.

6. **Cholesterol** is excreted into the intestine.

How are vitamins absorbed?

Fat-soluble vitamins, such as vitamins A, D, E and K, are absorbed in the chylomicrons along with lipid nutrients (Fig. 53-3).

The **water-soluble vitamins,** such as the B and C vitamins, cross the mucosa by diffusion and by association to **specific membrane transporter proteins**. Vitamin B_{12} (cyanocobalamine) is the largest of the vitamins, and its transfer utilizes a specific transport mucoprotein called **intrinsic factor** (see Chapter 52).

How are calcium and phosphate absorbed?

We consume 1000 mg (25 mmol) Ca^{2+} per day. However, we absorb only 400 mg Ca^{2+} totally. The **net-absorption is only 250 mg,** because we secrete 150 mg Ca^{2+} day^{-1} to the intestine (Fig. 67-1). The net-absorption is saturable, since it depends on available **Ca^{2+} binding protein** in the brush border and in the cytosol of the enterocyte. The synthesis of this protein, and thus intestinal Ca^{2+} absorption, is induced by active vitamin D and by the parathyroid hormone. Steroid hormones like vitamin D exert their major effects after binding to nuclear receptors and stimulating the synthesis of mRNA that codes for **cytosolic Ca^{2+} binding protein**. The basolateral membrane contains two transporters of Ca^{2+}: A **Na^+/Ca^{2+} exchanger**, which is more effective at high intracellular [Ca^{2+}], and a **Ca^{2+}-ATPase**, which is the major mechanism at low levels of intracellular [Ca^{2+}]. The duodenum–jejunum can concentrate Ca^{2+} against a 10-fold concentration gradient. About 750 mg (19 mmol) Ca^{2+} must be excreted in the faeces every day (Fig. 67-1).

The amount of phosphate absorbed, and its concentration in plasma, is determined by the amount available through the diet, but the active transport is somewhat dependent on vitamin D. High plasma [Ca^{2+}] and [phosphate], promote **bone formation** in children. The children increase the precipitation of Ca-hydroxyapatite in their **bone matrix**.

Are Ca^{2+} and phosphate important?

The [total calcium] in the blood plasma of healthy persons is **2.5 mM** or 100 mg l^{-1}. Ca^{2+} has critical roles in neuromuscular function and coagulation. The [Ca^{2+}] is regulated by **PTH** – see Chapters 66 and 67.

Intracellular phosphate is an essential component of nucleic acids, high energy molecules, co-factors, regulatory phosphoproteins, and glycolytic intermediates.

Further reading

Johnson, L.R. (Ed.) (1991) *Gastrointestinal Physiology*, 4th edn. Mosby-Year Book, St Louis.

Stevens, B.R. *et al.* (1990) Intestinal brush border membrane Na^+/glucose cotransporter function *in situ* as a homotetramer. *Proc. Natl Acad. Sci.* **87**: 1456–60.

Sullivan, S.K. and M. Field (1991) Ion transport across mammalian small intestine. In: M. Field and R.A. Frizzel (Eds) *Handbook of Physiology*, Sect. 6, Vol. 4., Am. Physiol. Soc., Bethesda, MD.

53. Multiple Choice Questions

Each of the following five statements have True/False options:

A: Hexoses and amino acids require Na^+ for active transport into the enterocyte.

B: A person with lactase deficiency cannot digest lactose, so undigested lactose from a milky diet would enter the colon.

C: The Na^+-K^+-pump is essential for intestinal Na^+ absorption.

D: All lipase proteins are lipid soluble.

E: Cytosolic peptidases from the enterocytes and brush border peptidases cannot cleave small peptides into single amino acids.

53. *Case History*

A well-known alcoholic male, with hepatic insufficiency, is brought to the intensive care unit of a hospital in a hepatic coma.

Normally ammonia is formed in the gastrointestinal tract as a product of protein digestion and bacterial action. The liver usually removes a major portion by converting ammonia into urea. Hereby, the toxic ammonia is eliminated.

The impaired liver function of this patient has led to the development of collateral venous shunts with oesophageal varicosities. Large quantities of blood from the gut, with a high [NH_4^+], are transported directly into the systemic veins and the brain of the patient. His blood [NH_4^+] is drastically increased, and his blood [glucose] is 2 mM (hypoglycaemia).

1. *What is hepatic coma? What is causing the unconsciousness?*
2. *Explain his condition in terms of abnormal glucose metabolism.*

Try to solve the problems before looking up the answers in Chapter 74.

SECTION VI.

The Kidneys and the Body Fluids

The kidneys **excrete water and solutes**, whereby they regulate the volume and composition of the body fluids within an extremely narrow range; they also **excrete most of our metabolic end-products**. The kidneys are important **endocrine organs** that trigger the **renin**-angiotensin-aldosterone cascade, and produce erythropoietic factor, as well as 1,25-dihydroxy-cholecalciferol, kinins and prostaglandins.

Chapter 54.
Body Water and Fluid Compartments

On average, our **total body water** comprises 60% of the body weight (Fig. 54-1). The so-called **lean body mass**, which means a body stripped of fat, has a water content of 0.69 parts of the body weight. Such high values are observed in newborn babies and in extremely fit athletes. Babies have a 10-fold higher water shift per kg of body weight than adults.

Is our water content sex dependent?

Female population groups seem to contain less water on average compared to males. Such differences show **sex dependency**, but the important factor is the **fraction of body fat,** since fat tissue contains significantly less water than other tissues (only 10% water).

Sedentary, overweight persons contain 50–55% water dependent on the body fat content, regardless of sex.

How much of the body water is present in cells?

The **intracellular fluid volume (ICV)** comprises 26–28 kg out of the total 42 kg water in a 70-kg person (Fig. 54-1).

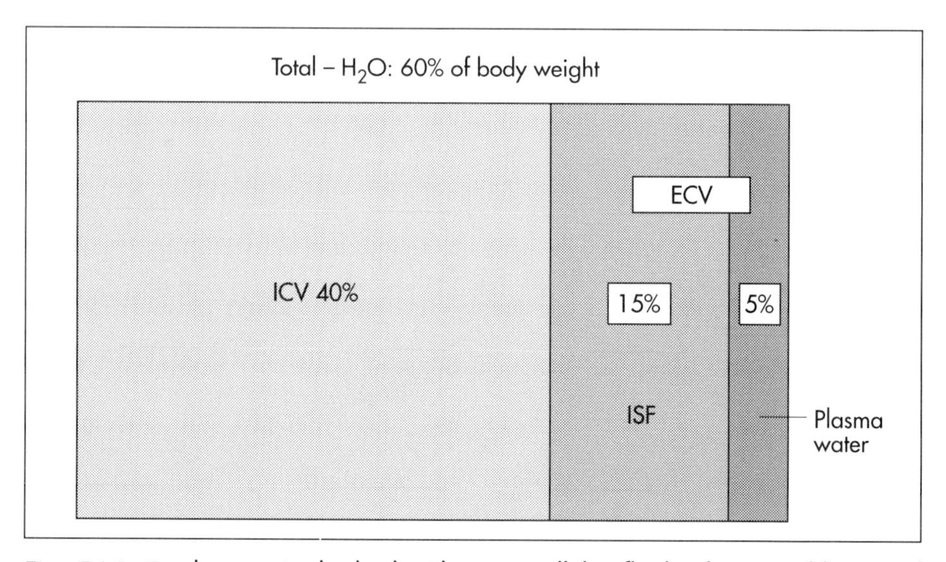

Fig. 54-1: Total water in the body. The extracellular fluid volume is abbreviated ECV, the intracellular volume is abbreviated ICV, and the interstitial fluid volume is termed ISF (ISV).

The **extracellular fluid volume (ECV)** compartment comprises the remaining water (14–16 kg) with most of the water in **tissue fluid (interstitial fluid or ISF)** and 3 kg water in plasma (Fig. 54-1).

Do our body waters vary with age and with body weight?

The **fraction of body fat** rises with increasing age and body weight, and the **relative mass of muscle tissue** becomes less.

Consequently, the **body water fraction falls with increasing body weight and age.** Ageing and fat accumulation imply that the cells lose more water, but the ECV is remarkably constant through life and under disease conditions.

When a dead person is cremated, the remains are sodium, potassium, calcium and magnesium. What is the approximate content from a 70 kg body?

Each dead body contains 4 mol of both sodium and potassium. A minor fraction of the potassium is radioactive. The calcium and magnesium content is 25 and 1 mol, respectively.

Radioactive measurements are linked to an important concept called specific activity (SA). What is SA?

SA is the **concentration of radioactive tracer** in a fluid volume divided by the **concentration of naturally occurring mother-substance**. SA is equal to **activity/mass unit** (radioactive/non-radioactive isotope). Radioactivity is measured as the number of degradations per second in becquerel or Bq l^{-1} and the mother-substance traditionally in mol l^{-1}. **One degradation per second equals 1 Bq**. Following even distribution, the SA for a certain substance must be the same all over the body. SA is preferably measured in plasma (with scintillation counters or other equipment).

Is it possible to measure total body potassium?

Our natural body potassium is ^{39}K, but we also contain traces of naturally occurring radioactivity (0.00012 or 0.012% is ^{40}K with a half-life of $(1.3 \cdot 10^9)$ years). When using this natural tracer, injection of radioactive tracer is avoided.

To measure the **total amount of the tracer ^{40}K in the body (*S* Bq)**, we must place the person in a sensitive whole body counter.

SA is the number of Bq ^{40}K per mol ^{39}K in the whole body. We can calculate all ^{39}K or **total body potassium**: *S*/SA mol/whole body – when SA is known to be 0.012%. The total body potassium of a healthy person is 4000 mmol. The SA of ^{40}K implies a $^{40}K/^{39}K$ ratio of 0.48/4000 (= 0.00012).

How do we measure the exchangeable body potassium?

An exchangeable ion pool in our body is the dynamic or exchangeable part of the total ion content. The balance is fixed in insoluble salts in the bones. The dynamic character implies the use of a dilution principle to measure such a pool.

We must inject a **radioactive tracer, such as ^{42}K with a physical half-life of 12 h (12.4 h)**. The total tracer dose given must be adjusted for by the loss of tracer in the urine and by the radioactive decay. Two urine samples are obtained and examined for tracer and for natural potassium. The first urine sample is from the first 12 h, and the second sample is covering 12–24 h. The tracer is assumed to distribute just as natural potassium after 12–24 h. When the tracer is distributed evenly in the **exchangeable potassium pool**, its SA must be the same in urine, plasma or elsewhere in the pool.

The **exchangeable body potassium** is equal to:

(Injected – eliminated)/SA.

We know the specific activity for the tracer (SA Bq mol^{-1}) from the plasma measurements. In this way we measure the **exchangeable body potassium**. The normal values are **41 mmol ^{39}K kg^{-1}** body weight for females, and **46 mmol kg^{-1}** for males.

How is exchangeable sodium (^{23}Na) measured?

This is an easy procedure using the dilution principle and a minimum of equipment.

Our natural non-radioactive body sodium is ^{23}Na. We administer the **radioactive tracer, ^{24}Na, with a physical half-life of 15 h**. We use a mixing period of 30 h to secure even distribution in the ECV.

The total tracer dose given must be adjusted for by the loss of tracer in the urine, and the radioactive decay of ^{24}Na (see the decay law in Chapter 1).

The **exchangeable sodium** is equal to:

(Injected – eliminated)/SA.

We know the specific activity for the tracer (SA Bq mol^{-1}) from the plasma measurements, therefore calculation of the **exchangeable ^{23}Na** is easy. The normal value for **exchangeable sodium** is 40 mmol kg^{-1} of body weight. In a patient with a body weight of 75 kg the exchangeable sodium is $(75 \cdot 40) = $ **3000 mmol**. The non-exchangeable sodium is fixed in the bones.

How do we measure the total body sodium?

The **total body sodium** is measured following discrete radiation called **neutron activation analysis**. The whole body of the patient is exposed to radiation with neutrons. A small fraction of the natural ^{23}Na now becomes radioactive sodium (^{24}Na) by uptake of an extra neutron.

A **sensitive whole body counter** records the radiation from ^{24}Na. Now we can calculate the **total body sodium**.

Normally, the total body sodium is 1000 mmol larger than the **exchangeable sodium** due to the fixed sodium content of the bones (1000 + 3000 mmol = **4000 mmol ^{23}Na**).

Do we have the same osmolality in plasma, ISF and in muscle cells?

Water permeable membranes separate these compartments, so that they contain almost the same number of osmotically active particles. The compartments have the **same osmolality**. They are isosmolal or have the same concentration expressed as mOsmol kg^{-1} water or the same freeze point depression.

None of these membranes can carry any important hydrostatic gradient.

The sum columns of electrolyte concentrations in muscle cells are essentially higher than the extracellular sum columns, because cells contain proteins, Ca^{2+}, Mg^{2+} and other molecules with several charges per particle (Fig. 54-2).

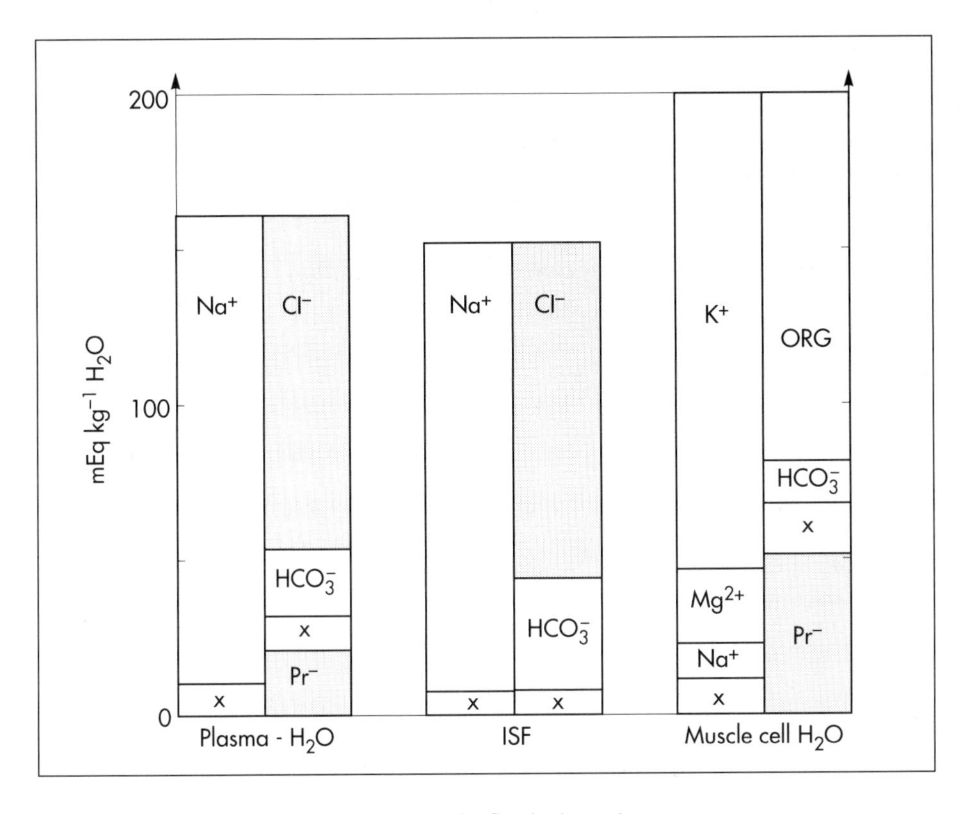

Fig. 54-2: Body fluid electrolytes.

The above columns show the ionic composition per kg of water, so we have 150 mmol Na per kg plasma water (Fig. 54-2). Out of 1000 g plasma 940 g is water, and the rest consists of plasma proteins and small ions. The **fraction of water in plasma is 0.94 (F_{water}).**

The exciting Donnan effect is described in Chapter 1. Calculations on ion concentrations are shown in Chapter 58 and below.

What is the ionic composition of normal human plasma? Calculate the concentration of Na^+ and Cl^- in a ultrafiltrate of plasma

The following concentrations are found in normal plasma: **[Na^+]** 140–141, **[K^+]** 3–4, **[Cl^-]** 103–104, **[bicarbonate]** 24, and **[Ca^{2+}]** 2.5 **mmol l^{-1} plasma.**

The [Na^+] and [Cl^-] depend upon a 5% uneven distribution due to the **Donnan effect**, and upon the **fractional content of water in plasma (0.94).**

$$[Na^+] = 141 \cdot 0.95/0.94 = \mathbf{143\ mmol\ l^{-1}\ ultrafiltrate}.$$

Based on the Donnan effect alone, this result should be less than 141. The Donnan effect is simply more than compensated by the **protein volume effect** or fractional content of water in plasma (0.94).

$$[Cl^-] = 103 \cdot 1.05/0.94 = \mathbf{115\ mmol\ l^{-1}\ ultrafiltrate}.$$

Based on the Donnan effect this result should be greater than 103 and the protein volume effect contributes further.

Such an ultrafiltrate is present in the kidneys and in ISF.

What is understood by the indicator dilution principle?

Mass conservation is always the underlying principle. The indicator M mol distributes in X l of distribution volume.

We measure the concentration C_p in mol l^{-1}, following even distribution, and calculate X: $X = M/C_p$.

Are there any errors involved?

Uneven distribution of indicator introduces a systematic error. A **non-representative concentration of indicator** in plasma makes it insufficient to correct for plasma proteins alone. **Loss of indicator** to other compartments is inevitable.

Elimination or synthesis of indicator in the body are frequently occurring errors. The indicator may be toxic or in other ways **change the size of the compartment** to be measured.

How is total body water measured?

We must use the dilution principle. **Tritium-marked water** is a good tracer. The equilibrium period is 3–6 h.

M mol indicator divided by C_p mol indicator l^{-1} is equal to the distribution volume for the indicator.

Healthy adolescents and children have normal values around 60% of the body weight assuming 1 l of water to be equal to 1 kg. Adult males and females with a sedentary life-style and larger fat fractions contain 55–50% water.

Is it possible to measure the extracellular fluid volume (ECV)?

We administer a **priming dose of inulin intravenously, and then infuse** inulin to maintain constancy of the plasma concentration of inulin (C_p).

The patient then urinates, and the infusion is stopped. For the next 10 h the patient collects their urine, which makes it possible, at the end of the infusion (M mol), to measure all the present body inulin.

Dividing M with C_p gives the volume of distribution after correcting for the difference in protein concentration between plasma and ISF.

Chromium-ethylene-diamine-tetra-acetate (^{51}Cr-EDTA) is a chelate with a structure that cannot enter into cells. The chelate molecule contains radioactive Cr, making it easy to measure. The ^{51}Cr-EDTA distributes and eliminates itself in our body just as inulin. We inject a single dose intravenously, and draw blood samples every hour for 5 h. **ECV is the renal blood flow (RBF) multiplied by the mean transit time.** The clearance of ^{51}Cr-EDTA is independent of C_p and a good estimate of glomerular filtration rate (GFR) just like the **inulin clearance.** Such methods – including renal lithium reabsorption – are important during renal function studies. Normal values for ECV are approximately 20% of the body weight or 14–17 kg.

Chronically ill patients with **debilitating diseases** often retain their ECV in spite of marked reductions in the cell mass of their body.

How is plasma volume measured?

Here, also, the dilution principle is used. The indicator for plasma volume can be **Evans Blue (T_{1824})** which binds to circulating plasma albumin. A small dose of albumin, marked with radioactive iodine, is also a good indicator (^{131}iodine has a physical half-life of 8 days).

We measure the indicator concentration in plasma (C_p) every 10 min for an hour after the administration. Then we plot the log of C_p with time. Extrapolation to the time zero determines the maximum concentration of indicator in plasma. This corrects for the biological loss, while the indicator distributes itself in the plasma phase. The tracer dose divided by C_p at time zero provides us with the **intravascular plasma volume**. Normal values for the plasma volume are close to 5% of the body weight.

In **diabetics and hypertensive patients** the tracer is lost more readily through the capillaries to the ISF than in healthy persons (**increased transcapillary escape**).

How is an elimination rate constant defined?

The rate constant (k) is the fraction of the total amount of a given substance in the distribution volume of the body eliminated per unit time. The **elimination rate constant (k)** for the renal excretion of a substance eliminated solely by glomerular filtration and distributed in ECV, is equal to **GFR/ECV**.

Further reading

Bruun, N.E., M. Rehling, P. Skøtt and J. Giese (1990) Enhanced fractional Na reabsorption in the ischaemic kidney revisited with lithium as a probe. *Scand. J. Clin. Lab. Invest.* 50: 579–85.

54. Case History A

A female patient (age 22 years; weight 71 kg) is in hospital suspected of having a potassium imbalance. She has taken diuretics for 2 years. She is tired and sleepy; her legs are paretic. The ECG shows prolongation of the Q-T interval, depression of the S-T segments and flattening of the T-waves. Her blood pH is 7.57 and the serum K^+ concentration is 2.9 mM. One morning she receives an intravenous injection of a solution containing the radioactive isotope of potassium (555 000 Bq, of $^{42}K^+$ with a physical half-life of 12 h). Following the injection her urine is collected in two periods, 0–12 and 12–24 h. The first urine collection contained 40 mmol K^+ ($^{39}K^+$) and 4144 Bq $^{42}K^+$. The second urine specimen contained 40 mmol K^+ and 2220 Bq $^{42}K^+$. Both urine specimens were analysed for radioactivity exactly 24 h after the injection, where the specific activity of her plasma was 55.5 Bq $mmol^{-1}$. The $^{42}K^+$, retained after the first 12 h, distributes in her body just like all other exchangeable K^+. The body contains traces (0.012% of the total) of naturally occurring radioactivity (^{40}K) with a half-life of $1.3 \cdot 10^9$ years.

1. *Calculate the exchangeable K^+ pool of her body after the 12-h distribution period. Is the result normal?*
2. *Calculate the elimination rate constant (k) for exchangeable K^+ in her body, and the biological half-life for this K^+ in hours. Calculate the ratio between the physical and the biological half-life of K^+.*
3. *What is the cause of her disease?*
4. *Describe a method for measurement of her total body potassium.*

54. Case History B

Two groups of substances are evenly distributed in the ECV of a healthy 25-year-old man. His weight is 70 kg, and his ECV is 14 l. Both groups of substances disappear solely by excretion through the kidneys. His GFR is 120 ml min^{-1}, and his renal plasma flow (RPF) is 700 ml min^{-1}.

1. *Inulin is representative for one family of substances. Inulin is only ultrafiltered in the kidneys. What fraction (k_1) of the total amount of inulin in the body is maximally excreted in the urine per minute?*
2. *The other substances are not only ultrafiltered, but they are also undergoing tubular secretion to such an extent that they totally disappear from the blood during the first passage. What is the elimination rate constant (k_2) for these substances?*

Try to solve the problems before looking up the answers in Chapter 74.

CHAPTER 55. GLOMERULAR FILTRATION

I. RENAL ANATOMY. URINE FORMATION

The urine is formed by three renal processes. Define these processes

The three processes are all defined as vectors transporting substances according to arrows 1, 2 and 3 on the nephron shown in Fig. 55-1:

1. **Glomerular filtration** which is due to a hydrostatic/colloid osmotic pressure gradient – the **Starling forces**.
2. **Tubular reabsorption** (active or passive) which is the movement of water and solute from the tubular lumen to the tubule cells and to the peritubular capillary network.
3. **Tubular secretion** (active or passive) which represents the net addition of solute to the tubular lumen.

The final excretion flux of the substance s in the urine (J_s) is shown in Fig. 55-1.

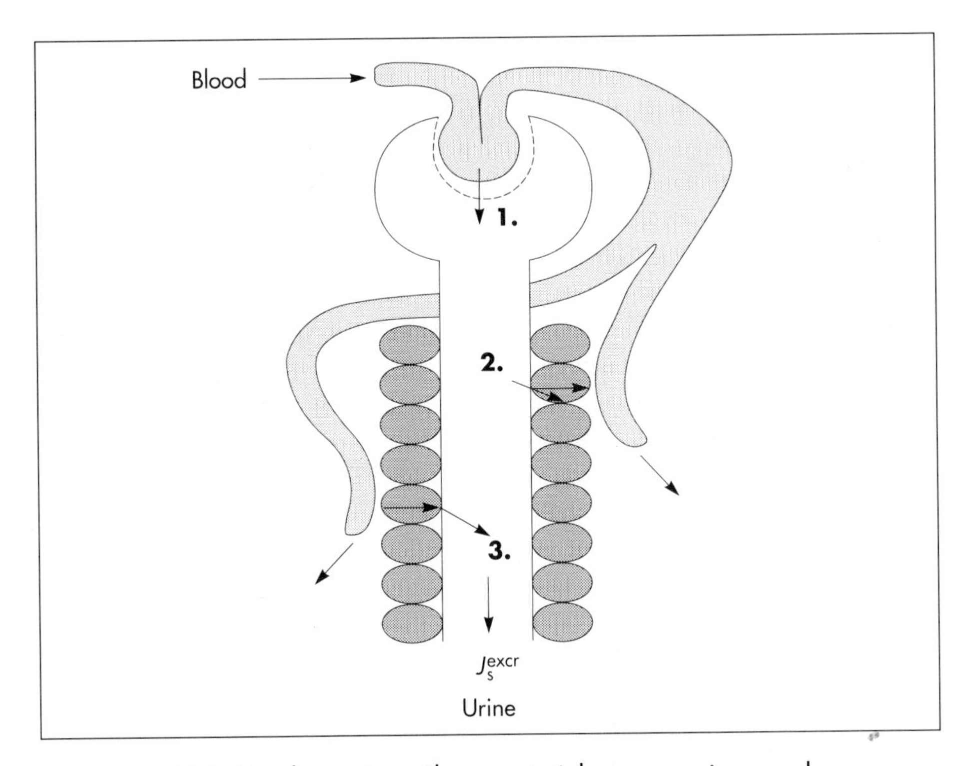

Fig. 55-1. Renal transport. Three vectorial processes in a nephron.

What is the functional unit of the kidney, and how is this structure localized in the kidney cortex and the medulla?

The functional unit is the **nephron**. Each human kidney contains **1 million units** at birth. Each **nephron** consists of a **glomerulus** (i.e. lots of glomerular capillaries in a Bowman's capsule), a **proximal convoluted tubule** ending in a straight segment (**pars recta**) still containing isotonic fluid. The **loop of Henle** is a regulating unit composed of the descending thin limb with its hairpin turn, the ascending thin limb and the ascending thick limb reaching the juxtaglomerular apparatus in the cortex (Fig. 55-2).

The tubule system is convoluted from the macula densa of the juxtaglomerular apparatus, and this is defined as the **distal tubules** ending in collecting ducts. Functionally we have a **distal tubule system** consisting of the **loop of Henle**, and the **distal tubule** (Fig. 55-2). The illustration shows a **collecting duct**, which receives urine from many nephrons. Such a unit is a **kidney lobulus**. Several collecting ducts join to empty through the duct of Bellini into a **renal cup** or **calyx** in the renal pelvis.

What is the difference between a cortical and a juxtamedullary nephron?

The **cortical nephron** represented on the left side of Fig. 55-2 A, does not reach the inner zone of the medulla, because its loop of Henle is short. The small, cortical nephrons have a smaller blood flow and GFR than the **juxtamedullary nephrons** (which are located close to the medulla and comprise 15% of all nephrons). The total inner surface area of **all the glomerular capillaries is approximately 1 m^2**. The capillaries are

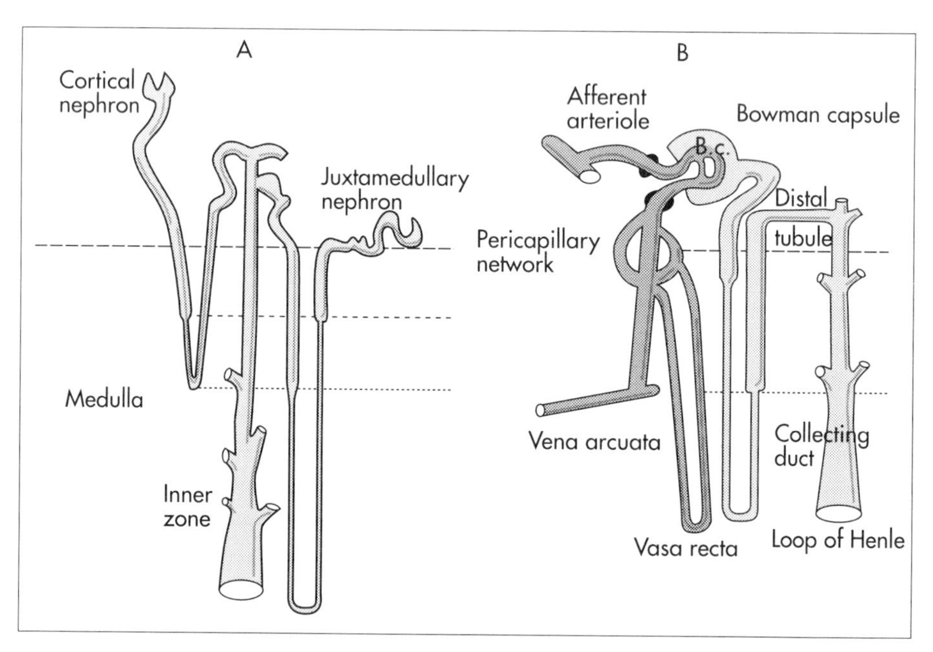

Fig. 55-2. A, A cortical and a juxtamedullary nephron leading to the same collecting duct. B, A juxtamedullary nephron with related blood vessels.

covered with epithelial cells with foot processes (so called **podocytes**). **Mesangial cells** in the glomerulus secrete **prostaglandins** and exhibit **phagocytosis.**

The **proximal tubules have an inner area of 25 m^2**, because they have characteristic microvilli or brush borders, which have carboanhydrase activity. The juxtamedullary nephron has a **long, U-shaped Henle loop**. The bottom of this loop extends towards the peak of the papilla (apex papillae) at the outlet of the collecting duct (Fig. 55-2). The juxtamedullary nephrons have large corpuscles with relatively large bloodflow. These nephrons also have efferent arterioles with large diameters, and afferent arterioles with small diameters. The juxtaglomerular apparatus is located close to the afferent arteriole and to the **macula densa** at the beginning of the distal tubule (Fig.59-2). When the blood has passed the juxtamedullary glomeruli it goes on to a primary capillary network and continues in the vasa recta to the medulla. The blood collects in vena arcuata, vena interlobaris and vena renalis.

What are the major functions of the kidneys?

1. The renal control of **body fluid osmolality** maintains the normal cell volume (ICV). The regulation of **ECV**, including **total blood volume (TBV)**, is essential for the normal function of the circulatory system. These functions are accomplished by renal excretion of water and NaCl. Normally, we excrete **1500 (1200–1800) ml of water and 2–5 g Na$^+$** (= 5–12 g NaCl) daily.
2. Regulation of K^+-balance. The daily intake of K^+ is matched by the renal K^+-excretion. Our daily urine contains **2–5 g K$^+$.**
3. Acid–base balance. The pH of the ICV and the ECV is maintained within narrow limits (many metabolic processes are sensitive to pH). The acid–base balance is accomplished by co-operative action of the kidneys and the lungs.
4. Renal excretion of waste products. Urea from amino acids is excreted with about 30 g or 0.5 mol of urea per day. The daily renal excretion of uric acid, creatinine, hormone metabolites and haemoglobin derivatives matches their daily production.
5. The daily renal excretion of metabolic intermediates and foreign molecules (drugs, toxins, chemicals, pesticides) is carefully matched to the intake or production.
6. Renal secretion of hormones. The kidneys secrete **erythropoietic factor**, renin, kinins, prostaglandins and **1,25-dihydroxy-cholecalciferol.**

What solutes contribute to the urinary osmolarity? How much can the osmolarity vary?

Half the osmolarity of urine is due to **urea**, and the other half is due mostly to **NaCl**. The osmolarity varies tremendously (from 50 to 1300 mOsmol l^{-1}).

II. CLEARANCE

In 1926 Poul Brandt Rehberg – an associate of August Krogh – found creatinine extremely concentrated in human urine (C_U mg ml^{-1}) compared to plasma (C_P mg ml^{-1}). He also measured the urine production per minute or **minute diuresis**. Square brackets around a substance denote concentration as the symbol C does.

Thus, the C_U/C_P ratio is great for creatinine. Multiplying this ratio with the **minute diuresis** (or $\mathring{V}_U$ ml min^{-1}) yields a result $((C_u \cdot \mathring{V}_u)/C_p)$ greater than similar results derived for most other substances. Brandt Rehberg used this result as his measure of **renal filtration rate**. The idea of a **filtration–reabsorption type of kidney** was developed. Rehberg was the first to realize that reabsorption in the proximal tubules controls filtration. A few years later Rehberg's **renal filtration rate** was called **creatinine clearance** and used as a measure of the **GFR**.

What is the renal plasma clearance for a certain substance?

Clearance is a **cleaning index for blood plasma** passing the kidneys. The efficacy of this cleaning process is directly proportional to the excretion flux $(C_u \cdot \mathring{V}_u)$ for the substance, and inversely proportional to its C_P. **Clearance** or the **cleaning index** is the ratio between excretion flux and plasma concentration.

The classical definition of **clearance** is: **volume of arterial plasma completely cleared of the substance in 1 min**. Of course, this is not a real volume, since no single ml of plasma has all molecules of the substance completely removed in one renal transit.

We developed the clearance formula above:

$$\textbf{Clearance} = (C_u \mathring{V}_u)/C_p \; [(\text{mg ml}^{-1})(\text{ml min}^{-1})/(\text{mg ml}^{-1})].$$

Clearance can also be thought of as the number of ml arterial plasma containing the same amount of substance as contained in the **minute diuresis**.

What is renal ultrafiltrate compared to arterial plasma?

Ultrafiltrate is plasma minus proteins – so-called **plasma water**. The fraction of plasma that is pure water is 0.94. Thus, the concentration of many substances in the ultrafiltrate, C_{filtr}, is equal to $C_p/0.94$.

What is the GFR?

GFR is the **volume of glomerular filtrate produced per minute**. In man the GFR is about 180 l per day (125 ml per min).

Why is inulin an ideal indicator for GFR measurements?

1. Inulin **passes freely** through the glomerular barrier.
2. All ultrafiltrated inulin molecules pass to the urine. In other words, they are neither reabsorbed nor secreted.
3. Inulin is **without effect on GFR**, it is non-toxic and easy to measure. Inulin is uncharged and not bound to proteins in plasma. Inulin crosses freely most capillaries and yet does not traverse the cell membrane. Since 1 l of plasma contains 0.94 l of water, the ultrafiltrate concentration of inulin is $C_p/0.94$.
   All of the above properties satisfy the criteria for an ideal indicator for GFR measurements.

How are GFR and ECV measured by use of the clearance for inulin?

1. The main idea is to measure the **amount of inulin excreted in the urine during a time period where the plasma [inulin] is constant**. This can be done by infusing

inulin intravenously until the plasma concentration is constant. The subject urinates first to eliminate any urine accumulated before infusion. The infusion continues, and after a while the subject urinates again, and the urine volume and [inulin] of this sample is measured. The amount of inulin filtered through the glomerular barrier per minute is: (**GFR** $\cdot$ **$C_p/0.94$**). All inulin molecules remain in the preurine until the subject urinates. Thus, the amount excreted ($C_u \cdot \mathring{V}_u$) is equal to the amount filtered:

$$\text{GFR} \cdot C_p/0.94 = C_u \cdot \mathring{V}_u.$$

$$\textbf{GFR} = (C_u \cdot \mathring{V}_u / C_p) \times 0.94 = \textbf{CLEARANCE}_{\textbf{inulin}} \times 0.94.$$

2. **The ECV** can be measured if all inulin molecules are collected in the urine for a couple of hours after the inulin infusion stopped. The **inulin distribution volume** is more or less identical to the ECV.

$$\textbf{ECV} = \textbf{Amount of inulin excreted}/(C_p/0.94).$$

What is the size of the inulin clearance and GFR in normal men?

The **inulin clearance** is 180 l per 24 h for young, healthy men or 125 ml min^{-1}. Thus the **GFR** is $(125 \cdot 0.94) = 118$ ml min^{-1}.

Are these values sex and age dependent?

The inulin clearance is 10% lower for young females than for young males, except when their body weights and body surface area are equal. The values decrease with age to **70 ml min^{-1} after the age of 70.**

Is the endogenous creatinine clearance a measure of GFR?

At the normally low plasma concentrations of creatinine the modest secretion of creatinine is detectable resulting in up to 15% over-estimation. The creatinine clearance is therefore only slightly higher than the inulin clearance, so the creatinine clearance is clinically **acceptable as a GFR estimate,** when corrected for the 15% over-estimation.

What principles are used for estimating the renal filtration capacity?

1. **The endogenous clearance method** for creatinine obviates the need for intravenous infusion.

   We must collect the urine for 6–24 h and measure the **creatinine excretion flux.** The creatinine production is from the creatine metabolism in muscles and the production is proportional to the muscle mass. Creatinine is produced at a relatively constant rate (1.2 mg min^{-1} or 1730 mg per day in a 70 kg person). A venous blood sample for C_p is also necessary:

$$\textbf{GFR} = (\textbf{CLEARANCE}_{\textbf{creatinine}} \cdot 0.94)/1.15.$$

   Note that we have corrected for the 15% creatinine secretion.

2. **Serum [creatinine]** and **serum [urea].** These concentrations depend upon both **protein turnover** and **kidney function.** The serum [creatinine] and [urea] are large after intake of meals rich in meat.

3. **The elimination rate constant, *k***, for compounds eliminated as **inulin** (the **inulin family**: ^{51}Cr-EDTA, ^{57}Co-marked vitamin B_{12}, ^{14}C-marked inulin, ^{3}H-marked inulin). Inulin is uncharged and unbound in plasma.
   The ECV is the distribution volume for inulin. One of these indicators is infused continuously until C_p is constant. When the infusion stops, the fall in C_p is followed with time. The elimination is exponential – i.e. the fraction (k) of the remaining amount in the body that disappears per time unit is constant. Since the **inulin family** of substances is eliminated solely by filtration, the elimination depends upon GFR, and the distribution volume of inulin (ECV). Thus, k = **GFR/ECV** for the inulin family of substances. $T_{\frac{1}{2}} = 0.69/k$.
4. **Inulin clearance** is a precise experimental measure, but inulin must be infused intravenously which limits its clinical use.

What is the passage fraction and the absorption fraction for a certain substance?

The fraction of the ultrafiltrated amount of substance passing a cross-section of the nephron is the **passage fraction**. The **reabsorption fraction** is the reverse of the passage fraction (1 minus the passage fraction).

How does the passage fraction for inulin vary through the nephron?

The passage fraction for inulin **does not vary at all.**

This fraction is **1** and remains so. After glomerular filtration not a single inulin molecule – theoretically – reaches the tubular fluid from the blood, or escapes from the tubular fluid from the blood.

What is the variation of the passage fraction for glucose in the proximal tubules?

The proximal tubules rapidly reabsorb glucose, so the **passage fraction** is 0 half way down the tubules.

For water and for NaCl?

For water and for NaCl the **passage fraction** is about 0.3 at the outlet of the proximal tubules.

For urea and for bicarbonate?

For urea the **passage fraction** is about 0.5, and for bicarbonate only 0.1 at the outlet of the proximal tubules.

What is the excretion fraction (EF) for a substance?

The excretion fraction (EF) is the fraction of its glomerular filtration flux, which passes to and is **excreted** in the urine.

$$\mathrm{EF} = J_{excr}/J_{filtr};\ \mathrm{EF} = C_u \mathring{V}_u/(\mathrm{GFR}\, C_{filtrate}).$$

$C_{filtrate}$ is the concentration of the substance in the ultrafiltrate. The **excretion fraction for inulin** is 1.

Substances with an EF above 1 have a **net secretion**. Substances with an EF below 1 are subject to **net reabsorption**.

How is the extraction fraction (E) for plasma defined?

The extraction fraction (**E**) is the fraction **extracted** from the total substance flux in the renal blood.

$$E = J_{filtr}/J_{total} = (C_a - C_{v.r.})/C_a.$$

C_a and $C_{v.r.}$ are the concentrations in the renal artery and the renal vein blood, respectively.

Substances with an **E** of **1** are cleared totally from the plasma during their first passage of the kidneys. Inulin has an extraction fraction of 0.2.

What is the glomerular filtration fraction (GFF)?

The glomerular filtration fraction (**GFF**) is the fraction of the plasma flowing to the kidneys that is ultrafiltered (**GFR/RPF**). GFF is normally 0.20. The remaining 80% of the blood plasma is returned to the systemic circulation. The GFF is reduced during acute glomerulonephritis (see Chapter 62).

Further reading

Rehberg, P. Brandt (1926) Studies on kidney function: I. The rate of filtration and reabsorption in the human kidney. *Biochem. J.* **20**: 447.

Vander, A.J. (1991) *Renal Physiology*, 4th edn. McGraw-Hill, New York.

55. Case History

A male patient with a body weight of 100 kg receives a bulk dose intravenously of inulin mixed with a substance named X. A saline solution with both substances is continuously infused to keep their plasma concentrations constant.

The average plasma concentrations are 225 mg inulin l^{-1} and 1 mg X l^{-1} in a period from 60 to 180 min. Some of the molecules of X binds to plasma proteins, but 90% are free.

The minute-diuresis is 2 ml urine min^{-1}, and the inulin concentration is 20 mg ml^{-1} urine. The concentration of X is 400 μg ml^{-1} urine.

1. *Calculate the clearance for inulin and for X. Is the inulin clearance normal?*
2. *Calculate the ultrafiltrated flux of X.*
3. *Calculate the mass of X excreted in the 120-min period.*
4. *How is the substance X treated in the tubules? Calculate the difference between the filtration flux and the excretion flux.*

Try to solve the problems before looking up the answers in Chapter 74.

Chapter 56. Tubular Reabsorption

The passage fraction for both Na^+ and water falls to 0.3 through the proximal tubules as described in Chapter 55. Thus **70% of the filtrate volume** is reabsorbed in the proximal tubules of healthy kidneys. The filtrate consists mainly of Na^+, anions and water, so the filtrate must remain isotonic in the proximal tubules. A constant percentage (fraction) of salt and water is thus reabsorbed in the proximal tubules regardless of the size of GFR. This effect is termed the **glomerulo-tubular balance**.

How are water and ions transferred through the epithelial layer of the proximal tubules?

The **Na^+-K^+-pump** is located in the **basolateral exit-membrane** of the cell (see Chapter 1). The **primary active ion-transport** provides energy for the secondary **water absorption**. The Na^+-K^+-pump in the **exit-membrane** drives the luminal transport across the **entry membrane** (see Chapter 1: Fig. 1-1).

This transport of NaCl and water is nearly isotonic. The bulk flow can take place against a large osmotic gradient, and increases in diluted solutions. The **entry membranes**, which have the **leaky junctions**, are highly permeable to water. Zeuthen's hypothesis concerning the ionic mechanism driving water through an exit-membrane is presented in Chapter 1. The Na^+-K^+-pump builds up a high electrochemical gradient for K^+ and indirectly for Cl^-. Zeuthen couples the water outflux to the outward transport of K^+ and Cl^-. Each ion passes with **at least 400 water molecules** to the ISF and blood (Fig. 1-1).

The ISF receives ions and glucose, causing its osmolarity to increase. The osmotic force will transfer water from the tubular fluid through the cells and tight junctions to the ISF, where the hydrostatic pressure will rise. This hydrostatic force transfers the **bulk of water, ions and molecules** through the thin-walled, tubular capillaries to the blood. When excess of water (solvent) passes through tight junctions, they lose part of their **tightness** and **water drags many Na^+/Cl^- ions out** by **solvent drag**. The mechanism is analogous to the transport mechanisms in the small intestine, salivary glands and sweat glands.

What substances change their concentrations in the fluid passing the proximal tubules?

The **passage fraction** for NaCl and K^+ is **0.3** at the outlet – the same as for water. Since the concentrations of these substances do not change much here, they must be reabsorbed in the same proportion.

1. The co-transport of glucose with Na^+ clears the tubular fluid of glucose already in the first part of the proximal tubules. The maximal reabsorption capacity (T_{max}) for this **glucose transport protein** is given below.
2. The water reabsorption increases the [urea] in the tubule fluid. Therefore, tubules reabsorb urea passively.

Since **urea diffuses quite easily**, it will diffuse from the proximal tubules to the blood. This results in a passage fraction of 0.5 with a rise in concentration of **67%** (0.5/0.3 = 1.67) at the end of the proximal tubules.

3. If **inulin and para-amino hippuric acid (PAH) molecules** are present, their concentrations will rise. The [inulin] increases by a factor of **3.33** (1/0.3), and that of PAH by much more. The passage fraction of PAH increases to 5 due to proximal secretion. Since the tubular volume falls to 0.3, the concentration of PAH must rise by a factor of **16.7** (5/0.3).

What is the transport mechanism for glucose and other organic substances through the tubule cells?

The substances belonging to the **glucose family** comprise **glucose, amino acids, acetoacetates, ascorbic acid, β-hydroxybutyrate, lactate, phosphate, sulphate and carboxylate.** The tubule cell membranes contain **special transport proteins** which reabsorb each substance coupled with Na^+.

There is an electrochemical Na^+ gradient (inside negative), which provides just enough electrochemical energy to transport glucose against its concentration gradient.

The electrical force transfers a **positive net charge (Na^+)** from the lumen to the cytoplasm of the cell.

What is Tmax for glucose?

This is the maximum reabsorption flux (J_{reabs}) for glucose in the proximal tubules (Fig. 56-1). The optimal value for this **glucose transporter** is 320 mg min^{-1} or 320/180 =**1.78 mmol min^{-1}** for healthy, young subjects with a body weight of 70 kg.

How does the reabsorption and the excretion flux for the glucose family vary with increasing plasma concentration?

For the **reabsorption family**, the excretion flux is zero at first because all the filtered load is reabsorbed (all glucose is reabsorbed, see Fig. 56-1). The excretion flux increases then linearly with the high filtration flux (Fig. 56-1). For the reabsorption family of compounds, the excretion flux is equal to the filtration flux minus T_{max}: $J_{excr} = J_{filtr} - T_{max}$. Dividing all three expressions by C_p provides us with: $J_{excr}/C_p = J_{filtr}/C_p - T_{max}/C_p$. Thus the clearance for the **reabsorption family** is equal to the **inulin clearance – T_{max}/C_p**. The **inulin clearance** or J_{filtr}/C_p is also equal to **GFR/** 0.94.

What is the renal threshold for glucose?

The **appearance threshold** is the blood [glucose] at which the glucose can be first detected in the urine (normally 1.5 g l^{-1} or 8.3 mM). This occurs when most nephrons are saturated (Fig. 56-1).

The actual **saturation threshold**, the point where all nephrons are saturated, is much higher (normally 2.2 g l^{-1} or 12.2 mM). The concentration difference (12.2 – 8.3 = 3.9 mM) represents a similar flux difference (1.78 – 1.22 mmol min^{-1}) called **splay**. The reabsorption capacity for glucose in the proximal tubule cells becomes saturated at these high blood concentrations (Fig. 56-1).

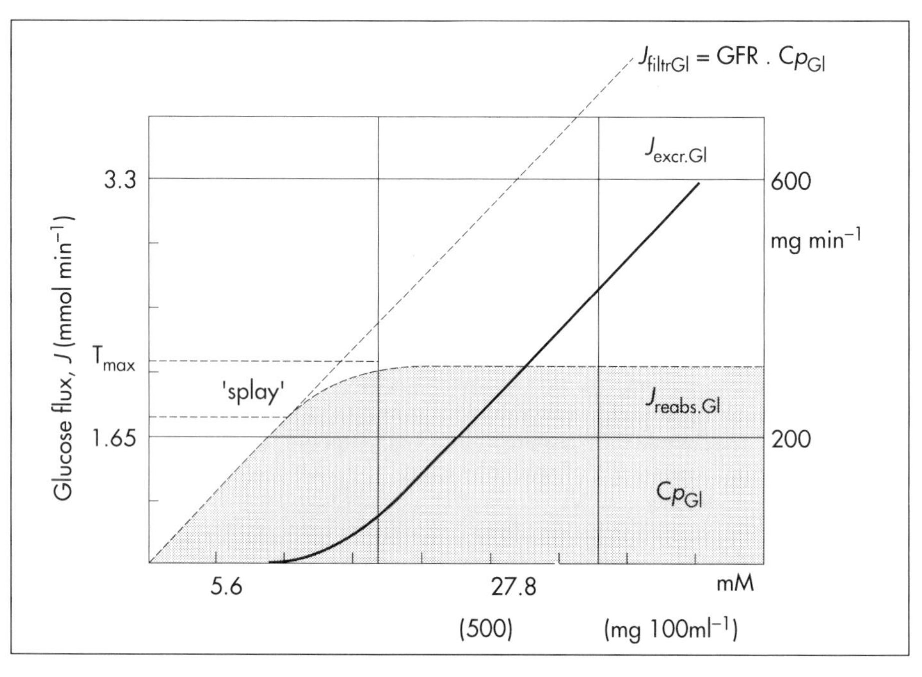

Fig. 56-1. Renal glucose fluxes as a function of the plasma concentration (C_p).

Are the proximal tubules involved in osmotic diuresis?

The presence of **impermeable substances** (sulphate, mannitol, sucrose, inulin, glucose above the threshold) in the tubular fluid will **inhibit the tubular reabsorption of water and Na^+**. Since 70% of the **bulk reabsorption** takes place in the proximal tubules this has severe consequences. Therefore, more than half the glomerular filtrate volume will be found in the urine (i.e. GFR=120 and $\mathring{V}_u$ =60 ml min^{-1}) instead of 1 ml min^{-1}.

How great is the osmolarity of the renal medulla?

The renal cortex is isotonic with the plasma. There is a graded increase in the osmolarity of the medullary interstitium going from the outer zone of the medulla to the tip of the papilla. The dominant solutes are urea, Na^+ and Cl^- (Fig. 56-2). The total osmolarity in the medullary interstitial tissue can be as high as **1400 mOsmol l^{-1} close to the papillae** in situations where the urine is very concentrated (Fig. 56-2).

When the preurine in the Henle loop passes down through the hypertonic medulla, water moves out into the medullary interstitium, due to the medullary osmotic gradient, making the urine concentrated. In man the maximal urine osmolarity is 1300–1400 mOsmol l^{-1}, which in a daily urine volume of **0.5 l** corresponds to a daily solute loss of 700 mOsmol. When ADH is absent, the preurine leaving the distal tubules remains hypotonic. **Large amounts of hypotonic urine** then flow into the renal pelvis (down to 50 mOsmol l^{-1}). A daily solute loss of 700 mOsmol, under this circumstance, implies a daily water loss of **14 l.**

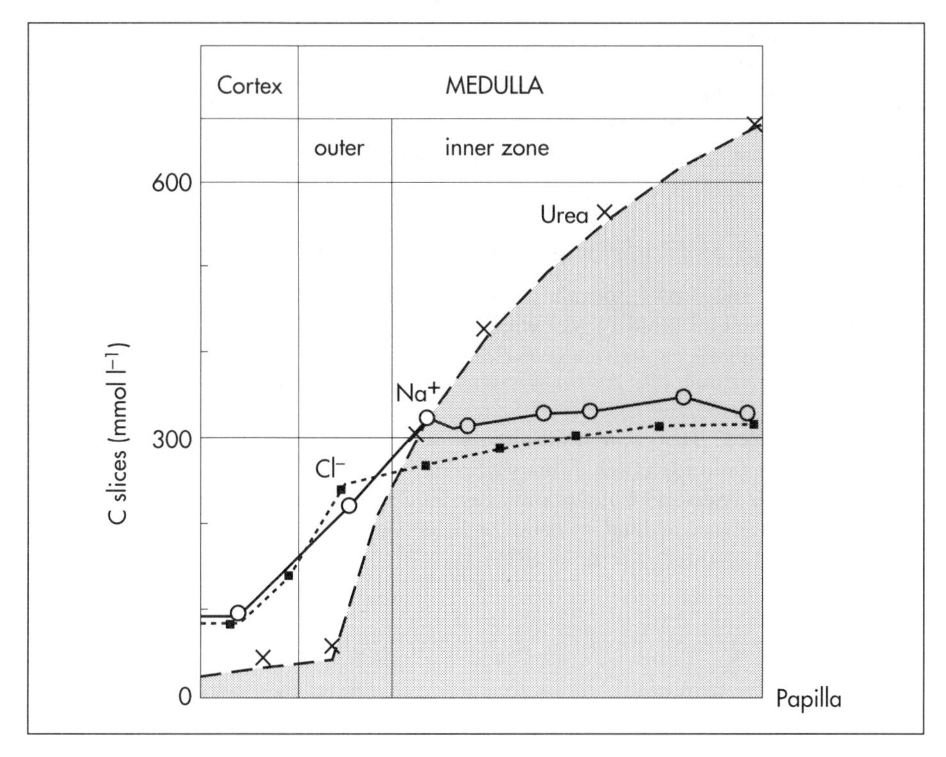

Fig. 56-2. Concentrations of solutes in slices of kidney tissue measured by microcryoscopy.

Describe the mechanism by which inorganic ions are reabsorbed in the thick ascending limb of Henle

A co-transport of solutes (Na^+, K^+, 2 Cl^-) occurs through the **luminal entry-membrane.** This co-transport is part of the transepithelial **single effect** at each level of the thick ascending limb. Intracellular [Na^+], [K^+] and [Cl^-] increase. The electrochemical energy for the function of the **basolateral Na^+-K^+-pump** is provided by its Na^+-K^+-ATPase. The pump throws Na^+ into the peritubular fluid. The K^+ and Cl^- ions leak out passively. The **basolateral Na^+-K^+-pump** can be inhibited by **loop diuretics** (furosemide) which abolish the entire osmolar gradient in the outer renal medulla (see diuretics p. 455).

What is the single effect in the loop of Henle?

We have many **solute pumps** at each level along the **thick ascending limb of Henle** which is found in the outer medulla. Such a solute pump and related co-transport is a transepithelial **single effect** (from german **Einzel-effect** describing the concentration gradient established at each level). Normally, the solute pump provides a concentration gradient of 200 mOsmol l^{-1} between the two limbs of the Henle loop at a given level.

The **thick ascending limb is impermeable to water**, and its cells actively transfer Na^+ to the interstitial fluid establishing the (200 mOsmol l^{-1}) gradient at each level.

The **descending limb** is highly permeable to water but only slightly permeable to solutes. There is no active transport through these thin epithelial cells. Water diffuses passively from the descending limb to the elevated osmolarity of the ISF of the outer medulla.

How are the loops of Henle able to act as a counter current multiplier?

When isosmotic urine flows from the proximal tubule through the Henle loop, we have a **single effect** of 200 mOsmol l^{-1} at each level of the thick ascending limb. The input to the descending loop in the outer medulla is 300 mOsmol l^{-1} and the output to the distal tubule is 100 mOsmol l^{-1}. At the bottom of the Henle loop the concentration can increase to at least 1300 mOsmol l^{-1}. The total osmotic gradient along the system is thus (1300 − 100) =**1200**. This is **6 multiples** of the **200 mOsmol l^{-1} single effect.** The **counter current fluid exchange** of the vasa recta reduces the washout of solute from the ISF. The **NaCl is reabsorbed again and again in the thick ascending limb of the Henle loop**. The counter current fluid exchange and the counter current NaCl reabsorption are combined in the **counter current multiplier**.

How is the concentration gradient dependent upon the vasa recta?

The vasa recta are involved in the concentration of solutes in the renal medulla. They receive blood from the efferent arterioles and consequently have an elevated colloid osmotic pressure and reduced hydrostatic pressure. The net force in these vascular loops favours **net fluid reabsorption**. The blood is then passed on in the direction of increasing medullary osmolarity. After the U-turn the vasa recta conduct blood in the direction of falling osmolarity.

Accordingly, the input blood must supply water to the hyperosmolar, interstitial fluid and passively reabsorb solutes. The reverse occurs in the ascending limb. **Solute** (NaCl/urea), that enters the vasa recta with water, subsequently diffuses down the medullary concentration gradient from the ascending into the descending vascular limb, thus minimizing solute loss from the medulla. The blood leaving the vasa recta absorbs water osmotically and delivers solutes to the interstitium.

The gross effect is a **water shunt** passing the medullary tissue. This avoids washing out the medullary concentration gradient (Fig. 56-3). The vasa recta operate as a **passive countercurrent exchanger**. The meagreness of the medullary bloodflow, and its reduction by ADH, also contribute to the maintenance of the medullary concentration gradient.

Does urea promote water reabsorption in the inner medulla?

Water enters the medullary interstitium by two routes – the **descending limb of the Henle loop** and all three sections of the **collecting duct**. The effective osmotic pressure across the inner medullary collecting duct depends on the difference in the [NaCl] between the interstitium and the tubular fluid. Urea does not promote this water reabsorption, because the **inner medullary duct** is permeable to urea in the presence of ADH.

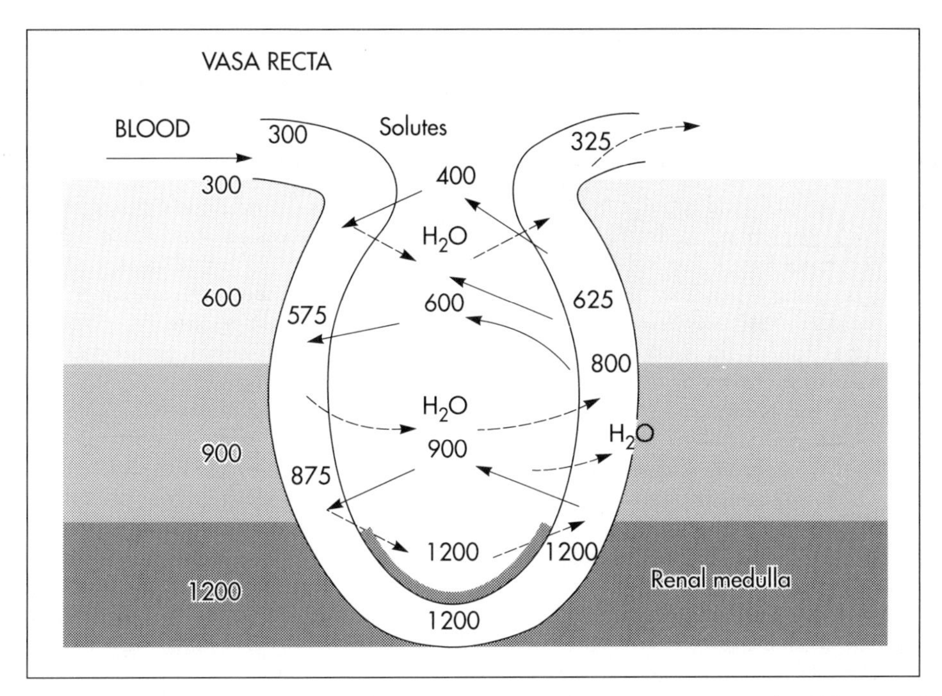

Fig. 56-3. Counter current exchange occurs in the vasa recta, with diffusion of solutes along thick arrows. Passive osmotic flux of water from the blood to the hyperosmolar interstitial tissue occurs along stippled arrows (numbers are mOsmol l^{-1}).

Is urea essential for the medullary concentration gradient in the kidneys?

Energy is necessary to establish a concentration gradient. The energy is from the **basolateral Na^+-K^+-pump** of the **thick ascending limb of Henle**. The thick ascending limb is **impermeable to urea and water**.

Urea constitutes 50% of the maximum concentration gradient. The urea contribution to the medullary concentration gradient is based on energy from the basolateral Na^+-K^+-pump of the thick ascending limb. The kidney reuses urea by recirculation in the **intrarenal urea circuit**.

The cortical and outer medullary collecting ducts are **not permeable to urea** whether or not ADH is present. Here, any water reabsorption (ADH dependent) increases the concentration of urea in the tubular fluid. Later, the **inner medullary collecting ducts reabsorb urea**. They are **leaky** for urea, so urea passes into the interstitium thus building up the medullary gradient. Part of the urea in the ISF diffuses back to the urine through the **thin ascending limb**.

Is a high concentration gradient vital?

Without the gradient we are unable to produce a concentrated urine, and we would quite easily **die** of water depletion.

A high osmolality in the medullary interstitium allows passive water reabsorption when ADH is present. ADH increases the concentration of solutes in the collecting ducts, and reduces the loss of water.

What are the consequences of tubular Na^+ reabsorption?

1. The extent of the **tubular Na^+ reabsorption** determines the magnitude of the ECV. **Na^+ depletion** leads to shrinkage of ECV, hypotension or shock. **Na^+ accumulation** leads to expansion of ECV, circulatory overload and oedema formation. When the filtration flux of Na^+ is increased, both the proximal and the distal reabsorption increase, thus minimizing the increase in urinary excretion. Expansion of the ECV inhibits tubular Na^+ reabsorption (mainly proximal), and leads to **increases in urinary Na^+ excretion**. This response is aided by increases in GFR and reductions in aldosterone output, but **natriuretic peptides** (inhibiting the tubular reabsorption) may contribute as well. Redistribution of the blood to shorter cortical nephrons with smaller reabsorptive capacity will also have the same effect.
2. The **Na^+ reabsorption** in the **thick ascending limb** of the Henle loop is critical for the extremely high medullary osmolarity. It is important for **water reabsorption**.
3. **The proximal Na^+ reabsorption** is coupled to any one of the following secondary reabsorption processes: **glucose, amino acids, Cl^- and H^+ secretion**. A special **glucose transport protein** (GLUT 2) transfers glucose and Na^+ to the tubule cells. The filtered load of Na^+ per day is $(140 \cdot 180\ \text{l day}^{-1}) = 25\ 000\ \text{mmol day}^{-1}$. Only 100–200 mmol Na^+ are excreted per day.

What characterizes the tubular K^+ reabsorption?

The total body potassium is 4 mol, and only a few per cent of the K^+ are outside the cells and subject to control. As seen above more than 99% of the filtered Na^+ is reabsorbed, and less than 1% is excreted in the urine. However, K^+ excretion is normally 20 to 30% of the filtered load. The main K^+ transfer is passive and paracellular through **tight junctions** of the proximal tubules. Moreover **K^+ excretion** can vary over a wide range from almost complete reabsorption of filtered K^+ to excretion fluxes in excess of filtered load, i.e. **net secretion of K^+. Normally the passage fraction is larger at the inlet of the collecting ducts than at the outlet, showing a net reabsorption here. In conditions with low serum [K^+], a K^+-pump in the medullary collecting ducts seem to reabsorb K^+.**

The **excretion of K^+** by overload is almost entirely determined by the **extent of distal secretion**. Any rise in serum [K^+] immediately results in a marked rise in K^+ secretion. The **passive part of this secretion** is driven by a favourable electrical gradient, since the tubular lumen is negative with respect to the outside surface of the distal tubules and collecting ducts. The **secretion of K^+** is coupled to the distal **Na^+ reabsorption**, since the electrical gradient is established by distal Na^+ reabsorption. This transport mechanism is controlled by **aldosterone** (see Chapter 70).

Describe four factors affecting distal tubular handling of K^+

1. **Body stores of K^+**. Low K^+ diet enhances K^+ reabsorption, whereas acute or chronic K^+ administration enhances K^+ secretion.
2. **Distal Na^+ load**. High distal Na^+ load enhances K^+ secretion, and conversely maximal Na^+ conservation limits K^+ secretion.

3. **Acid–base balance**. Alkalosis enhances while acidosis inhibits K^+ secretion. This is because a high intracellular H^+ concentration will compete with K^+ in the secretion mechanism, whereas a low $[H^+]$ will allow more K^+ to be secreted.
4. **Aldosterone**. Aldosterone enhances both Na^+ reabsorption and K^+ secretion in the distal tubules (Chapter 70).

What characterizes the proximal tubular reabsorption?

1. The Na^+ reabsorption is active, and 70% of the filtered Na^+ and water is reabsorbed here.
2. The reabsorption of fluid is **isosmotic.**
3. The Cl^- reabsorption is **passive.** This ion follows Na^+ in order to maintain electrical neutrality.
4. Reabsorption of **water is passive** as the result of the osmotic force created by the absorption of NaCl.
5. The descending limb of the Henle Loop has a **high water permeability** but is **much less permeable to solutes (Na^+ and urea).**
6. About 30 g of plasma albumin pass through the glomerular barrier each day. Fortunately most of this albumin is absorbed through the brush border of the proximal tubules by pinocytosis. Inside the cell the protein molecule is digested into amino acids, which are then absorbed by **facilitated diffusion** through the basolateral membrane.

What characterizes the distal reabsorption?

1. The **Na^+ reabsorption** in the thick ascending limb and the distal convoluted tubule is active.
2. The Na^+ and water reabsorption here are not tightly coupled. The **thick ascending limb, the distal convoluted tubule and the cortical collecting ducts** are much less permeable to water than to solutes, so that active Na^+ reabsorption is not accompanied by proportional amounts of water. This is what may reduce the $[Na^+]$ in the tubular fluid left behind.
3. The permeability to water in the late distal tubule and the cortical collecting ducts is minimal in the absence of ADH, but **importantly, is increased by the presence of ADH** (vasopressin) in the blood.
4. The **amounts of Na^+ reabsorbed** in the distal system are **much less** than in the proximal.
5. The Na^+ reabsorption in the distal system is increased by the adrenocortical hormone, **aldosterone.**

What is the function of vasopressin (ADH) in the kidneys?

See Chapters 60 and 64.

What is the function of vasopressin in the rest of the body?

Vasopressin is a **universal vasoconstrictor** causing blood pressure to rise.

Has aldosterone any function in the kidneys?

Aldosterone is a mineralocorticoid, which **promotes the reabsorption of Na^+ (and thus Cl^-) and the secretion of K^+ (and H^+) in the distal tubules.**

What is the function of aldosterone outside the kidneys?

Like the glucocorticoids, aldosterone also **increases the working capacity** of muscle fibres. Aldosterone **promotes the reabsorption of Na^+ (and thus Cl^-) and the secretion of K^+ (and H^+) in the collecting ducts of sweat and salivary glands** just as in the distal tubules of the kidney. Aldosterone-antagonists (spironolactone) inhibit all aldosterone effects.

Further reading

Berne, R.M. and M.N. Levy (1993) *Physiology*, 3rd edn. Mosby Year Book, St Louis.

Jamison, R.L. (1987) The renal concentrating mechanism. *Kidney Int.* **32** (Suppl. 21): S43–S50.

Zeuthen, T. (1992) From contracting vacuole to leaky epithelia. Coupling between salt and water fluxes in biological membranes. *Biochim. Biophys. Acta* **1113**: 229–58.

56. Case History A

A man with a body weight of 70 kg receives a dose of 1 mmol inulin (MW 5000) intravenously. Inulin is continuously infused at a rate of 50 mg min^{-1}. Half an hour later the patient urinates. The plasma [inulin] is 0.4 mg ml^{-1} from a sample of blood. Thirty minutes later, the patient has an inulin concentration of 0.5 mg ml^{-1} plasma, and he again empties his bladder. This urine sample has a volume of 70 ml with concentrations of inulin and glucose of 12 mg ml^{-1} and 250 mM (45 mg ml^{-1}), respectively. The mean glucose concentration in plasma is 21.11 mM or 3.8 g l^{-1}.

1. *Calculate the inulin clearance for this man. Is the calculated value normal?*
2. *Calculate the maximal (proximal) reabsorption capacity for glucose (T_{max}). Is the calculated value normal?*
3. *Does the above calculated T_{max} value suggest a functional damage? What is the relation between the glomerular and the tubular function?*

56. Case History B

A male with a total body water phase of 42 kg combusts 100 g of protein daily. He produces an equivalent amount of urea (molecular weight: 60) and he is in urea balance. Urea is evenly distributed over the total water phase. Urea is excreted by the kidneys at a rate of 60 ml plasma totally cleared of urea per minute.

1. *Calculate his renal excretion of urea in grams per 24 h.*
2. *Calculate the maximal fraction of urea eliminated per minute (the elimination rate constant, k) of the total urea prevailing in the body.*

Try to solve the problems before looking up the answers in Chapter 74.

CHAPTER 57. TUBULAR SECRETION

How does the secretion and excretion flux for the PAH family vary with increasing plasma concentration?

Substances secreted like PAH constitute the **secretion or PAH family**. The filtration flux (J_{filtr}) increases in direct proportion to the rise in C_p (Fig. 57-1). The excretion flux ($J_{excr} = \mathring{V}_u \cdot C_u$) is equal to the filtration flux plus the secretion flux: $J_{excr} = J_{secr} + J_{filtr}$. Dividing the excretion flux with C_p provides us with the **PAH clearance**. The clearance is the slope of the excretion flux curve, and it is obviously a constant value that is almost independent of C_p (Fig. 57-1). The secretion flux approaches a maximum (Fig. 57-1). The excretion flux of PAH must reach high levels of [PAH] before the plasma protein binding effect is negligible, so the excretion flux curve is probably not linear before the [PAH] is high (Fig. 57-1).

The secretion of organic acids and bases is restricted to the proximal tubules. What is the mechanism?

Organic acids and bases secreted in the proximal tubules include endogenous substances and drugs. The endogenous substances include adrenaline (Ad), bile salts, cAMP, creatinine, dopamine, hippurates, NA, oxalate, prostaglandins and urate. The drugs comprise acetazolamide, amiloride, atropine, chlorothiazide, cimetidine, diodrast,

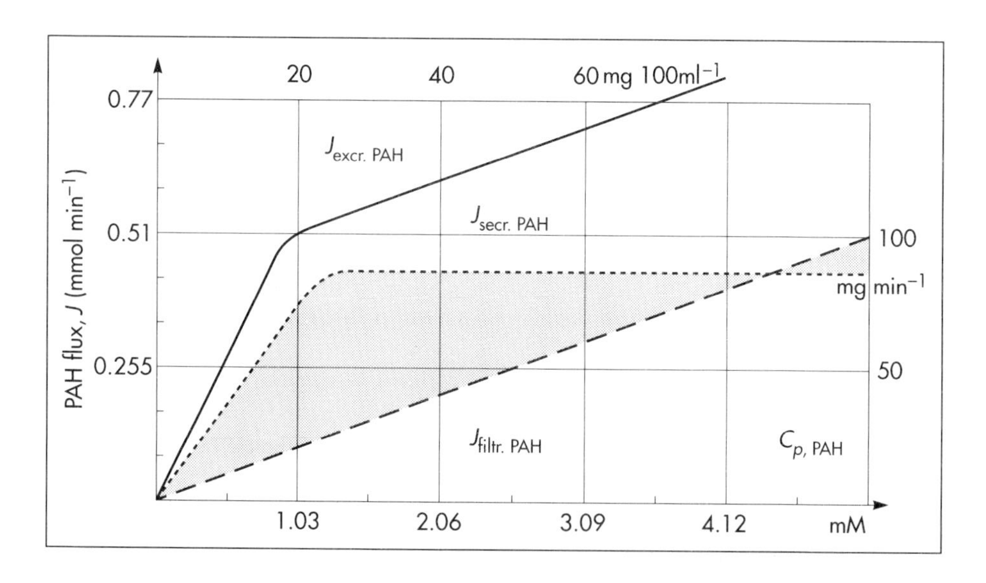

furosemide, hydrochlorothiazide, morphine, para-aminohippuric acid (PAH), penicillin, probenecid, and acetylsalicylic acid. All these substances have varying but high affinity to an **organic acid–base secretory** system in the proximal tubule cells showing saturation kinetics with a T_{max}. These molecules leave the blood of the tubular capillaries and bind to basolateral receptors with symports on the tubule cell. These **channels** are driven by energy from the **basolateral Na^+-K^+-pump** transporting the molecules against their chemical gradient across the basolateral membrane. Inside the cell the molecules accumulate until they can diffuse towards the luminal membrane. Here, an **antiport** transfers the ions into the tubular fluid. All these molecules compete for transport, so intake of the drug probenecid can reduce the penicillin secretion.

The luminal membrane **contains specific receptor proteins** for **nutritive mono- and dicarboxylates**. These receptor functions are also coupled to **Na^+ transfer**.

What is understood by T_{max} for PAH?

T_{max} is the maximum secretion flux for PAH in the tubules (Fig. 57-1). Normally the T_{max} is 0.40 mmol min^{-1} (80 mg min^{-1}) for PAH. For the **secretion or PAH family** at T_{max}, the excretion flux is equal to J_{filtr} plus T_{max}. Dividing all three expressions by C_p provides us with: $J_{excr}/C_p = J_{filtr}/C_p + T_{max}/C_p$. The expression J_{filtr}/C_p is actually inulin clearance or GFR/0.94. Thus, the clearance for the secretion family is equal to (GFR/0.94 + T_{max}/C_p).

At low PAH concentrations in the blood (Fig. 57-1), the slope of the excretion flux curve is high (the clearance for PAH is high). Here the PAH clearance is an acceptable estimate of RPF because the blood is almost cleared by one transit.

The secretion flux is maximal, when the blood [PAH] is high enough to achieve **saturation kinetics**. The weak organic acids and bases mentioned above are similarly secreted into the proximal tubule, and have secretory T_{max} values just like PAH (Fig. 57-1). In humans of average size (with an average body surface area of 1.7 m^2), the T_{max} for diodrast and phenol red average 57 and 36 mg min^{-1}, respectively.

What can be inferred from the clearance value of a substance?

If the clearance of a substance has the **same value as the inulin clearance** for the person, then the substance is only subject to **net-ultrafiltration**. Theoretically, reabsorption might balance tubular secretion and give the same result. In general, substances showing net-ultrafiltration are neither reabsorbed nor secreted.

If the clearance of a substance is **greater than inulin clearance**, then clearly this substance is being added to the urine as it flows along the tubules; in other words, it is being **secreted**.

Similarly a clearance which for any substance is **less than inulin clearance** means that the substance is being **reabsorbed** at a higher rate than any possible secretion.

Describe the tubular secretion of H^+

There is a **secondary active secretion of H^+** in the proximal tubules. In the tubule cell the H^+ production depends upon the amount of carbon dioxide and the presence of carbonic acid anhydrase. A H^+ inside the tubule cell binds with the carrier protein in the brush border, while a Na^+ binds with the other end of the same carrier. Then the Na^+ enters the cell, while the H^+ is forced outward. This is called **luminal Na^+-H^+-counter transport**, and the H^+-secretion is electroneutral.

The cortical collecting ducts contain special epithelial cells termed **intercalated cells** that secrete H^+ by **primary active secretion** against a concentration gradient. The amounts of secretion are precisely determined by the $[H^+]$ in the ECV.

Describe the tubular secretion of K^+

The secretion of K^+ into the tubular system is linked to the Na^+ reabsorption. This process is controlled by aldosterone, which activates the **Na^+-K^+-pump**. This is the essential control mechanism for $[K^+]$ in the ECV.

The secretion of K^+ takes place mainly in the **principal cells** of the late distal tubules and in the cortical collecting ducts. Secretion only occurs when the $[K^+]$ in the ECV is higher than a critical level (about 5 mmol l^{-1}).

The Na^+-K^+-pump in the basolateral membrane draws Na^+ out into the ISF, and K^+ into the **principal cells**. These cells have a special luminal membrane, which is permeable to K^+. Therefore, K^+ rapidly diffuses to the tubular fluid.

How is urate excreted?

The renal tubules have a capacity of both actively reabsorbing urate ions and actively secreting them.

The **active reabsorption** of urate ions occurs in the proximal tubules by a Na^+-co-transport. The tubular reabsorptive capacity is normally far greater than the amount in the glomerular filtrate. Above a critical concentration in the ECV of about 420 μmol l^{-1}, the urate precipitates in the form of uric acid crystals, provided the environment is acid. Precipitation in the joints is termed **gout (arthritis urica)**, often affecting several joints. Urate ions are accumulated in the ECV of gout patients, and also in patients with uraemia (Chapter 62). High doses of probenecid compete with urate about the proximal reabsorption mechanism. Use of this drug to patients with acute gout increases the excretion of urate in the urine. See also Chapter 53.

The **active secretion** of urate ions occurs from the blood to the tubular fluid by the **organic acid–base secretory system,** which has a low capacity for urate.

Describe the tubular secretion of creatinine

Essentially all creatinine in the glomerular filtrate passes on and is excreted in the urine. The molecule is larger than that of urea, so none of it is reabsorbed.

Contrary, creatinine is secreted into the proximal tubules, so that the [creatinine] in the urine increases more than 100-fold.

Further reading

Guyton, A.C. (1991) *Textbook of Medical Physiology*, 8th edn. W.B. Saunders, Philadelphia.

Larson, T.S. *et al.* (1989) Renal handling of organic compounds. In: S.G. Massry and R.J. Glassock (Eds) *Textbook of Nephrology*. Williams & Wilkins, Baltimore.

57. Case History

A 63-year-old woman (weight 70 kg) with a chronic kidney disease is in hospital due to frequent, massive urination and fast deterioration. She is examined with a test where inulin and PAH molecules are infused intravenously until constant plasma concentrations are achieved. The patient's urinary bladder is catheterized. Following 150 min of continuous infusion with constant plasma concentrations, the patient's bladder is emptied, and the infusion is stopped. The 150-min urine contains 2.03 g of inulin and 18.78 g of PAH. The constant plasma concentration of inulin and PAH is 48 mg per 100 ml and 400 mg per 100 ml, respectively. The free fraction of PAH (F_{free}) is 0.75. The fraction of water in her plasma is 0.94.

1. *Calculate the renal inulin clearance for this patient. Is the obtained value normal?*
2. *Is the proximal, tubular function of this patient normal?*

Try to solve the problems before looking up the answers in Chapter 74.

CHAPTER 58. THE THREE CLEARANCE GROUPS

All substances treated by the kidneys can be divided into three groups or **families**, namely the **filtration group**, the **reabsorption group**, and the **secretion group**.

I. THE FILTRATION GROUP

The kidney treats this group of substances just like **inulin**, which is a fructose derivate from Jerusalem artichokes. Inulin has a spherical configuration and a molecular weight of 5000.

Inulin filters freely through the glomerular barrier, and the tubuli neither reabsorb nor secrete inulin. Thus, under steady-state conditions, the flux of inulin leaving the Bowman's capsules (J_{filtr}) must be exactly equal to the flux of inulin arriving in the final urine (J_{excr}).

The **inulin family** consists of **mannitol, raffinose, sucrose, thiocyanate, thiosulphate, radioactive iothalamate and radioactive-labelled EDTA**. These substances are more or less evenly distributed in the ECV.

The **elimination rate constant** (k) for any substance is the constant **fraction of the total amount of substance in the distribution volume of the body eliminated per time unit**. The inulin family consists of substances which are eliminated by glomerular ultrafiltration (by GFR), without being secreted or reabsorbed. Such substances are distributed in the ECV. The elimination rate constant (k) for the **inulin family** is roughly equal to **GFR/ECV**.

How does the excretion (J_{excr}) and filtration flux for inulin (J_{filtr}) vary with increasing plasma concentration (C_p)

The J_{filtr} increases in direct proportion to the rise in C_p (Fig. 58-1). A simple mathematical deduction shows the same:

$$J_{filtr} = \mathring{V}_u C_u = J_{excr}.$$

Dividing all three expressions with C_p provides us with three expressions of **inulin clearance**. The clearance is the slope of the curve, and it is obviously a constant value that is independent of C_p. Since 0.94 parts of normal plasma are water, the inulin C_p must be equal to $(0.94 \cdot C_{filtr})$.

GFR is equal to $(\mathring{V}u \cdot Cu)/C_{filtr}$. Substitution in the inulin clearance expression yields the following relation:

INULIN CLEARANCE = GFR/0.94 (Chapter 55).

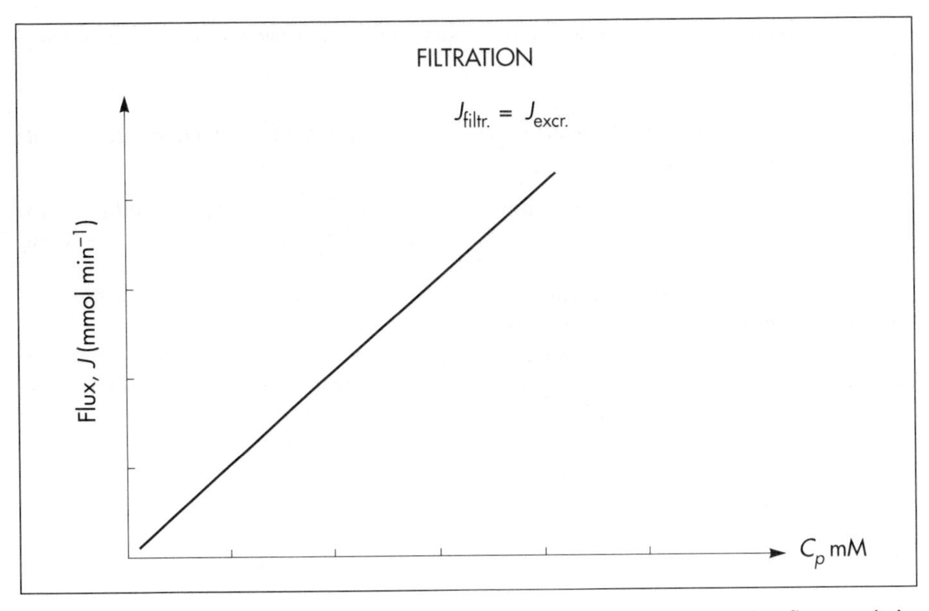

Fig. 58-1. The straight line shows a direct relationship between the flux and the concentration for the inulin family.

II. The Reabsorption Group

The **reabsorption group** or the **glucose family** contains many vital substances such as **glucose, amino acids, albumin, acetoacetates, β-hydroxybutyrate, ascorbic acid, other vitamins, lactate, pyruvate, Na^+, Cl^-, HCO_3^-, phosphate, sulphate and carboxylate.**

For the reabsorption family of compounds, the excretion flux is equal to the filtration flux minus T_{max}:

$$J_{excr} = J_{filtr} - T_{max}.$$

Dividing all three expressions by C_p provides us with:

$$J_{excr}/C_p = J_{filtr}/C_p - T_{max}/C_p.$$

Thus the clearance for the reabsorption family is equal to the inulin clearance $-T_{max}/C_p$. The inulin clearance or J_{filtr}/C_p is also equal to GFR/0.94 as shown above.

III. The Secretion Group

The secretion group comprises organic acids and bases such as PAH, penicillin, probenecid, X-ray contrast substances, sulphonamides, other therapeutics, salicylic acid and phenol red. Foreign substances are often distributed in the ECV, but some of them are also entering cells. At low concentrations their elimination rate constant (k) is roughly equal to RPF divided by ECV.

Is it possible to change the excretion flux curve for inulin to clearance by a simple mathematical procedure?

Yes. This is easily done by differentiating the excretion flux curve for the **inulin family** with respect to C_p. Thus, we automatically produce all the clearance curves for these substances.

For the **inulin family** the flux equals ($\mathring{V}_u \cdot C_u$), and by division with C_p we have the inulin clearance; a value which is always constant.

Since 0.94 parts of plasma are water, the concentration in the ultrafiltrate (C_{filter}) is $C_p/0.94$. By substitution into the equation **GFR/C_p = inulin clearance**, it follows that **inulin clearance is equal to GFR/0.94.**

For all substances belonging to the inulin family the excretion flux curves are linear, so the **rate of change** which is the **clearance** must be constant (Fig. 58-2):

$$\text{If } y = k \cdot x \text{ then the derivative } dy/dx = k$$

or $J = k \cdot C_p$, which by differentiation with respect to C_p produces:

$$\mathbf{d}J/\mathbf{d}C_p = \textbf{clearance} = k.$$

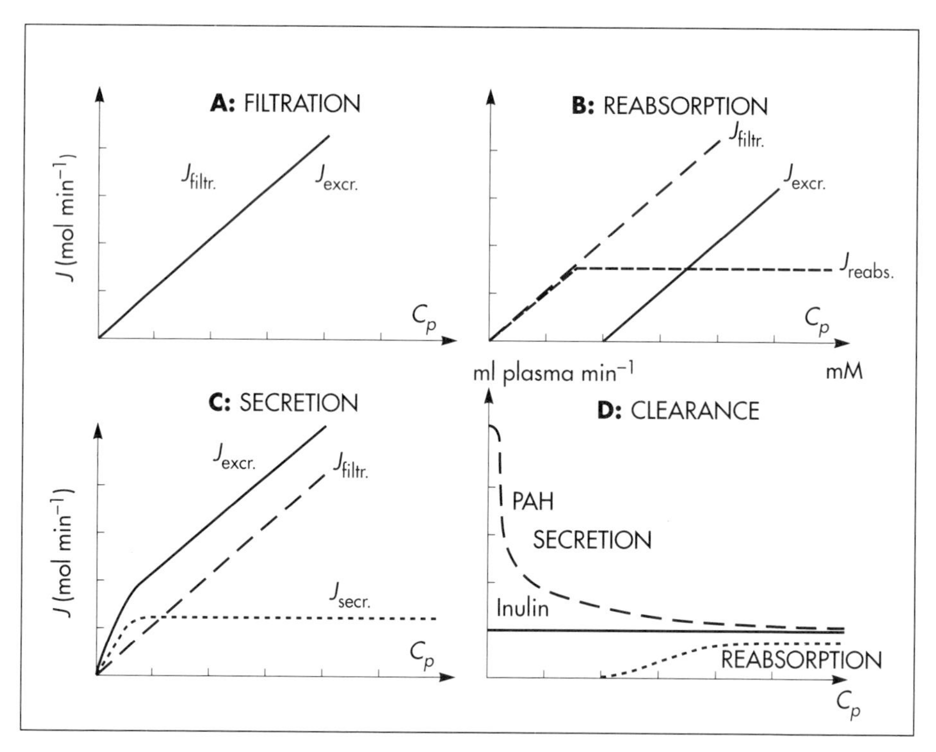

Fig. 58-2: A, B, and C are the filtration-, reabsorption- and secretion- families, respectively. D shows the clearance curves.

Describe the excretion flux and clearance curves

The three excretion flux curves, when differentiated (dJ_{excr}/dC_p), provide us with the same GFR in a given person. The results are plotted with C_p as the dependent variable (*x*-axis of Fig. 58-2).

For the **reabsorption family**, the clearance is zero at first, because the excretion flux is zero (Fig. 58-2 D). The clearance increases, and finally it approaches the inulin clearance (Fig. 58-2 D). The clearance for the reabsorption family is ($GFR/0.94 - T_{max}/C_p$). Therefore, the clearance is steadily increasing towards inulin clearance with increasing C_p.

For the **secretion family**, the clearance must be equal to the excretion flux divided by C_p. Thus, the clearance for the secretion family is equal to ($GFR/0.94 + T_{max}/C_p$).

At low [PAH] in the blood, the clearance for PAH is high and an acceptable estimate of RPF, because the blood is almost cleared by one transit. When the [PAH] increases, more and more PAH is eliminated by filtration, and the secretory elimination is suppressed (**auto-suppression**). The clearance for the secretion family falls with increasing C_p, and approaches that of inulin (Fig. 58-2 D).

Is it possible to calculate the concentration of a low molecular substance in the renal ultrafiltrate (C_{filtr})?

Yes, and the calculations for each of the following cases are:

1. **Uncharged, free molecules** like inulin:

$$C_{filtr} = C_p/0.94 \text{ (mol l}^{-1}\text{ ultrafiltrate)}.$$

This value is only dependent upon the fractional content of water in plasma ($F_{water} = 0.94$ l water/l plasma).

2. **Uncharged, protein bound molecules:**

$$C_{filtr} = C_p \cdot F_{free}/0.94 \text{ (mol l}^{-1}\text{ ultrafiltrate)}.$$

This value depends upon the fractional content of water in plasma and of the fraction of free, unbound molecules (F_{free}).

3. **Low molecular ions:**

$$Cl^- : \text{isosmolal } C_{filtr} > C_p \rightarrow C_{filtr} = C_p \cdot 1.05/0.94$$

$$Na^+ : \text{isosmolal } C_{filtr} < C_p \rightarrow C_{filtr} = C_p \cdot 0.95/0.94.$$

The concentration of low molecular ions in the ultrafiltrate is affected by the 5% Donnan effect (see Chapter 1), and the **fractional content of water** in plasma.

58. Case History

A 20-year-old female (with the normal average body surface area 1.7 m^2) has a normal cardiac output ($\dot{Q}$), and an oxygen consumption for the whole body of 250 ml min^{-1}. Her arterial oxygen concentration is 200 ml l^{-1}, and her mixed venous oxygen concentration is 150 ml l^{-1}.

The arterial haematocrit is 0.5, the RBF/$\dot{Q}$ ratio is 0.3, and her GFF (GFR/RPF) is 0.2. The urine flow rate is 0.75 ml min^{-1}.

1. *Calculate the RBF.*
2. *Calculate the GFR.*
3. *The inulin clearance is a generally accepted measure of GFR. Experimental evidence validates this. Write at least four arguments for the use of the inulin method.*
4. *Calculate the urine/plasma concentration ratio (C_u/C_p) for inulin.*

Try to solve the problems before looking up the answers in Chapter 74.

CHAPTER 59.
RENAL CIRCULATION

Measurement of renal bloodflow (RBF). What is the principle?

The **Fick's principle** (mass balance principle: $RPF = J_{excr}/C_p$) is used to measure the renal clearance at low plasma [PAH], since at low concentrations the blood is almost cleared by one transit.

The law of mass balance states that the infusion flux of PAH is equal to its excretion flux at steady state.

Only one passage through the kidneys **effectively** eliminates PAH from the venous blood plasma at **low [PAH]**. We can make a methodological short-cut and measure the [PAH] in the medial cubital vein only, instead of the true arterial [PAH] by arterial catheterization. Hereby is obtained an acceptable approximation called the **effective renal plasma flow (ERPF)**. In a healthy, resting person the ERPF is 600–700 ml of plasma per minute. The **ERPF principle** avoids complex invasive procedures. The **effective renal blood flow (ERBF)** is calculated with the help of a total body haematocrit (normally 0.45). If **ERPF** is measured to be 660 ml plasma min^{-1}, we can calculate **ERBF**: $660/(1 - 0.45) = 1200$ ml whole blood min^{-1} at rest. This is 20% of $\mathring{Q}$. The true **RBF** is 10% higher than the measured **ERBF**.

At high [PAH] the T_{max} for PAH is a valuable measure of the **secreting tubular** mass because the proximal tubule cells are saturated with PAH.

Are RBF and GFR sensitive to changes in blood pressure?

RBF is autoregulated within the physiological pressure range (Fig. 59-1). GFR follows the autoregulation pattern of the blood flow (explained below).

The RBF falls drastically below a **mean arterial pressure of 9.3 kPa (70 mmHg)**. The **medullary bloodflow** is always small in both absolute and relative terms. Any **severe RBF reduction** as in shock, easily leads to ischaemic damage of the medullary tissues resulting in papillary necrosis and total collapse of the kidney. During such pathophysiological conditions, **prostaglandins (PGE_2 and PGI_2)** are secreted from the mesangial cells due to sympathetic stimulation. These prostaglandins dilate the afferent and efferent glomerular arterioles and dampen the renal ischaemia caused by sympatho-adrenergic vasoconstriction.

How is RBF changed during exercise?

Sympathetic vasoconstriction reduces the resting RBF. At maximum exercise RBF falls to half the resting level. RBF also drops during emotional stress and during haemorrhage.

NA and circulating Ad from the adrenal medulla, constrict the afferent and efferent glomerular arterioles, when they are bound to α_1-adrenergic receptors. This constriction decreases both RBF and GFR.

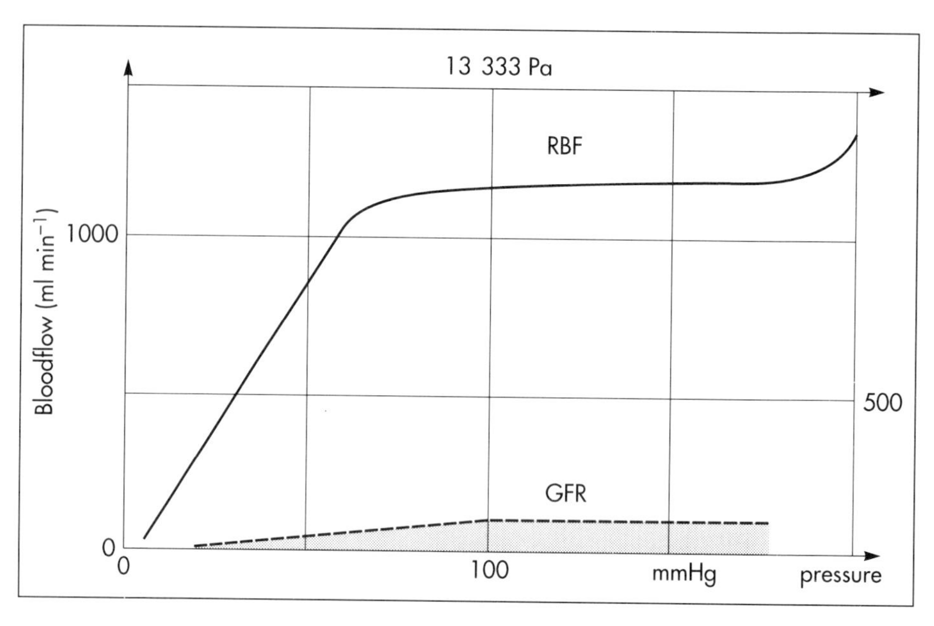

Fig. 59-1: Pressure–flow relations in the kidney. The RBF curve shows autoregulation, and GFR follows the bloodflow.

Is RBF redundant compared to renal oxygen consumption?

1. At rest the RBF is extremely high at **1200 ml blood min^{-1}**, and the renal oxygen uptake ($\mathring{V}_{O_2}$) is 18 ml min^{-1}. For human beings the **average arteriovenous oxygen difference** is 50 ml l^{-1}, while the **renal difference** is only 18/1.2 = 15 ml oxygen l^{-1}. Of a total $\mathring{V}_{O_2}$ of 250 ml min^{-1} for a resting person the kidney's use (18 ml·100)/250 = 7.2% oxygen.
2. The human kidneys form only a small fraction of the weight of a 70 kg body, that is (300 g · 100)/70.000 g or **0.43% of the body weight**.
3. Thus, kidneys consume much more O_2 than indicated by their relative weight. The kidneys have a unique nutritive supply. Their redundant RBF reflects the need of a **high filtering and reabsorption capacity** for the regulation of the body fluids.

Is the bloodflow higher in the cortex than in the medulla of the kidneys?

The 300 g of kidney tissue receive a total bloodflow (RBF) of 1200 ml min^{-1}, which is one-quarter of the cardiac output at rest. On average, RBF is **400 ml of blood min^{-1} 100 g^{-1} kidney tissue**. These units are called **flow units (FU)**. The renal blood flow/weight unit is higher than any other major organ in the body. **The medullary bloodflow** is **only** 200 FU and maximally makes up 10% of the 300 g total kidney weight (0.3 parts medulla as compared to 2.7 parts cortex). This blood flow is also relatively large, and is required for the delivery of large amounts of energy. **Energy requiring reabsorption of Na^+** and **reabsorption of large fluid volumes** takes place in the medulla. Within the renal medulla, the P_{O_2} falls off sharply toward the tip of the medulla, where the blood supply is at a minimum. The **counter current exchange of oxygen** in the vasa recta is a disadvantage to the renal papillae because their cells are fed last with oxygen by the blood. The inner cells meet their energy requirements primarily by anaerobic breakdown of glucose by glycolysis. The amount of energy obtained here is only one-tenth of the oxidative breakdown of 1 mol of glucose (2888 kJ free energy). The medullary bloodflow is small in absolute terms: (200 FU · 0.3) = 60 ml min^{-1} or 5% of the total bloodflow. The medullary bloodflow is reduced by vasopressin to 1%.

The cortical bloodflow is much larger than the medullary bloodflow (422 · 2.7 = 1140 ml min^{-1}). Here, one-fifth of the whole plasma stream passes the glomerular barrier by ultrafiltration and become **preurine**. Fortunately, we obtain the greater part of the energy required for cortical tubular transport by oxidative metabolism.

Is GFR related to the oxygen consumed by the kidneys?

Any change of RBF will change GFR.

1. A **reduced GFR** implies a **smaller tubular Na^+ reabsorption** and thus a **smaller O_2 demand**. When kidneys are perfused by anoxic blood the tubular reabsorption is blocked first, then the GFR is reduced.
2. When **more Na^+ filters** through the glomerular barrier, more reabsorption occurs in the tubules. As Na^+ reabsorption is the **main oxidative energy demanding activity,** a high GFR is correlated to a **high oxygen consumption** in the normal kidney.

What determines the size of GFR?

1. The **net filtration pressure (P_{net})** is as follows. This pressure is the capillary blood pressure (60 Torr) minus the pressure in Bowman's capsules (15 Torr), and minus the net colloid osmotic pressure (28–0 Torr). The values in brackets are true for young, healthy persons.
   In resting humans P_{net} is 10–20 Torr or **1.3–2.6 kPa.**
2. The **total area of the capillary barrier in the glomeruli** is **A**. In healthy humans A is about 12 m^2.
3. R is the resistance of the glomerular barrier.

$$\text{GFR} = P_{net} \cdot A/R.$$

The resistance per area unit is expressed as kPa min ml^{-1} of filtrate.

What determines the hydrostatic pressure in the proximal tubules?

1. The **propulsion pressure of the glomerular blood.**
   This pressure depends upon the haemostatic and the colloid osmotic pressure of the blood. The propulsion pressure is (60–28) = 32 Torr or **4.2 kPa,** according to the values given above. The glomerular capillaries form a **portal system,** and the potential for varying the glomerular capillary pressure is great, so the GFR can vary.
2. The **reabsorption flux of salts and water** in the proximal tubules.
   Sympathetic stimulation increases the proximal reabsorption flux. This causes the hydrostatic pressure to fall in the proximal tubules and Bowman's capsule thus causing GFR to increase. In reverse, **angiotensin II** secretion inhibits the **proximal reabsorption flux, increases the proximal pressure and reduces the GFR.**
3. The **total resistance** below the proximal tubules (i.e. in the distal system).
4. The **resistance of the glomerular barrier** (R – see above).

What is the tubuloglomerular feedback mechanism (the TGF response)?

An increase in the flow rate and in the [NaCl] in the early part of the distal tubules **triggers the tubuloglomerular feedback (TGF) response**. The macula densa probably senses the Na^+/Cl^- concentration of the passing distal fluid (Fig. 59-2). The macula densa elicits a constriction of the afferent arteriole and reduces the release of renin.

Leyssac and co-workers extended the TGF concept to include the effects of local **angiotensin II** on the efferent arteriole and on the rate of proximal reabsorption. They gave anaesthetized rats an intravenous saline infusion equal to 1% increase of body weight. They found a 30% rise in RBF, a 25% increase in GFR and in the proximal reabsorption rate, and a significant decrease in **plasma renin concentration**. Both the urinary flow rate and the solute excretion flux doubled. The TGF response controls the glomerular capillary pressure and the proximal tubular pressure by afferent vasoconstriction. The depressed renin release can dilate the efferent arteriole. The proximal reabsorption rate influences the size of the hydrostatic pressure difference across the glomerular membrane and thereby GFR.

The juxtaglomerular (JG) apparatus contains epithelioid JG cells which contain renin granules: the macula densa cell plug in the final part of the ascending Henle-loop, and the Lacis cells connecting the afferent, and the efferent arteriole (Fig. 59-2).

The TGF response thus also involves the **renin–angiotensin II-aldosterone cascade** (Fig. 61-1), but the response can also be modulated by Ca^{2+} and prostaglandins.

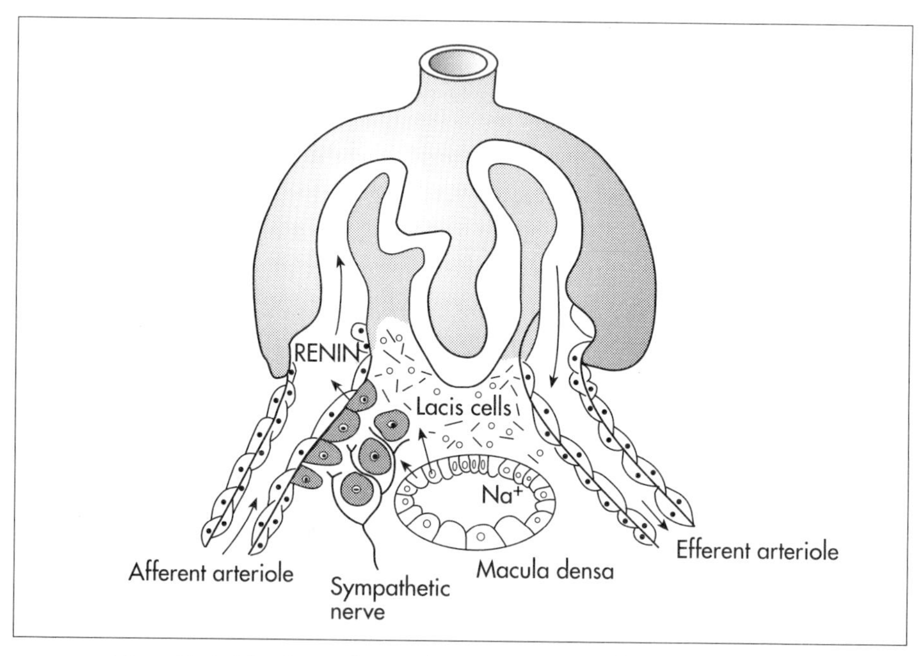

Fig. 59-2: The juxtaglomerular apparatus with renin secretion.

Can tubuloglomerular feedback explain renal autoregulation?

Yes. The macula densa probably senses a **rise in Na^+/Cl^- concentration** in the passing distal fluid. This elicits a constriction of the afferent arteriole and reduces the release of renin. The related fall in angiotensin II causes a rise in RBF and GFR.

When the distal fluid flow falls, the **[NaCl] at the macula densa drops** and the afferent arterioles dilate. More renin is released which causes a rise in **angiotensin II with arteriolar constriction** and thus a reduction in both RBF and GFR, which could be involved in the myogenic autoregulation (see below).

These renin responses tend to keep RBF and GFR normal (i.e. autoregulation). Aldosterone which is released simultaneously increases Na^+ reabsorption and K^+ secretion.

In reverse, a purely mechanical muscle phenomenon can explain renal autoregulation when the blood pressure waves include high frequences. A **dynamic** fall in blood pressure has been claimed elicits a **myogenic autoregulation**.

Although these hypotheses are reasonable, we have to consider the contribution of other mechanisms. Recently, **NO** has been related to autoregulation in general, and to relaxation of smooth muscle cells in the genito-urinary system (see Chapter 5).

Further reading

Leyssac, P.P., F.M. Karlsen and O. Skøtt (1991) Dynamics of intrarenal pressures and glomerular filtration rate after acetazolamide. *Am. J. Physiol.* **261** (*Renal Fluid Electrolyte Physiol.* 30): F169–F178.

Pollack, D.M. and R.O. Banks (1991) Perspectives on renal blood flow autoregulation. *Proc. Soc. Exp. Biol. Med.* **198**: 800–5.

59. Case History

A young woman (body weight 56 kg) with an inulin clearance of 125 ml of plasma min^{-1} is tested with PAH. The free fraction of PAH in the plasma is 0.80, and the rest binds to plasma proteins.

Her urine is collected over a period and the excretion flux of PAH is measured to 100 mg min^{-1}. The average concentration of PAH in plasma from the renal arterial and venous blood is 0.2 and 0.02 g l^{-1}, respectively. The haematocrit is 43%.

1. *Calculate the clearance for PAH.*
2. *Calculate the tubular secretion flux for PAH, which is equal to T_{max} for PAH at the blood plasma concentration concerned.*
3. *Calculate the RBF.*

   The patient collects the urine in a second period, and the average concentration of PAH in plasma from the arterial blood is 1 g l^{-1}.
4. *Calculate the excretion flux for PAH in the urine.*
5. *Calculate the new clearance for PAH.*

Try to solve the problems before looking up the answers in Chapter 74.

Chapter 60. Renal Regulation of Acid–Base Balance

Do we produce acid or base in the body?

Mostly, we produce **net-acid**. A sedentary person produces a large amount of CO_2, 18–24 mol day^{-1}. The combustion of fat and carbohydrate does not produce base. On the contrary, sulphuric acid and phosphate are produced, up to 100 mmol day^{-1}. Most of this acid is eliminated in the kidneys, and the balance of organic anions is oxidized. With a daily standard diet of 100 g of protein (corresponds to 16 g nitrogen), it is possible to produce about 1 mol of the acid NH_4^+ (pK 9.3), which dissociates to NH_3 and H^+ to a minimal extent at pH 7.4. Any surplus of amino acids is metabolized in the liver. Hereby, stochiometric amounts of bicarbonate and NH_4^+ is produced.

Since the NH_4^+ production reacts with a small part of the metabolic CO_2 to produce **urea**, the base production eliminates itself. The daily excretion flux of urea in the urine is 500 mmol or 30 g.

The liver controls the production of urea and glutamine. The high $[H^+]$ in acidosis stimulates the hepatic delivery of glutamine to the kidney and thus the excretion of NH_4^+. The low $[H^+]$ in alkalosis stimulates the hepatic production of urea.

See also Chapter 34.

How is bicarbonate treated in the tubules?

The tubule cells **reabsorb bicarbonate**. The cells secrete H^+ to the bicarbonate containing tubular fluid, and here they combine to form carbonic acid. Then the anhydrase containing **brush borders** of the proximal tubules catalyse the conversion of carbonic acid to carbon dioxide.

Most of the CO_2 diffuses back to the tubule cells and converts to bicarbonate.

Large doses of carbonic anhydrase inhibitors (eg. acetazolamide) alter the diuresis, the bicarbonate and the Na^+ elimination in the urine. How?

Inhibition of the **carbonic anhydrase enzyme** in the tubule cells will reduce the H^+ secretion and thus the bicarbonate reabsorption. We will then lose great amounts of bicarbonate in the urine. With every mol of bicarbonate lost, 1 mol of Na^+ is also lost. This extra loss of ions implies that additional amounts of water will be excreted. The end result of the administration of **carbonic anhydrase inhibitors** is **dehydration and loss of ions.**

Is the acidity of the urine a good estimate for the H^+ secretion in the tubules?

No. The maximal $[H^+]$ in urine is $\mathbf{10^{-5}}$ **mol** $\mathbf{l^{-1}}$. This is a negligible amount in the daily urine (1.5 l daily). Daily, almost 4500 mmol bicarbonate reacts with the H^+ secreted (in an equimolar relation), and the resulting CO_2 diffuses back to the tubule cells. Yet, there is a correlation between **titratable acidity** and pH.

How do we excrete an acid surplus?

1. We produce about 24 mol CO_2 daily, which is eliminated through the lungs. A small part of this CO_2 production (0.5 mol daily) reacts with ammonia:

$$2\,NH_3 + H_2CO_3 = CO(NH_2)_2 + 2\,H_2O.$$

   In this way the body base production almost (see below) eliminates itself as urea.
2. Of the 50–100 mmol daily acid production, 30 mmol are neutralized by the 30 mmol bases normally absorbed from the gut.
3. The **non-metabolizable acid surplus** is excreted by the kidneys, but their capacity is limited. The GFR of the two kidneys is the filtered volume per minute (see Chapter 55).

   Large amounts of bicarbonate are ultrafiltered by the renal glomeruli with a normal inulin clearance of 130 ml min^{-1} and an SB of 24 mM:

   $(\mathbf{SB \times Inulin\ clear \times 1440\ min\ day^{-1}}) = (24 \times 0.130 \times 1440) = \mathbf{4500\ mmol\ day^{-1}}$.

   The tubule cells reabsorb bicarbonate and other acid anions, so the total reabsorption of base to the body is larger than the ultrafiltration flux. The net acid excretion for a person on a normal protein diet is **50–100 mmol daily**.

   Approximately **30 mmol is excreted as NH_4^+**, and **30 mmol as titratable acids** (mainly primary phosphate).

How is bicarbonate reabsorbed in the renal tubules?

Already in the proximal tubules we reabsorb 90% of the filtered bicarbonate (4000 mmol daily) by means of **H^+ secretion**. Most of the H^+ secreted in the proximal tubules is derived from the **Na^+-H^+-exchange** and the rest is mediated by a **H^+-ATPase**.

The small amount of filtered bicarbonate that leaves the proximal tubules is reabsorbed in the distal system, so that only 1–5 mmol is found in the daily urine. Secreted H^+ combines with filtered bicarbonate and generates carbonic acid. This acid instantly dissociates into water and CO_2, because the brush border of the proximal tubules contains **carbonic anhydrase (CA)**.

What is the function of carbonic anhydrase (CA) and CA-inhibitors?

Water and CO_2 diffuse passively across the luminal membrane, and in the CA containing tubular cell they form **carbonic acid**, which then almost instantly form **H^+ plus bicarbonate**. Bicarbonate then moves across the basolateral membrane, and its reabsorption is finished.

Drugs called **CA-inhibitors** increase the renal loss of bicarbonate, Na^+, and water (diuretics). The mechanism is described above.

What are the two other buffer systems reacting with the H^+ secreted?

1. The H^+ in the tubular fluid react with secondary phosphate and form primary phosphate (this is the major part of the **titratable acidity**).
2. Glutaminase synthesizes ammonia from glutamine in the tubular cells, and during acidosis the production rate can increase 10-fold within 5 days. The delay is caused by a regulated entry of glutamine into the mitochondria. Because the pH of the tubular fluid is lower than that of plasma, there is a preferential diffusion of

ammonia into the preurine. The reaction, $NH_3 + H^+ \Rightarrow NH_4^+$, provides a **sink** for continued secretion of NH_3. The subsequent excretion of non-diffusible NH_4^+ produces the net loss of H^+ and leads to generation of new bicarbonate.

Which are the three most important buffer systems used for the elimination of protons with the urine?

1. As the bicarbonate reabsorption continues the [bicarbonate] and pH of the tubular fluid decrease. This makes the phosphate buffer more important.
2. In the tubular fluid the **secondary phosphate** reacts with H^+, and together they form the **primary phosphate** (a daily **titratable acidity of 30 mmol**). When we lose large amounts of acid, the pH of urine can approach 5 or even 4.5. When this happens we must mobilize a new buffer in order to increase the excretion of H^+.
3. The tubule cells contain glutaminase. This enzyme synthesizes ammonia from certain amino acids including glutamine. **Acidosis stimulates the hepatic glutamate delivery to the kidneys and thus ammonia synthesis.**

   Ammonia diffuses easily to the lumen, where it reacts with H^+ and forms NH_4^+ (pK = 9.2). Thus, ammonia exists in the body fluids (pH = 7.4) in form of NH_4^+. The NH_4^+ cannot diffuse back from the lumen, and it is therefore eliminated with the urine without further changes in pH.

   We normally eliminate 30–50 mmol NH_4^+ day^{-1}, but this capacity for eliminating H^+ can be increased tremendously during serious acidosis.

How is acidic urine produced?

The **secreted H^+ in the kidney tubules** has the following fate:

1. The **major part reacts with bicarbonate** from the ultrafiltrate, and produces CO_2 and water.
2. A **few ions remain as H^+** in the tubular fluid and leave as such in the urine (at pH 5, this is maximally 10^{-5} mol l^{-1}).
3. Some H^+ ions react with secondary phosphate from the glomerular filtrate and forms **primary phosphate.**
4. Some H^+ ions react with ammonia to form **NH_4^+ which remains in the urine.**

Further reading

Good, D.W. and M.A. Knepper (1985) Ammonia transport in the mammalian kidney. *Am. J. Physiol.* **248**: F459–F471.

Seldin, D.W. and G. Giebisch (1989) *The Regulation of Acid–Base Balance.* Raven Press, New York.

60. Case History

Two male patients with a normal arterial pH (pH_a = 7.40) are in hospital. One patient starts to vomit excessively, which increases his pH to 7.80. The other patient does a cardiopulmonary exercise test, which reduces his pH to 7.00.

1. *Calculate the three concentrations of H^+ (in nmol $l^{-1} = 10^{-9}$ mol l^{-1}) at these conditions.*
2. *The two changes in pH from the normal pH = 7.4 are similar. Why are the related changes in H^+ concentration not identical?*
3. *Describe other conditions with metabolic acidosis and metabolic alkalosis.*

Try to solve the problems before looking up the answers in Chapter 74.

Chapter 61. Regulation of Body Fluids

What happens to the urine production during dehydration?

Thirst is partly due to a high osmolarity in the blood. Sensors in the brain called **osmoreceptors** will cause **vasopressin** or **ADH** to be liberated from the neurohypophysis (Chapter 64). ADH has three renal effects:

1. **ADH increases the water permeability of the cortical and medullary collecting ducts** causing a small, concentrated urine volume (**antidiuresis**). ADH binds to receptors on the basolateral surface of the cells, where they liberate and accumulate cAMP, which diffuses to the luminal surface and increases its permeability.
2. **ADH stimulates renal NaCl reabsorption in the thick ascending limb of Henle.**
3. **ADH reduces the bloodflow through the vasa recta,** thereby protecting the concentration gradient.

How do we sense thirst?

Thirst is a subjective sensation and thus studied only in man. Thirst acts on a system composed of receptors, a cerebral integrator and a trigger. Humans have **cerebral or brain osmoreceptors.**

These receptors sense the osmolarity of the blood, and an increased osmolarity of only one per cent triggers the sensation of thirst, and vasopressin (ADH) is released.

The **cerebral integrator** includes the **subfornicate organ in the hypothalamus, the cingulate gyrus, the amygdaloid bodies (Fig. 11-2) and mesencephalon.**

The **effector** is the locomotor system, and water is consumed. When we are in a water deficient state, such an input control is appropriate.

What is the renal response, when a healthy person drinks 1 l of water or 1 l of physiological saline?

When a person rapidly drinks 1 l of water, the intestine absorbs water. Ions diffuse out into the intestinal lumen and the immediate blood osmolarity falls drastically. The **low osmolarity** causes a **block of the ADH secretion.**

The urine volume starts to rise after 30 min and approaches 20 ml min^{-1} after 90 min. One litre of isotonic saline implies an ECV expansion. There is no dilution of body fluids. The **volume and salt expansion** will not increase the urine volume much. The increased ECV can be sustained for 24 h.

How is the Na^+ composition of ECV maintained?

The dominating ion in ECV is Na^+, and its elimination depends upon three factors:

1. **The first factor** is the GFR, which is responsible for the filtered flux of Na^+.
2. **The second factor** is the **renin–angiotensin–aldosterone cascade,** starting with a

high renin output to the blood from the JG apparatus. Any factor increasing the sympathetic tone or reducing the renal artery pressure triggers the renin cascade.

3. **The third factor** is a **peptide or peptides** from the atrial tissue (**atrial natriuretic factor or peptide, ANF or ANP**) with natriuretic effect. An interesting study of this effect has been performed on healthy dogs with **split infusion** by Emmeluth *et al.* Split infusion is a technique by which a hypertonic NaCl solution is infused into the carotids and water into the jugular vein in amounts making the total load isotonic. This technique has revealed the existence of **concentration-sensitive receptors** in the brain which contribute to the control of renal Na^+ excretion possibly by modulation of the synthesis of the renal natriuretic peptide, **urodilatin**. Urodilatin has been isolated from human urine, and contains four amino acids more than ANP. The specific mechanisms linking the brain to this renal effect are under investigation.

Are the kidneys involved in the pathogenesis of oedema?

Oedema is a clinical condition where the ISF volume is abnormally large.

A **voluminous ISF** is due to **increased haemostatic pressure** (heart insufficiency), or a **reduced colloid osmotic pressure** (hypoproteinaemia) as predicted from Starling's law for transcapillary transport.

Reduced protein synthesis (liver disease) and **abnormal protein loss** with the urine (proteinuria) causes hypoproteinaemia. Thus protein losing kidneys are involved.

A normal person drinks 1200 ml of water daily and 1000 ml in food. He loses 2500 ml. Why is there an imbalance?

Water is lost in the urine (1500 ml), in the stools (200 ml), in sweat (400 ml), and through evaporation from the respiratory tract (400 ml).

The total loss of water is 2500 ml, and this corresponds perfectly to the intake plus a normal production of **300 ml metabolic water per 24 h.**

In what way is sodium excreted through sweat, and what is its concentration in sweat?

Sweat secretion takes place in perhaps three compartments just as the secretion in the pancreas and in the salivary glands (Chapters 48 and 50).

Normal sweat is a **hypotonic solution** because Na^+ is reabsorbed in the duct system. The $[Na^+]$ can increase up to 80 mmol l^{-1} with increasing sweat flow – due to the limited time for **Na^+-reabsorption**.

Is it possible to use the serum-Na^+ as an estimate of osmolarity in the whole of the ECV?

A normal $[Na^+]$ in ECV of 150 mmol kg^{-1} plasma water corresponds to a total osmolality of 300 mOsmol kg^{-1}.

Alterations in $[Na^+]$ (osmolarity in mmol l^{-1} plasma) will be followed by similar changes of the ECV osmolarity. Osmolarity is defined in Chapter 1.

How is it possible to estimate the size of an acute haemorrhage using a small blood sample?

Attempting to compensate the loss of blood, the resistance vessels contract following an acute bleeding. This contraction reduces the capillary pressure and drags ISF into the blood.

Consequently, both the **plasma proteins and the haematocrit fall** almost in proportion to the size of the blood volume loss.

How can an output control be used to stabilize the ionic composition of the ECV?

We use output control, when the body is overloaded with water.

The most important osmotically active solute in ECV is NaCl, because it only passes into cells in small amounts. Urea, glucose and other molecules with modest concentration gradients are of no importance.

During health, two primary control systems are used.

1. The **osmolality** (osmol kg^{-1}) or **ion concentration** controls elimination of water. **Water overload** results in low osmolality which blocks the ADH secretion resulting in an increased water excretion. An increased water loss increases the osmolality of ECV.

   Hyperosmolality elicits a linear increase in plasma ADH, which causes water retention until isosmolality is reached.
2. In healthy persons it is the increase in **pressure and volume** of the ECV that controls salt elimination, and not **osmolarity**.

   The salt excretion depends upon three factors as described earlier in this chapter: the **first factor** or GFR (the filtered mass is the primary event but subject to autoregulation); the **second factor** or the renin–angiotensin–aldosterone cascade is the major regulator; and the **third factor renal, natriuretic peptide** contributes.

   Only when the **arterial blood pressure falls drastically** will the body drop the protection of normal concentration. In such a disease state large amounts of **vasopressin** is released in an attempt to improve the pressure.

How does the renin–angiotensin–aldosterone cascade work?

Increased renal sympathetic tone or falling pressure in the renal artery trigger β-receptors on the JG cells of the juxtaglomerular apparatus and liberate renin from cells located in the afferent glomerular arteriole (Fig. 61-1). This takes place, when we lose ECF, move into the upright position and when the blood pressure falls.

Renin is a protease that separates angiotensin I from the liver globulin angioten–sinogen.

When angiotensin I passes the lungs or the kidneys, a dipeptide is separated from the decapeptide by an angiotensin converting enzyme (ACE). This process produces the octapeptide angiotensin II (Fig. 61-1).

Angiotensin II has two effects:

1. Angiotensin II stimulates the aldosterone secretion from the zona glomerulosa of adrenal cortex, and thus stimulates Na^+ reabsorption and K^+ secretion in the distal tubules (Fig. 61-1).

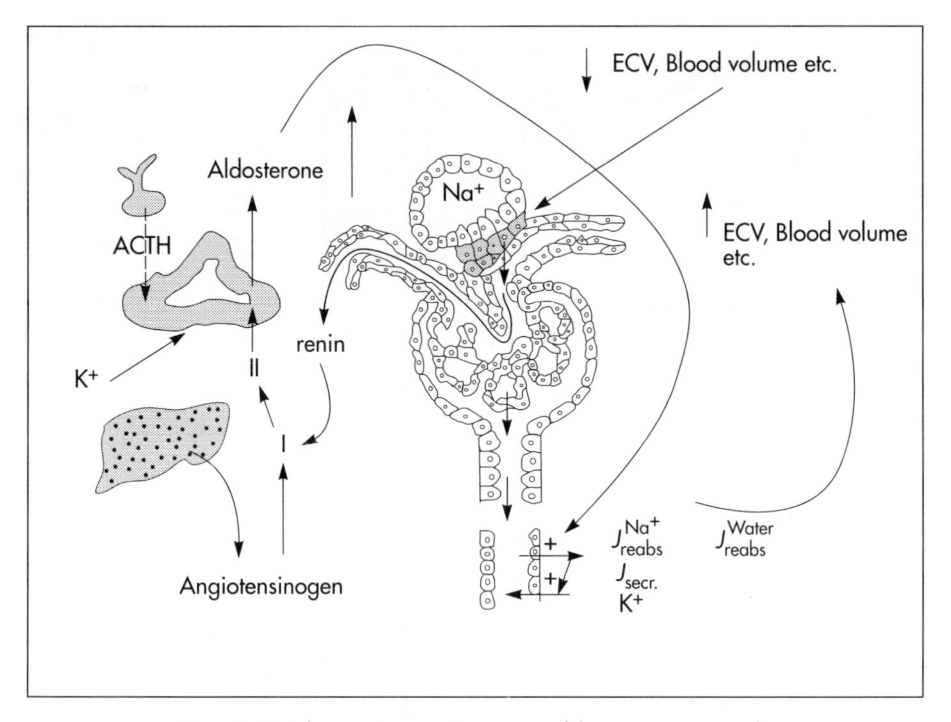

Fig. 61-1: The renin–angiotensin–aldosterone cascade.

2. Angiotensin II constricts arterioles (a strong effect on the afferent arteriole via sympathetic α-receptors) which reduces the renal perfusion and proximal reabsorption. Therefore GFR is also reduced. There is also a constricting effect on the efferent arterioles.

Aldosterone stimulates the Na^+ reabsorption and the K^+ secretion of the distal system. The combined effect of the whole cascade is ECF homeostasis.

The most likely trigger of the renin–angiotensin–aldosterone cascade is the **falling NaCl concentration** – especially Cl^- – of the fluid at the macula densa in the distal renal tubules. Other mechanisms are also involved in the release of renin: (1) **falling perfusion pressure** to the kidney; (2) **stimulation by prostaglandins**; (3) **inhibition by angiotensin II**; and (4) **sympathetic stimulation** of the renal nerves, which stimulates renin secretion directly via β-adrenergic receptors on the JG cells. β-Blocking drugs inhibit the renin secretion.

What is dehydration? Types of dehydration?

This is a reduction of total body water. When we lose more than 5% there are clinical consequences. The condition becomes life threatening if the patient loses 20%. Extracellular dehydration may be visualized by a bad skin turgor.

Isosmolal dehydration is a proportional loss of water and solutes. There is no concentration gradient over the cell membranes, and the loss is mainly from ECV (Fig. 61-2).

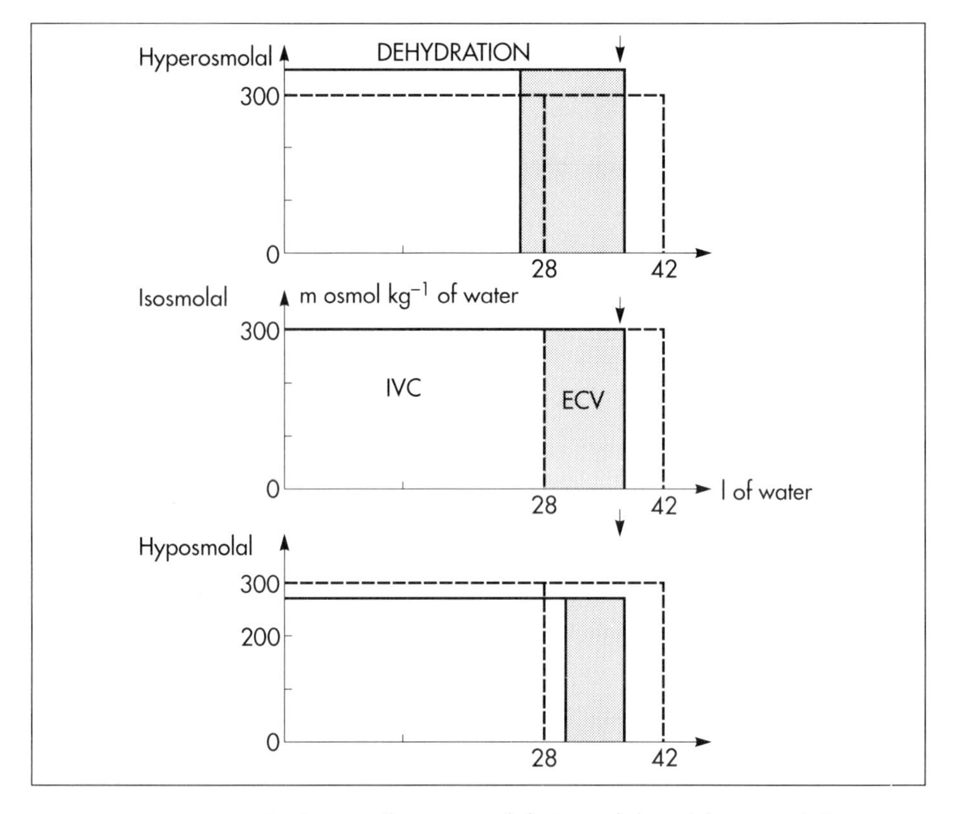

Fig. 61-2: Dehydration (hyperosmolal, isomolal and hyposmolal).

Hyperosmolal dehydration occurs in persons deprived of water. The hyperosmolal ECV drags water from ICV and dehydrates the cells (Fig. 61-2). The hyperosmolality liberates ADH to restrict the water loss. The patient excretes a very small urine volume.

Hyposmolal dehydration is seen in persons losing salt-containing fluid and drinking water only.

Water dragged into the cells further reduces the hyposmolal ECV (Fig. 61-2). The small ECV elicits a hyperaldosteronism, which is called secondary, because it is not initiated as primary hypercorticism in the adrenal cortex.

What is overhydration and what is the danger associated with hyperosmolal overhydration?

Overhydration is an abnormal increase in total body water frequently of iatrogenous origin.

Large quantities of infused saline increase the hyperosmolality of ECV.

Hyperosmolality drags water from the cells, so that the patient develops intracellular dehydration with hallucinations, loss of consciousness and respiratory arrest.

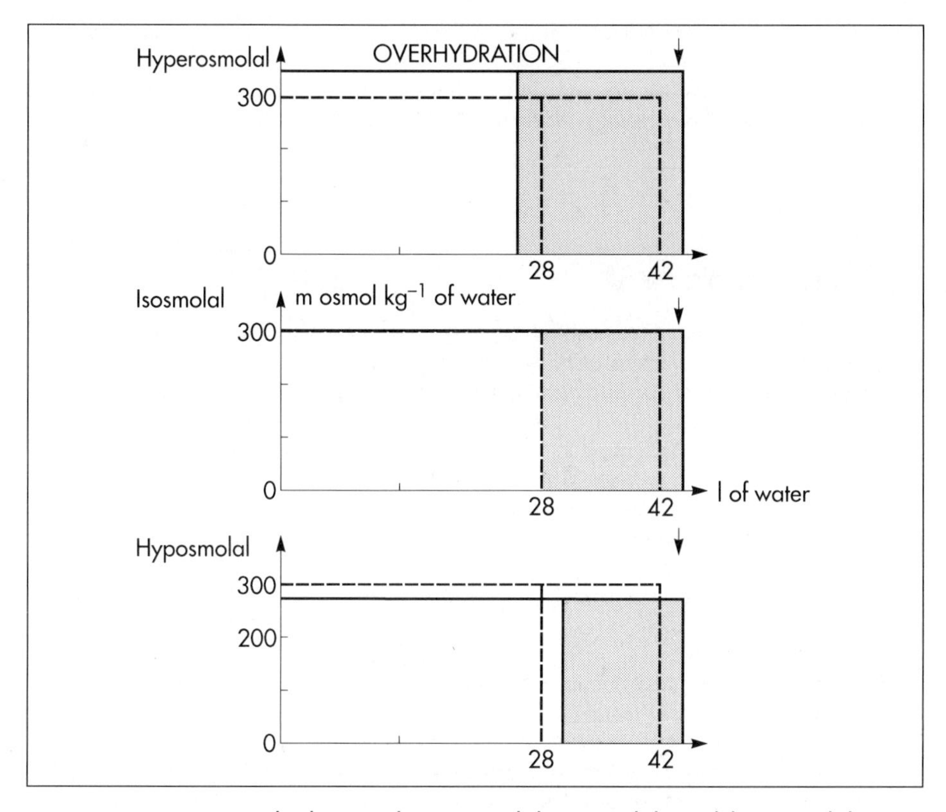

Fig. 61-3: Overhydration (hyperosmolal, isosmolal, and hyposmolal).

Is it possible to diagnose hyposmolal overhydration?

The patient is typically in fluid treatment and develops muscle cramps and disorientation. The skin turgor is normal. A low serum – [Na^+] confirms the diagnosis. The water overload in ECV is dragged into the cells (Fig. 61-3). In the brain and the muscles this intracellular overhydration causes headache, disorientation, increased spinal pressure, coma and muscle cramps. Both **hyposmolal** and **hyperosmolal** overhydration conditions are characterized by cerebral symptoms and signs.

Further reading

Emmeluth, C., C. Drummer, R. Gerzer and P. Bie (1992) 'Roles of cephalic Na^+ concentration and urodilatin in control of renal Na^+ excretion.' *Am. J. Physiol.* **262** (*Renal Fluid Electrolyte Physiol.* **31**): F513–F516.

Gibbons, G. H. (1984) Interaction of signals influencing renin release. *Ann. Rev. Physiol.* **46**: 291–308.

Krogh, A. (1965) *Osmotic Regulation in Aquatic Animals.* Dower, New York.

Skøtt, O. and B.L. Jensen (1993) Cellular and intrarenal control of renin secretion. *Clin. Sci.* **84**: 1–10.
Suki, W.N. (1989) Renal actions and uses of diuretics. In: S.G. Massry and R.J. Glassock (Eds) *Textbook of Nephrology*. Williams & Wilkins, Baltimore.

61. Case History A

I. A 62-year-old man is in hospital for surgery of an inguinal hernia. He arrives with a normal status of body fluid compartments. His ICV is 28 l, and his ECV is 14 l. Both compartments have an osmolality of 300 mOsmol kg^{-1} water.

The patient is anaesthesized in the semirecumbent position. Lidocaine is introduced through a catheter in the spinal channel. A sudden change in the operation programme leaves the patient waiting for 6 h in a warm room. During this time he loses 0.5 l h^{-1} of sweat containing a total of 200 mOsmol of NaCl in the 3 l.

1. *What is the rise in osmolality in ECV assuming the ECV to be reduced from 14 to 12.25 l?*

II. The patient restitutes well with water and food, but 3 days later he develops high fever and diarrhoea due to hospital bacteria. At this time he loses 3 l of intestinal fluid with 150 mmol Na^+ l^{-1}.

2. *Does the concentration gradient over the cell membranes change?*

III. While losing two more litres of jejunal fluid, and before rational treatment with infusion of physiological saline is carried out, the patient drinks water. By mistake, he receives intravenous infusions of isotonic glucose. When the water loss is compensated his total body water is again 42 l.

3. *Calculate the new osmolality and the new ECV.*

61. Case History B

A male patient (35 years of age; body weight 73 kg) with a negative calcium balance, is tested for sodium balance disturbances. One morning his blood plasma Na^+ is 146 mM. He receives intravenously a solution containing radioactive sodium (2812000 Bq of $^{24}Na^+$ with a physical half-life of 15 h). He collects all his urine for the next 30 h (2.5 l in 29 h and with a Na^+ concentration of 70 mmol l^{-1}). The last hour's urine production is 0.06 l with a Na^+ concentration of 90 mmol l^{-1}. The patient only loses $^{24}Na^+$ in the urine.

After exactly 30 h the urine portions 1 and 2 contain the radioactivity: 20350 and 18870 Bq l^{-1}, respectively, and the plasma of the patient has an SA of 210 Bq $mmol^{-1}$.

1. *Calculate the total exchangeable body Na^+ of the patient, and compare the result to a normal value of 40 mmol kg^{-1} body weight.*
2. *Calculate the fraction of the total exchangeable body Na^+ excreted in the urine per 24 h, and compare the result to a normal minimum of 5.5% per 24 h.*
3. *Does the above condition change the aldosterone secretion?*
4. *Calculate the volume of distribution for the exchangeable Na^+. The $^{24}Na^+$ distributes evenly in this volume, with the same concentration as in plasma water. F_{water} is 0.94. Is the calculated volume normal?*

The following year this patient is again examined following 'neutron activation analysis' (exposure to a discrete radiation with neutrons).

5. *Describe a method of measurement of his total body sodium using a whole body counter.*

Try to solve the problems before looking up the answers in Chapter 74.

CHAPTER 62. RENAL FUNCTION IN DISEASE

Renal insufficiency is a clinical condition, where the GFR is inadequate to clear the blood of nitrogenous substances classified as **non-protein nitrogen** (urea, uric acid, creatinine, and creatine). The retention of non-protein nitrogen in the plasma water is called **azotaemia**, and the clinical syndrome is called **uraemia**. The number of filtrating nephrons falls below one-third of normal, as determined by measurement of a GFR below 40 ml min^{-1}.

Acute renal insufficiency accompanies extremely severe states of circulatory shock. Such a serious disorder leads to progressive uraemia and **chronic renal insufficiency.**

Is the renal function reduced in acute glomerulonephritis?

The inflammation is an abnormal immune reaction often caused by repeated streptococcal tonsillitis. An insoluble antigen–antibody complex precipitates in the basement membrane of the glomerular capillaries. The cells of the glomeruli proliferate, and disease will of course reduce GFR and to some extent, the RBF (measured as PAH clearance). Thus the infection depresses the **Glomerular Filtration Fraction, (GFF)** (GFF=GFR/RPF).

What is pyelonephritis?

Pyelonephritis is a bacterial disease that **begins in the renal pelvis**, and then progresses into the renal medullary tissue.

The essential function of the medulla is to concentrate the urine during water depletion. Therefore, in patients with pyelonephritis, the ability to concentrate the urine is abolished (**isosthenuria**). The ability to dilute the urine deteriorates also. Thus, in isosthenuria the urine is always isotonic with the plasma.

What is uraemia?

Uraemia is a clinical condition of **acute renal failure with retention of urea and other waste products**. The end products of protein catabolism are uric acid, creatinine, creatine, NH_4^+, guanidine and urea. Uraemic patients generally exhibit **hyperkalaemia** (serum-$[K^+]$ above 5 mM) and **metabolic acidosis** (pH below 7.35 and a **base deficit**). This is due to the inadequate secretion of K^+, NH_4^+ and H^+. In complete renal shut-down the patient dies within 1–2 weeks without dialysis with the artificial kidney.

The nephrotic syndrome includes albuminuria and oedema. What is the basic problem?

The **nephrotic syndrome** refers to the marked increase in the permeability of glomerular capillaries to protein, resulting in **massive loss of albumin in the urine** (albuminuria), causing hypo-albuminaemia and generalized oedema. Oedema is visible in the face – especially around the eyes. A serious but rare complication may develop when a large volume of fluid accumulates in the abdominal cavity as **ascites**.

Why do diabetics develop leaky renal glomeruli?

Diabetic nephropathy includes **hypertension, persistent albuminuria, and a decline in GFR**. One-third of all insulin-dependent diabetics develop nephropathy. The mortality rate is high. The metabolic disturbance in diabetics causes hypertension and leaky renal glomeruli, but the mechanism remains uncertain (see below).

Is it possible to suppress the development of diabetic nephropathy?

Treatment with **angiotensin converting enzyme inhibitors** has an interesting effect on diabetic nephropathy. This therapy **reduces urinary albumin excretion**, but it does not change the GFR and the arterial blood pressure. The treatment also **postpones the development of diabetic nephropathy**. This is also true for insulin-dependent diabetics with persistent microalbuminuria (a minor albumin excretion between 30 and 300 mg per 24 h) and normal blood pressure. The effectiveness of this treatment suggests that angiotensin oversecretion may be the cause of diabetic nephropathy.

Hypertensive patients handle Na^+ just as healthy persons. How is that?

A long-term increase in sodium intake results in changes of the kidney function. Surprisingly, the changes are similar in hypertensive and normotensive humans! Almost all people increase their ECV and GFR without changing the absolute reabsorption flux of Na^+ and water in the proximal tubules. Therefore, the rise in filtration flux of Na^+ and water will reach the loop of Henle. Bruun *et al.* found arterial blood pressure and heart rate unaffected by the amount of sodium in the diet. The **plasma concentrations of active renin, angiotensin II and aldosterone decreased with increasing Na^+ intake, but ANP and cyclic GMP increased**. Arginine vasopressin (ADH) in plasma did not change.

These homeostatic reactions are all appropriate physiology in both healthy and hypertonic humans.

A patient suffering from a traffic accident has to have one destroyed kidney removed. What happens to the GFR immediately after the removal and later?

Immediately after the removal, the GFR of the patient falls to half its original value, because half the functioning nephrons have been removed.

Most individuals will soon increase their GFR towards normal values by compensatory **work hypertrophia** by the remaining kidney. The **hypertrophia factor** is not known. Each remaining nephron must filter and excrete more osmotically active particles than before.

Describe the hormonal condition in a patient immediately after extensive surgery

Exposure to stress and painful stimuli triggers the whole adrenal hormone production. The **sympatho-adrenergic system** is activated with release of catecholamines, and the **secretion of ACTH and glucocorticoids (cortisol) is increased**. ACTH stimulates the secretion of cortisol from the adrenal cortex. Exposure to fluid loss, reduced glomerular propulsion pressure, and the increased sympathetic activity release **renin** from the **JG cells**, so the **renin–angiotensin–aldosterone cascade** is triggered (Fig 61-1). If the patient is also dehydrated the **high plasma osmolality** leads to liberation of ADH.

What happens to a patient with an isolated damage of the sodium reabsorption?

An isolated damage of the Na^+ reabsorption (salt-losing nephritis) is a condition in which the disease processes are mainly due to dysfunction in the renal medulla. There is a **marked loss of Na^+ in the urine and seriously low ECV and blood volume** (hypovolaemia with threat of imminent shock). Thus the patient must have a high salt intake to prevent shock and stay alive.

Renal hypoxia causes increased erythrogenesis, whereas renal artery stenosis does not. Why?

When the kidneys are perfused with **hypoxaemic blood (low P_{aO_2})**, the JG apparatus in the walls of the afferent arterioles releases an enzyme (the **renal erythropoietic factor**) into the blood, but erythropoitin cannot be extracted from kidney tissue. Within minutes, the **renal erythropoietic factor** cleaves a plasma globulin from the liver, to produce large amounts of the glycoprotein hormone, **erythropoietin**. This is a circulating molecule with a strong effect on the **erythrogenesis** (from the haemopoietic stem cells in the red bone marrow). The stem cells are stimulated to produce proerythroblasts, which speed up the production of new red cells after a few days. The increased erythrogenesis improves tissue oxygenation, which decreases erythropoietin production and the balance is re-established. **Glucocorticoids, sex hormones and thyroid hormones** also increase erythropoietin production.

Stenosis of the renal artery often implies a small renal bloodflow, a small glomerular filtration and a small NaCl reabsorption with a related small oxygen consumption. As long as the renal oxygenation is sufficient, the erythropoietin production is normal.

Describe the renal element in chronic congestive heart failure

1. **At rest the GFR, the RBF, and the $\mathring{Q}$ are close to normal.** The cardiac patient has an abnormally high Na^+ retention and the accompanying high water retention increases his or her ECV. This leads to oedema and increased venous return to the heart with a small rise in $\mathring{Q}$.
2. **During exercise the rise in $\mathring{Q}$ is insufficient.** The insufficient rise in bloodflow and blood pressure, elicits (via the baroreceptors) a marked vasoconstriction of the renal circulation. Thus the RBF must decline. Other vascular beds also constrict markedly during exercise.

The key factor to the **abnormal Na^+ retention** is the **reduction in RBF**.

The major Na^+ reabsorption normally takes place in the proximal tubules. Therefore the distal tubular fluid contains a small load of Na^+, allowing only a small K^+ secretion here. The patient with serious heart failure may develop **secondary hyperaldosteronism** and eventually become **K^+-depleted.**

This is in contrast to a patient with hyperplasia of the adrenal zona glomerulosa (**primary hyperaldosteronism** or **Conn's disease**). Here aldosterone works directly on the highly Na^+ loaded distal tubules, so the patient become **severely K^+-depleted.**

What are diuretics?

Diuretics are therapeutic agents that increase the production of urine. Diuretics are employed to enhance the excretion of salt and water in cases of cardiac oedema or arterial hypertension.

Diuretics act at different points on the nephron.

1. The so-called **loop diuretics** (bumetamide and furosemide) inhibit primarily the reabsorption of salt in the thick ascending limb of the loop of Henle, probably by blocking the (Na^+, K^+, $2Cl^-$)-cotransport process in the luminal entry membrane.
2. Aldosterone antagonists (e.g., spironolactone) competes with aldosterone for receptor sites in the distal tubules. As aldosterone promotes Na^+-reabsorption and H^+/K^+ secretion, spironolactone causes a natriuresis and reduces urinary H^+ and K^+ excretion. Aldosterone antagonists are relatively weak diuretics but useful in reducing K^+ excretion produced by other diuretics.
3. Thiazide diuretics (bendrofluazide, hydrochlorothiazide) act on the early part of the distal tubule by reducing the Na^+, Cl^--cotransport process. Thiazides cause urate retention, glucose intolerance and hypokalaemia. Thiazides are contraindicated by gout.
4. Potassium-retaining diuretics (e.g., amiloride) inhibit Na^+ reabsorption in the collecting duct and reduce renal K^+ excretion.
5. Carbonic anhydrase inhibitors (e.g., acetazolamide) act on the anhydrase in the brush borders of the proximal tubules.

What is non-ionic diffusion in the renal tubules?

Non-ionic diffusion is a passive tubular reabsorption of weak organic acids and bases, that are lipid soluble in the undissociated or non-ionized state. In this state these compounds penetrate the lipid membrane of the tubule cell by diffusion. The tubule cells, however, are practically impermeable to the dissociated form of these compounds. Therefore, the **ionic form of the weak acid or base** is fixed in the tubular fluid and **favoured for urinary excretion.**

A weak organic acid is mainly undissociated at low urinary pH, whereas an organic base is more dissociated. In acid urine the reabsorption flux of weak organic acids is **increased,** whereas the reabsorption flux of weak organic bases is **reduced.** In alkaline urine the opposite situation prevails.

Examples of acids showing this phenomenon are phenobarbital and procain (both with pK about 7), NH_4^+, acetylsalicylic acid, and many other therapeutics. Weak bases are the doping substance, amphetamine, and many therapeutics.

In cases of poisoning with weak bases, the patients are treated with infusions of **amino acid–HCl solutions,** which acidify the urine. In cases of poisoning with weak acids, the patients receive **infusions of bicarbonate** solutions, whereby alkalization of the urine is instituted.

Further reading

Bruun, N.E., P. Skøtt, M.D. Nielsen, S. Rasmussen, H.J. Schütter, A. Leth, E.B. Pedersen and J. Giese (1990) Normal tubular response to changes of sodium intake in hypertensive man. *J. Hypertens.* **8**: 219–27.

Mathiesen, E.R., E. Hommel, J. Giese and H.H. Parving (1991) Efficacy of catopril in postponing nephropathy in normotensive insulin dependent diabetic patients with microalbuminuria. *Brit. Med. J.* **303**: 81–7.

Rose, B.D. (Ed.) (1987) *Pathophysiology of Renal Disease,* 2nd edn. McGraw-Hill, New York.

62. Case History

During her working hours a 22-year-old nurse delivered an arterial sample for blood gas tensions. She had no symptoms or signs of disease, but doubted that an arterial sample could be taken without causing pain. The sample was taken from a radial artery with a fine needle following local anaesthesia. To her surprise she experienced no pain. The arterial values were: P_{CO_2} 24 Torr, P_{O_2} 102 Torr, pH_a 7.36, and standard bicarbonate 15.8 mmol l^{-1}. The nurse had been starving for 24 hours.

1. *What was the explanation of her condition?*
2. *What was the rational treatment?*

Try to solve the problems before looking up the answers in Chapter 74.

SECTION VII.

Endocrine Glands in Humans

The **endocrine** and the **nervous systems** coordinate the functions of the other organ systems. The endocrine system exerts its influence through blood-borne substances (hormones) produced in glands without secretory ducts (i.e. endocrine glands). The endocrine glands comprise the **hypothalamo-hypophyseal axis**, which regulates the function of the **thyroid, parathyroid, adrenal and reproductive glands**, as well as the **pancreatic islets** of Langerhans.

CHAPTER 63. ENDOCRINOLOGY

What is the definition of hormones?

Hormones are blood born **messenger** molecules. **Endocrine hormones** are secreted into the blood and their target cells are equipped with **receptors that recognize the hormones**. They co-ordinate the activities of different cells. Hormone molecules form a large **signal family** together with neurotransmitters, autocrine and paracrine substances.

Paracrine and autocrine hormones are secreted into the ISF surrounding the cells and their actions are restricted either to **nearby cells or to the cell of origin**.

Neurotransmitters or **neurocrine hormones** (peptides, amines, amino acids) exert a type of paracrine action, since they are released in the synaptic region.

Describe historical principles for hormonal function analysis

The scientists of history used extirpation, substitution and transplantation to get insight in the classical part of our present knowledge.

Removal of the pancreas produced diabetes in animals. Removal of the pituitary gland, followed a few days later by removal of the pancreas, produced no symptoms of diabetes. This is because the adenohypophysis produces a hormone that is antagonistic to the effect of insulin in glucose metabolism. The hormone is human growth hormone (GH). Houssay received a Nobel Prize for this work in 1947.

How are lipophobic hormones secreted to the blood?

The water-soluble hormones (peptides and catecholamines) are packed in the Golgi complex in secretory granules that migrate to the cell surface.

Exocytosis of the granule contents to the ISF and diffusion through fenestrae to the capillary blood is a common method. The secretory cells are first stimulated by chemical or electrical signals.

All reactions in the cell linking stimulation and secretion together are termed **stimulation-secretion couplings**. Stimulation-secretion coupling involves depolarization of the cell membrane, or opening of Ca^{2+} channels, so that Ca^{2+} can diffuse into the cell and combine with its binding protein, calmodulin. Intracellular $[Ca^{2+}]$ is necessary for the exocytosis.

How are hormones transported with the blood?

Protein-binding protects small hormone molecules (such as the thyroid hormones) from metabolism.

Protein-binding also eases the transportation of the lipid-soluble steroids, and maintains an equilibrium with a small free pool of hormone, so the concentration of free hormone is constant.

How are hormones eliminated from the blood?

Metabolic elimination

Proteolytic enzymes inactivate peptide hormones both in the specific target organs and in other organs, in particular the liver, where they undergo transformation and conjugation.

Excretory elimination

1. The **kidneys** excrete some hormones or their metabolites in the urine.
2. The **liver** excretes other hormones in the bile after coupling them to glucuronic acid or sulphate, but part of these hormones are reabsorbed in the **entero-hepatic circuit**.

Identify three chemical types of hormones

Peptides and proteins include neuropeptides, pituitary and gastrointestinal hormones.

Steroids consist of adrenal and gonadal steroids and vitamin D, which is converted to a hormone. Steroids are lipid-soluble (lipophilic).

Monoamines (modified amino acids) comprise catecholamines, serotonin, melatonin, and thyroid hormones. Catecholamines (dopamine, NA and Ad) are derived from tyrosine – and serotonin/melatonin from tryptophan – by a series of enzymatic conversions. Monoamines and amino acid hormones are water-soluble just as peptides. Thyroid hormones are iodinated derivatives of tyrosine, and thyroid hormones are lipophilic.

What are hormone receptors?

These are proteins, to which hormones bind. They are present in cell **membranes, cytoplasm** and **nucleus**, and serve two functions. Firstly, they are required for selectivity. Secondly, they are connected to an effector mechanism in the cell (Fig. 63-1).

In response to **hormone-binding** the receptor conformation is changed, and this activates a specific enzyme system that serves as an amplifier. In the cytosol, multiple **second messengers** have evolved to serve such purposes, whereas in the nucleus, the hormone-receptor complex binds to DNA and regulates gene expression (Fig. 63-1). The **effector domain of the membrane receptor** is tightly coupled to the **regulatory portion** of the effector enzymes (such as **adenylcyclase**). These effector enzymes control ion fluxes, membrane transport systems, the production of cyclic nucleotides, and the breakdown of phospholipids. The effector systems are also coupled to **high-energy phosphorylation** of proteins including the receptor itself.

This phosphorylation is critical for cell viability (synthesis, transport and metabolism of vital molecules). Many hormones initiate a series of reactions when bound to membrane receptors. A family of coupling molecules called **G-proteins** links receptors to nearby effector molecules (see Chapter 5: Fig. 5-1 and 5-2).

Where do steroids and thyroid hormones act in a cell?

Both are **lipophilic** and therefore pass easily through the cell membrane by diffusion.

Steroids binds to specific **cytosol-receptor proteins** that are then translocated into the cell nucleus where they reversibly bind to DNA (Fig. 63-1). Some unbound receptor proteins may even exist in the nucleus.

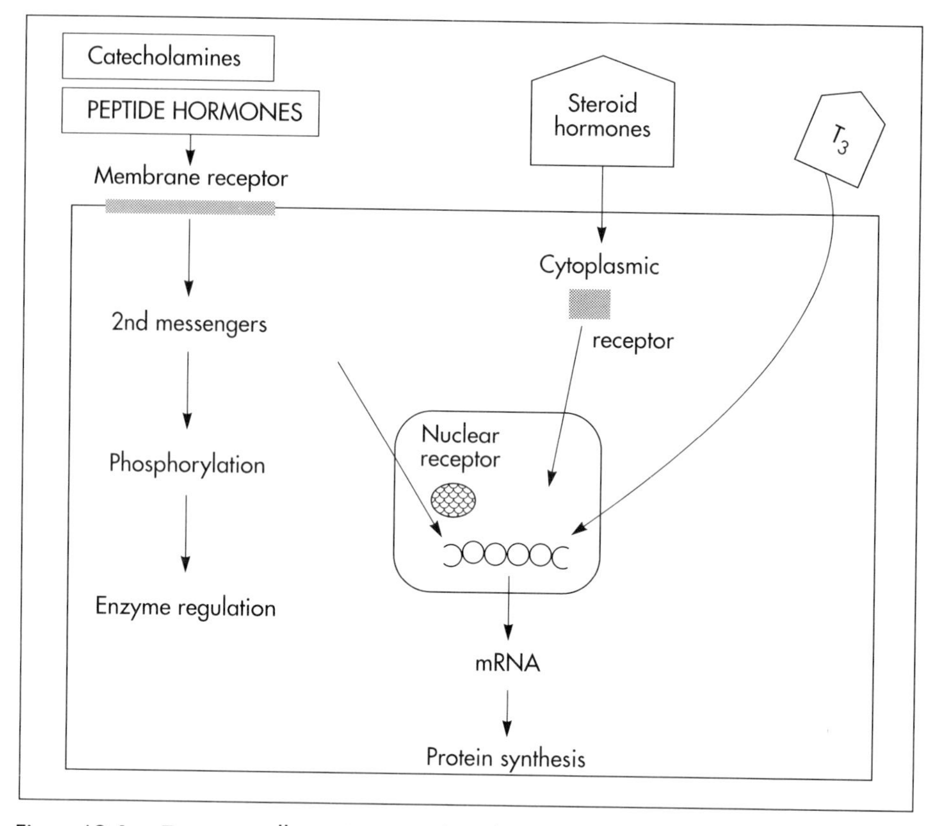

Fig. 63-1: Target cell activation by hormones acting at **membrane**, **cytoplasmic**, and **nuclear** receptors.

The binding of the steroid–receptor complex to the specific gene enhances mRNA transcription and gene expression.

Tri-iodo-thyronine (T_3) binds to **nuclear receptor proteins**, which then attach to a thyroid response unit in the gene in a manner similar to that of steroid receptors. The result is increased mRNA formation (Fig. 63-1).

Steroids and thyroid hormones frequently work in conjunction with each other (**potentiated amplification of gene expression**).

What is the mechanism of action for lipophobic hormones (monoamines, amino acids and peptides)?

Such water-soluble hormones (**first messengers**) bind to hormone receptors on the lipid-rich plasma membrane. Peptide hormone and catecholamine receptors are **membrane receptors** with a **binding domain** located extracellularly and an **effector domain** intracellularly (Fig. 63-1).

The **second messengers** involved are cAMP, cGMP, IP_3, Ca^{2+}, DAG, etc. The Ca^{2+} ion is an important second messenger. The Ca^{2+} influx to the cytosol is controlled by **hormone receptor binding, neural stimuli or modified by other second messengers.**

Sutherland discovered cAMP and demonstrated its role as a second messenger in mediating body functions (Nobel Prize 1971).

Is cAMP involved in fight-or-flight reactions in humans?

Increased activity of the sympathetic nervous system including release of Ad triggers **fight-or-flight reactions.** In the heart, Ad molecules diffuse to the myocardial cells, where they bind to membrane β-receptors. A stimulatory signal is hereby transmitted to an associated enzyme called **adenylcyclase.** This enzyme catalyses the conversion of ATP to **cAMP** (see Chapter 5, Fig. 5-1). The importance of cAMP is that it activates **protein kinase A** that, among many other functions, **phosphorylates** the **Ca^{2+}-channel protein.** This activation is correlated with an increase in the magnitude of the Ca^{2+}-influx, the force of contraction, and the heart rate.

The parasympathetic system counteracts the sympathetic by slowing the heart rate and decreasing the force of contraction. ACh is bound to another set of specific membrane receptors located on the heart cell membrane. ACh reduces the Ca^{2+}-influx that was increased by Ad. This reduction is related to a reduction in the cAMP levels or to an increase in **cGMP.**

How is cAMP liberated from the cell membrane to the cytosol?

Stimulating and inhibiting protein receptors in the target cell membrane control the cAMP liberation. Binding of a stimulating or inhibiting hormone to these receptors results in stimulation or inhibition of a membrane-bound enzyme called **adenylcyclase.** This enzyme catalyses the endergonic production of cAMP from ATP.

Most hormones have a blood concentration of approximately 10^{-10} mol l^{-1}. One molecule bound to a cell receptor, releases 10 000 times more cAMP in the cell. Hence, cAMP works as an amplifier of the hormone signal.

Phosphodiesterase (PDE) destroys cAMP. PDE enhances hydrolysis of cAMP to the inactive 5′-AMP by a highly exergonic process.

Inhibitors of the PDE (theophylline and caffeine) act synergistically with hormones that use cAMP as a second messenger.

What is the effect of cAMP?

cAMP stimulates catabolic processes such as lipolysis, glycogenolysis, gluconeogenesis, and ketogenesis. The cAMP also stimulates amylase liberation in the saliva by the parotid gland, the HCl secretion by the parietal cells, the insulin release by the β-cells in pancreas, and the increased ion permeability of cell membranes.

What second messengers are involved in insulin secretion?

When the [glucose] increases in the arterial blood and close to the β-cells of the pancreatic islets of Langerhans, the **high [glucose] triggers an increase in Ca^{2+} flux to the cell interior** via voltage-sensitive Ca^{2+} channels (Fig. 5-2).

The initial surge in insulin secretion is caused by calmodulin-dependent **protein kinases.**

The high cytosolic [Ca^{2+}] activates the membrane phospholipases A_2 and C. Phospholipase A_2 releases arachidonic acid (AA) which stimulates insulin secretion.

Phospholipase C catalyses the formation of IP_3 and DAG (Fig. 5-2). The IP_3 releases more Ca^{2+} from the ER, and DAG activates protein kinase C (Fig. 5-2).

The decrease in insulin secretion after the initial surge and its subsequent increase can be explained by the action of protein kinase C.

Initially, the active protein kinase C reduces cytosolic [Ca^{2+}] and thus reduces the initial calmodulin-dependent insulin secretion. Later, protein kinase C stimulates the formation of cAMP and amplifies the induction of calmodulin-dependent protein kinase thereby causing a gradual increase in insulin secretion. Prolonged glucose stimulation probably leads to downregulation of protein kinase C. An abnormally prolonged glucose stimulation may render β-cells **glucose blind** and thus spoil their function (see Chapter 73).

Insulin secretion is not only stimulated by **glucose**, but also potentiated by ACh via phospholipase C and by glucagon via activation of adenylcyclase. β-Agonists stimulate β-receptors on the glucagon producing α-cells, whereas α-agonists inhibit insulin secretion via α_2-receptors on the β-cells. ACh and glucagon react by activating **protein kinase C** and cAMP-dependent **protein kinase A**, respectively. Both mechanisms potentiate the Ca^{2+}-triggered insulin secretion.

How are peptide hormones synthesized?

Transcription in the cell nucleus produces a **precursor messenger RNA** molecule complementary to part of a DNA. The precursor is processed into messenger RNA and transported through the nuclear membrane into the cytoplasm. Messenger RNA carries the genetic information in triplet codons. Messenger RNA binds to ribosomes and **transfer RNA** molecules synthezise peptides (**ribosomal translation**).

Translation produces big precursor molecules (**pre-prohormones**). Precursors have a **signal peptide** that contains processing information to ensure that the protein enters the rough ER. Here enzymes split the precursor into a signal molecule and a prohormone. Finally, peptide hormones undergo **post-translational processing** (e.g. TSH and gonadotropins are glycosylated; insulin forms a zinc-complex). The hormones reach the Golgi complex, where they are packed into secretory granules that migrate to the cell surface.

Roger Guillemin synthesized brain peptides that regulate the pituitary secretion *in vitro*. He received the Nobel Prize in 1977.

What is the principle in radio-immunoassays (RIA)?

Radio-immunoassay (RIA) is the measurement of hormone concentrations by utilizing their immunological effect. First a specific antibody is produced. A known amount of this antibody is placed in a series of test tubes containing known amounts of the hormone to provide a standard curve. A known amount of antibody is also placed in a series of test tubes containing unknown concentrations of the hormone. Now it is easy to compare the response of the unknown sample to the standard curve. Rosalyn S. Yalow and Saul Berson developed the RIA method. Rosalyn Yalow received the Nobel Prize in 1977.

1. An **antigen–antibody complex** is formed, and with sufficient antibody, the overshoot in each test tube is measurable by addition of known amounts of radioactive hormone in excess of the overshoot.

2. The radioactivity is measured in the **supernatant** containing the excess of free, radioactive hormone or in the **precipitate**.

Recent variations of the RIA technique include **immunoradiometric, chemiluminescent, and enzyme-linked immunosorbent assays**. In the **radioreceptor assay** a hormone receptor is substituted for the antihormone–antibody in RIA.

Are physical–chemical principles still advantageous for hormone analysis?

High pressure liquid chromatography (HPLC) is a sensitive and excellent analysis used for many hormones and biochemical key molecules.

Colorimetry/spectrophotometry uses absorption of monochromatic light which is a precise method.

What is receptor modulation?

Cell membrane and intracellular receptors can change their affinity and number. A specific ligand for a receptor is able to modulate the total number of this receptor. Increasing the concentration of the ligand (hormone, neurotransmitter, drug) often reduces the number of receptors and vice versa. **Maximal effects of hormones are generally observed at a receptor occupancy of less than 50%**. The myoepithelial cells (myometrium and breast) contain oxytocin receptors. Their number is upregulated by oestrogens and downregulated by progesterone. The cardiac muscle contains **noradrenergic receptors** (β_1). Both affinity and number of receptors is increased by **thyroid hormone stimulation** (T_3/T_4).

What is endocrine cybernetics?

Endocrine cybernetics is a control system discipline. We all use this concept when explaining and representing humoral regulation using simple feedback systems.

An endocrine feedback system is a system whereby the first hormone controls the secretion and liberation of the second. The second hormone acts by feedback and modulates the secretion of the first.

A negative feedback system contains at least one step of inhibition. The total effect is to minimize any external change introduced to the system. Almost all hormone systems maintain homeostasis by negative feedback.

A positive feedback system is a system where an external change leads to increased secretion of hormone 1, which also leads to a secondary rise in hormone 2's concentration. This is an auto-accelerating phenomenon and a rarity. The most important example in humans is the **steep rise in blood [oestradiol] in the middle of the menstrual cycle**. High [oestradiol], when maintained for longer than 35 h, stimulates by positive feedback the luteinizing hormone (LH) and follicle-stimulating hormone (FSH) secretion from the adenohypophysis. By contrast, moderate plasma [oestradiol] levels, which are present during the other parts of the cycle, provide negative instead of positive feedback.

Long feedback systems act on the hypothalamo-pituitary system from remote target organs.

Short feedback systems use a short distance feedback, such as the influence of the hypophysis back to the hypothalamus. **Auto-feedback** refers to the action of a liberated hormone that was secreted on the cell from where it came, thereby modulating its own secretion.

What is internalization?

Internalization is the **transport of hormone–receptor complex into the cell by an endocytotic vehicle.** This is primarily a means of **terminating** the action of the hormone. After destruction of the hormone by lysosome enzymes, the receptor returns to the surface and is reused.

How is it possible to estimate the function of a hormone system in the clinical situation?

1. The **hormone concentration** in the blood is commonly used. It can be measured by taking advantage of the new methods described above.
2. The **secretion flux** of T_3 and T_4 from the thyroid gland.
3. The **metabolic rate** or the **absorption rate** of ^{131}I (radioactive iodine) in the thyroid gland. The physical half-life of ^{131}I is 8 days or 192 h. The **elimination rate constant (*k*)** of a substance is the amount eliminated per unit time divided by the total amount present in the distribution volume. The variable ***k*** is easy to calculate:

$$T_{\frac{1}{2}} = \ln 2/k = 0.693/k.$$

   The value of ***k* for iodine** is 0.693/192 or 0.0036 h^{-1}.
4. **The elimination rate.** Abnormal amounts of catecholamines or **VMA (vanillyl mandelic acid)** in a 24-h urine sample suggest the presence of a **catecholamine producing tumour** (pheochromocytoma).
5. **Stimulation test.** Stimulation with ACTH without a substantial rise in plasma [cortisol] suggests **primary, adrenocortical atrophia.**
6. **Suppression test.** Dexamethasone (a cortisol synergist) is administered to a Cushing suspect in the evening. The next morning a measurement of plasma [cortisol] shows suppression in normal persons and in patients with a **primary, adrenocortical hyperfunction.** Hypothalamic/pituitary Cushing is never suppressed by cortisol.
7. **The glucose tolerance test.** A load of glucose normally triggers an **increased rate of insulin production.**

Describe some widely used clinical applications of hormones

Distribution of oestrogens and progesterone in **contraceptive** pills is world wide. Oestrogens are widely used to relieve **postmenopausal discomfort.** Now some females with **osteoporosis** are treated experimentally with calcitonin.

Insulin is a life-saver for **diabetics,** and it is produced and distributed as pure human insulin.

In the affluent areas of the world many **women deliver their babies** following an oxytocin infusion.

Oestrogens and gonadotropins are used in treatment of **sterility and menstrual disturbances.**

Huggins received the Nobel Prize in 1966 for the introduction of a new form of **cancer therapy in which sex hormones are used to retard cancer growth.** He used androgens for breast cancer and oestrogens for prostate cancer.

Further reading

Guillemin, R. (1985) Physiological studies with somatocrinin, a growth releasing factor. *Ann. Rev. Pharmacol. & Toxicol.* **25**: 463–85.

Sutherland, E.W., Jr *et al.* (1966) The role of cAMP in response to catecholamines and other hormones, *Pharmacol. Rev.* **18**: 145–61.

Yalow, R.S. (1987) Radioimmunoassay: Historical aspects and general considerations. In: *Handbook of Experimental Pharmacology*.

63. Case History

The adenohypophysis of a 23-year-old woman contains approximately 300 μg of TSH with a molecular weight of 31 000. TSH has a $T_{\frac{1}{2}}$ in plasma of 55 min and a concentration of 100 pmol l^{-1} plasma. The haematocrit of the patient is 0.5. The woman secretes thyroid-stimulating hormone (TSH) to her TBV, which is 4 l.

1. *Develop an equation for the calculation of her TSH secretion (J mol h^{-1}). The rate constant k can be used.*
2. *Calculate the secretion of TSH from her adenohypophysis.*
3. *What fraction of her total TSH store is secreted per 24 h?*

Try to solve the problems before looking up the answers in Chapter 74.

CHAPTER 64. THE HYPOPHYSIS AND THE HYPOTHALAMUS

What is special about the pituitary gland?

The pituitary gland consists essentially of two parts both controlled by the hypothalamus.

The **glandular** part is the adenohypophysis or anterior lobe, and the **neural** part is the neurohypophysis or posterior lobe.

The **adenohypophysis** develops ectodermally from the primitive mouth cavity. The **neurohypophysis** develops from the neuroectoderm. The two parts combine to form one body called the adeno-neurohypophysis that weighs about 0.5 g.

Describe three compartments of the hypothalamo-hypophyseal system

1. The neurosecretory neurons passes from the supraoptic and paraventricular nuclei of the hypothalamus to the neurohypophysis.
2. **The hypophysiotropic zone** in the median eminence of the hypothalamus is connected to the adenohypophysis. Both releasing and inhibiting peptides are transported along the portal circulation.
3. **The neuroregulatory peptides** are **endogenous opiates** (endorphins and encephalins), **β-lipotropin, neurotensin, substance P, VIP**, etc. Many of these peptides are cut off from a big mother molecule: pro-opio-melanocortin (POMC). These peptides may exhibit a **permissive effect** on other hormones (ACTH, GH) related to behaviour and autonomic responses. During exercise (a Cooper test which lasts 12 min) the plasma [β-endorphin] and [ACTH] increases by 200–300% from the normal resting averages of 1.7 and 2.2 pmol l^{-1}, respectively. There is an increase in plasma [ACTH] and its accompanying neuroregulatory peptides during prolonged **stress** such as exhaustive exercise and chronic disease. The endogenous opiates affect **stress adapting behaviour**, e.g. the euphoria observed in the chronically ill.

I. THE NEUROHYPOPHYSIS

Where are the hormones of the neurohypophysis formed?

Neurosecretory neurons (which have nuclei in the hypothalamus and axons that lead to the posterior lobe of the hypophysis) and **peptidergic neurons** (spread in the nervous system) produce and liberate peptides in much the same way. We store the neurosecretory hormones (vasopressin and oxytocin) in the neurohypophysis, but the **production starts** in the **hypothalamus**. First mRNA are formed in the cell nucleus of the neuron bodies in the supraoptic and paraventricular nuclei. These mRNA are produced by transcription from DNA molecules. The mRNA diffuse through the nuclear membrane and bind to protein synthesizing ribosomes. They can translate the genetic code to several **transfer RNA (tRNA) molecules** and build the respective protein or peptide molecule. In this way we synthesize **high molecular pre-prohormones** in the ribosomes close to the nucleus. The signal peptide is cut off in the ER. The precursors are then packed in secretion granules in the Golgi complex. The **big pre-prohormone** consists of four subunits: a signal peptide, the hormone, neurophysin and a glycopeptide (in the case of vasopressin).

What is axoplasmic transport?

The secretory **granules travel through the axons** of the neurosecretory neurons, that form the supraoptico-hypophysial tract, with high velocity (more than 100 μm h^{-1}). This tract runs through the pituitary stalk and ends in the neurohypophysis. The transfer is known as the **axoplasmic transport**. The neurohypophysary terminals are located close to the capillary blood. During transport the prohormone splits into its subunits. The two hormones are then released. These hormones are built of nine amino acids, and the mean molecular weight for these acids is 120. Accordingly, these two nonapeptides must have a molecular weight around $(9 \cdot 120) = 1080$. This estimate is approximately equal to the precise molecular weight for vasopressin (1084 Da).

The nerve endings of this tract in the neurohypophysis is the **storage area** for these two neurosecretory hormones; secretion to the blood takes place through fenestrated capillaries. The secretion granules release their content by regulated exocytosis. Exocytosis is triggered when the neurosecretory neuron is depolarized and an action potential is transferred to the terminals.

When is vasopressin released and what is its actions?

Even a small rise in the osmolarity of plasma stimulates osmoreceptor cells, located close to the neurosecretory cells in the hypothalamus.

The osmoreceptors stimulate both production and secretion of vasopressin (ADH) in the neurosecretory cells. The plasma [ADH] will then rise from the basal level which is 2 pmol l^{-1}. The normal secretion flux is 10^{-13}mol ADH kg^{-1} body weight min^{-1}, and the biological half-life in human plasma is 18 min. The plasma [ADH] in some females is increased in the premenstrual phase.

1. ADH eases the renal reabsorption of water in the **cortical collecting ducts** (and not in the outer medulla but in the **inner medulla**) – leading to antidiuresis.
2. ADH probably stimulates the **active solute reabsorption** (NaCl) in the **thick ascending limb** of the renal Henle loop. Thus, ADH helps maintain the concentration gradient in the kidney.
3. Vasopressin is a **universal vasoconstrictor**. Vasopressin reduces the small, medullary blood flow through the vasa recta along the Henle loop.

   ADH acts on the basolateral membrane of the cells, and the result is a rise of [cAMP] in the cytosol (Fig. 5-1). The cAMP diffuses to the luminal side, where it causes vesicular structures to develop and fuse with the luminal membrane. Hereby, the membrane receives a large number of water channels, so the membrane becomes highly water permeable. Water diffuses through the cell to the basolateral membrane and into the ISF.

What is known about inappropriate ADH secretion?

ADH producing tumours in the hypophysis or in the lungs cause the **syndrome of inappropriate ADH secretion.** Water retention, concentrated urine, hyposmolar plasma, and muscle cramps characterize this syndrome.

Another type of disturbed secretion is **ADH insufficiency**, which causes diabetes insipidus. A total lack of ADH can result in a diuresis of 25 l per 24 h.

When is oxytocin released and what are its actions?

Stimulation of **tactile receptors in the mammary nipple** causes the neurosecretory neurons to release oxytocin through a neuroendocrine reflex.

The latency between the stimulus and milk ejection is due mainly to the transport of oxytocin in the blood from the neurohypophysis to the milk ducts (20–30 s). Oxytocin stimulates the **myoepithelial cells in the milk ducts** of the lactating breast so that milk is ejected to the baby.

Oxytocin also stimulates the **myoepithelial (myometrial) cells of the uterus** which is believed to satisfy some women sexually during breast feeding. Oxytocin is in fact the drug of choice to induce labour and to augment labour when the uterus muscle is not functioning adequately.

What is diabetes insipidus?

The true form of **diabetes insipidus** is caused by **ADH deficiency**. There are two types of diabetes insipidus. The primary or idiopathic type, which is probably due to a genetic defect that blocks the hormone production, and the secondary type, where the hypothalamo-hypophysary system is damaged.

The renal tubule cells are rarely insensitive to ADH, and the condition is called **renal diabetes insipidus**.

The symptoms and signs are mainly large diuresis and a tremendous thirst (polydipsia).

II. The Adenohypophysis

What characterizes the five hormones secreted from the adeno-hypophysis?

These **five** hormones are tropic hormones – they regulate the hormone secretion of other cells. They include:

1. **Thyrotropin** or TSH, which is produced in thyrotropic cells, stimulates synthesis and secretion of thyroid hormones and maintains the integrity of the thyroid gland.
2. **Gonadotropins (FSH and LH)** are from gonadotropic cells. FSH stimulates the growth of the ovum, the Grafian follicles in females and spermatogenesis in males. LH stimulates ovulation and the development of the corpus luteum. LH regulates testosterone production in males. FSH and LH stimulate secretion of oestrogens through the menstrual cycle, and of progesterone in the second half of the cycle.
3. **Corticotropin (ACTH)** from corticotropic cells stimulates secretion of adrenal cortical hormones (mainly glucocorticoids), and maintains the integrity of adrenal cortex.
4. **Somatotropin (human growth hormone (HGH))** from somatotropic cells regulates growth, partly directly and partly through evoking the release of somatomedins from the liver and elsewhere. Protein synthesis, lipolysis and blood [glucose] are increased by this hormone.

5. **Prolactin** or **mammotropin** produced in mammotropic cells, promotes development of mammary tissue during pregnancy and stimulates milk production in the post-partum period.

What is special about the blood supply to the adenohypophysis?

The carotid artery supplies blood to the hypophysis through the superior hypophysary artery. The blood spreads out into a **primary capillary network** in the **eminentia mediana hypothalami**.

From this network the venous blood is transferred into long, portal veins to a **secondary network** of highly permeable capillaries or **sinusoids** in the adenohypophysis.

[A **third capillary network**, between the neurohypophysis and the median eminence of the hypothalamus, allows **short loop feedback** from the hypophysis to the hypothalamus.]

The adenohypophysis produces tropic, peptide hormones. What is characteristic of their molecular structure?

1. One group contains **glycoproteins** with two peptide chains (α96-β-glycoproteins). There is a special, biologically active β-chain for each of the three hormones TSH, FSH and LH, although the inactive α-chain is the same for all of them.
2. Somato-mammotropins are **single-chain peptides** containing 200 amino acids of almost the same sequence.
   HGH and prolactin (PRL) are probably simple gene duplicates from the same prohormone molecule.
3. **POMC peptides are neuroregulatory hormones**: ACTH, endogenous opiates, β-endorphin, β-lipoprotein, α-MSH and β-MSH.

Histamine plays an important role in pituitary hormone secretion (Knigge and Warberg). It stimulates the secretion of ACTH, β-endorphin, α-MSH, and PRL. Histamine participates in the release of these hormones during prolonged **stress** and possibly in the suckling- and oestrogen-induced PRL-release.

The release of GH and TSH are predominantly inhibited by histamine.

Histamine increases the secretion of LH in females – mediated by GnRH. Histamine probably affects the cell bodies in the supraoptic and paraventicular nuclei, stimulating the formation of arginine vasopressin and oxytocin.

Describe clinical pictures of abnormal pituitary hormone production

Panhypopituitarism is due to total destruction (lesions or tumour invasion) of all hormones in the hypothalamo-hypophysary system. Lack of GH and somatomedins result in a dwarf without normal sex development (lack of LH and FSH). This dwarf has also hypothyroidism (lack of TSH), and a Cushing-like hypercorticism (no ACTH secretion).

Hyperpituitarism is often caused by prolactin producing microadenomata which cause abnormal milk production. This leads to disturbance of the menstrual cycle and infertility. Other pituitary adenomas produce large amounts of GH leading to gigantismus in childhood and to acromegaly in adults.

Further reading

Bowman, W.C. and M.J. Rand (1990) The Endocrine System and drugs affecting Endocrine Function. In: *Textbook of Pharmacology*, 2nd ed. Blackwell Scientific Publications, London.

Cohen, S. and G. Carpenter (1987) Receptor for epidermal growth facor and other polypeptide mitogens. *Ann. Rev. Biochem.* **56**: 881–914. (Nobel Prize to Cohen 1986).

Knigge, U. and J. Warberg (1991) The role of histamine in the neuroendocrine regulation of pituitary hormone secretion. *Acta Endocrinol.* (Copenhagen) **124**: 609–19.

Levi-Montalcini, R. *et al.* (1989) Nerve growth factor mRNA and protein increase in hypothalamus in a mouse model of aggression. *Proc. Natl Acad. Sci., USA* **86**: 8555–9, 1989.

64. Case History

A woman (24 years of age; height: 1.70 m; weight: 60 kg) is in hospital due to a tremendous thirst, and she drinks large amounts of water. Since she is producing 10 or more litres of urine each day, the doctors suspect the diagnosis to be diabetes insipidus. The vasopressin concentration in plasma (measured by a RIA method) is 10 fmol l^{-1}. Her secretion of vasopressin is only 5% of the normal flux of 10^{-13} mol min^{-1} kg^{-1} body weight. The normal plasma [vasopressin] is averagely 2 pmol l^{-1}. The ECV is 20% of her body weight.

Vasopressin is injected intravenously on several occasions. A dose of 3 μg vasopressin is the minimum necessary to normalize her diuresis for 4 h. Before the injection her diuresis is 6 ml of urine min^{-1}, but within 25 min her urination is constantly around 0.5 ml min^{-1}.

1. *Calculate the secretion of vasopressin (in μg h) from the neurohypophysis of a normal 60 kg person and of this patient.*
2. *Calculate the distribution volume for vasopressin, which is 20% higher than ECV.*
3. *Assume the 3 μg vasopressin injected to be distributed evenly immediately after the intravenous injection. Calculate the rise in vasopressin concentration in the distribution volume.*
4. *Estimate the relation between this concentration and that of a healthy individual.*
5. *Does this ratio have implications for the interpretation of her special type of diabetes insipidus?*
6. *Is it dangerous to lose 10 l of urine per day?*

Try to solve the problems before looking up the answers in Chapter 74.

CHAPTER 65.
THE THYROID GLAND

The thyroid gland maintains the metabolic level of almost all cells in the body by producing, in its follicular cells, two thyroid hormones: **tri-iodothyronine** (T_3), and **tetra-iodothyronine** (T_4) or **thyroxine**. Iodine (I_2) has an atomic weight of 127 and a molecular weight of 254; T_4 has a molecular weight of 777 Da of which 508 is iodide.

Thyroid hormones are essential for **normal neural development, linear bone growth** and **proper sexual maturation.**

How are the thyroid hormones synthesized?

The synthesis depends on an adequate supply of dietary iodine (at least 75 μg daily is required to prevent goitre formation).

1. The **iodine** is reduced to **iodide** (I^-) in the gastrointestinal tract and is rapidly absorbed.
2. Follicular cells have an active **iodide pump** that acquires and concentrates iodide (I^-) from the blood against an electrochemical gradient of more than 50 mV (inside negative). The pump is dependent on the function of a Na^+-K^+-ATPase. The absorption of iodide is inhibited by ouabain and negative ions (thiocyanate, nitrate).
3. Iodide is rapidly oxidized by a peroxidase to iodine (I_2) and incorporated into tyrosyl residues in **thyroglobin** which is located **in the colloid**. This molecule is produced in the Golgi complex by **organification** and packed into cellular vesicles.
4. The multiple tyrosyl units on the thyroglobulin molecule are iodinated, either at one or two positions, forming mono-iodinethyronin (MIT), and di-iodinethyronin (DIT) respectively. When MIT **couples** to DIT it forms T_3, and two DIT molecules form T_4 or **thyroxine**. Coupling depends on the conformation of thyroglobin and is catalysed by **thyroid peroxidase.**

We store iodinated thyroglobin containing both T_3/T_4 molecules as well as MIT and DIT by **exocytosis** of vesicle contents into the central colloid fluid. Each of the thyroid follicles consists of a single epithelial layer and a lumen that is filled with colloid. Some follicle cells are only separated from the capillary blood by a thin basement membrane and fenestrations in the endothelial walls of the capillaries.

How are the thyroid hormones liberated from the colloid?

The follicular cells absorb **colloid with thyroglobulin which contains thyroid hormones** by **endocytosis** of small colloid vesicles. These vesicles fuse with lysosomes forming **phagolysosomes** which contain proteases called **lysozymes** that liberate T_3 and T_4. These hormones are eventually secreted to the blood through the basement membrane. TSH stimulates almost all processes involved in thyroid hormone synthesis and secretion.

Do we have vehicles for the transport of T_3 *and* T_4 *in the blood?*

In the blood we have only small amounts of **thyroxine-binding globulin** (TBG; approximately 10 mg l^{-1}), but the affinity for T_4 is high. The total T_4 is 10^{-7} mol l^{-1} equal to 77.7 μg l^{-1} of blood serum, because 777 g of T_4 equals 1 mol. Out of the total, 90% binds to TBG, and the rest to **thyroxine-binding prealbumin** (TBPA) and **thyroxine-binding albumin** (TBA; minimal amounts). Oestrogens stimulate the synthesis of TBG. T_3 binds to TBG and albumin, but not to TBPA. The T_3 hormone is eliminated quickly (half-life 24 h), because it has the lowest degree of protein-binding. The thyroxine (T_4) molecule has a half-life of 7 days, almost equal to the physical half-life of the radioactive isotope ^{131}I (8 days).

T_4 is likely to be a prohormone, which is deiodinized by **mono-deiodinase** to the more potent T_3 just before it is used in the cells. Thus T_3 is probably the final hormone, although it is present only in a very low concentration (10^{-9} mol l^{-1}).

In what way are the thyroid hormones broken down?

The thyroid hormones are broken down by **deiodination, deamination** and **decarboxylation.**

Small parts of these molecules are coupled to sulphate and glucuronic acid in the liver and end in the bile. Bile is released into the **enterohepatic circuit.**

What happens when thyroid hormones enter the cells?

Thyroid hormones are lipid soluble and pass through cell membranes easily. T_3 binds to specific **nuclear receptors** with an affinity that is 10-fold greater than for T_4. The information alters **DNA transcription** into mRNA, and the information is eventually **translated** into many **effector proteins**. Important cellular constituents are stimulated by T_3: the mitochondria, Na^+-K^+-ATPase, myosin ATPase, adrenergic β-receptors and many enzyme systems.

What are the effects of thyroid hormones?

Thyroid hormones **stimulate oxygen consumption** in almost all cells.

They stimulate the rate of (1) hepatic glucose output and peripheral glucose utilization; (2) hepatic metabolism of fatty acids, cholesterol and triglycerides; and (3) the synthesis of important proteins.

The many rate-stimulating effects are summarized in an overall increase in oxygen consumption. This slow – but long lasting – calorigenic and thermogenic effect is confined to the **mitochondria**.

The thyroid hormones and the catecholamines work together in metabolic acceleration. Thyroid hormones increase the number of β-adrenergic receptors. Thyroid hormones modulate the secretion of sex hormones (sex development), growth hormone (growth), and nerve growth factors (CNS development).

The high basal metabolic rate raises the core and shell temperature, so that the peripheral vessels dilate. This vasodilatation forces the cardiac output ($\mathring{Q}$) to increase. A circulatory shock develops, if the high $\mathring{Q}$ is insufficient – so-called **high output failure.**

A human body overloaded with thyroid hormones for a prolonged period (**hyperthyroidism**) will suffer from muscle atrophia, bone destruction and hunger damage, due to increased catabolism of cellular proteins and fat. Eventually

hypothyroidism may develop due to suppression. Thyroid hormones increase the absorption of carbohydrates in the intestine and the excretion of cholesterol.

How are the thyroid hormones regulated?

Without the thyroid gland the metabolic rate slows down. **TSH (thyrotropin) from the thyrotrope cells of the anterior pituitary is the only known regulator of thyroid hormone secretion in humans.** The iodide uptake by the thyroid gland increases within some hours after injecting TSH. Synthesis of MIT, DIT, T_3 and T_4 is enhanced, and secretion increases in proportion to the increases in pinocytosis of colloid.

TSH activates adenylcyclase bound to the cell membranes of the follicular cells and increases their cAMP. **T_3 has a strong inhibitory effect on TRH secretion,** as well as on the expression of the gene for the TRH precursor.

Mutations of a gene located on chromosome 10 can produce a mutation in receptor tyrosine kinase proto-oncogene associated with **thyroid medullary carcinoma.**

What is hypothyroidism?

Adult hypothyroidism is a condition characterized by the accumulation of mucopolysaccharides and by cold, dry skin almost like in **oedema**, but of the non-pitting type (**tortoise-skin**) often called **myxoedema.**

Iodide deficiency in childhood may result in a **hypothyroid dwarf** or **cretin.**

1. **Thyroid gland hypothyroidism** is characterized by high TSH. A test dose of TSH to a patient with thyroid hypothyroidism will not stimulate the thyroid gland.
2. The condition can also be a **pituitary or a hypothalamic hypothyroidism.** A test dose of TRH to a hypothalamic myxoedema patient will result in a normal TSH response.

   A test dose of TRH will result in an **increased TSH response** in **thyroid gland hypothyroidism** and decrease in **hyperthyroidism.** This is due to the negative feedback of thyroid hormones on the hypophysis. Hypothyroid females often have excessive and frequent menstrual bleedings (menorrhagia and polymenorrhoea).

What is hyperthyroidism?

Patients with hyperthyroidism or **thyrotoxicosis** have overwhelmingly high metabolic rates. The patient is extremely nervous. The patient trembles, sweats and has a high heart rate and basal metabolic rate. Erroneous treatment with thyroid hormone can kill the patient by causing vasodilatation and **cardiac output failure.**

Graves disease or **Morbus Basedowii** is the combination of **thyrotoxicosis, struma and exophthalmus. Hyperthyroidism** is probably caused by **TSH-receptor antibodies,** which bind to the thyroid follicle cells and stimulate the gland to secrete T_3 and T_4. The rise in thyroid hormone concentration will **suppress TSH secretion.**

Lean hyperthyroid females – like female distance runners – have small fat stores and greatly reduced menstrual bleedings (oligomenorrhoea) or even amenorrhoea.

What is struma?

Struma is a **visible or palpable enlargement of the thyroid.** Struma is due to **iodine deficiency, increased iodine demand** or **strumagens.** Any prolonged TSH stimulation results in an enlarged thyroid.

Diseases in the thyroid gland including struma are caused by malfunction in the gland itself or by **hypothalamo-pituitary defects.**

What are the different effects of antithyroid substances?

The **iodide trap** is inhibited by monovalent anions and ouabain.

Thiocarbamide inhibits the **iodination of tyrosyl residues**.

Sulphonamides inhibit **thyroid peroxidase** which oxidizes iodide to iodine. Large doses of iodide inhibit the **TSH receptors** on the thyroid gland.

Calcitonin and bone minerals

The thyroid gland perifollicular chief cells (C-cells) secrete another hormone, **calcitonin**, which has important effects on our calcium balance. Calcitonin is a single-chain peptide with a disulphide ring, containing 32 amino acids. Calcitonin has a MW of Da. **Calcitonin is secreted from the thyroid gland in response to hypercalcaemia and it acts to lower plasma [Ca^{2+}] (as opposed to the effect of parathyroid hormone (PTH)).**

Administration of calcitonin leads to a rapid fall in plasma [Ca^{2+}]. **Calcitonin is the physiological antagonist to PTH and inhibits Ca^{2+} liberation from bone** (i.e. inhibits both osteolysis by osteocytes and bone resorption by osteoclasts). But calcitonin reduces plasma phosphate just as PTH.

Calcitonin probably inhibits reabsorption of phosphate in the distal tubules of the kidney, but calcitonin also inhibits the renal reabsorption of Ca^{2+}, Na^{+} and Mg^{2+}. Calcitonin may inhibit gut absorption of Ca^{2+} and promote phosphate entrance into bone.

Calcitonin deficiency does not lead to hypercalcaemia, and **excess calcitonin from tumours** does not lead to hypocalcaemia. Therefore, most effects of calcitonin are evidently off-set by appropriate regulation through the actions of PTH and vitamin D.

What is the physiological importance of calcitonin?

Calcitonin causes important **bone remodelling.**

Calcitonin in plasma declines with age and is lower in women than in men. Low levels of calcitonin is involved in accelerated bone loss with age and after menopause (**osteoporosis**).

Calcitonin **protects the female skeleton** from the drain of Ca^{2+} during pregnancy and lactation.

Calcitonin is a **neurotransmitter** in the hypothalamus and in other CNS locations.

Calcitonin is administered to postmenopausal females in order to prevent osteoporosis.

Further reading

Boulieu, E.-E. and P.A. Kelly (1990) *Hormones. From Molecules to Disease*. Hermann Publ. and Chapman & Hall, London.

Griffin, J.E. and S.R. Ojeda (1992) *Textbook of Endocrine Physiology*. Oxford University Press, London.

Hofstra, R.M.W. *et al.* (1994) A mutation in the RET proto-oncogene associated with multiple endocrine neoplasia type 2B and sporadic medullary thyroid carcinoma. *Nature* **367**: 375–7.

65. Case History

A 49-year-old woman (weight 52 kg; height 1.69 m) is in hospital and is being examined for thyroid disease. Her distribution volume for iodine is 12 l and her renal clearance is 36 ml plasma min^{-1}. In a period where her iodine intake equals her output, she is subjected to the following test. In the morning she receives a small dose of the radio-active isotope ^{131}I, and 3 h later she urinates. From that moment she collects her urine for the following 2 h. The urine collection has a volume of 0.2 l and an iodine concentration of 65 μg l^{-1}. The total urine radioactivity is $1.6 \cdot 10^7$ Bq. During the 2-h test period, her plasma concentration of ^{131}I falls from 38 300 to 26 100 Bq l^{-1}. The radioactivity in the thyroid gland (measured with a scintillation counter) increases during the test by 77 500 Bq.

1. *Calculate the concentration of iodide in her plasma at the start of the test and at the end of the test.*
2. *Calculate the uptake of iodide in the thyroid during the 2-h test and compare the result with a mean value of 2.4 μg h^{-1} for healthy persons.*
3. *Calculate the thyroid plasma clearance for iodide and compare the result to the expected value of 10 ml min^{-1}.*
4. *Calculate the elimination rate constant for iodide.*
5. *Calculate the biological half-life for iodide in its distribution volume and compare the result to the physical half-life of ^{131}I (8 days).*

Try to solve the problems before looking up the answers in Chapter 74.

CHAPTER 66.
THE PARATHYROID GLANDS

For many years we overlooked the existence of the parathyroid glands until we realized the consequence of their surgical removal. **Without the parathyroids, a person develops tetanic cramps due to a fall in the concentration of** [Ca^{2+}] in blood plasma. In 1909, MacCallum treated this condition successfully with calcium salts.

Characterize the parathyroid hormone (PTH)

PTH is a single-chain peptide containing 84 amino acids with a mean MW of 113, so that PTH is (113 · 84), about 9500 Da. Humans carry a single PTH gene on chromosome 11. The **chief cells** (C cells) of the parathyroids produce PTH and **parathyroid secretory protein** of unknown function. In man, the four parathyroid glands are located just behind the thyroid gland. Ectopic tissues sometimes develop in the mediastinum or in the neck. Removal of such tissues in order to cure hyperparathyroidism is an extremely difficult task.

How does PTH work?

PTH binds to **membrane receptors** on all target cells. The binding activates adenylcyclase and this raises the [cAMP], which interacts with protein kinase A. The enzyme then catalyses the **phosphorylation of effector proteins**. PTH is secreted in response to hypocalcaemia, in particular, a low [Ca^{2+}] (or low [Mg^{2+}]). The major effects of PTH are to **increase plasma [Ca^{2+}] and decrease plasma [phosphate]** through its effect on three target organs: bone, kidney and gut.

1. PTH accelerates the **removal of Ca^{2+} from bone** by several processes. PTH stimulates the osteolysis by surface osteocytes causing the release of Ca^{2+} for rapid equilibrium with the ECV. After 12 h, the delayed effect of PTH, which stimulates the osteoclasts to resorb mineralized bone, sets in.
2. PTH reduces the **reabsorption of Ca^{2+} from the proximal tubules** of the kidneys and increases the reabsorption of Ca^{2+} from the distal tubules. This often results in an increased net loss of Ca^{2+} in the urine. PTH binds to the basolateral membrane of the tubule cell, stimulates cAMP, which in turn diffuses through the cell to the luminal membrane. Here cAMP activates a Ca^{2+} reabsorption port.
   Another drastic action of PTH is on the proximal tubules, where PTH **inhibits phosphate reabsorption** so efficiently that its excretion in the urine increases within 5 min.
3. The PTH action on the gut is indirect. PTH stimulates renal formation of biologically **active vitamin-D** (1,25-dihydroxy-D_3). This vitamin **directly stimulates the active absorption of Ca^{2+} and phosphate across the gut mucosa** and potentiates the action of PTH on bone resorption (Fig. 67-2).

What are the symptoms and signs of hypoparathyroidism?

The cause is often iatrogenous (caused by thyroid surgery). The result is a dramatic fall in plasma [Ca^{2+}] and a rise in [phosphate], leading to **hypocalcaemic cramps** and – without therapy – death.

What are the symptoms and signs of hyperparathyroidism?

Excessive secretion of PTH leads to: bone resorption, high [Ca^{2+}] in plasma, high Ca^{2+} excretion flux in the kidneys with **renal stone formation, bone lesions** and **metastatic calcification.**

Further reading

See **PTH/calcitonin**: Chapter 65. **PTH/vitamin D**: Chapter 67. **PTH/growth**: Chapter 68.

Auerbach, G.D., S.J. Marx and A.M. Spiegel (1985) Parathyroid hormone, calcitonin, and the calciferols. In: J.D.Wilson and D.W. Foster (Eds) *Textbook of Endocrinology*, 7th edn, pp. 1137–1217. W.B. Saunders, Philadelphia.

66. Case History

A man, 49 years of age (height 1.86 m; weight 62 kg) is in hospital due to the following symptoms and signs. He is nervous and has a diffuse struma. A characteristic, blowing sound is heard from the thyroid gland with a stethoscope. The blood pressure is 145/70 mmHg. An attack of cardiac arrhythmia is recorded with an ECG. A P-wave frequency above 400 min^{-1} is present during the attack. The concentration of thyroxine in blood serum is 180 nM. The distribution volume for radioactive thyroxine is 8 l. The elimination rate of this thyroxine is 14% of the total content per 24 h.

1. *Calculate the serum [thyroxine] in $\mu g\ l^{-1}$.*
2. *How much thyroxine is eliminated per 24 h expressed in $\mu g\ day^{-1}$?*
3. *Calculate the thyroid plasma clearance of iodide, when the concentration of free iodide in plasma is 4 $\mu g\ l^{-1}$.*
4. *Explain the condition of this patient.*

Try to solve the problems before looking up the answers in Chapter 74.

CHAPTER 67.
THE CALCIUM AND PHOSPHATE BALANCE

What causes the stiffness and the elasticity in bones?

Bones are like iron and concrete. The **elasticity** is due to the collagen (iron) network, and the **stiffness** (concrete) is due to calcium salts (Ca – hydroxyapatite complex). Bone tissue consists of two compartments, bone cells (metabolically active) and the bone matrix (metabolically inert extracellular compartment). The organic part of the bone matrix consists of **collagen fibres and ground substance**, and the inorganic part consists of **calcium-phosphate-hydroxyapatite**.

Are calcium and phosphate balance related?

The atomic weight of calcium is 40, and we consume 1000 mg (25 mmol) day^{-1}. However, we absorb only 400 mg in total. The **net absorption is only 250 mg**, since we secrete 150 mg day^{-1} to the intestines. Thus, 750 mg (19 mmol) must be excreted in the faeces every day (Fig. 67-1). The amount of phosphate absorbed and its concentration in

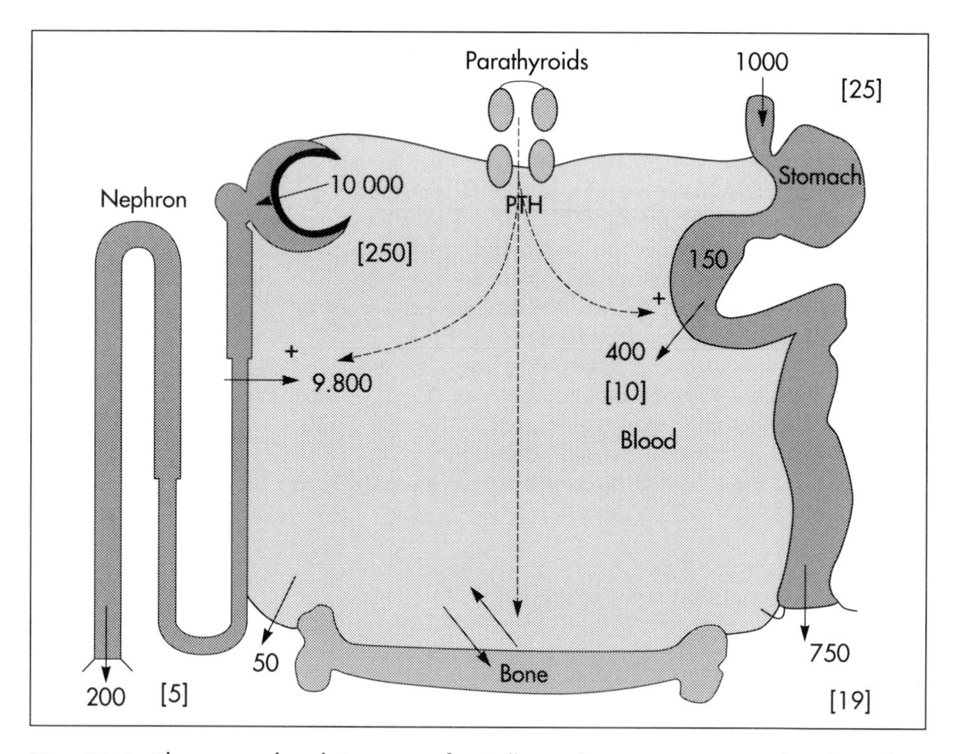

Fig. 67-1: The normal calcium transfer (all numbers are in mg day^{-1} and in [mmol day^{-1}]).

plasma, are determined by the amount available in the diet, but the active transport is somewhat dependent on vitamin D.

High plasma [Ca^{2+}] and [phosphate] promote bone formation in children, i.e. they increase the precipitation of Ca-hydroxyapatite in their bone matrix.

The [total calcium] in the blood plasma of healthy persons is **2.5 mM** or 100 mg l^{-1}. Half the calcium binds to plasma proteins and 5% binds to a citrate–phosphate–bicarbonate complex. A **significant portion (45%) is free ionized calcium ([Ca^{2+}]); this part has** critical roles in neuromuscular function and coagulation. If the plasma [Ca^{2+}] falls to half its normal level, the body develops increased neural excitability with **tetanic cramps**. On the other hand, an increase in [Ca^{2+}] leads to **calcification** of soft tissue with heart, kidney and intestinal diseases. The [Ca^{2+}] is the one variable regulated by PTH – see Fig. 67-1. For important information on Ca^{2+} and phosphate, see the last part of Chapter 53.

Describe the distribution of Ca^{2+}

The normal adult contains 27 500 mmol or 1100 g of calcium and 600 g of phosphate. The major part exists in the bones (1080 g fixed as hydroxyapatite and only 4 g or 100 mmol as **exchangeable bone calcium**). We exchange this 4 g of calcium five times a day. The flux between the bones and the ECV (16 l) is approximately 500 mg or 13 mmol day^{-1}. This fast exchange regulates [Ca^{2+}] in plasma. But we also have a small amount of daily bone formation and resorption (**bone remodelation**).

In our soft tissue cells 99% of all calcium is complexed with phosphate in the **mitochondria (10^{-5} mM)**, or bound to membranes and the ER. The cytosol contains an extraordinarily low [Ca^{2+}]: 10^{-7} mol l^{-1}. The intracellular [Ca^{2+}] is of extreme physiological importance as **regulator** of enzymatic reactions and secretion, as well as a **secondary messenger** for peptide hormones.

Intracellular phosphate is an essential component of nucleic acids, high energy molecules, co-factors, regulatory phosphoproteins, and glycolytic intermediates.

Describe the elimination of Ca^{2+}

The glomerular filtration of Ca^{2+} is easy to calculate, since approximately half the total plasma concentration is free and filtrable (2.5/2 mmol l^{-1}). Since 0.94 parts of plasma is water, the [Ca^{2+}] in the ultrafiltrate is (1.25/0.94) 1.33 mmol l^{-1}. A person with an average **GFR** of 0.131 l min^{-1}, will produce a 24-h ultrafiltrate of 188.6 l.

Thus, a total Ca^{2+} flux of (1.33 · 188.6) 250.9 mmol day^{-1} or **more than 10 000 mg day^{-1}**, will pass the glomerular **ultrafilter** (Fig. 67-1).

Fortunately, almost all Ca^{2+} is reabsorbed in the kidney tubules (about 67% is reabsorbed in the proximal, and the reabsorption of the balance in the distal tubules is regulated by PTH). We only excrete 100–200 mg day^{-1} (or 2.5–5 mmol day^{-1}) in the urine and 50 mg or 1.1 mmol through the skin. This is a maximum of **250 mg (or 6 mmol) Ca^{2+} excretion flux from the body per 24 h** (Fig. 67-1).

Does the [H^+] in plasma influence the [Ca^{2+}]

Both ions are in equilibrium with calcium-proteinate:

$$[\textbf{Calcium} - \textbf{proteinate}] + [\textbf{H}^+] \Rightarrow [\textbf{Ca}^{2+}] + [\textbf{HProt}^-].$$

Here, the negative protein ions have taken up some H^+ and formed $HProt^-$.

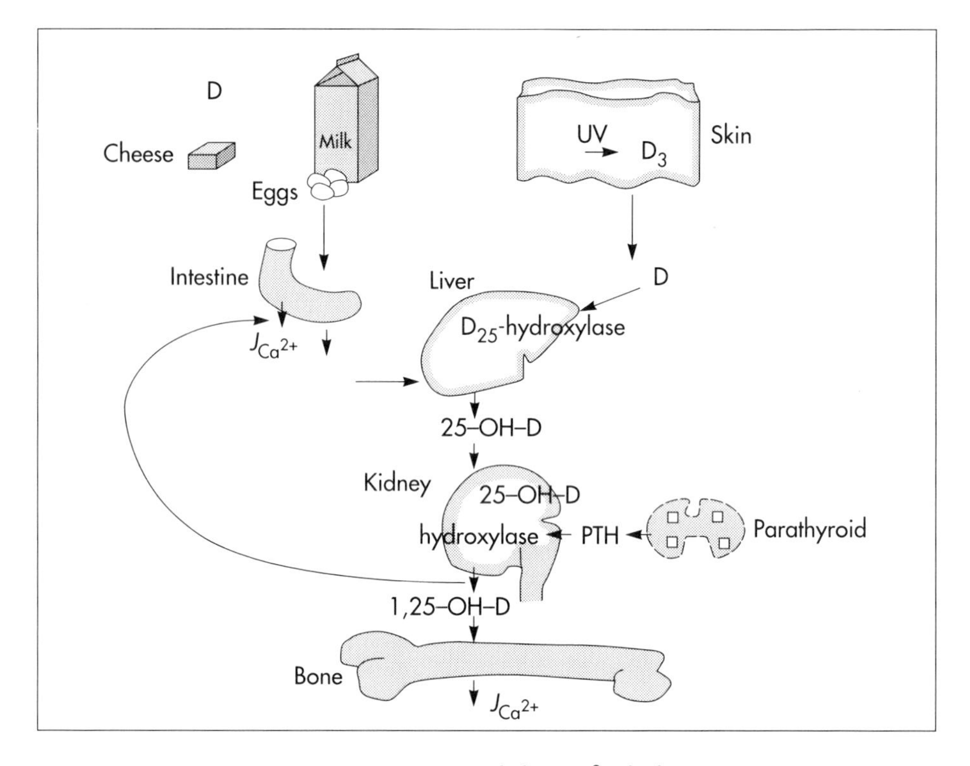

Fig. 67-2: Vitamin D and the Ca^{2+} balance.

A hyperventilating patient eliminates carbonic acid, and thus reduces $[H^+]$ and $[Ca^{2+}]$ in plasma, because more Ca^{2+} binds to protein. Low $[Ca^{2+}]$ in plasma leads to tetany (tetanic muscle cramps). The treatment is simply to reinhale the exhaled air (rebreathing) resulting in carbonic acid accumulation.

Vitamin D and bone minerals

Vitamin D_3 (**cholecalciferol**) from the skin, and vitamin D_2 (**ergocalciferol**) from vegetables are concentrated in the liver. Both are produced in the skin by ultraviolet irradiation (Fig. 67-2). In the liver microsomal oxidase simply hydroxylates the molecules to the weakly active **25-hydroxy-cholecalciferol (25-OH-D)**. This substance is transferred with the blood to the kidneys, where it is further hydroxylated at the C-1 position to the most potent form **1,25-dihydroxy-cholecalciferol (1,25-OH-D) (Fig. 67-2)**. This is what makes vitamin D a potent **steroid hormone**. Vitamin D is stored in muscle and adipose tissue and circulate in plasma bound to **vitamin D-binding protein**.

High PTH, low plasma [phosphate], and oestrogens stimulate the renal hydroxylation to active vitamin D. The active vitamin D metabolite stimulates the Ca^{2+} transport across the cell and mitochondrial membranes and has the following two effects:

1. Enhanced effect on **gut absorption of Ca^{2+}**, so that plasma $[Ca^{2+}]$ increases.
2. Enhanced effect of **PTH on bones**.

Is gravity an important stimulant for bone formation?

Ca^{2+} loss continues for months following space flight. Stimulation of bone deposition requires the stress of muscular activity as when a person is working against a gravity field. Appropriate exercise during space missions reduces the total Ca^{2+} loss.

Further reading

Stewart, A.F. and A.E. Broadus (1987) Mineral metabolism. In: P. Fegil *et al.* (Eds) *Endocrinology and Metabolism*, 2nd edn. pp. 1317–453. McGraw-Hill, New York.

67. Case History

A female (52 years of age; height 1.68 m; weight 62 kg) is in hospital due to her third attack of kidney stone pains. The first routine examination with arterial blood analysis reveals the following. Her blood pH is 7.21 and her plasma [Ca^{2+}] is 2 mM in ionized form. Her [Ca^{2+}] constitute 62% of the total calcium concentration in plasma. The inorganic phosphate concentration (total) is 0.84 mM plasma. The patient excretes 2–3 l of urine per day. $pK_2 = 6.8$ for H_3PO_4.

1. *Calculate the fractions of primary ($H_2PO_4^{2-}$) and secondary (HPO_4^{2-}) phosphate in her plasma. Compare the results to the normal mean value of 1 mM of inorganic phosphate with 20% primary and 80% secondary phosphate at pH = 7.40.*
2. *Calculate the total plasma [calcium] of this patient and compare the result to the normal mean value of 2.5 mM.*
3. *Why is the ionized calcium fraction much higher than normal (0.45)?*
4. *Could this condition be the result of a classical endocrine disease?*
5. *Why did this patient develop kidney stones? Was her diuresis normal? If not explain why.*

Try to solve the problems before looking up the answers in Chapter 74.

CHAPTER 68. GROWTH

GROWTH FACTORS

Most growth factors (GH, insulin, T_3, prolactin, etc.) do not pass through the placenta. How can the foetus grow?

1. **GH** produced in the placenta differs from the pituitary GH by a few amino acid residues. **Placental GH** suppresses release of maternal, pituitary GH during pregnancy. **Placental GH** stimulates maternal metabolism and foetal cell proliferation and hypertrophia.
2. **Foetal thyroid hormones** stimulate brain development.
3. **Foetal insulin** stimulates foetal growth, cellular glucose uptake and glucose utilization.
4. Paracrine and autocrine growth factors are also important for foetal growth: **insulin-like growth factor-II (IGF-II), nerve growth factors (NGF), epidermal growth factor (EGF), and platelet derived growth factor (PDGF).**

How is postnatal growth controlled?

The **pituitary GH** has an endocrine effect on the production of the growth factor **somatomedin C (or IGF-I)** and its specific binding protein in the liver. Hepatic IGF-I circulates in plasma bound to **IGF-binding protein I (IGF-BPI)**. The binding proteins are important, not only as a vehicle in plasma, but also for the final binding of the hormone to its cell membrane receptor. **Hepatic IGF-I** stimulates bone formation, and GH stimulates the precondrocytes directly. **Locally released IGF-I** stimulates the condrocytes in healthy states, but it also has a regenerative function in damaged tissues.

Pituitary GH also stimulates the production of receptors for other growth factors. GH serves to commit a precursor cell to a specific pathway, and IGF-I enhances its growth and replication. Somatomedins inhibit the secretion of GH from the somatotropic cells of the adenohypophysis. Somatomedins also stimulate the hypothalamic secretion of **somatostatin (growth hormone inhibiting factor (GHIH) or somatotropin releasing inhibiting factor).**

Do we have tissue specific growth factors?

Such factors are produced in damaged tissues, where they are important for regeneration. Nervous tissues produce **NGF**, epidermal tissues produce **EGF**, thrombocytes produce **PDGF**, and fibroblasts produce both **fibroblast growth factor (FGF)** and **transforming growth factors (TGF-α, and TGF-β).**

Hepatocyte proliferation after liver damage is induced by both **hepatocyte growth factor (HGF)** and by **hepatocyte stimulating substance (HSS)**. Growth factors are used in regenerative therapy.

Are the gonads important for early pubertal growth?

The relative large growth of the testes at puberty shows the importance of **sex hormones** for early pubertal growth (Fig. 68-1). In general, the growth spurt in early puberty is caused by increased production of **sex steroids** (stimulated by LH (ICSH) and by FSH), because the sex hormones stimulate hypothalamus to release **more GH**.

What is the action of GH?

GH and insulin are probably the most important **anabolic hormones** in the body; however, GH has many insulin-antagonistic effects. Human GH stimulates protein synthesis, mitosis in cells, chondrogenesis, ossification, and phosphate balance, while increasing glycolysis (i.e. anaerobic breakdown of glycogen). GH stimulates hepatic glucose production (glycogenolysis) but not its gluconeogenesis. GH enhances RNA synthesis, accelerates glucose uptake and antagonizes the lipolytic effect of adrenaline. GH is produced in spikes during the day (hunger spikes) and during deep sleep (EEG stage III and IV); the half-life of GH is 20 min.

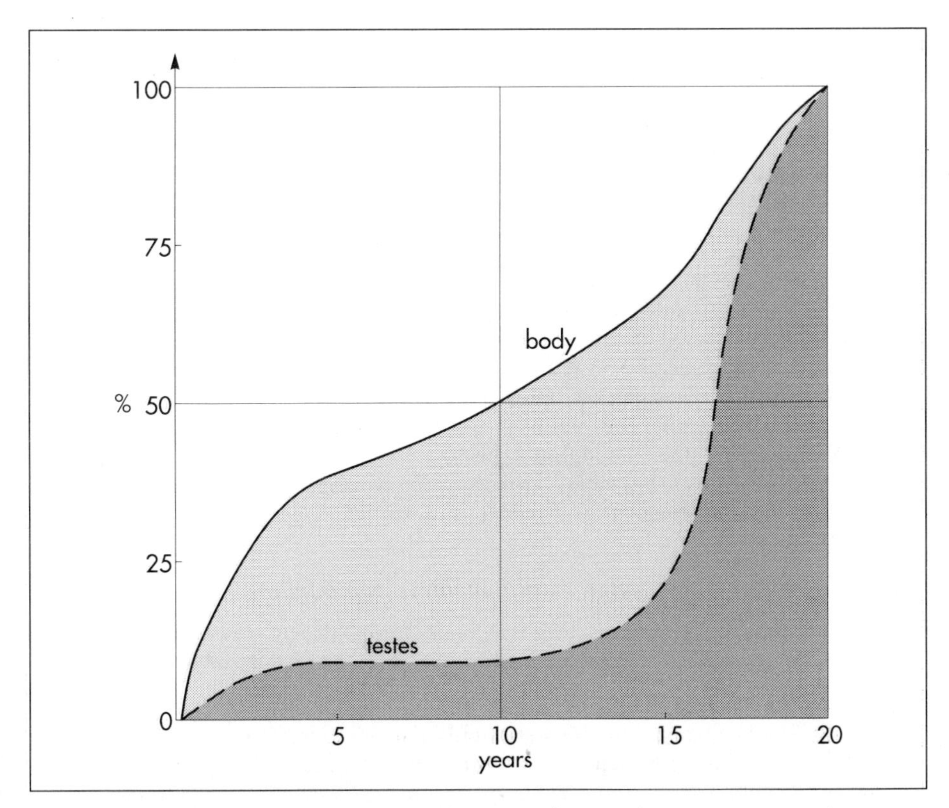

Fig. 68-1: Growth curves in boys (and a similar pattern in girls).

What hormones are essential for growth and development?

Children with **pan-hypopituitarism** become **infantile dwarfs**. They become **juvenile dwarfs**, as they do not develop sexually.

Children with **hyperpituitarism** become giantesses or giants. The clinical diagnosis is **gigantismus. Pituitary hormones** are essential to growth.

Hypothyroid children (cretins) also become dwarfs. The **thyroid hormones** are necessary or **permissive** for normal growth and development. **Cretins** are **mentally retarded, hypothyroid dwarfs.**

Sex steroids in high concentrations close the epiphyseal lines. If this occurs early in life, the child also becomes an infantile dwarf, but such an individual is often sexually active. The clinical diagnosis is **precocious puberty** or **pubertas praecox**. Without proper **insulin** treatment, children with diabetes (mellitus) become dwarfs (diabetic, infantile nanismus), because of the intracellular hypoglycaemia. Poorly regulated diabetics also develop **osteopenia** (bone decalcification).

Primary hormones for bone remodelling are **GH, PTH, calcitonin and 1,25-dihydrocholecalciferol (1,25-D_3)**. Secondary hormones for bone remodelling are glucocorticoids, sex hormones, thyroid hormones, prostaglandins (PGE_2), insulin and **IGF-I–III (somatomedins).**

What is dwarfism?

Dwarfism (nanismus) is used as a classification of adult females with a total height below 1.40 m, and adult males with a height below 1.50 m (in Scandinavia and other Anglo-Saxon countries).

Apart from hormones, what other factors are essential for optimal growth and development?

Genetic factors are essential as shown by the strong correlation between the height of the parents and the final height of the child. Tall people are tall before the pubertal growth spurt, which is more or less the same for tall and shorter people. Persons with only one sex chromosome (X,0) show retarded growth from birth, whereas persons with an extra sex chromosome (XXY; XYY) become tall (Fig. 69-2).

Optimal growth depends upon **optimal nutrition** (essential amino acids, vitamins, minerals, and fatty acids) and optimal neuroendocrine, metabolic control. Optimal growth also depends upon an **optimal health**. Most disease states and all immunodefence threatening treatments retard growth or imply weight loss. Such disorders cause catabolic hormones to dominate and induce anorexia.

Describe the primary hormones acting in bone remodelling

Bone remodelling is a balance between bone formation by osteoblasts and bone resorption by osteoclasts and mononuclear cells. This balance involves the following hormones besides **GH**:

1. **1,25-D_3** is a D vitamin, but also a **steroid hormone**. Kidneys produce 1,25-D_3 when stimulated by PTH. The **steroid hormone** increases the calcium absorption from the gut and mobilizes Ca^{2+} from the bones. This increases the [Ca^{2+}] and [phosphate] in plasma. The 1,25-D_3 also increases the renal reabsorption of Ca^{2+}.

2. **PTH** also increases the plasma [Ca^{2+}]. PTH mobilizes Ca^{2+} from the bones and increases the reabsorption of Ca^{2+} in the distal, renal tubule cells while inhibiting phosphate reabsorption in the proximal tubules of the kidneys.
3. **Calcitonin** from the thyroid inhibits bone resorption by blocking the PTH receptors on the osteoclasts. Thus plasma [Ca^{2+}] and [phosphate] fall.

Is human calcitonin of biological importance?

Cyclic treatment with calcitonin or combined treatment with 1,25-D_3 and PTH **improve bone formation in osteoporosis** (i.e. bone atrophia involving both minerals and matrix). Postmenopausal osteoporosis is treated successfully with calcitonin. Calcitonin is important in **bone remodelling** and in **treatment of osteoporosis**.

Immune Defence Systems

We have a cellular and a humoral immune defence system.

Cellular immune reactions are exquisitely antigen-specific just like the humoral system. The major organs of the **reticuloendothelial system (RES)** (bone marrow, lymph nodes, spleen, and thymus) receive sympathetic efferents – in particular the T-lymphocyte regions (see below). Hereby, the CNS (via the hypothalamus) modulates the intensity of immunoreactions.

Endotoxins from the normal bacterial flora of the intestine constantly enter the blood. Ordinarily they are inactivated by phagocytic activity of the RES mainly in the liver. Macrophages not only inactivate endotoxins, they also release hydroxylases, proteases, certain coagulation factors and AA derivatives (prostaglandins, thromboxanes, leukotrienes and monokines).

Monokines are control proteins that modulate metabolism (Chapter 42), temperature control (Chapter 44), hormone secretion (Chapter 64), and the immune defence systems. The important role of RES in haemorrhagic and endotoxic shock is described in Chapter 27.

Lymphocytes produced in the thymus during foetal life mediate our cellular immunity. Those so-called **T-lymphocytes** remain in the thymus, where they proliferate until they are mature and then migrate to the blood. T-lymphocytes constitute the majority of blood lymphocytes. The lymphocytes proliferate at first contact between antigen and T-lymphocytes. Some new cells bind the antigen in an antigen–antibody reaction and destroy the antigen. **Killer T cells** is the proper name for these cells, but the destruction of antigen requires the co-operation of **helper T cells**. Helper T cells stimulate the proliferation and differentiation of killer T cells to increase their number. A subgroup of **effector T cells** can suppress antibody formation by **B-lymphocytes** and inhibit other effector T cells – **suppressor T cells**.

Cellular immunity is a delayed form of immunity. The response reaches a peak after 2 days. Delayed immunity reaction encompasses the rejection of transplants, contact allergies and defence reactions against certain viruses and fungi. The T-cell number is deficient in AIDS victims (**Acquired Immune Defence Syndrome**). **Humoral immunity** is due to circulating antibodies in the blood. In foetal life, cells from the bone marrow pass through the gastrointestinal lymph nodes. Here the resting cells become immunologically **active B-lymphocytes.**

The cells re-enter the blood and migrate to the foetal spleen, liver and other lymph nodes. When an antigen binds to receptors on these cells, the lymphocytes divide. Helper T cells activate **resting B-lymphocytes**, so they differentiate to **plasma cells** or to **active B-lymphocytes**. Some new cells develop to plasma cells and remain in the lymph nodes. A few of the new cells enter the blood as **memory cells** – these comprise only a small fraction of the blood lymphocytes.

The **plasma cells are enormous** (10–20 μm in diameter) with an overwhelming surplus of **protein-producing ER**. These plasma cells produce large quantities of antibodies and release them into the blood. The antibodies form part of the **gamma-globulin fraction**. They are the **immunoglobulins** of the blood plasma.

How do lymphocytes recognize all the different antigens?

Burnet received the Nobel Prize in 1960 for his **clonal selection theory**. Stem cells differentiate into millions of different B-lymphocytes. Without exposure to all the antigens each of these B-lymphocytes have inherited the ability to divide into a clone of plasma cells. The first contact with the specific antigen starts the clone production. The clone of plasma cells produces the specific immunoglobulin. This understanding of the immunoreaction made transplantation possible.

Thomas made the first **transplantation** of kidneys in 1956, and received the Nobel Prize in 1990 for his contribution to science and therapy. The second successful transplantation was the transplantation of bone marrow to treat **leukaemia** (i.e. uncontrolled proliferation of **impotent leucocytes**).

Do we contain enough genes to inherit the ability to produce millions of specific immunoglobulins?

At first it seems incredible. However, Tonegawa has shown how only a few genes can be rearranged to produce millions of different antibodies in an individual (Nobel Prize 1987).

Is nitric oxide a killer molecule?

Macrophages produce high amounts of nitrite and nitrate when immuno-stimulated. In fact, the endogenous nitrate production mainly occurs in the body, and not in intestinal microorganisms as previously believed. The immuno-stimulated macrophages have killer activity and the molecule responsible for that activity is the unstable **nitric oxide (NO).**

NO is synthesized from one of the guadino nitrogens of L-arginine by the enzyme **nitric oxide synthase**. Several synthases have been purified and cloned. The enzymes represent a new family that contains a haeme moiety.

NO, produced in large quantities, destroys microorganisms and cells which become cancerous.

Further reading

Alberts, B., D. Bray, J. Lewis, M. Raff, K. Roberts and J.D. Watson (1989) Molecular biology of the cell, 2nd edn. Garland, New York.

Änggård, E. (1994) Nitric oxide: mediator, murderer, and medicine. *Lancet* **343**: 1199–206.

Tonegawa, S. (1985) The molecules of the immune system. *Sci. Am.* **253**: 122–31.
Thomas, E.D. and R. Storb (1983) Allogeneic bone-marrow transplantation. *Immunol. Rev.* **71**: 77–102.

68. Case History

A 22-year-old medical student is treated with an intravenous dose of PTH. This changes his renal excretion flux of two substances and their plasma concentrations.

1. *Describe the alterations and explain the mechanisms.*
2. *What is the diagnosis?*
3. *Describe the most likely symptoms and signs of this patient before treatment.*

Try to solve the problems before looking up the answers in Chapter 74.

CHAPTER 69.
SEXUAL SATISFACTION AND REPRODUCTION

Describe the physiological basis of the sexual drive

We feel the **sexual drive (libido)** when **sex related areas** in the higher brain centres are stimulated. These centres include the limbic system, stria terminalis and the preoptic region of the hypothalamus.

The sex drive of females is variable – for some it increases near the time of ovulation, when **oestradiol** secretion is increasing, while others experience a peak drive near menstruation.

Male libido is increased by **testosterone** secretion. The CNS cells involved (see above) must contain **sex hormone receptors**. Sex hormones are steroids. They are lipid soluble and pass the cell membrane easily. After binding to **cytoplasmic receptors**, the receptor–hormone complex translocates to the cell nucleus. Here the information is transcribed and translated. The result is release of new proteins, with the same information, into the cytosol, where the physiological response is triggered.

Castration is assumed to reduce female libido minimally, but male libido is most often lost. Removal of one testis need not change the male libido. These clinical observations reflect psycho-socio differences, and not necessarily a different libido mechanism in the two sexes. Hypothyroid persons lose their sex drive. The sex drive is stimulated by a multitude of **sense impressions** (visual, auditive, olfactory, and psychological). **Potency** refers to the ability to engage in intercourse.

Describe different types of sexual satisfaction

The **brain** is an important sex organ. Obviously, any **natural body contact** can be considered part of a healthy sex life – including the penetration of the penis in the vagina.

Sexual satisfaction is synonymous with **orgasm** in western cultures. **Orgasm** is the psychological climax or the culmination of **total commitment** in a sexual act that is accompanied by a series of physiological reactions.

Sexual enjoyment covers several phenomena, e.g. the **fetishistic satisfaction** of wearing the clothes of the opposite sex. This is the important part of a transvestite's sex life. Some transvestites become asexual in the general sense of the term, since they do not need partners. Some individuals prefer **masturbation (onany)** as a substitute for partnership. Many individuals prefer heterosexual contacts, others prefer homosexual activities, while bisexuals may prefer either sexes depending on the circumstance. Sexual activities can vary. Besides homosexual activity, oral sex, anal sex and many other variants are not uncommon.

Sex before birth

Normal sexual development in the embryo involves several processes. The sperm, which can be an X or a Y sperm, decides the genetic sex or sex genotype (Fig. 69-1). The genetic sex is independent of the ovum.

If the ovum is fertilized by an X spermatozoa (22 + X chromosomes) the offspring is XX, a female. If the ovum is fertilized by a Y spermatozoa (22 + Y chromosomes) the offspring is XY, a male (Fig. 69-1).

Sex differentiation in the embryo usually harmonizes with the sex genotype, but hormonal disturbances can lead to abnormalities. Proliferation of non-germinal and germinal cells in the **genital ridge** create the gonadal primordia, which develops into a cortex surrounding the medulla. Until the 7th week of gestation, each sex has a bipotential system. Around the 7th week, the medulla of the primitive gonad begins to differentiate into **testes if a Y chromosome is present**. Around the 12th week, the cortex of the indifferent gonad differentiate into **ovaries if no Y chromosome is present**. When the testes are present, the Wolffian ducts develop into the male reproductive tract (epididymis, vas deferens and seminal vesicles), whereas the Müllerian ducts regress. Conversely, in the female, the Müllerian ducts develop into the female reproductive tract (the uterine tubes, uterus and the upper vagina), and the Wolffian ducts degenerate. The Leydig cells produce testosterone and dihydro-testosterone, in response to **human chorionic gonadotropin (hCG)**. Secretion of these two androgens peaks at the 12th week. The Sertoli cells produce **Müller inhibiting hormone (MIH)**. Visible differentiation of the gross anatomy does not appear until late in the second month of embryonic life. Testosterone causes the **differentiation** of the foetus to a male. The foetal genital tract will always develop into female genitals (a phenotypic female), if unexposed to embryonal testicular secretion. If testosterone is present, male external sex organs develop and the genital tubercle elongates to form the male phallus. If testosterone is absent, female organs develop instead. It is the action of **5-α-dihydrotestosterone** on the

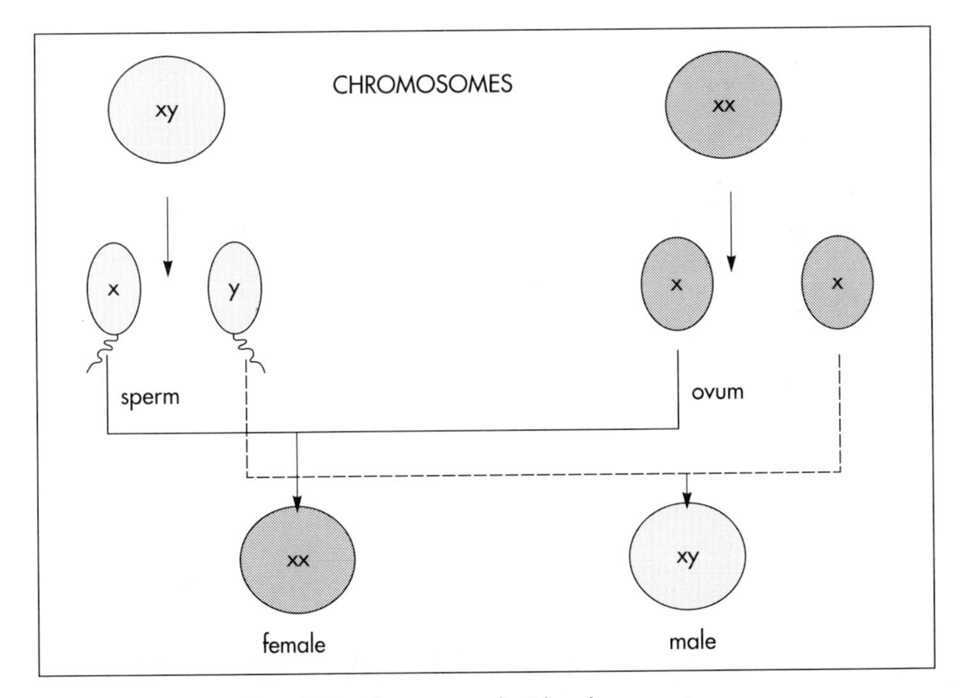

Fig. 69-1: The sperm decides the genetic sex.

urogenital sinus that is behind the normal development of the male, external sex organs. The **urogenital sinus** develops into the external genitals in both females and males.

The presence of normal ovaries or testes determines the **gonadal sex**. Without normal ovaries or testes any genetic sex will develop into an **apparent female**. In 1949 Barr found a densely coloured body in the periphery of the nucleus (**the Barr body** or **sex chromatin**) of the buccal mucosa in women. The Barr body is also present in other individuals with two or more X chromosomes in each cell.

Individuals with one sex chromatin (Barr body) also have a **drum stick** attached to a small fraction of their leucocytes (Fig. 69-2). We find sex chromatin and drum sticks in cells, whether they divide or not. Chromosomes are only visible in dividing cells. The **maximum number of sex chromatins and drum sticks is always one less than the number of X chromosomes** (Fig. 69-2).

What can cause abnormalities in human sex development?

Aberration of sex development can arise from two different causes. The **sex chromosomes** can create **genetic sex disturbances**, and **hormones** can disturb our **sex differentiation**.

In 1938 Turner described a syndrome in small persons, retarded in growth and in

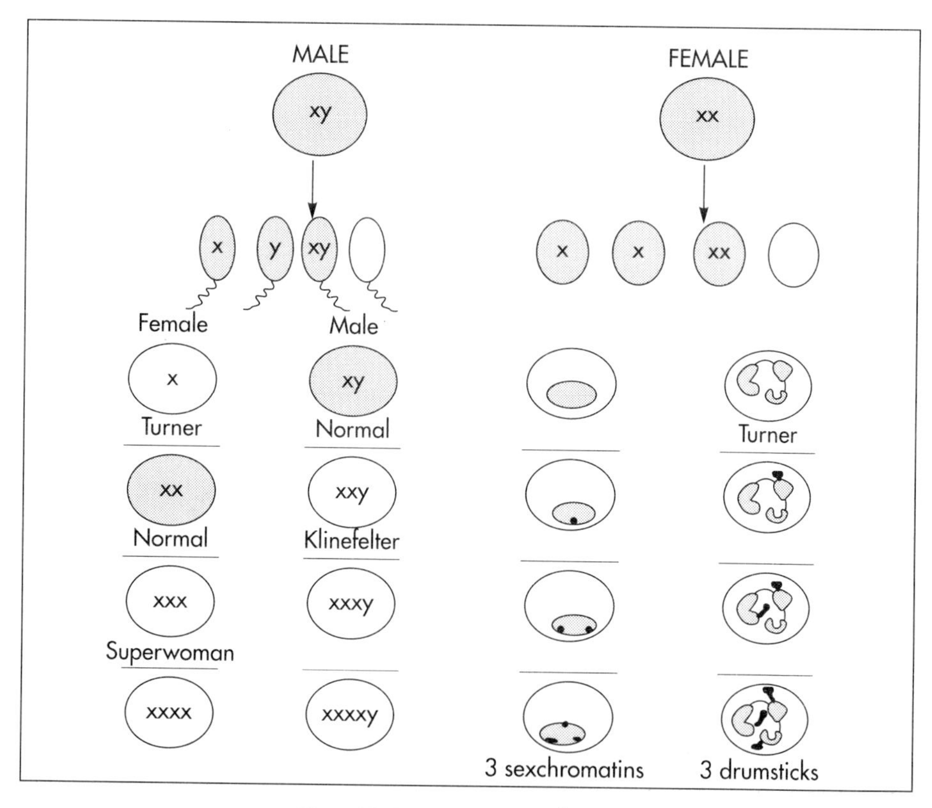

Fig. 69-2. Intersex syndromes.

sexual development. They are apparent females with small or no ovaries. Since they have only one sex chromosome (X), their total chromosome number is 45. They have no sex chromatin and no drum stick (Fig. 69-2).

In 1942 Klinefelter described a syndrome in persons appearing as men. They are tall, have small testes, some have female breasts (**gynaecomastia**), and they are sterile. Their cells contain XXY chromosomes (47 instead of the normal 46). Thus they must have one sex chromatin and one drumstick just like normal women (Fig. 69-2). These phenotypic XXY-males have significantly higher LH and FSH, and lower blood [testosterone] than matched XY controls. The XXY-males do not show more feminine behaviour than matched controls. A similar group of tall males with XYY chromosomes are not extraordinarily masculine. Some XYY-males have significantly higher [testosterone] in their blood than matched XY controls. Some small **super women** have an extra X chromosome: XXX, making a total of 47 chromosomes. We expected them to have two sex chromatins and two drumsticks, and this has been confirmed (Fig. 69-2).

Apparent men with XXXY (48) chromosomes have Klinefelter characteristics, and also two sex chromatins and two drumsticks (Fig. 69-2).

Individuals with four X chromosomes are extremely rare, e.g. apparent females with XXXX (48), and apparent males with XXXXY (49). Cells with four X chromosomes contain a maximum of three sex chromatins (Barr bodies) and three drumsticks, regardless of whether the cells come from apparent females or males (Fig. 69-2).

A very small number of individuals end up being of indeterminate gonadal sex (i.e. having both ovarial and testicular tissues present). Some persons have an ovary on one side and a testis on the other – a **true hermaphrodite**. In Greek mythology **Hermaphroditos** was the child of Hermes and the beautiful Aphrodite.

Pseudo-hermaphrodites have external genitals from both sexes, but only one gonadal sex. Males have normal XY chromosomes, but small testes with poor sperms (poor spermiogenesis). Some of these genetic (XY) boys are born as apparent girls. An enzyme defect that blocks the conversion of testosterone to **5-α-dihydro-testosterone** disturbs the development of the external genitals. Female hermaphrodites have ovaries, female ducts, XX chromosomes, and varying degrees of masculine differentiation of the external genitals. The virilizing effect of testosterone on the urogenital sinus causes this **adrenogenital syndrome** (with **adrenal hyperplasia** caused by enzyme defects).

What are sex identity and sex roles?

Sex identity is the individual's **perception** of herself or himself as a female or a male. Sex identity is established early, and is not lost by castration. Both psychological and social factors can interfere with normal sexual development. An imminent urge to change sex (operative sex shifts, etc.) characterizes **trans-sexual persons**.

The **sex role** is the **social behaviour or cultural role** played by or forced upon each individual. Some male homosexuals wish to express their femininity while other males clearly signal that they are men. **Transvestites** love to dress like the opposite sex. Transvestites are heterosexual, homosexual or asexual just as others.

What is the most drastic change at puberty?

Young children have low blood [gonadotropin] from birth. Through childhood they develop pulsatile secretion of gonadotropins which peak at night. Puberty is probably triggered by gonadotropin releasing hormone (GnRH) in a sufficiently mature CNS.

The **hypothalamic gonadal axis** switches from inactivity to a mature feedback system. Hormones are then produced at high rates, and the secondary sex characteristics then develop.

THE FEMALE

Describe the menarche and the menstrual cycle

The menarche is the **first menstrual bleeding**. It often occurs between the 12th and the 14th year.

The menstrual cycle starts at the first day of bleeding. The bleeding is due to decrease in oestrogen and progesterone secretion. The FSH and LH secretion start to rise before the bleeding and stimulate the growth of several follicles – in particular following the bleeding. One of these – by mysterious means – will ripen by selection and grow so fast that the follicle will protrude more than 10 mm from the surface of the ovary. Oestrogen from the granulosa cells of this selected follicle binds to specific, **cytoplasmic receptors** in the endometrial and other uterine cells. Oestrogens activate and stimulate formation of **intracellular progesterone receptors.**

Oestrogen increases the **thickness of the endometrium**, the **size of the myometrial cells** and the **number of gap junctions**, thus allowing the myometrium to work as a unit. The **oestrogen phase** is also called the **proliferative phase**. The concentration of sex hormones in plasma (C_H) is shown in Fig. 69-3. Oestrogens work synergistically with progesterone to release gonadotropins by **positive feedback** just before ovulation (Fig. 69-3).

Following the rupture of the follicle (ovulation), the **luteal corpus (corpus luteum)** produces increasing amounts of progesterone in addition to oestradiol also from a new developing follicle (Fig. 69-3).

Owing to the **priming effect** of oestrogen on progesterone receptors, both hormones stimulate the growth of the endometrial glands, so that they curl like a helix. The progesterone effect in particular provides the endometrial/myometrial tissues with their **high secretion and blood perfusion**, so the uterus is prepared to receive the fertilized ovum. During sexual stimulation the vaginal fluid secretion increases, as does the bloodflow to the organs involved.

If fertilization does not occur, the level of oestradiol and progesterone switches off both gonadotrophins. The corpus luteum, with no LH to support it, fades out and degenerates (Fig. 69-3).

The ovarian hormones almost cease to flow, and the uterus is deprived of their stimulating action. Therefore the **uterus shrinks and sheds its swollen lining** endometrium. The lack of progesterone and oestrogen (by negative feedback) increases the release of LH and FSH just before the menstrual bleeding.

On the first day of the menstrual bleeding, the low progesterone and high prostaglandin level probably **releases enough Ca^{2+} to start spontaneous contractions** of the myometrial cells. Ca^{2+} ions enter myometrial cells and stimulate their activity in the secretory (progesterone) phase.

The **gap junctions synchronize** these contractions, so that they include the whole myometrium. This can make excretion of blood and necrotic cells (containing prostaglandins) extremely painful. Prostaglandins dominate in menstrual fluid and

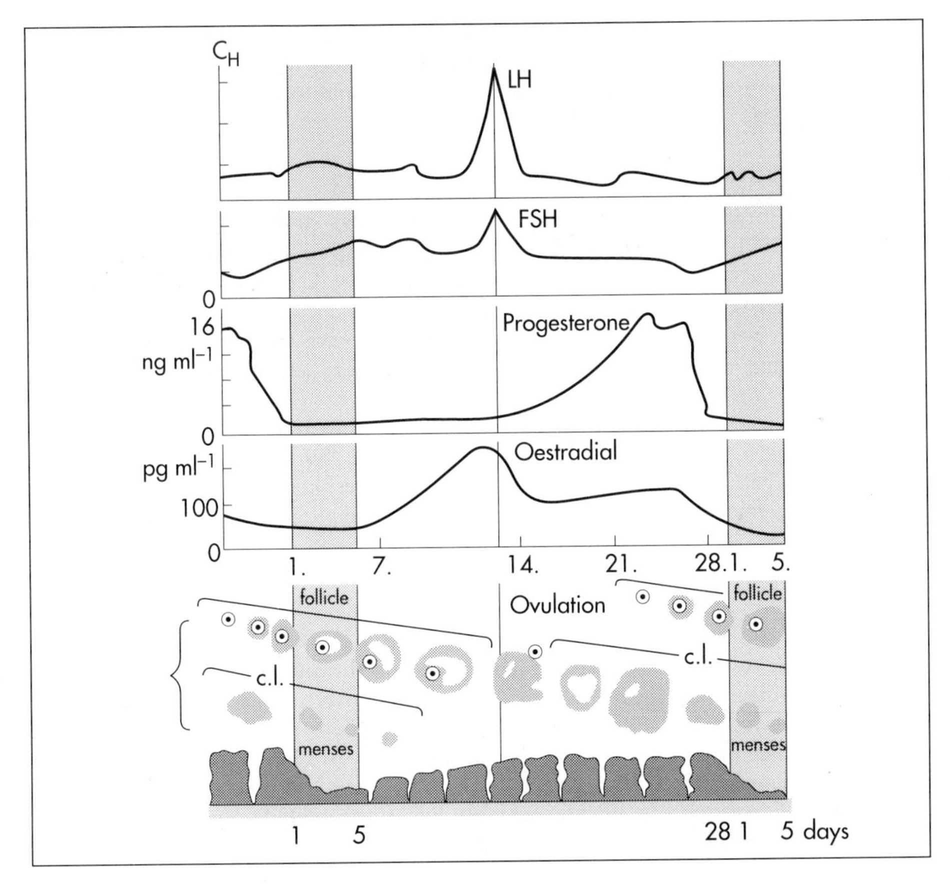

Fig. 69-3: The menstrual cycle in a female. c.l., Corpus luteum.

stimulate the spontaneous activity of the human myometrial cells. A **normal bleeding** corresponds to a **loss of 50 ml whole blood**. The mixture of vaginal fluid and menstrual blood produces a pH close to that of normal blood. The average cycle length is 28 days.

ADH (vasopressin) secretion from the neurohypophysis can cause **premenstrual tension** and an unpleasant increase in body fluid volume.

What causes ovulation?

At a certain level oestradiol can increase FSH output (by **positive feedback**) whereas at lower levels it is a potent inhibitor of GnRH secretion and thus of FSH output (**negative feedback**).

The first phenomenon is called **the positive feedback release ovulation**. The negative feedback forms the basis for the ovulation-inhibition by contraceptives. The primary inhibitor of FSH secretion is the peptide, **inhibin**, that is secreted by the ovary (and testis), and blocks the effect of GnRH.

The [oestradiol] increases sharply in the last part of the follicular phase, while the [LH] also increases. The sharp rise in LH and a modest rise in FSH coincides with ovulation. **The LH not only causes rupture of the follicle**; it continues to act on the follicular cells, turning them into a yellow endocrine organ, the corpus luteum.

Fertilization – essentials for female orgasm

The **time for foreplay** including **clitoral and multifocal stimulation** is important for most females. A clitoral orgasm in the foreplay often triggers more female orgasms later during the intercourse. Female orgasm is released from the spinal cord reflexes via sympathetic signals in the pudendal nerves.

Two persons with a simultaneous sexual drive must have the necessary time for the sexual act. If they are also in love, it is natural to explore and use all means to satisfy each other.

Years ago, when the Kinsey report was made, the average duration of sexual intercourse was 17 s in the US. American males able to ejaculate even faster were assumed to be particularly virile. Today, such a short performance is considered a male disease called **premature ejaculation** due to its possible implication for the female orgasm and fertilization.

Conception – describe the essentials

The **autonomic moving** spermatozoa pass through the uterus while prostaglandins (PGE) inhibit their spontaneous activity. The spermatozoa can keep their **vitality for more than 4 days.**

They lose their protective cover in the uterine tube. The **head of the spermatozoa swells** and liberates **proteolytic enzymes**. These enzymes dissolve the corona radiata around the egg (oocyte). The **oocyte can only live 14 h** without conception. Fusion of the two sex cell membranes forms the zygote, and the mitosis is complete within 24 h.

The zygote passes into the uterine tube within a few days, protected against other spermatozoa by an increased permeability for K^+, so that the zygote membrane hyperpolarizes. **Peristaltic movements of the tube** and ciliary motion conduct the zygote to the uterine cavity while undergoing **cleavage division**. Each cell is capable of developing into a complete human being up to the **eight-cell stage**. At the **morula stage,** the cells start to develop into the inner cell mass or **blastocyst**, and the trophoectoderm or **trophoblast**, which give rise to the **extra-embryonic tissues**. Seven days after conception, the blastocyst loses the zona pellucida and **implants in the wall of the uterus (nidation)**. Nidation depends on prior conditioning of the endometrial stromal cells by progesterone. The stromal cells accumulate nutrients and swell or **decidualize** around the blastocyst. **Histamine and prostaglandins** are involved in the high permeability of the vessels around the nidation site. The **blastocyst** produces **hCG,** and hCG acts like LH and binds to the LH receptors. This is what keeps the **luteal corpus** in being, and the pregnancy continues.

During pregnancy hCG thus conserves the corpus luteum, taking over the role of LH 1–2 weeks following conception.

At the end of pregnancy the **uterus is sensitized by oestrogen**. After a high peak in progesterone secretion the **progesterone output falls** (Fig. 69-4). This fall in progesterone allows the uterus to respond to **oxytocin**, whose release is the **final trigger for parturition**.

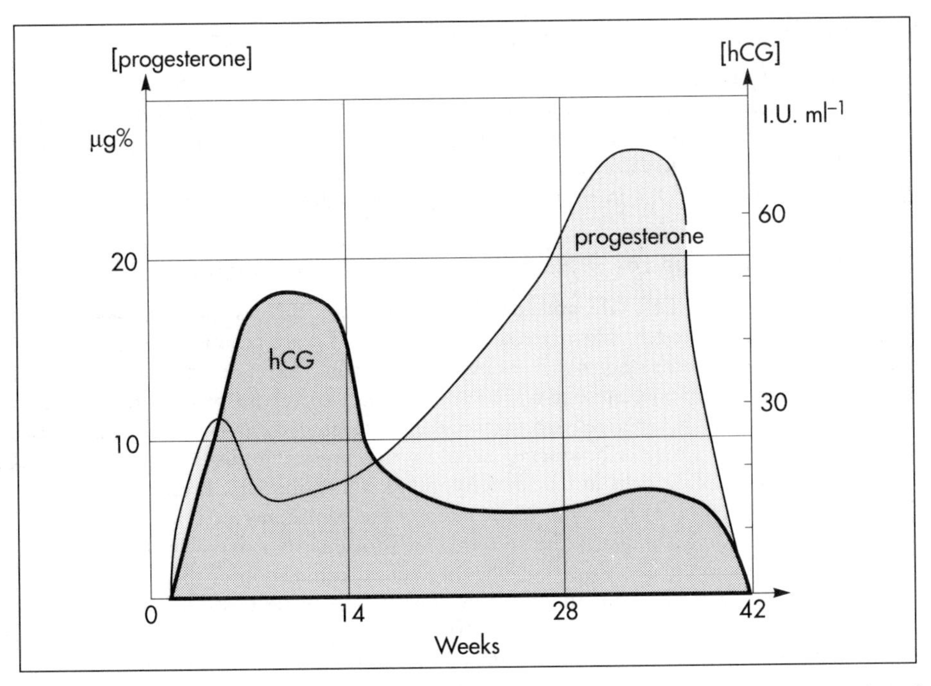

Fig. 69-4: Variations in blood [progesterone] and in [hCG] during 42 weeks of pregnancy.

How is pregnancy diagnosed?

The β**-group of hCG** is specific and found in the blood by specific antibody methods even before the first menstrual bleeding fails to appear. The hCG is detectable in the urine 8–12 days after the first missing vaginal bleeding.

What are the changes of the progesterone and hCG in the blood of a woman during pregnancy?

The [hCG] reaches a peak value after **10 weeks of pregnancy**, when the cytotrophoblast count is maximum (Fig. 69-4).

The first peak on the [progesterone] curve is progesterone produced by the luteal corpus. The placenta takes over the progesterone production in the remaining pregnancy period ending with a peak concentration before birth (Fig. 69-4).

What is a foeto-placental unit?

The **foetus and the placenta** form a foetoplacental unit. It produces all the hormones necessary for a successful pregnancy. The placental progesterone blocks the menstrual cycle of the mother. Pregnant women therefore develop amenorrhoea. Aside from **progesterone** and **oestrogens** the unit also produces **peptides** such as mammotropin (**human placental lactogen, hPL**) and **hCG**.

Can a female with occluded uterine tubes become pregnant?

From the start of the menstrual cycle the woman is given **FSH to stimulate her ovaries** before ovulation. On the 12th day she is given **hCG**. When ovulation occurs (after 30–35 h), **egg cells are sucked out**, placed in a tissue culture and **exposed to spermatozoa**. After 48 h some eggs fertilize into the 4–8 cell stage. A few of these **fertilized eggs are placed in the uterus**. One in four of these eggs will nidate (*in vitro* fertilization).

What hormones develop the breasts during puberty, pregnancy and lactation?

During puberty **FSH, LH, GH and insulin** are important besides the obvious effects of progesterone and oestrogen. The thyroid hormones (T_3/T_4) are **permissive**. At the end of pregnancy there are other hormonal events. Progesterone secretion reaches a peak and then falls. This fall in progesterone allows the pituitary to release **prolactin** (LTH).

Prolactin acts on the enlarged mammary glands turning them into **milk producers**. Prolactin develops the milk-producing acini in the breasts during pregnancy and is formed in the acidophilic cells in the adenohypophysis just like GH. **Prolactin inhibiting factor** (PIF or **dopamine**) from the brain inhibits the prolactin secretion.

The baby's suckling stimulates the secretion of prolactin and oxytocin, but oestradiol and sexual stimulation is also involved. The mechanical stimulation of the breast increases the secretion of prolactin from the pituitary, but the response is strikingly reduced by alcohol.

Prolactin is important for the **development of the mammary gland tissue**; oxytocin, however, governs the **ejection of milk** during lactation. **Oxytocin** causes contraction of the myoepithelial cells in the milk ducts (just as it does in the myometrial cells).

Mother-milk contains long chain fatty acids that are essential for brain development. Suckling babies are protected against juvenile diabetes in comparison to non-suckling babies (Chapter 73). Cow's milk contains much more protein and less lactose than human milk.

How is labour initiated?

When the foetus has reached a **critical size**, the myometrial fibres are stretched, which increases their contractility.

The foetal pituitary–adrenal axis signals to the placenta a decrease in the **progesterone–oestrogen ratio** acting on the myometrium. This increases myometrial contractions that are mediated by prostaglandins (PGE_2 and PGE_2-α). The density of oxytocin receptors in the myometrium increases throughout gestation and particularly at term.

The role of the stable plasma concentration of **maternal oxytocin** at parturition is an enigma. This fact does not exclude an important role of oxytocin in normal human parturition. Therapeutic doses of oxytocin initiate labour in most cases at the end of gestation.

The foetal cortisol production prepares the foetus to adapt to extrauterine life by stimulating lung maturition, by increasing the hepatic glycogen stores, and by promoting closure of the **ductus arteriosus** (Fig. 25-2).

Relaxin is a proinsulinlike polypeptide produced by the corpus luteum. The hormone relaxes pelvis articulations and softens the uterine cervix in order to facilitate passage of the foetus.

These, and several other factors, are involved in human labour but the exact trigger mechanism remains unclear.

What happens at the menopause?

Menses terminate and reproductive capacity is lost at an average age of 51 years, because the follicles disappear, and thus reduce the ovarian oestrogen secretion substantially.

The loss of oestrogen increases the gonadotropin levels by negative feedback.

Adrenal and ovarian stromal cells secrete androgen precursors that are converted to oestrogens by **aromatase** in adipose tissues. This is why menopausal females with sufficient adipose tissue suffer less from oestradiol deprivation than lean females. Vascular flushing and sweating, emotional lability, **osteoporosis**, and increased frequency of heart diseases are related to oestrogen deficiency.

How may lactating women be protected against pregnancy?

The suckling baby stimulates the mother to liberate prolactin, oxytocin, ACTH, GH and inhibin. **Inhibin** blocks **GnRH** secretion, and thus FSH and LH secretion thereby reducing fertility, but lactating women can still become pregnant.

THE MALE

Where are the male sex hormones produced?

The **Leydig interstitial cells** of the testes produce androgens and small amounts of oestrogens. The seminiferous tubules must be stimulated by adequate concentrations of two hormones – FSH and testosterone – to produce spermatozoa. The **Sertoli cells** produce **inhibin** as do the granulosa cells in females. Inhibin inhibits FSH but not LH secretion.

The **human hypophysis** produces four sex related hormones FSH, LH, prolactin and oxytocin. LH is also called **interstitial cell stimulating hormone** (ICSH) in the male, because it stimulates the interstitial cells that produce testosterone, which in turn specifically inhibits LH secretion. Removal of the male pituitary causes complete loss of all testicular functions; administration of FSH and ICSH then restores these functions completely.

How does the sex feedback system work in the male?

At the onset of puberty a timing device in the brain triggers the gonadotropin producing machinery in the **hypothalamic–pituitary–testicular axis**. Negative feedback control operates both before and after puberty, but the output of FSH and ICSH from the adenohypophysis is more than 100 times greater in young adults than in boys.

Circulating inhibin is the primary inhibitor of FSH secretion by negative feedback on the pituitary gonadotropes. FSH stimulates Sertoli cells to produce inhibin (Fig. 69-5).

A gonadotropin releasing hormone is formed in the **hypothalamus** (a decapeptide **GnRH** = LHF = LHRH), and stimulates in a pulsatile manner **pituitary gonadotropes** to secrete FSH and LH (= ICSH). Circulating testosterone regulates ICSH secretion by negative feedback primarily on the eminentia mediana hypothalami. The blood [testosterone] is highest during the night and in the morning (circadian rhythm) but there is virtually no seasonal rhythm with testosterone secretion in humans. In disease or

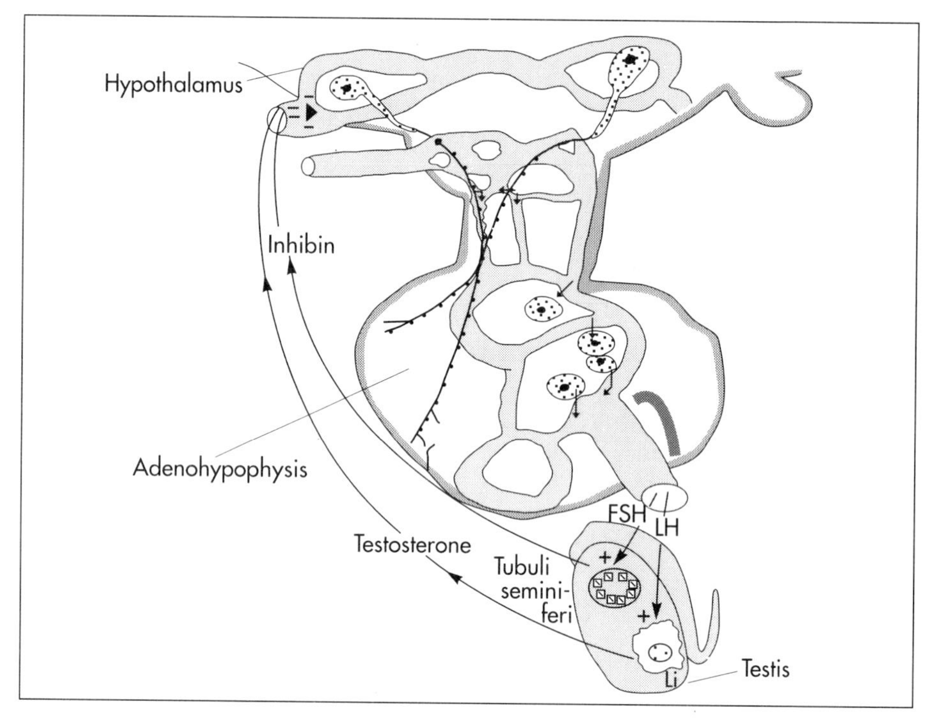

Fig. 69-5: Testicular control systems.

old age the seminiferous tubules may cease functioning, but the **sexual capacities** (other than fertility) are well maintained as long as testosterone is produced.

Describe the most important efferent activity during coitus

The activity in males is described as an example. The typical sequence of efferent events in the male include: (1) **erection** (penile rigidity and elongation due to parasympathetic vasodilatation); (2) **emission** of semen (sympathetic contraction of smooth muscles in epididymis, ducts and glands) driving the fluids into the posterior urethra; and (3) **ejaculation**. Rhythmic contractions of skeletal muscles expel the semen from the urethra.

1. **Erection:** Psychological factors trigger penile rigidity, and sexual thoughts can cause erection, emission and ejaculation. The penis contains erectile tissue located in two dorsal **corpora cavernosa** and in a single ventral **corpus spongiosum**. All the cavernous spaces of the three penile corpora receive blood from thick-walled arteries ending centrally in each corpus. The blood leaves the cavernous spaces through thin-walled veins starting peripherally. Tactile stimuli, especially from the very sensitive **glans penis** activates sensory, somatic fibres in the pudendal nerve, whereby impulses reach the **sacral plexus**. Parasympathetic impulses (S_2–S_4) from the sacral plexus elicit dilatation of the arteries and constriction of the veins in penis. The cavernous spaces are hereby filled with blood under high (arterial) pressure within seconds, causing the penis to become hard and elongated. Erection occurs quite normally during the REM phases of sleep. The new-formed sperm, produced by a

complex sequence of cell divisions during which the **chromosome number is halved**, move slowly along a coiled tube, the epididymis, where the **spermatozoa are stored and develop motility**.

2. **Emission:** Oxytocin ejects sperm into semen. Two exocrine glands near the neck of the bladder (the seminal vesicles and the prostate) secrete fluids which nourish the sperm and transport them through the urethra during the sexual act.
3. **Ejaculation:** Ejaculation is a sympathetic response. Signals from the glans penis reach the lumbar region of the spinal cord through afferent fibres in the internal pudendal nerves. Filling the posterior urethra with semen triggers sensory impulses that travel through the pudendal nerves to the spinal cord. The spinal cord transmits rhythmic signals to the **skeletal ejaculation muscles** (the ischio-and bulbo-cavernous muscles and those of the pelvis). These rhythmic signals stimulate rhythmic contractions that expel the semen from the urethral meatus into the female genitals.

What is chryptorchism?

Chryptorchism means **hidden orchids** (testes). The flower orchid (French orchidé) has a root which is actually shaped like a testis.

If the testes do not descend from the abdominal cavity to the scrotum, heat destroys the sperm-producing seminiferous tubule cells. Heat does not harm the Leydig (testosterone-producing) interstitial cells.

What are the effects of testosterone?

The hormone acts on many tissues, including the CNS. Testosterone is responsible for the growth, maturition and maintenance of the **primary sex structures** (seminiferous tubules, epididymis, prostate, seminal vesicles and penis). The hormone also stimulates the development of **secondary sex characteristics** (beard, body hair, sebaceous glands in the skin with waxy secretion, larger larynx, bones and muscles than females and castrates, male behaviour and attitude). **Acne during puberty** is due to **testosterone**, but in the female **adrenocortical androgens** are involved. Testosterone promotes protein synthesis (anabolic effect). Anabolic steroids that have a powerful anabolic action but only a modest androgenic action have been synthesized. These artificial hormones are still used to produce short-term **super-athletes**. Such a misuse of medicine for doping purposes often results in addiction, which has serious psychological, social and physical effects.

Further reading

Theilgaard, A. (1984) A psychological study of the personalities of XYY- and XXY-men. *Acta Psychiat. Scand.* **69**, Suppl. 315: 1–134.

Volpi, R., P. Chiodera, D. Gramellini, C. Cigarini, C. Papadia, G. Caffari, G. Rossi and V. Coiro. (1994) Endogenous opiod mediation of the inhibitory effect of ethanol on the prolactin response to breast stimulation in normal women. *Life Sci.* **54** (11): 739–44.

Yen, S.S.C. and R.B. Jaffe (Eds) (1986) *Reproductive Endocrinology. Physiology, Pathophysiology and Clinical Management*, 2nd edn. W.B. Saunders, Philadelphia.

69. Case History A

A 24-year-old female is going through her last menstrual cycle before pregnancy.

1. *Summarize schematically the most important events in her menstrual cycle.*
2. *Summarize schematically the most important events during continued pregnancy and delivery.*

69. Case History B

A pregnant woman delivers oxygen to her foetus. Her A-haemoglobin (A = adult) is functionally different from that of her foetus (F-haemoglobin).

1. *Why is this difference important? How are the two dissociation curves related?*
2. *FSH and LH are important for this woman. Describe why. Describe the function of the two hormones in her husband.*
3. *Following birth the mother breastfed her baby and experienced a feeling of sexual pleasure including uterine contractions. Describe the mechanism.*

Try to solve the problems before looking up the answers in Chapter 74.

CHAPTER 70.
THE ADRENAL CORTEX

The cortex produces three types of hormones: (1) the **gluco-corticoids** (cortisol and corticosterone); (2) a **mineralo-corticoid** (aldosterone); and (3) a minimal amount of **sex steroids** (androgens and oestrogens).

The adrenal cortex of the normal human has three layers: the **outer** zona glomerulosa is narrow, the **middle** zona fasciculata is wide, and the **inner** zona reticularis contains a reticulum of interconnected cells.

How are the adrenal corticosteroid hormones built and synthesized?

All three types of hormones represent chemical modifications of four-ring structures (A–D) forming the **cyclopentano-perhydro-phenantren**.

The precursor for these hormones is **cholesterol** absorbed from the blood HDL and LDL fractions by the cortex cells.

Most of the synthetic reactions involve mixed oxygenases belonging to the **cytochrome P-450 enzymes**.

The synthesis of **glucocorticoids** occurs in the **zona fasciculata** with a small contribution from zona reticularis. **Aldosterone** is produced in no other region of the cortex than the **zona glomerulosa**. The synthesis of **sex hormones** occurs mainly in the **zona reticularis**. During ACTH stimulation the size and number of cells in the zona fasciculata and zona reticularis increase, mainly because the cortisol and the sex hormone production increases. The mitochondria, central ribosomes, vesicular cristae and ER increases in these cells.

A microsomal desmolase removes C_{20-21} from the precursors, pregnenolone and progesterone (C_{21} steroids). The residues are dehydroepi-androsterone and androstenedione (C_{19}). These androgens are weak and are converted to a more potent form, testosterone, in peripheral tissues.

In the zona reticularis, testosterone is converted to oestradiol (C_{18}), due to removal of a CH_3 group by aromatase. The gene for the human androgen receptors is found on the **X chromosome**. The receptor protein has a molecular weight of 98 kDa and is found both in the cytosol and the nucleus.

Do CRH and ACTH influence the adrenal cortex secretion?

The hypothalamic **corticotropin releasing hormone (CRH)** stimulates the secretion of ACTH. Stress stimulates not only the sympatho-adrenergic system with catecholamine release, but also the CRH/ACTH release with increased secretion of neuroregulatory peptides and cortisol. This is important, since **small amounts of cortisol** have **permissive effects** on catecholamines, while inhibiting TSH. Stress also releases **GH** that stimulates glycogenolysis/glycolysis, and **prolactin**, both from the acidophilic cells in the adenohypophysis. Prolactin released by stress is possibly mediated by **hypothalamic histaminergic neurons**. ACTH binds to the cells of the zona fasciculata and activates adenylcyclase which results in a rise in the cAMP level. The major effect of ACTH –

through increased cAMP level – is to stimulate the **conversion of cholesterol to pregnenolone** by desmolase. This is the rate-limiting step in the production of cortical steroids. Plasma levels of **cortisol (hydrocortisone), adrenal androgens and their precursors** rise within 3 min of intravenous ACTH injection. Cortisol inhibits the release of CRH and ACTH by **negative feedback**.

What are the effects of cortisol?

1. **Physiological effects:** Glucocorticoids are insulin-antagonists, since they increase the blood [glucose] and act as protein catabolic hormones.
   Cortisol stimulates **hepatic glucose production** (glycogenolysis and gluconeogenesis). Cortisol stimulates **lipolysis, formation of FFA and of ketone bodies**. Cortisol **inhibits the glucose uptake in insulin target tissues** (GLUT 4 in muscle, heart and adipose tissue). The gene for the human glucocorticoid receptor is present on **chromosome 5**. The receptor protein (94 kDa) is found in both the cytosol and the nucleus.
2. **Pharmacological effects:** Supraphysiological doses of glucocorticoids are used to treat more diseases than any other group of drugs. **Glucocorticoids are anti-inflammatory, anti-allergic, and they stimulate haemopoiesis. Unfortunately, glucocorticoids also delay the healing of wounds and destroy tissue proteins by increased gluconeogenesis.**

Kendall isolated cortisone, and Hench and Reichstein directed the first administration of cortisone and ACTH to patients with rheumatoid arthritis. The result was a dramatic improvement. They shared the Nobel Prize in 1950. The serious side effects were recognized later.

How is cortisol transported and eliminated?

In plasma, cortisol binds to **transcortin** or **corticosteroid-binding-globulin (CBG)**, and a small amount binds to albumin.

The CBG binding means that cortisol has a long $T_{\frac{1}{2}}$ (90 min). Oestrogens stimulate the production of CBG. Patients with liver and kidney diseases do not produce enough CBG.

Cortisol is converted into tetrahydrocortisol, cortisone or to 17-ketosteroids. Conjugation with glucuronic acid or sulphate forms water soluble products. The fraction excreted in the bile is released to the **enterohepatic circuit.**

What is the relation between blood hormone concentration and the hormone binding globulin concentration?

A low concentration of **hormone specific globulin** implies a low concentration of the **bound** hormone fraction, but the **free** hormone concentration can still be the same.

How is aldosterone transported and eliminated?

Only a small fraction of the aldosterone binds to proteins. This makes its half-life **short** (20 min).

Aldosterone is reduced to **tetrahydro-aldosterone** in the liver and is conjugated with glucuronic acid.

How does aldosterone function?

Aldosterone **promotes the reabsorption of Na^+ and increases the secretion of K^+ and H^+ in the distal tubular system** (especially the cortical collecting ducts and the connecting segment). A rise in normal serum-[K^+] or a fall in serum-[Na^+] liberates aldosterone.

What mechanism makes aldosterone function?

Aldosterone, like all other steroids, is lipophilic thus it passes easily through the cell membrane. The human mineralocorticoid receptor is found primarily in the cytoplasm, and its gene is present on **chromosome 4**. This receptor has been cloned revealing a molecular weight of 107 kDa. When aldosterone is bound to the receptor, the receptor reveals a **DNA-binding site**.

The receptor–aldosterone complex translocates from the cytosol to the nucleus, where it binds to chromatin by means of the DNA-binding site. This process activates the transcription of the specific gene producing mRNA, which is then translated into specific proteins. Following complete transmission, the **receptor–aldosterone complex** dissociates from chromatin and from each other. The receptor returns to the cytosol with a **hidden DNA-binding site**.

By this process aldosterone promotes the synthesis of new proteins, that may **stimulate the Na^+-K^+-pump or facilitate Na^+ entry into the tubular cell through integral Na^+ channel proteins in the luminal membrane**.

Why is the aldosterone feedback control of [Na^+] in ECV important?

Increased Na^+-reabsorption from the tubules due to aldosterone causes a simultaneous reabsorption of water and thus an increased ECV, with increased arterial blood pressure. The pressure rise leads to increased GFR. The rapid filtration flow overrides the high reabsorptive effect of aldosterone, which downregulates the size of ECV.

What effect has ACTH on the secretion of aldosterone?

The ACTH effect is only a moderate stimulation in acute situations. The action of **angiotensin II** on aldosterone secretion is much more important.

The adenohypophysis probably produces an **aldosterone stimulating factor**. Dopamine (prolactin inhibiting hormone, PIH) is an **aldosterone inhibitor**.

How is the renin–angiotensin–aldosterone cascade released?

Thus far, this question has not been resolved.

The most likely trigger of the **renin–angiotensin–aldosterone cascade** during shock is the falling NaCl concentration – **especially [Cl^-]** – of the fluid at the macula densa.

Other mechanisms are also involved in the release of renin at falling arterial pressure: (1) falling perfusion pressure to the kidney; (2) stimulation by prostaglandins; (3) **angiotensin II decrease** releases renin by negative feedback; and (4) the nerve supply to the kidney, which stimulates renin secretion directly via **β-adrenergic receptors** on the JG cells.

Describe hypofunction of the adrenal cortex

Hypocorticism causes **Addison's disease**. This is a life-threatening condition with loss of Na^+ and thus also of ECV. Symptoms and signs include: reduced blood pressure, reduced blood [glucose], tiredness and skin pigmentation caused by MSH, whose level is increased concurrent with the **overproduction of ACTH, due to the decreased negative feedback**.

Describe hyperfunction of the adrenal cortex

Cushing's syndrome is adrenal hypercorticism – a condition dominated by cortisol effects.

Often the primary disease is a **secreting tumour** in the cortical zona reticularis and fasciculata. An excessive secretion of ACTH in the adeno-hypophysis can also cause Cushing's syndrome.

The patient may suffer from **muscular and bone atrophy** due to the **catabolic effect** of the cortisol surplus. There is a **delayed healing of lesions and wounds** due to the slow fibrocyte formation. Sodium and water accumulate in the body, whereas potassium is lost. The nitrogen balance is negative.

The arms and legs are thin and weak due to the protein catabolic effect, but fat collects in the back and neck regions (**bulls neck** and **buffalo hump**). Fluid accumulation leads to a special look called **moon face** or **tomato face**.

The high blood [cortisol] also increases the blood [glucose]. A **diabetic condition**, which is resistant to even large doses of insulin, develops. The patient suffers from **hypertension** and rapidly develops **arteriosclerosis**.

How is the adrenal cortex function studied in a person in whom Addison's disease is suspected?

1. **A single plasma ACTH level** is often diagnostic, as it will be elevated if hypocorticism originates in the **adrenal gland** (primary) and normal or low in **hypothalamic–pituitary hypocorticism** (secondary).
2. Differentiation between the hypothalamic and the pituitary hypocorticism can be accomplished by **injection of CRH**, which stimulates the pituitary directly.
3. **Insulin-induced hypoglycaemia** normally recruits the combined hypothalamic–pituitary axis. Hypoglycaemia is a prolonged stress which will normally increase ACTH and glucocorticoid secretion. If there is adrenal gland insufficiency then this increase in glucocorticoid secretion is not observed. **Hypothalamic and pituitary failure** is detected when a dose of CRH restores glucocorticoid release.
4. **A single ACTH-injection** along with measurement of plasma [cortisol] is a simple screening procedure for adequate endogenous ACTH secretion. Synthetic ACTH is injected intravenously. If plasma [cortisol] is normal 30 min later, the disease is **not a primary adrenocortical atrophia**.
5. **The metyrapone test** is used in patients who are suspected to have Addison's hypocorticism, but who have reacted normally to the ACTH-test. [Cortisol] and [11-deoxycortisol] in plasma is measured two mornings in succession, and the

adrenal ll-hydroxylase is inhibited with 500 mg metyrapone every second hour during 24 h. The plasma [cortisol] must fall while [11-deoxycortisol] accumulates. The fall in cortisol biosynthesis increases the CRH and ACTH secretion by negative feedback, if the **hypothalamo-pituitary system is intact**.

How do you follow up when Cushing syndrome is suspected?

1. The best assessment of increased ACTH production is by measuring the **24-h urinary cortisol excretion.**
2. **Suppression** of ACTH and cortisol is performed with a synthetic glucocorticoid (**dexamethasone**), which is potent enough to suppress the ACTH production, without influencing the blood [cortisol].

The principle of the **suppression test** is to **challenge the hypothalamo-pituitary feedback system.** If the system is intact a fall in blood [cortisol] is expected, and if not (hypothalamo-pituitary Cushing) the high [cortisol] is maintained.

In healthy persons such a suppression will **inhibit ACTH secretion**, and lead to a reduced blood [cortisol]. Such a normal observation is also typical for patients with a **primary adrenocortical hypertrophy**. Dexamethazone does not suppress the high cortisol level of patients with a **damaged feedback system.**

Further reading

Hench, P. (1950) The present status of cortisone and ACTH in general medicine. *Proc. Royal Soc. Med.* **43**: 769–73.

Skøtt, O. and B.L. Jensen (1993) Cellular and intrarenal control of renin secretion. *Clin. Sci.* **84**: 1–10.

70. Multiple Choice Questions

Each of the following five answers have True/False options:
ACTH stimulates the liberation of:

A. Hydrocortisone or cortisol.
B. Adrenaline.
C. Aldosterone.
D. Adrenal androgens.
E. Dopamine.

Try to solve the problems before looking up the answers in Chapter 74.

CHAPTER 71.
THE ADRENAL MEDULLA

Describe the sympatho-adrenergic system

This system is a functional and phylogenetic unit of the sympathetic system and the adrenal medulla.

The **medulla** is a **modified sympathetic ganglion**.

The **sympathetic system** consists of short preganglionic fibres, which have tropic centres in the lateral horn of the spinal cord from the first thoracic segment to the third lumbar segment.

The myelinated nerve fibres form ramus communicans albus and pass to the paravertebral or **lateral** sympathetic ganglion chain. Here most of the fibres end on the postganglionic cell bodies. Some preganglionic fibres pass the sympathetic chain and reach **collateral ganglia** such as the **solar plexus**, and the **ganglion mesentericus superior and inferior**.

In these lateral and collateral relay stations each preganglionic fibre is in contact with many cell bodies. These cells have long, postganglionic, unmyelinated fibres serving the different organs.

The preganglionic fibres to the adrenal medulla pass all the way to the **chromaffine cells** in the adrenal medulla. These chromaffine cells are embryological analogues to postganglionic fibres and the synapse is cholinergic. (Further information in Chapter 15).

What are the main functions of this system?

Stress comprises severe emotional and physical burdens (fear, pain, hypoxia, hypothermia, hypoglycaemia, hypotension, etc). Stress mobilizes the catecholamines of the adrenal medulla and of the sympathetic nervous system. The sympatho-adrenergic system give rise to the **fright, flight or fight** reactions which are all **acute** stress situations.

What are catecholamines and where are they synthesized?

Catecholamines are substances consisting of catechol (an aromatic structure with two hydroxyl groups) linked to an amine.

The dietary amino acid **tyrosine** is absorbed as tyrosine from the gut (Fig. 71-1). Tyrosine is also produced in the liver from dietary phenyl alanine. The L-tyrosine is hydroxylated (tyrosine hydroxylase), and the product is L-dihydroxy-phenylalanine (DOPA). **Tyrosine hydroxylase** limits the DOPA production. DOPA is decarboxylated to DOPAMINE, catalysed by DOPA-decarboxylase. The three important catecholamines in humans are **dopamine, epinephrine or adrenaline (Ad), and norepinephrine or noradrenaline (NA)**. DOPA is converted to melanine in the melanocytes of the skin, stimulated by **melanocytic stimulating hormone** (MSH in Fig. 71-1).

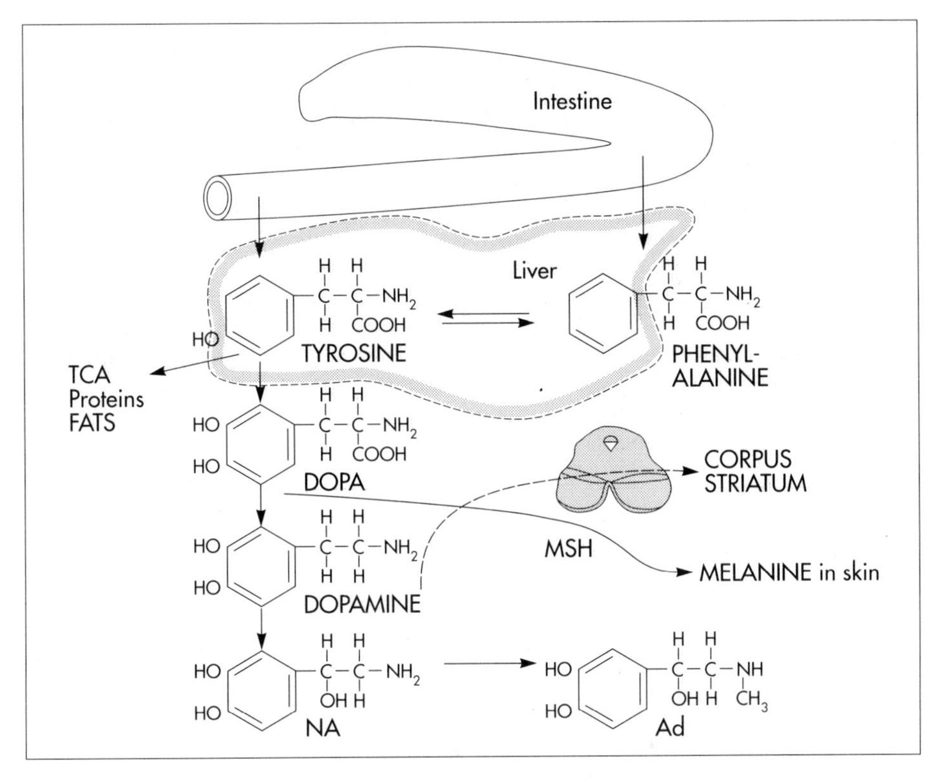

Fig. 71-1: Synthesis of catecholamines.

The neurons of the substantia nigra produce dopamine, which by axonal transport reaches the striatum (corpus striatum in Fig. 71-1). Here, dopamine is stored in the granules of the terminals – ready for liberation in the synapses (Fig. 12-3). In the sympathetic ganglion cells, dopamine is just an intermediary step, which is oxidized (dopamine oxidase) to form NA.

Catecholamines are formed in **adrenergic, noradrenergic and dopaminergic neurons.**

In the adrenal medulla, the process continues for one more step. NA is converted into **Ad** by methylation of the N-atom. In the newborn the primary end-product of the medulla is NA, but with advancing age there is a dramatic rise in the Ad/NA ratio. This conversion depends upon cortisol. Euler identified NA as the chemical transmitter from adrenergic nerves, and Axelrod demonstrated the reuptake of the transmitter after its release from nerve terminals. In 1970 they shared the Nobel Prize with Katz, who explained the role of calcium in synaptic transmission.

What are the actions of catecholamines?

Catecholamines increase the heart rate and the cardiac output ($\dot{Q}$) by stimulation of the **adrenergic β_1-receptors** in the myocardium. **Noradrenergic nerve fibres** innervate vessels **all over the body,** and this system usually has some tonic, vasoconstrictor activity. **The α_1-receptors** are located on the surface of **vascular smooth muscles.**

Catecholamines **dilate** the bronchial airways by stimulation of their **adrenergic β_2-receptors**. They increase both **tidal volume and respiratory frequency**. The result is an increased ventilation with an increased $\mathring{Q}$.

Catecholamines relax the **smooth muscles of the digestive tract** (β_2-receptors), but **contract the sphincters** just like the sympathetic nerve system. Catecholamines stimulate metabolism (T_3) and lipolysis. Ad stimulates hepatic glycogenolysis.

Finally, Ad stimulates the **ascending reticular system** (i.e. the **RAS**) in the brainstem, keeping us alert and causing **arousal reactions** with desynchronization of the EEG.

Catecholamines support the sympathetic system in modifying the circulation during exercise. During exercise the blood is directed to the working muscles from other parts. The resistance vessels of the striated muscles in hunting predators (and perhaps in humans) are also innervated by another system. This is the **cholinergic, sympathetic vasodilatator system**. It is capable of a rapid and appropriate bloodflow response during hunting. **The fall in the α-adrenergic tone in the muscular arterioles** is probably the most important exercise response in humans.

When are catecholamines liberated?

Prolonged **stress** liberates ACTH via hypothalamic signals. **ACTH stimulates the glucocorticoid secretion through cAMP. Small amounts of glucocorticoids** are **permissive** for the actions of catecholamines. **Acute stress** activates the splanchnic nerves and liberates large amounts of Ad from the medulla. Diabetics who are developing acute hypoglycaemia, secrete large amounts of catecholamines. Acute muscular activity starts a large catecholamine secretion in exercising persons.

How are catecholamines eliminated?

Plasma catecholamines are rapidly removed from the blood and have a half-life in plasma of **less than 20 s**. This is the combined result of **rapid uptake by tissues** and **inactivation** in the liver and vessel walls.

Enzymes inactivate catecholamines by methylation or by oxidation. **Catechol-O-methyl-transferase** (COMT) on the surface of the target cells catalyses methylation. **Monoamine oxidase** (MAO) in the mitochondria catalyses oxidative removal of the amino group. The final product in the urine is **vanillyl mandelic acid (VMA)**.

Does hypersecretion occur in the adrenal medulla?

Some patients have attacks of severe hypertension. These attacks are in a few cases caused by a **medullary tumour** (a pheochromocytoma of chromaffine cells), which liberates large amounts of catecholamines (Chapter 15).

What substance acts as the neurotransmitter in sympathetic ganglia?

ACh is the transmitter between the pre- and the postganglionic fibres, not only in the sympathetic system, but also in the parasympathetic system.

Describe the adrenergic receptors

Adrenergic receptors are membrane-receptors, divided into two groups, α- and β-receptors, on the basis of **blocking drugs** (see Chapter 15, Table 15-1).

Describe the cholinergic receptors

Cholinergic receptors are described in Chapter 15 (see Table 15-2).

What are the major differences between the sympathetic and the parasympathetic system?

See Chapter 15.

Further reading

Axelrod, J. and Hirati (1980) Phospholipid methylation and biological signal transmission. *Science* **209**: 1082–90.

DeQuattro, V. (1989) Catecholamines and adrenal disorders. In: L.J. DeGroot (Ed.) *Endocrinology*, 2nd edn, pp. 1717–800. W.B. Saunders, Philadelphia.

Dolan, L.M. and R.M. Carey (1989) In: E.D. Vaughan and R.M. Carey (Eds) *Adrenal Disorders*, pp. 81–145. Thieme Medical, New York.

Katz, B. and R. Miledi (1970) Further study of the role of calcium in synaptic transmission. *J. Physiol.* **207**: 789–801.

71. Multiple Choice Questions

Each of the following five statements have True/False options:

A. Muscarinic receptors are blocked by atropine, and nicotinic receptors by d-tubocurarine.

B. β-Receptors are blocked by propranolol.

C. Catecholamines have a half-life in plasma of more than 20 min.

D. The three important catecholamines in humans are Ad, NA, and DOPA.

E. β_1-Receptors are stimulated by adrenaline and located in the myocardium.

Try to solve the problems before looking up the answers in Chapter 74.

CHAPTER 72. THE ENDOCRINE PANCREAS

What is the function and structure of this organ?

The **endocrine pancreas** or the **pancreatic islets** are synonyms for the production site of four polypeptide hormones: glucagon, insulin, somatostatin, and pancreatic polypeptide (PP).

The **1 million islets of Langerhans** are discrete structures scattered throughout the pancreas (meaning **all meat**), but which only comprise 1% of its total weight. The islets are arranged along fenestrated capillaries, so that the hormones can pass easily to the portal blood. The islets of Langerhans receive both **sympathetic** (adrenergic) and **parasympathetic** (cholinergic) fibres.

The membranes of the islet cells contain **gap junctions** between neighbour cells, so hormones from one cell can act on its neighbor (paracrine action). Gap junctions allow passage of small molecules from one islet cell to its neighbour. In many pancreatic lobules, the α- β- and δ-cells form a **paracrine syncytium**.

Where is glucagon produced?

The α-cells are the source of glucagon (containing 29 amino acid residues). Glucagon stimulates **adenylcyclase** in the hepatocytes. This enzyme activates phosphorylase, which breaks down glycogen. Actually, glucagon triggers a **glycogenolytic cascade**, so that considerable amounts of glucose are released in response to falling blood [glucose]. In addition, glucagon stimulates the hepatic production of glucose (**gluconeogenesis**) from glycerol, alanine and lactate. Glucagon is a direct antagonist to insulin being catabolic in its actions (**gluconeogenetic, glycogenolytic, lipolytic and ketogenic, and deaminating amino acids**).

Where is insulin produced and how does it relate to glucagon?

The source of insulin is the β**-cells** which contain 51 amino acid residues in two chains (MW 5734). Insulin reduces the blood glucose for the following reasons:

1. Insulin increases the cell uptake of glucose (and potassium) in most tissues (adipocytes, heart and other muscle tissue). The exceptions are the brain, kidney and erythrocytes. The **uptake capacity for glucose in hepatocytes** is so large, that any insulin effect is immaterial.
2. Insulin increases the rate of **glycogen synthesis** in the liver and muscles and inhibits the rate of gluconeogenesis.
3. Insulin is a direct antagonist to glucagon being **anabolic** in its actions (increased glucose entry to cells, increased glycogen and lipid synthesis, decreased protein catabolism and ketogenesis).

How is insulin released from β-cells?

Glucose-evoked insulin secretion is the result of a chain of events in the pancreatic β-cell. The sequence is developed from material provided by Jørgen Frøkjær:

1. The **glucose uptake** takes place through a specific transporter protein (GLUT-2). The pancreatic β-cell membrane contains several K^+ channels, of which two are directly involved. This is the **K^+-ATP channel** and the **maxi-K^+ channel** (Fig. 72-1).
2. **Hyperglycaemia** accelerates the glucose metabolism and thus increases the ATP/ADP ratio.
3. **The K^+-ATP channels** are closed by increased [ATP], so the cell **depolarizes (hypopolarizes)**. During hypopolarization from the normal resting membrane potential of −70 mV, a threshold is reached at −50 mV, where the **voltage dependent Ca^{2+} channels** open (Fig. 72-1).
4. The Ca^{2+} influx triggers exocytosis of **insulin containing granules** following vesicular fusion with the cell membrane.
5. Normally, depolarization is stopped by the **maxi-K^+ channel** and other K^+ channels (Fig. 72-1). When intracellular [Ca^{2+}] and [K^+] has increased, it opens the **maxi-K^+ channel** (Fig. 72-1). The K^+ efflux restores the resting membrane potential (−70 mV) towards the equilibrium potential of K^+ (−100 mV).

Do islet hormones regulate pancreatic exocrine secretion?

Endocrine glandular tissue is localised in the Langerhans islets, which produces insulin in the β-cells, glucagon in the α-cells, somatostatin (GHIH) plus gastrin from the δ- and G

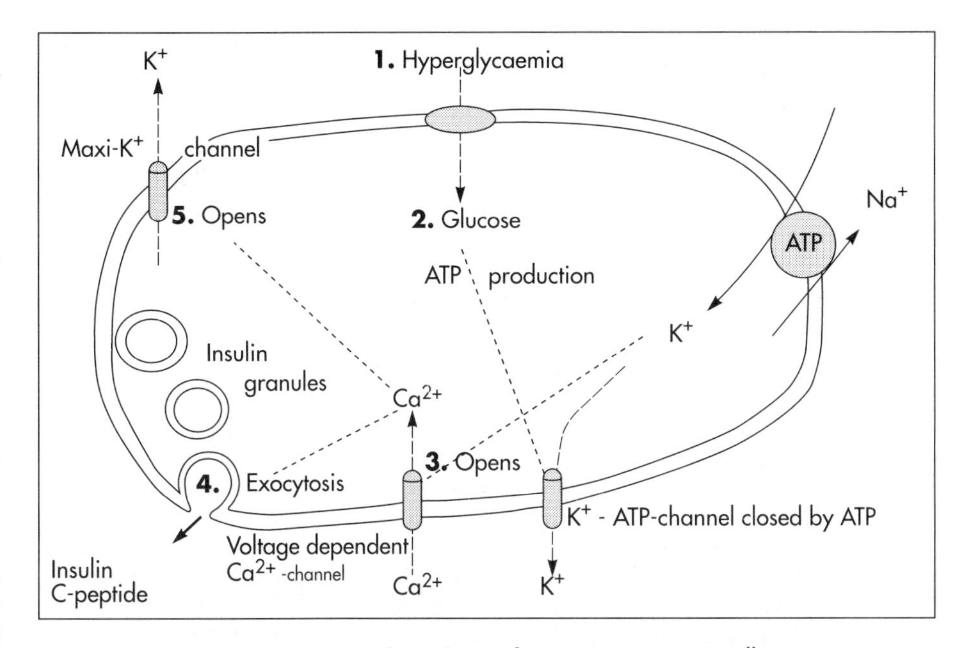

Fig. 72-1: Insulin release from pancreatic β-cell.

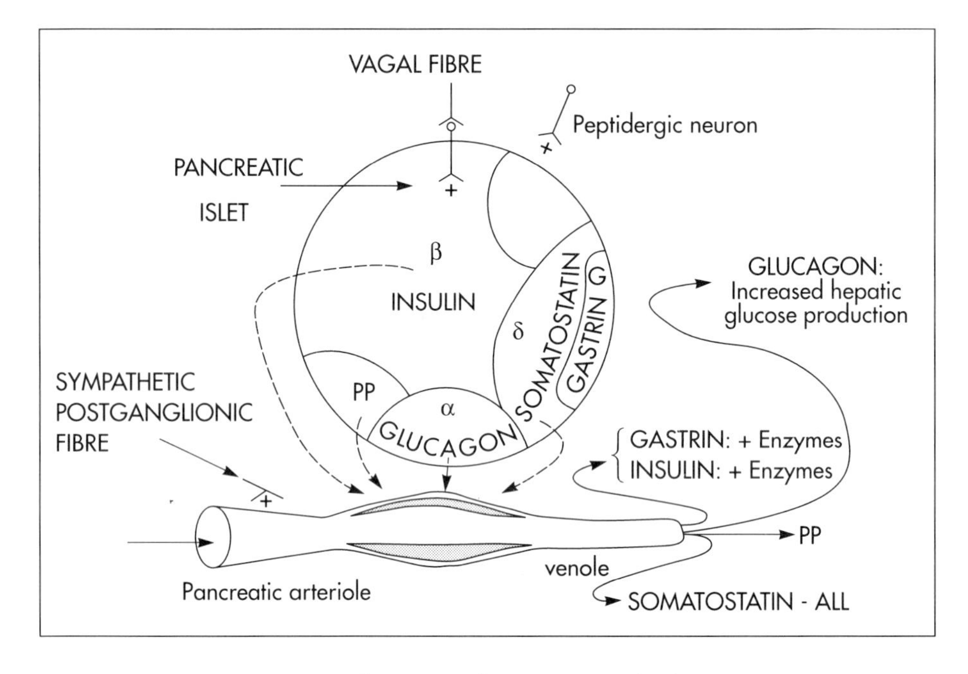

Fig. 72-2: Liberation of pancreatic islet hormones.

cells, and pancreatic polypeptide (PP) from the P cells (Fig. 72-2). Bombesin, galanin, and neuropeptides are present in pancreatic neurons and act as transmitters.

Stimulation of vagal fibres to the pancreas enhances the rate of enzyme secretion into the pancreatic juice (Fig. 72-2). **Stimulation of sympathetic fibres** reduces bloodflow to the pancreas, and thus inhibits pancreatic secretion (Fig. 72-2). **Gastrin enhances enzyme secretion** and insulin potentiates the effect, whereas **somatostatin inhibits secretion** from both acinar and duct cells (Fig. 72-2).

Tumours in the pancreas can secrete large amounts of gastrin, which stimulates gastric acid secretion and causes peptic ulcers (Zollinger–Ellison syndrome).

Describe the role of somatostatin

δ-Cells are the source of **somatostatin,** a potent and multipotent hormone inhibitor. Somatostatin contains a disulphide bridge and 14 amino acid molecules. Somatostatin produced in the islets **inhibits the local secretion of the other islets hormones**, while glucagon stimulates the local release of insulin and somatostatin. Somatostatin is also produced in the hypothalamus, where it functions as the **growth hormone inhibiting hormone (GHIH)**. Pancreatic somatostatin is released in response to high blood [glucose] and [alanine]. Somatostatin inhibits the secretion of the gastrointestinal tract (but not its motility) and functions as a synaptic transmitter in the CNS. Persons with somatostatin-producing tumours develop **diabetes and gallstones**.

Describe the role of pancreatic polypeptide

The cells responsible for pancreatic polypeptide (PP) secretion are particularly abundant in the head of the pancreas. PP contains 36 amino acid residues in a linear polypeptide. The plasma [PP] **increases** markedly after a **protein-rich meal,** but it is not released by alanine infusion. The PP secretion is increased by **exercise** (with a high plasma [alanine]), by **fasting** and by **hypoglycaemia**. The plasma [PP] is **suppressed by glucose infusion.** PP inhibits the **exocrine pancreas** and reduces the **gallbladder contractions**. This is an appropriate response during exercise and fasting, where any reduction in blood [glucose] would trigger a PP release.

Patients with **pancreatic islet cell neoplasms** have **elevated plasma [PP]**.

Further reading

Dunne, M.J. (1991) Block of ATP regulated K^+-channels by phentolamine and other α-adrenoreceptor antagonists. *Brit. J. Pharmacol.* **103**: 1847–50.

Felig, *et al.* (Eds) (1987) *Endocrinology and Metabolism.* 2nd edn. McGraw-Hill, New York.

72. Case History

A 19-year-old male, body weight 80 kg, suddenly complained of fever during his work and was ordered home to bed. The patient lived alone. Fortunately, a colleague visited him the next morning. He had to break the door down and found the patient unconscious. The patient arrived at the hospital in a deep coma. A blood sample from the radial artery showed the following results: β-cell antibodies, P_{aCO_2} 27 mmHg, pH 7.21, actual bicarbonate 10.5 mmol l^{-1}, O_2 saturation 0.96 and [glucose] 32 mmol l^{-1} (5.75 g l^{-1}). The actual values correspond to a base deficit (−15 mM) in the ECF volume (see Fig. 34-2). His urine contained glucose and ketone bodies.

The patient's breathing was deep and fast, his heart rate was 115 beats min^{-1}, and his blood pressure was 90/55 mmHg. The mucous membranes of the mouth were dry and the tonsils were enlarged and infected. The rectal temperature was 39.9°C.

1. *Explain the condition of the patient concerning thermobalance, carbohydrate metabolism, acid–base balance and fluid balance.*
2. *Describe a rational treatment of the four homeostatic disturbances in (1).*
3. *Following 8 h of treatment the blood pH was 7.41 and P_{aCO_2} 42 mmHg but the patient was still hyperventilating. Explain why.*
4. *Following 24 h of treatment all blood gas values were normal and the patient was restituted. How was the hyperventilation eliminated?*

Try to solve the problems before looking up the answers in Chapter 74.

CHAPTER 73.
BLOOD GLUCOSE REGULATION

This chapter covers the control of the blood glucose concentration in the **fed state** as well as in the **fasting states**. The third part of the chapter presents the **clinical aspects** of these control mechanisms.

I. GLUCOSE REGULATION IN THE FED STATE

How is blood glucose regulated following a meal?

In the **absorptive state** after a balanced meal, nutrients enter the blood and lymph from the gastrointestinal tract (as monosaccharides, triglycerides and amino acids). All the blood passes directly to the liver, which converts most of the other monosaccharides into **glucose**. Much of the absorbed carbohydrate **enters the liver cells**, but little of it is oxidized; instead it is stored as glycogen. Absorbed glucose, which did not enter hepatocytes but remained in the blood, is stored as glycogen by muscle cells, or it may enter into adipose tissue. A large fraction is oxidized to CO_2 and water in the various cells of the body. Glucose is the major source of energy during the absorptive state. Homeostatic mechanisms maintain the plasma [glucose] within narrow limits in healthy humans, so that a person's current energy needs during the postabsorptive state can be met by stored fuel.

A high glucose intake results in a high blood [glucose] or extracellular **hyperglycaemia**. Hyperglycaemia increases insulin secretion from the β-cells and inhibits glucagon secretion from the α-cells of the pancreatic islets. These hormones block **hepatic glucose production by glycogenolysis and gluconeogenesis. Insulin secretion** dominates over all **insulin-antagonists** (GH, glucagon, cortisol and some catecholamines).

Banting and Macleod shared the Nobel Prize in 1923 for their work in identifying the role of insulin in the carbohydrate metabolism. Their research led to the practice of insulin therapy for diabetes.

How does a meal potentiate insulin secretion?

The sight and the smell of a meal triggers **cephalic insulin secretion**. When the meal reaches the intestine, several peptides of the **incretin family** are released; this is the **intestinal secretion phase**. Typical representatives of the incretin family are **gastric inhibitory peptide (GIP or glucose-dependent insulin-releasing peptide), glicentin (intestinal glucagon), and glucagon-like peptides (GLP-1 and -2)**. Incretin strongly potentiates the insulin secretion induced by the rising blood [glucose]. The **incretin effect** causes a much larger insulin secretion than the i.v. administration of glucose, even at the same rise in blood [glucose]. This extra insulin secretion is called the **incretin effect**. The insulin released following a meal increases the storage rate of glucose-related energy in the liver, muscles and fat tissues. The storage effect is much larger than when glucose is administered intravenously.

How is glucose transferred through the intestinal cells?

Glucose is absorbed through the luminal membrane of the intestinal cells in **glucose-Na^+ transporter proteins.** The two substances pass through the basolateral membrane via separate routes: glucose passes in a special glucose-transporter, and Na^+ is transferred by the **Na^+-K^+-pump** (Fig. 53-1).

Glucose transport proteins and insulin receptors are described in Chapter 1.

What are the sources of glucose in the fasting state?

In the fasting state hepatic **glycogenolysis** produces most of the glucose and the remaining glucose is produced by **gluconeogenesis. Hepatic glucose** is produced by **glycogenolysis** (glycogen breakdown to glucose) and by **gluconeogenesis** (glucose formation from amino acids and other sources). The gluconeogenetic substrate is mainly lactate, glycogenic amino acids, glycerol, and small amounts of pyruvate. Muscle glycogen cannot deliver glucose to the blood, since muscle tissues lack **glucose-6-phosphatase.**

How is glucose eliminated?

In the fasting condition a healthy adult has a **blood [glucose] of 4–5 mM.** With an average [glucose] of 4 mmol l^{-1} in 15 l of ECV, therefore, the total glucose content in ECV is 60 mmol. This amount is equal to the glucose eliminated in 1 h at rest (60 mmol h^{-1}).

The CNS and the erythrocytes neither synthesize nor store glucose which is their primary fuel. Any surplus of glucose is deposited as liver and muscle **glycogen.** The liver cells contain an especially **efficient glucose transporter** (GLUT 2), and its **glucose uptake rate** cannot be increased further by insulin or by other hormones.

Is it normal to have glucose in the pre-urine or urine?

The filtration flux for glucose (mmol min^{-1}) increases proportionally to the concentration in the blood (as for all other filtratable substances).

Normally, all glucose is reabsorbed in the first part of the proximal renal tubules with a **T_{max} of 1.8 mmol min^{-1} or 320 mg min^{-1}.** In other words, the passage fraction falls from 1 to 0 already halfway through the proximal tubules.

The excretion flux for glucose is 0 in healthy humans.

Glucose appears in the urine of diabetics who have a blood [glucose] exceeding the **appearance threshold (10 mmol l^{-1}).**

Reabsorption of glucose over the luminal membrane of the proximal tubule cell takes place through the **glucose-Na^+ transporter** (Fig. 53-1).

How is insulin synthesized?

Pre-proinsulin is the precursor of insulin. When pre-proinsulin reaches the ER, enzymes separate the molecule from the signal molecule, to form proinsulin. In the **Golgi apparatus** enzymes cleave proinsulin to **insulin** (51 amino acids in two chains: A and B) and the **C peptide (connecting peptide).** Insulin and C peptide are wrapped in the same secretion granule. The content of these secretion granules is expelled from the cell by exocytosis with one C peptide for each insulin molecule.

The blood [glucose] increases after a meal. Increasing [glucose] is a strong stimulus to the β-cells of the pancreas. Glucose enters these cells through GLUT 2. The cells empty their granules into the ECV, and the granules dissolve immediately after entering the blood (Fig. 72-1:4). This sequence of events supplies the blood with insulin, C-peptide and proinsulin in the ratio 19:19:1.

Insulin is a vital hormone. Blood from the pancreas passes through the liver, where insulin promotes the production of glycogen from the recently absorbed glucose. The liver destroys a substantial amount of the insulin, whereas the C-peptide passes the liver undisturbed. The **plasma [C-peptide]** is thus a good estimate of **insulin secretion**. Insulin can now be synthesized from genetically modified microorganisms.

What are the major effects of insulin?

Insulin is an **anabolic hormone**.

Insulin reduces the blood [glucose] because it increases glycogen synthesis in the liver and muscles. Insulin increases the uptake of glucose through GLUT 4 (in adipocytes, heart and skeletal muscles). Insulin inhibits the gluconeogenesis from glycogenic amino acids in the liver.

Insulin promotes the storage of energy, the synthesis of glycogen, mRNA and proteins. Insulin thus reduces ureagenesis.

Insulin promotes lipogenesis in the fat stores; however, it inhibits lipolysis. It may be noted that the glycerol portion of the triglyceride molecule is a derivative of glucose.

Insulin increases the synthesis of cholesterol in the liver, in particular the rate of VLDL formation (**very low density lipoprotein**). The dangerous cholesterol fraction is LDL (**low density lipoprotein**).

Insulin increases the GLUT 4 transfer of glucose and K^+ into the muscle cell interior.

Certain major tissues (kidney, brain, intestine) are **insensitive** to the direct action of insulin.

What is glucagon, where is it synthesized and how is it liberated?

Intestinal glucagon (glicentin) is built up from 69 amino acids, whereas the glucagon from the α-cells of the pancreatic islets only contains 29 of the 160 amino acid residues in **proglucagon**. Conditions where there is **intracellular lack of glucose** (hunger, insulin deficiency, protein-rich meals, and amino acid infusion) liberate glucagon from the α-cells of the pancreatic islets to the pancreatic vein and then to the portal vein. Glucagon stimulates ketogenesis (formation of ketone bodies). High blood [glucose] and [FFA] inhibit glucagon secretion.

What are the actions of glucagon?

Glucagon and glicentin are **hepatic insulin-antagonist**.

Glucagon stimulates **hepatic glucose production** by glycogenolysis in the hepatocytes and thus increases the blood [glucose].

Glucagon also stimulates **gluconeogenesis from glycogenic amino acids** in the liver and thus increases **ureagenesis**. Glucagon stimulates ketogenesis (formation of ketone bodies). In addition to the ketogenic effect intestinal glucagon or glicentin is a potent stimulator of insulin secretion – as are other members of the **incretin family**. Incretins act by **increasing cAMP** in the β-cells (Fig. 72-2).

Do other hormones influence the glucose metabolism?

Somatotropin – human growth hormone (HGH) – is an **insulin-antagonist**, but together with insulin is probably the most important **anabolic hormone.**

Conditions where energy sources are lacking (hypoglycaemia, hunger, fasting state, exhaustion, stress) trigger the release of GRH from the hypothalamus, which in turn stimulates the release of GH from the hypophysis. This hormone has a **tropic effect** on the **α-cells of the pancreatic islets.** GH releases glucagon from these cells, just as sympathetic stimulation from the hypothalamus does.

GH increases blood [glucose] by **increasing hepatic glucose production** (glycogenolysis but not its gluconeogenesis) and by inhibiting the insulin sensitivity of the muscle cells and thus reduces their glucose uptake. GH also has the same effect on fat cells, mobilizing fatty acids and glycerol. GH stimulates protein synthesis, mitoses, chondrogenesis, ossification, and phosphate balance, while increasing glycolysis (i.e. anaerobic breakdown of glycogen).

Glucocorticoids are insulin-antagonists. They stimulate the **hepatic glucose production** (glycogenolysis and gluconeogenesis) but inhibit the cellular glucose uptake. Glucocorticoids are **permissive and potentiating** for catecholamines and glucagon.

Catecholamines (Ad and NA) are **insulin-antagonists.** Ad **stimulates hepatic glucose production** (glycogenolysis).

Catecholamines also stimulate lipolysis. The increase in mitochondrial oxygen uptake by T_3/T_4 is potentiated by catecholamines.

Somatostatin is **growth hormone inhibiting hormone (GHIH).** GHIH is synthesized both in the hypothalamo-pituitary system and in the pancreatic islets. The δ-cells of the pancreatic islets of Langerhans produce GHIH, which controls the function of the other islet cells by paracrine action. Somatostatin is a multipotent hormone inhibitor. Somatostatin inhibits not only somatotropin (GH) but also TSH, insulin, and **glucagon.** Somatostatin blocks the **gastrin secretion** in the gastric antrum. Somatostatin inhibits the secretion of digestive fluids, but increases gastrointestinal motility.

Meals rich in protein and fat release **PP** (from the PP-cells). **PP** inhibits both **enzyme secretion** from the pancreas and the **emptying of bile** into the small intestine. This leads to a **delay in the absorption of nutrients** including **glucose.**

Does the hypothalamo-pituitary system affect blood [glucose]?

Glucose-sensitive neurons in the hypothalamus (**the glucostatic centre**) react to **hypoglycaemia** by releasing **glucagon from the pancreatic α-cells** and **catecholamines from the adrenal medulla** by action of the sympathetic system. The **glucostatic centre** also reacts to **hyperglycaemia** to release insulin from pancreatic β-cells, and to activate **hepatic glycogen synthesis** by vagal stimuli (Fig. 43-1). Insulin promotes the entry of glucose into tissues, including the neurons of the **hypothalamic glucostatic centre** (but in no other brain neurons). A balanced blood [glucose] is achieved by sympathetic signals stimulating **hepatic glucose production** (Fig. 42-4). This **balance theory** is called the **glucostatic theory.** In the glucostatic theory the hypothalamus is considered a glucostat and the liver is a unique **glucose exchanger,** due to the **portal system** and the **hepatic glucose-6-phosphatase.**

Since the hypophysis hormones ACTH and GH are **insulin-antagonists** the net effect of the **hypophysis, when not balanced by a normal pancreas,** is a reduced glucose tolerance.

II. Blood Glucose Regulation in the Fasted States

How do we keep the fasting blood glucose constant?

Glucose production (gluconeogenesis and glycogenolysis) must equal **glucose combustion** (Fig. 42-4). Thus a precise relation must exist between the secretion of insulin and glucagon from the pancreatic islets.

Any **small fall in blood [glucose]** releases more glucagon (Fig. 42-4). During fasting **gluconeogenesis, glycogenolysis and lipolysis** are dominant. If a normal person does not eat for 24–48 h the CNS cells revert to combustion of ketone bodies, and a reversible condition that is similar to diabetes develops (**hunger diabetes**).

How is the blood glucose regulated during exercise?

We keep our blood [glucose] surprisingly constant around the fasting level, considering the wide variety of daily activities.

At rest a normal adult uses **60 mmol glucose h^{-1}**, but during maximum exercise that same person may use **five times more glucose**.

The exercise stress on the hypothalamus **increases the activity of the sympathoadrenergic system** which includes increased secretion of Ad from the adrenal medulla. Sympathoadrenergic activity **inhibits the insulin secretion** from the β-cells. Sympathetic activity also increases hepatic glucose production (glycogenolysis and gluconeogenesis from glycogenic amino acids such as alanine, glycerol, lactate, etc.).

We tend to increase glucagon secretion only if the blood [glucose] falls. A slight fall in blood [glucose] can occur both during exercise bouts and during prolonged exercise.

During exercise the blood [glucose] is maintained fairly constant by a **bihormonal interplay between insulin and glucagon**.

Generally, an increasing demand of more energy elicits increased glycogenolysis, lipolysis, and increased gluconeogenesis caused by insulin-antagonistic hormones (catecholamines, glucagon, cortisol, and GH).

Ad inhibits the insulin and stimulates the glucagon secretion, so the blood [glucose] increases (Fig. 42-4).

III. Clinical Aspects

What happens in the body during insulin-induced hypoglycaemia or insulin shock?

A high blood [insulin] will cause tissues to **store away the available glucose rapidly**, mainly through muscular GLUT 4, and stop simultaneously **the production of new glucose**.

A low blood [glucose] elicits a large secretion of glucagon to the portal blood. **Glucagon** is the most important insulin-antagonist. Glucagon increases hepatic glucose production (enhancing glycogenolysis, gluconeogenesis and protein breakdown). Low glucose levels trigger glucagon production, even from denervated, pancreatic α-cells; hence, they must be **glucose sensitive**. An increased **catecholamine secretion** from sympathetic nerve endings (NA) and **Ad** from the adrenal medulla (elicited from the hypothalamic glucostat via the sympathetic nervous system) help within minutes to compensate, as catecholamines stimulate glycogenolysis, increase lipolysis and inhibit peripheral glucose uptake. Hours later, cortisol and GH also contribute. An appropriate rise of plasma [cortisol] in response to insulin-induced hypoglycaemia documents an intact **CRH–ACTH–adrenal axis,** and this is the most widely used **stress test.**

The hypoglycaemic patient is warned by **adrenergic effects** such as trembling fingers, tachycardia, and muscular stiffness. The glucose consumption by the heart and brain continues. The lack of glucose in the brain makes the patient uneasy at first, he or she is then insecure, anxious and has cold sweat. Later in hypoglycaemia, the patient becomes confused and angry, and refuses with slurred speech to take glucose. Blood [glucose] below 2.5 mM elicits **hypoglycaemic shock** with loss of consciousness (somnolence, sopor or coma), universal cramps and respiratory stop. Intravenous injection of glucose (50%) is the rational therapy for **hypoglycaemic coma**. The patients wake up almost immediately, and are then often in a hyperglycaemic state.

What is diabetes mellitus?

Diabetes mellitus (DM) is a collective term for a multitude of metabolic disorders where **lack of insulin (type I)**, or **insulin resistance (type II diabetes)** dominates. **Insulin resistance** is defined as insufficient sensitivity to insulin. DM is characterized by **hyperglycaemia, protein depletion** and **increased lipolysis with deposition of lipids in vascular walls (atherosclerosis in arteries of the brain, the muscles, the heart, the eyes and the kidneys).**

The diabetic condition is characterized by **abnormal glucose tolerance** (see below).

The majority of patients with diabetes can be classified into two groups, those who are **(1) insulin-dependent diabetes mellitus (IDDM)** and those who are **(2) non-insulin-dependent (NIDDM).** Heredity and peripheral resistance to insulin seem to be a prominent feature of both IDDM and NIDDM.

1. **IDDM** is also known as **type I diabetes.** This is a serious, life-threatening metabolic disease, where continuation of life depends upon insulin treatment. The first treatment with insulin took place in Canada in 1922. Until then these patients died within half a year in ketoacidotic coma. Persons, often with hereditary predisposition, are suddenly attacked by autoimmune destruction of **all β-cells in the pancreatic islets,** which results in the complete absence of insulin. This autoimmune destruction occurs more often in populations where breast feeding is unpopular, and protein-rich cow's milk is used generally. Lack of insulin leads to **extracellular hyperglycaemia, and increased lipolysis.**
2. **NIDDM.** The much more frequent **type II diabetes** (maturity-onset) is the result of **insulin resistance and β-cell defects.** Type II diabetes also occurs in younger persons, especially in persons with a **high fat–low muscle mass.** A strong genetic element is always present, but **inactivity and stress** (an inactive life style with a low $\mathring{V}_{O_2}$) seems to be involved in the development of type II diabetes. Lack of exercise predisposes one to obesity, a condition which greatly decreases **insulin sensitivity of the target cells** (adipocytes, heart and skeletal muscle tissues). Reduced glucose combustion creates hyperglycaemia. The hyperglycaemia elicits insulin secretion from defective β-cells in some patients, resulting in **raised serum [insulin].** Since insulin is present, the acute complications such as **ketonaemia and metabolic**

acidaemia (often found with IDDM) are rare in these patients. The high serum [insulin] may further downregulate the activity of their **insulin receptors**. The insulin secreted in NIDDM patients does not increase the uptake of glucose as in normal persons. Many NIDDM patients need much more insulin for a given test effect than IDDM patients and healthy people.

Is it possible to explain the reduced glucose combustion in NIDDM? Yes. NIDDM can be caused, theoretically, by (1) **β-cell defects** including genetic defects, resulting in abnormal insulin production; or by (2) **target cell defects** including receptor failure. The possible defective sites in (1) and (2) have one common denominator. They are all **key proteins (hormone, receptors and transporters)**. Muscular activity is required to stimulate the normal production of **key proteins**. NIDDM relates to inactive life style.

The basic problem is therefore possibly a genetically and activity dependent **defect in key protein production** in the cell interior. Actually, a genetic defect has just been demonstrated at certain steps of insulin action in a subset of patients with late-onset NIDDM.

What is wrong in the β-cells of type 2 diabetics?

The β-cell defects are insufficiently described. The following answer is given by Jørgen Frøkjær and concerns only **stimulation–secretion coupling** and **exocytosis of insulin.**

Type 2 diabetics do not produce sufficient levels of ATP in the pancreatic β-cells to completely **block the K^+-ATP channels** (Fig. 72-1). Thus, the β-cell do not hypopolarize adequately in response to hyperglycaemia (Fig. 72-1). Therefore, the **voltage dependent Ca^{2+} channels** are insufficiently activated, and intracellular [Ca^{2+}] does not increase enough to trigger the insulin exocytosis needed (Fig 72-1). **Sulphonylurea compounds** close the **K^+-ATP-channels** and thus help to treat type 2 diabetes.

How is a glucose tolerance test performed?

The test is performed orally or intravenously.

Oral test. The patient drinks a glucose solution containing 75 g glucose within 4 min (WHO). The blood [glucose] is followed over 3 h by blood sampling. Normal individuals start from a low fasting [glucose] (5.5 mM or less). The blood [glucose] peaks after 1 h and **returns to normal within 2 h** (less than 6.7 mM in Fig. 73-1).

The diabetic patient typically starts from a high fasting [glucose] and increases to a very high level (Fig. 73-1). **The blood [glucose] is not back to normal within two hours. This is the clinical criterion of diabetes (Fig. 73-1).** A patient with hyperthyroidism (Graves disease or Morbus Basedowii) has a rapid intestinal absorption and a rapid combustion of glucose. The myxoedematous patient has a slow absorption and a slow combustion of glucose (Fig. 73-1).

Intravenous (i.v.) test. We inject 25 g glucose i.v. over a period of 4 min. We then measure the blood [glucose] every 10 min for at least 1 h in order to determine the half-life ($T_{\frac{1}{2}}$) from a semilogarithmic plot (Fig. 73-2). The metabolic combustion rate for glucose is exponential, so it is easy to calculate the metabolic rate constant (k) expressed in percentages.

Note that the **metabolic rate constant k** here is the amount of glucose combusted divided by the total amount of glucose in a mainly extracellular distribution volume. The ($T_{\frac{1}{2}}$) is equal to $0.693/k$.

All glucose combustion rates **above 1.2% per min** are normal (Fig. 73-2).

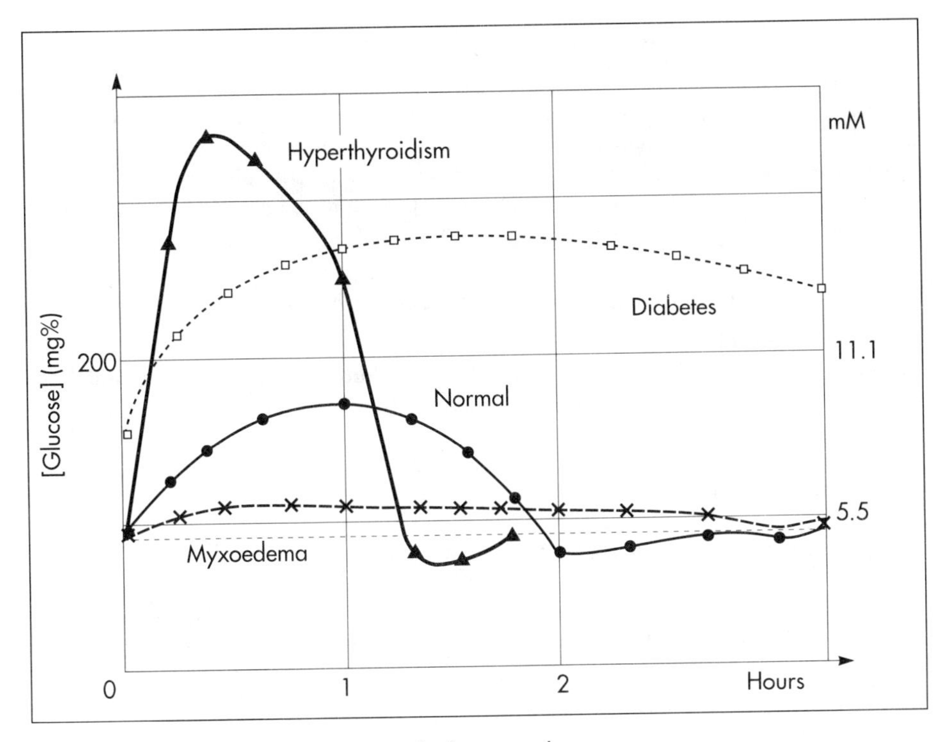

Fig. 73-1: Oral glucose tolerance curves.

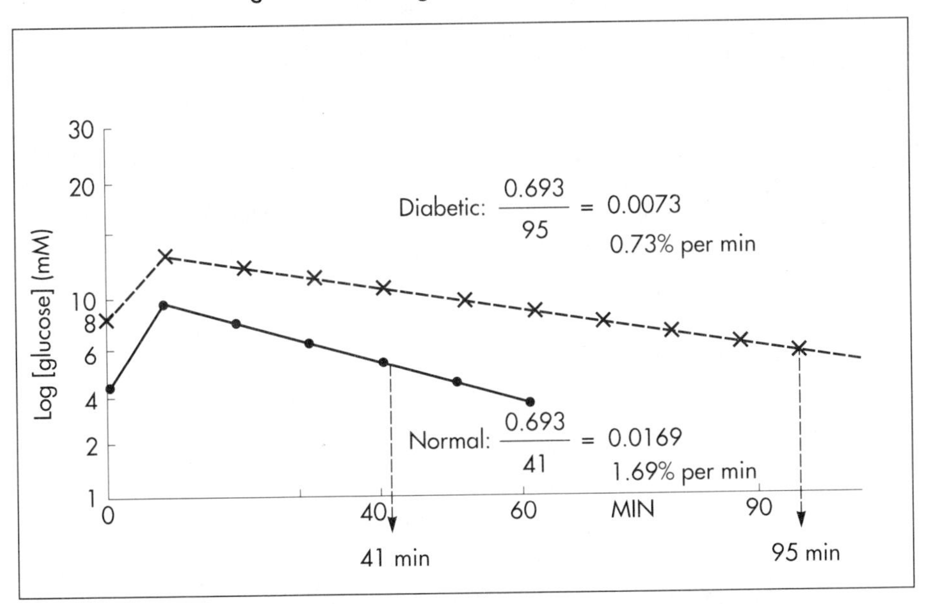

Fig. 73-2: Results of i.v. glucose tolerance tests from a normal person and from a diabetic.

How is an ideal insulin treatment administered?

A normal person with three meals per day will have **three peak concentrations of glucose and insulin in his or her blood**. It is possible to obtain such a time profile in a diabetic person by the following strategy. Inject a fast-acting insulin three times a day just before meals and a slow-acting insulin at night. This is **the physiological principle**. The aim of this procedure is to reduce the number of acute and chronic complications for diabetics.

What are the essential malfunctions in the diabetic person?

A poorly controlled diabetic condition leads to **extracellular hyperglycaemia, glucosuria, metabolic acidosis, polyuria (osmotic diuresis), dehydration and polydipsia**. The osmotic diuresis leads to the excretion of Na^+ and water, which results in Na^+ and ECV depletion.

Intracellular lack of glucose activates glycogenolysis in the liver and muscles, and accelerates muscular proteolysis and lipolysis. This liberates FFA, which are converted to ketone bodies.

A patient with hyperglycaemia above 25 mM loses consciousness to such a degree that contact is impossible (i.e. coma).

The increased rate of cholesterol production increases the occurrence of **atherosclerosis** and of diabetic nephropathy (Chapter 62). **Diabetic nephropathy** is characterized by albuminuria, hypertension and low GFR.

Further reading

Almind, K., C. Bjørbæk, H. Vestergaard, T. Hansen, S. Echwald and O. Pedersen (1993) Aminoacid polymorphisms of insulin receptor substrate-1 in non-insulin-dependent diabetes mellitus. *The Lancet* **342**: 828–32.

Ashcroft, F.M. and S.J.H. Ashcroft (1992) *Insulin*. IRL Press at Oxford University Press, Oxford.

Banting, F.G. and C.H. Best (1922) Internal secretion of pancreas. *J. Lab. and Clin. Med.* 7: 251–326.

Flatt, P.R. (1991) Nutrient regulation of insulin secretion. Portland Press, London.

Klip, A., A. Marette, D. Dimitrakoudis, T. Ramlal, A. Glacca, Q. Zhi and V. Mladen (1992) Effect of diabetes on glucoregulation. From glucose transporters to glucose metabolism *in vivo*. *Diabetes Care* **15** (4): 1747–66.

Thomas, H.M., A.M. Brant, C.A. Colville, M.J. Seatter and G.W. Gould (1992) *Biochem. Soc. Transact.* **20**: 1991–93.

73. Case History

A 49-year-old woman (height 1.52 m) is in hospital due to obesity and related problems. Her weight is 74 kg. She has developed skin mycoses and multiple boils. In the morning (fasting state) a blood sample shows a blood [glucose] of 7.4 mM, but glucose is not found in the urine .

At the hospital her total body water is measured following intravenous injection of radioactive water ($5.5 \cdot 10^7$ Bq tritium water). Her bladder is emptied at the time of injection. Two hours later the bladder is drained for 95 ml of urine with a concentration of 1 598 400 Bq l^{-1}.

At this time the indicator is evenly distributed in the total body water with a concentration of 1 520 700 Bq l^{-1}. The amount of indicator lost in the urine is two-thirds of the total loss.

1. *What is the principle for estimation of total body water?*
2. *Calculate the total body water in litres and in fraction of her body weight.*
3. *Is this a normal result?*
4. *The patient is obese, but is this a serious overweight?*
5. *Does she have symptoms and signs of complications?*

Try to solve the problems before looking up the answers in Chapter 74.

CHAPTER 74. ANSWERS

1. Membrane Transport

1. $\pi = R\,T\,(C_1 - C_2)$; $\phi = 8.83 \cdot 293 \cdot (150 - 50) = 243\,483$ [J/(K mol) K mol m^{-3}]. $\pi = 243\,483$ J m^{-3} = **243 483 Pascal (Pa)**, since 1 J = 1 Nm, and 1 Pa = 1 N m^{-2}.
2. A hydrostatic pressure of 243 kPa must be applied **over the sucrose solution** in order to stop the net water transport and create osmotic equilibration. This is the net osmotic pressure across this **ideal selective membrane**.

1. Case History

1. *In an ultrafiltrate of plasma at 310 K, with an osmolality of 0.315 Osmol kg^{-1}, the P_{osmot} is:*

$$P_{osmot} = 0.315\ (\text{Osmol kg}^{-1}\ \text{or}\ l^{-1}) \times 0.082\ (l\ \text{atm Osmol}^{-1} K^{-1}) \times 310\ (K) = 8.01\ \text{atm}$$

or ***a column of water 80 m high****.*

2. *The molecular weight of glucose is 180, and the plasma glucose concentration is 720 mg per 100 ml. This corresponds to 7.2 g l^{-1} or 7200 g m^{-3}. Since 1 mol of glucose equals 180 g, the plasma [glucose] is 7200/180 =* ***40 mol m^{-3}****. Actually, the concentration unit mol m^{-3} is always equal to mmol l^{-1} or mM. Since the plasma [glucose] of this young patient is almost eight-fold larger than normal, we have a severe case of diabetes. The first appearance is here, life-threatening* ***unconsciousness****.*
3. *The large output of water and glucose in the urine is related. Glucose has a strong osmotic effect, so the patient becomes* ***dehydrated*** *and the osmolality of his or her plasma ultrafiltrate is increased.*

2. Multiple Choice Questions

Answers A, B and D are true, whereas C and E are false statements.

3. Multiple Choice Questions

Answers A, B, C, and E are true statements, whereas D is false. During the early part of the AP the cell membrane is not **relatively** refractory, but **completely** refractory.

4. Multiple Choice Questions

Answers B, D, and E are true statements, whereas A and C are false.

5. Multiple Choice Questions

Answers A, B, D, and E are false statements. C is true. Catecholamines and peptide hormones are bound to **membrane receptors on the cell surface**.

6. Case History

I.

1. *The total muscle mass is assumed to represent 40% of the body weight 70 kg: (70 0.40) = 28 kg muscle tissue. The ATP hydrolysis at maximal muscle activity demands (20 × 28) = 560 mmol s^{-1}. The enthalpy of the ATP reaction is 42 kJ mol^{-1}. Accordingly, the total heat energy liberated must be (42 × 0.56) =* ***23.52 kJ s^{-1}***.
2. *The specific heat content of the body is 3.47 kJ $kg^{-1°}C^{-1}$. In 1 min the total energy liberated is (23.52 × 60) = 1411 kJ. The theoretical rise in body temperature in 1 min is 1411/(3.47 × 70) =* ***5.81°C****, provided heat dissipation could be totally blocked. Of course, this is not realistic, but the calculation illustrates the great rise in cellular metabolism at maximal exercise.*

II.

3. *The* ***Na^+-molality (mmol kg^{-1} of water)*** *is calculated as follows: The weight of the intracellular water phase is (954 − 193) = 761 mg; it contains (35 − 19) = 16 μmol of Na^+. The molality is 16/761 = 0.021 μmol mg^{-1} of water or* ***21 mmol kg^{-1}.***
4. *The half-life of NaCl: k* = 0.20/3 = 0.06667. $T_{\frac{1}{2}}$ = 0.693/*k* = **10.39 days**.
5. *The athlete must be in* ***salt balance****, with a salt intake equal to the salt elimination each day (so-called steady state).*

7. Case History

1. *The geometrical argument for the volume of a cylinder is used to calculate the volume of all arterioles: V = π radius² × length. Conversion factors are also used: 1 μm = 10^{-6} m; 1 μ**m²** = 10^{-12} m².*

   3.17 × 10^2 × ***10^{-12}*** *10^{-3} × 4 = 12.68 × 10^{-13} m³. The volume in all arterioles is (12.68 ×* ***10^{-13}*** *× 10^8) = 12.68 ×* ***10^{-5}*** *m³ =* ***126.8 ml of blood***

2. *The volume in all arterioles at the reduced radius is: 3.17 × 5^2 ×* ***10^{-12}*** *×* ***10^{-3} × 4 × 10^8 = 317 × 10^{-7} m³ = 31.7 ml of blood****.*
3. *Arterioles are resistance vessels, so arteriolar constriction increases the vascular resistance. Thus the arterial blood pressure is increased during stress.*

8. Multiple Choice Questions

Answers A, C, and D are true statements, whereas B and E are false.

9. Case History

The lesion of this patient must be localized in the ***lateral spinothalamic tract*** *of the spinal cord on the right side. The precise location is the thoracic segment, which has the upper affected skin area as dermatome. This is probably the 10th thoracic segment to the navel.*

9. Multiple Choice Questions

Answers A, D, and E are true statements, whereas B and C are false.

10. Multiple Choice Questions

Answer B is true, whereas A, C, D, and E are false.

11. Case History

1. *The defect in language function is* ***aphasia (word deafness)****.*
2. *The difficulties in understanding spoken and written language is called* ***sensory aphasia. Damage to Wernicke's area in the dominant hemisphere*** *(left hemisphere in 95% of all persons) explains the findings.*

11. Multiple Choice Questions

Answers A, C, D, and E are true statements, whereas B is false.

12. Case History

1. *Bleeding has interrupted the corticospinal tract as it traverses the* ***internal capsule****. The block of the excitatory pathways to the spinal cord result in* ***severe contralateral paresis****.*
2. ***Spasticity*** *means a motor condition dominated by increased tonic and phasic stretch reflexes, and it involves damage of the reticulospinal tract.* ***Foot clonus*** *is a repetitive pattern of violent contractions when the achilles tendon is tapped by the reflex hammer.*
3. *The innervation of the frontalis and the orbicularis oculi muscles is* ***bilateral****. Accordingly, these muscles function normally in a patient with a* ***unilateral central paresis of the facial nerve****. A patient with a* ***peripheral paresis*** *of the facial nerve, can only knit his brows and screw up his eyes at one side.*

12. Multiple Choice Questions

Answers A, D, and E are true statements, whereas B and C are false.

13. Case History

1. *The patient has **facultative** hypermetropic eyes. In order to foveate diverging light she is forced to accommodate relatively more than a normal eye during near vision. Thus, the ciliary muscles are fatigued, and she gets **eyestrain**.*
2. *The accommodative power is calculated as follows: 1/F – 1/N = 4 – (–3) = **7 D**. This result is exactly the same as that of a normotropic person of 30 years.*
3. *A fine text can be read further away and until the near point, 0.33 m in front of the eyes. Within the range of 0.2 m, she cannot obtain a sharp picture on the retina (foveation).*
4. *Hypermetropic persons usually have **too short eyeballs** – the saggital diameter is too short.*
5. *The fluid outflow at the iridocorneal junction is reduced, because such flat eyes implies a **small chamber angle**. This leads to increased intraocular pressure or **glaucoma**, a frequent disorder in hypermetropic patients.*

14. Case History

1. *The threshold pressure for a healthy person is 20 μPa or $2 \cdot 10^{-5}$ Pa (p_0). Thus, the relation is:*

$$dB, SPL = 20 \log (p/p_0).$$

$$dB, SPL = 20 \log (p/(2 \times 10^{-5})).$$

*Accordingly the sound pressure in the power plant is **2 Pa**, and the **ratio is 10^5.***

2.

$$26\ dB, SPL = 20 \log (p/(2 \times 10^{-5})).$$

The threshold pressure for the patient is (3.99×10^{-4}) Pa. This pressure is 20 times as high as that of the normally hearing person.

15. Case History

1. The **vagal nerve is traumatized**.
2. This is a **bilateral lesion** of the vagal nerves – a bilateral vagal paresis.

16. Case History

1. *Castle demonstrated lack of **intrinsic factor** in patients with pernicious anaemia as early as in 1929. This factor is normally secreted by the parietal cells of the gastric mucosa together with HCl. The case described is classical pernicious anaemia with all severe symptoms and signs. The diagnosis is confirmed by the megalocytic anaemia, with lack of HCl and low [vitamin B_{12}] in the serum. The lack of vitamin B_{12} in the liver and the red bone marrow inhibits the **methyl-malonyl Co-A mutase** and spoils the **purine–pyrimidin–DNA synthesis**. The*

inhibition of these two processes leads to the ***neurological*** *and the* ***haematological*** *disorders in pernicious anaemia, respectively.*

2. *The oxygen capacity of the patient's haemoglobin is* $(1.34\ ml\ g^{-1} \times 63\ g\ l^{-1}) =$ ***84 ml oxygen per litre of blood****.*
3. *The cell-rich bone marrow is filled up with immature stages of leucocytes, platelets and erythrocytes. They remain immature because of lack of* ***maturity factor*** *(vitamin* B_{12}*). The leucopenia was causing frequent infections, and the thrombocytopenia was causing the bleeding tendency of the patient.*
4. *Physiological adaptations to anaemia: (1) The falling red cell count leads to falling viscosity of the blood. The* ***reduced viscosity*** *can reduce the TPVR to less than half of the resting value, which is an appropriate event in order to ease the bloodflow. (2) A slight fall in systemic arterial pressure reduces the stimulus of the arterial baroreceptors. This is the reason for the* ***rise in heart rate and cardiac output****. (3) The low oxygen capacity of haemoglobin is compensated by an increased coronary bloodflow at rest, but during stair climbing the patient felt* ***precordial pain*** *(angina pectoris) caused by hypoxia. (4) The myocardial tissue suffers during* ***long lasting, severe anaemia****. This results in* ***cardiac failure*** *with* ***oedema, large sore liver, and stasis of the neck veins****. (5) The severely anaemic patient has an* ***increased respiration and metabolic rate*** *due to the large cardiac work, and a* ***chronic rise in temperature*** *is typical.*

17. Case History

1. *The linear mean velocity of the blood is related directly to the bloodflow and inversely to the cross-sectional area.*
2. *The volume of a cylinder equals (height · luminal cross-section area) according to the classical geometric argument. Height corresponds to* $\bar{v}$*. The area is* $(\pi \times 1^2) = \pi\ cm^2 = 3.14 \times \mathbf{10^{-4}}\ m^2$*. The* ***volume corresponds to the volume rate 80*** $\mathbf{cm^3/0.3\ s = 267\ cm^3\ s^{-1}}$ ***or*** $\mathbf{2.67 \times 10^{-4}}\ m^3\ s^{-1}$*. From this value, the* ***linear mean velocity*** *can be calculated and the question analysed.* $\bar{\mathbf{v}}$ = ***volume rate/area. Using this formula*** $\bar{\mathbf{v}} = 2.67/3.14 = 0.85\ m\ s^{-1}$*. Thus, the linear mean velocity in the aortic ostium of this person is more than twice the critical mean velocity* $(\bar{v} = 0.4\ m\ s^{-1})$ *for turbulence.*
3. $\bar{v} = \mathbf{2.67 \times 10^{-4}}\ m^3\ s^{-1}$ *per cross-sectional area.* $\bar{v} = 2.67 \times 10^{-4}/0.1 = \mathbf{2.67 \times 10^{-3}\ m\ s^{-1}}$*. Thus, the linear mean velocity of the blood is high in the aorta, and low in the capillary bed, where the cross-sectional area is 0.1* m^2*.*
4. *The blood pressure increases, so the* δ *decreases. Since they are inversely related the* ***pulse wave velocity increases****. In contrast, the* σ *is not changed appreciably.*
5. *The radius of the capillary is small, and the wall tension (T) becomes minimal. For a long, thin-walled cylinder* r_2 *goes towards infinity, therefore the Laplace-formula achieves the following form:*

$$\Delta P_t = T/r + T/\text{infinity} = T/r \text{ or } \mathbf{T = \Delta P \times r.}$$

$$\mathbf{T = [2666 \times 5 \times 10^{-6}] = 1.33 \times 10^{-2}\ Pa \times m\ (=N\ m^{-1}).}$$

This small tension is enough to carry the high transmural pressure.

6. **Poiseuille's law**: $\mathring{Q} = \Delta P/TPVR$ ($l\ min^{-1}$). $TPVR = \Delta P/\mathring{Q}$. $\mathring{Q}$ is 80 ml/0.3 s or 267 $ml\ s^{-1}$. This value is equal to $2.67 \times 10^{-4}\ m^3\ s^{-1}$.

TPVR = 13 330/2.67 × 10^{-4} = 0.5 Pascal seconds m^{-3} (PRU).

18. Case History

1. There are (7 × 1 440) l of whole blood per day, or (7 × 1440 × 0.55) = 5544 l of plasma per day. Thus, the fraction is: **20/5544 = 0.0036.**
2. Three litres of lymphatic fluid are produced every day, the interstitial phase is supplied with (20 × 5) g of protein day^{-1}, but only 90 g net, since 10 g returns to the blood via the capillaries.
   Thus, the **lymphatic mean concentration of protein** is **90/3 = 30 g l^{-1}.**

18. Multiple Choice Questions

Answers A, B, and D are true statements, whereas C and E are false.

19. Case History

1. $$\mathring{V}_{O_2} = (D_{LO_2} \cdot \Delta P_{O_2});\ \mathring{V}_{O_2} = (22 \times 12) = \mathbf{264\ ml\ STPD\ min^{-1}}.$$

   The oxygen tension gradient can rise to a maximum value of 70 Torr during exercise.
2. The rise in the tension gradient from 12 to 70 Torr is caused by a rise in P_{AO_2} and a fall in $P_{\bar{v}O_2}$ during exercise. This rise alone is insufficient without a corresponding rise in the **lung diffusion capacity**. Thus the **lung diffusion capacity** must also increase to reach 4900. It must increase exactly to **70 ml STPD min^{-1} $Torr^{-1}$** (70 × 70 = 4900). The number of open capillaries increases, and they dilate. $\mathring{Q}$ increases by a factor of 5–8. All these changes increase the total diffusion of oxygen, the diffusion barrier area and reduce the diffusion distance in the expanded lungs.
3. The renal bloodflow is 1200 ml min^{-1} and C_{aO_2} is 200 ml l^{-1}. The renal oxygen consumption is 15 ml min^{-1}. The renal oxygen influx is (200 × 1.2) = 240 ml min^{-1}. The renal oxygen outflux is (240 − 15) = 225 ml min^{-1}. The $C_{\bar{v}O_2}$ is 225/1.2 = 187.5 ml l^{-1}. Thus, the renal oxygen content difference is (200 − 187.5) = **12.5 ml l^{-1}**.
4. The **small oxygen content difference** obtained in question (3) is reasonable, because the high renal bloodflow has a clearance purpose. The oxygen delivery to the kidney tissues is obviously redundant.

20. Case History

1. *The hydrostatic pressure of a 900-mm blood column is: (900/13.6) = 66 mmHg or Torr (=10.1 kPa).*
   The hydrostatic component is maximal, since the muscular venous pump is passive. The venous pressure in the muscles is thus 66 plus the 10 mmHg at heart level, a total of ***76 mmHg****.*
2. *In the Poiseuille law the radius is in the denominator in the 4th potency. Thus the vascular resistance is reduced by a factor of 3^4, that is a* ***resistance of only 1/81 of the arteriolar resistance at rest****.*
3. *The MAP is (70 + 100/3) = 103 mmHg and the venous pressure is reduced to 20 Torr. Thus the driving pressure is (103 − 20) =* ***83 mmHg****.*
4. *The* ***total muscle bloodflow is*** *(100 × 30)/170 =* ***17.6 l blood min^{-1}****.*
   The relative bloodflow is 17600/300 = ***59 FU****.*
   The rise in muscle bloodflow is thus: 59/3 = ***19.7-fold****.*

21. Case History

1. *The work rate = $(\Delta P \cdot \mathring{Q})$; $\mathring{Q}$ = sv · 60. Since the work rate must be expressed in watts, the cardiac bloodflow per second is used: (68 × 60/60 s) = 68 ml s^{-1} or (68×10^{-6}) m^3 s^{-1}. This result must be multiplied by 13300 N m^{-2} =* ***0.9 W for the pressure–volume work rate of the left ventricle****.*
   The right ventricle has the same bloodflow, only the pressure is (2.67/13.33) 20% or one-fifth, so the ***right ventricular pressure–volume work rate is (0.9/5) = 0.18 W*** *or 180 mW.*
2. *The heart frequency is 60 beats min^{-1}, which indicates a basic* ***resting condition****. All other variables are also typical for the resting condition.*
3. *The work rate = 20 kPa · $(25 \times 10^3 \times 10^{-6})/60$ m^3 s^{-1} =* ***8.3 W****.*
4. *Work = (work rate · time) = $[(0.9 \times 0.999 + 8.3 \times 10^{-3})\ 2 \times 10^9]$ (J) = $(1.8148 \cdot 10^9)$J = 1814.8 MJ.*
5. *The mechanic efficiency of the heart is 10%. Thus 10-fold as much energy must be provided: 18148 MJ.*
6. ***At rest*** *the ventricle ejects (60 × 68) = 4080 ml blood in 1 min. Actually, the heart performs better: (60 × 0.40) = 24 s ejection time, i.e. 170 ml s^{-1} or 1.7×10^{-1} l s^{-1} or 1.7×10^{-4} m^3 s^{-1}. The linear mean velocity ($\bar{v}$) is: 1.7/3.8 = 0.447 m s^{-1}. The kinetic work rate: $[\frac{1}{2}m\ (l\ s^{-1} \times kg\ l^{-1}) \cdot \bar{v}^2]$. The unit is N m s^{-1}.*

$$[[\tfrac{1}{2} \times (1.7 \times 10^{-1} \times 1.06) \times 0.447^2]$$

$$[(0.18 \times 10^{-1})]\ W = \mathbf{0.018\ W\ or\ 18\ mW}.$$

During exercise *the ventricles eject 25000 ml of blood in (60 × 0.55) = 33 s, i.e. 758 ml s^{-1} or (7.58×10^{-4}) m^3 s^{-1}. Now $\bar{v}$ is: 7.58/3.8 = 1.99 m s^{-1}. The kinetic work rate:*

$$[[\tfrac{1}{2} \times (7.58 \times 10^{-1} \times 1.06) \times 1.99^2]$$

or

$$[(15.91 \times 10^{-1})]]\ W = \mathbf{1.6\ W}.$$

22. Case History

1. $$J_f = 0.075 \times [(3.1 - 0.133) - 0.9(3.3 - 3.3/7)];$$

$$J_f = 0.075\ [2.97 - 2.55] = (0.075 \times 0.42)$$

$$= \mathbf{0.0315\ ml\ min^{-1}\ and\ per\ 100\ g\ of\ tissue}.$$

2. *In 30 kg of tissue: (300 × 0.0315) = 9.45 ml min^{-1} or a daily production of* ***13 608 l.***
3. *The distance corresponds to the pressure of a water column: (1.2 m × 1000 kg m^{-3} × 9.81 m s^{-2}) = 11 772 Pa (or 90 Torr). To this value must be added the supine mean capillary pressure of 3.3 kPa. The expected total value is* ***15 kPa.***
4. *The partial compensation is due to (1) arteriolar constriction; (2) the muscular venous pump reducing the venous pressure considerably; and (3) a very small filtration coefficient in the foot capillaries, compared to less gravity loaded capillaries.*

23. Multiple Choice Questions

Answers A, B, and D are true statements, whereas C and E are false.

24. Case History

1. *Since the circulating TBV is 5400 ml and $\mathring{Q}$ is 5400 ml min^{-1}, then $\mathring{Q}_s$ must be 5400/60 = 90 ml s^{-1}. Thus $\bar{t}$ = 5400/90 =* ***60 s.***
2. *The method is shown in 1: t = 650/90 =* ***7.2 s.***
3. *Flow units (FU) are ml of blood per 100 g tissue and per minute. The calculation goes like this: (3 × 35 000/100) =* ***1050 ml min^{-1}****. Bloodflow per second: 1050/60 =* ***17.5 ml s^{-1}****.*
4. $\bar{t} = \mathring{V}/\mathring{Q}_s$ *or* $\mathring{V} = \bar{t} \cdot \mathring{Q}_s$*; thus* ***V = (5 × 17.5) ml s^{-1} = 88 ml.***

25. Multiple Choice Questions

Answers A, B, and D are true statements, whereas C and E are false.

26. Case History

1. *TBV = 6/(1.5 − 0.17) =* ***4.51 l of blood****. Dimensional analysis: TBV = mmol/(mmol l^{-1} of blood) = l of blood.*
   TEV = 6/(1.33/0.43 l of erythrocytes) = ***1.94 l of erythrocytes.***
2. *Healthy resting males have a TBV of 75 and a TEV of 30 ml per kg body weight. Accordingly, the predicted values of this man are: (75 · 60) =* ***4500 ml of blood, and (30 · 60) = 1800 ml of erythrocytes.***
3. *TH = [CO tracer volume in ml STPD/(1.34 · ($S_{CO,2} - S_{CO,1}$))]. TH = (6 × 22.4)/(1.34 · (0.166 − 0.016)) =* ***669 g haemoglobin.***

The predicted TH for a healthy adult is (4.5 × 150 g l^{-1}) = 675 g haemoglobin, so the result is normal.

4. *An unknown fraction of the CO absorbed by the body is bound to* ***non-circulating haemoglobin in the spleen and liver sinusoids****. An unknown CO volume is bound to myoglobin.*

27. Case History

1. *The most likely explanation is a bleeding peptic ulcer caused by gastric hypersecretion of HCl. Actually, a large bleeding lesion was found at the cardia, but the ulcer was located in the bulbar part of the duodenum.*
2. *The hypotension, arteriolar constriction and falling venous pressure during the severe haemorrhage* ***lower the hydrostatic pressure*** *in the capillaries. This promotes a net reabsorption of ISF into the blood. Hereby, the blood [haemoglobin] is diluted to half its original value within an hour. Thus the blood loss must* ***approach half the total blood volume (5 l) or 2.5 l****.*
3. *Endotoxins from the intestinal flora are not inactivated by the* ***depressed macrophage system of the shock patient****. He has lost about 50% of his circulating macrophages. As a consequence the patient develops fever and his haemorrhagic shock is aggravated.*
4. *The massive oedema of 8 l is clearly an* ***overdose of physiological saline*** *– so-called iatrogenous oedema or overhydration. This is usually avoided by daily use of a bed weight for body weight control and daily serum electrolyte measurements.*

28. Case History

1. *Original volume of metabolic ratemeter:* $V_{STPD} \cdot 760/273 = n \cdot R = V_{ATPS} \cdot (P_B - P_{water})/(273 + t)$
   $V_{STPD} \cdot 760/273 = 50 \cdot (760 - 18)/(293)$
   V_{STPD} = ***45.5 l STPD.***
2. *Safety period until 50% of the oxygen was used: The metabolic ratemeter contained at the start: (45.5 × 0.2093/2) = 4.76 l STPD of oxygen. Safety period: 4.76/ 0.333 =* ***14 min.***

29. Case History

1. *For the small alveolus the transmural pressure must be:*
   $\Delta P = 2T/r = (2 \times 0.07)/0.00004$ = ***3500 Pa.***
   With the double radius the transmural pressure necessary to avoid collapse is only: 3500/2 = ***1750 Pa.***
2. *Small alveoli must empty into larger alveoli and then collapse because they all communicate. The distending pressure is essentially the same in the 300 million basically spherical alveoli of different size. This implies that the* ***alveolar wall tension*** *must change commensurate with the change in* ***alveolar radius****, otherwise the small alveoli will collapse.*

3. *Yes, normally we produce **surfactant** from the type 2 cells of the alveoli. Surfactant can change the alveolar wall tension (the part that is due to surface tension). Let us assume the same surfactant density in the liquid film covering all alveoles. During expiration, surfactant molecules are packed tightly together, so that the surfactant separates the water molecules and **reduces the total alveolar wall tension**. When the radius is large (or during inspiration), surfactant is scattered, and thus the attractive forces between water molecules will increase **(increasing surface tension)**.*
4. *0.5/3 = **0.167 l BTPS cm^{-1} of water at FRC**. The ratio to FRC is: 0.167/2.5 = **0.067. The normal value is 2 ml Pa^{-1} or 0.2 l BTPS per cm of water at FRC.***
5. *The pressure applied for sufficient effect must be at least 1 l divided by the total compliance of this person. $1/C_{total} = 1/C_{lung} + 1/C_{chest\ wall} = 1/0.167 + 1/0.15$ = **12.65 cm of water/l BTPS.***

30. Multiple Choice Questions

Answer C is true, and A, B, D, E are false statements.

30. Case History

1. *$RQ = \dot{V}_{CO_2}/\dot{V}_{O_2} = 1$. $\dot{V}_{CO_2} = \dot{V}_{O_2}$ = 10 000/20 = 500 l STPD day^{-1} or 500/1440 = 0.347 l min^{-1}. Since $\dot{V}_{CO_2} = (\dot{V}_A \cdot F_{ACO_2})$, it follows that $\dot{V}_A = \dot{V}_{CO_2}/F_{ACO_2}$ = **0.347/0.056 = 6.2 l STPD min^{-1}.***
2. *$\dot{V}_D + \dot{V}_A = \dot{V}_E$; $\dot{\mathbf{V}}_{\mathbf{D}} = f \cdot V_D$.*
   Since the person is in a steady state her RQ equals her R value = 1; thus, since her inspired and expired volumes are the same, her fractions of nitrogen must be the same, namely 0.79: $(F_{EN_2} \cdot \dot{V}_E) = (F_{IN_2} \cdot \dot{V}_I)$ and $(F_{AN_2} \cdot \dot{V}_A) = (F_{IN_2} \cdot \dot{V}_{AI})$. F_{EO_2} = 120/(P_B – 20) = 0.162. The F_{ECO_2} = (1 – 0.79 – 0.162) = 0.048.
   *$\dot{V}_{CO_2} = (\dot{V}_E \cdot \mathbf{F_{ECO_2}}) = (\dot{\mathbf{V}}_{\mathbf{A}} + \mathbf{f} \cdot \mathbf{V_D})\ \mathbf{F_{ECO_2}}$. **Anatomic dead space = (0.347 – 6.2 0.048)/(12 0.048) = 0.085 l STPD.***
3. *During triple hyperventilation her F_{ACO_2} must be: 0.347/(3 6.2) = 0.0187. Thus, F_{AO_2} is (1 – 0.78 – 0.0187) = 0.1913 and P_{AO_2} is (0.2013 713) = **144 Torr or 19.2 kPa.***
4. *$\dot{Q} = \dot{V}_{O_2}/(C_{aO_2} - C_{\bar{v}O_2})$ = 347/(200 – 160) = **8.68 l min^{-1}.***

31. Multiple Choice Questions

Answers A, B, D, and E are true statements, whereas C is false.

31. Case History

1. *Calculate $\dot{V}_{CO_2}$ and cardiac output. $R = \dot{V}_{CO_2}/\dot{V}_{O_2}max$; $\dot{\mathbf{V}}_{\mathbf{CO_2}} = 0.9\ \dot{V}_{O_2}max$ = **4.05 l STPD min^{-1}.***
   Since R is 0.9 the ratio: $(C_{\bar{v}CO_2} - C_{aCO_2})/(C_{aO_2} - C_{\bar{v}O_2})$ must be the same. Thus, (650 – 500)/(200 – $C_{\bar{v}O_2}$) is 0.9. Accordingly, $C_{\bar{v}O_2}$ is 33.3 ml STPD l^{-1}.
   *$\dot{Q}$ = 4500/(200 – 33.3) = **27 l of blood min^{-1}.***

2. *Calculate the coronary bloodflow.*
   Coronary bloodflow = 420/(200 − 30) = ***2.47 l min^{-1}.***
3. *During exercise the myocardial energy resources switch from mainly β-oxidation of fatty acids and only 30% carbohydrate combustion, to mainly carbohydrate utilization including lactate combustion and only 30% β-oxidation of fatty acids.*
4. $\dot{Q} = \dot{V}_{O_2} \cdot (C_{aO_2} - C_{\bar{v}O_2})$ *or* $\mathbf{5450 = 273 \cdot (200 - C_{\bar{v}O_2})}$.
   Thus ***$C_{\bar{v}O_2}$ equals 150 ml STPD l^{-1}****, and the O_2 difference equals* ***50****. These results are typical for the resting condition.*

32. Case History

1. ***RQ*** *= (542 − 500)/(200 − 150) =* ***0.84.***
2. *The calculated RQ value corresponds to mixed diet.*
3. $\dot{V}_{CO_2}$ *= (10 450/20.6) = 507.3 l STPD daily =* ***0.352 l min^{-1}.***
4. $\dot{V}_{CO_2}$ *= ($\dot{V}_{O_2}$ · RQ) = (0.352 · RQ) =* ***0.296 l min^{-1}.***
   $F_{ACO_2} = 5.3/(101.3 - 6.25) = 0.056$; $[F_{ACO_2} = (\dot{V}_{CO_2}/\dot{V}_A)]$
   $\dot{V}_A = \dot{V}_{CO_2}/0.056 = 5.28$ l min^{-1}. $\dot{V}_E = \dot{V}_{CO_2}/F_{ECO_2} = 0.296/[4.4/(101.3 - 6.2)] = 6.43$ l min^{-1}. $\dot{V}_D = (\dot{V}_E - \dot{V}_A) = 1.15$ l min^{-1}. $\mathbf{V_D = 1.15/14 = 0.082}$ **l.**
5. $\dot{Q} = \dot{V}_{O_2}/(C_{aO_2} - C_{\bar{v}O_2}) = 352/(200 - 150) =$ ***7.04 l min^{-1}.***
6. *The cause of unconsciousness is neither respiratory nor cardiovascular. By exclusion the most likely cause is* ***brain damage*** *from the traffic accident.*

33. Multiple Choice Questions

Answers A, B, C, and E are true statements, whereas D is false.

33. Case History

1. $F_{ACO_2} = P_{ACO_2}/(101.3 - 6.2)$. $\mathbf{P_{ACO_2} = 5.32}$ **kPa.**
2. $C = (\alpha \cdot P)$
   α is the Bunsen solubility coefficient. Its size must ***decrease with increasing temperature****, because the gas escapes with thermic movements.*
3. *22.4 ml = 1 mmol.* $F_{ACO_2} = 0.056$ *and* $(101.3 - 6.2) = 95.1$ *; Physically dissolved* $C_{aCO_2} = (0.51 \times 0.056 \times 95.1 \times 1000)/101.3 =$ ***26.8 ml STPD l^{-1} or 26.8/22.4 = 1.2 mM.***
   Physically dissolved $C_{aO_2} = (0.022 \times 13.3 \times 1000)/101.3 =$ ***2.89 ml STPD l^{-1} or 2.89/22.4 = 0.13 mM.***
4. $P_{\bar{v}CO_2}$ is 6.1 kPa and her $P_{\bar{v}O_2}$ is 6.0 kPa.
   Physically dissolved $C_{\bar{v}CO_2} = (0.51 \times 6.1 \times 1000)/101.3 =$ ***30.7 ml STPD l^{-1} or 30.7/22.4 = 1.37 mM.***
   Physically dissolved $C_{\bar{v}O_2} = (0.022 \times 6.0 \times 1000)/101.3 =$ ***1.3 ml STPD l^{-1} or 1.3/22.4 = 0.06 mM.***

34. Multiple Choice Questions

Answers B, C, D, and E relate to metabolic alkalosis and are true, whereas A is false.

34. Case History

1. *pH = 6.1 + log([bicarbonate]/(0.225 · P_{CO_2})).*
   pH = 6.1 + log([500/22.4]/(0.225 × 4.9) = 6.1 + log 20.246 = ***7.406****. This patient is in a normal state between attacks of dyspnoea.*
2. *pH = 6.1 + log([bicarbonate]/(0.225 × P_{CO_2})).*
   pH = 6.1 + log([21]/(0.225 × 2.6) = 6.1 + log 35.897 = ***7.655****. During emergency conditions the patient has developed an* ***acute respiratory alkalosis*** *by hyperventilation.*
3. *The diagnosis is probably hyperventilatory tetany.*

35. Case History

1. ***Standard affinity*** *is the binding force between two molecules, when half of the binding sites are occupied (i.e. at 50% saturation). In the case of oxyhaemoglobin the P_{50} is used. Here, standard affinity is* ***1/P_{50}.***
2. *The dry CO-fraction (F_{ACO}). All the binding sites for CO and oxygen in the haemoglobin of this person are described by the following volume: (148 · 1.34) = 198.3 ml STPD l^{-1}. Thus, all binding sites are occupied by CO and oxygen. The following equation is always true for CO-haemoglobin (HbCO) and oxy-haemoglobin (HbO_2):*

   $$[HbCO]/P_{aCO} = \text{Standard affinity} \cdot [HbO_2]/P_{aO_2}.$$

   Using the data given:
   28.3/P_{aCO} = 260 × 170/100. P_{aCO} = (28.3 × 100)/(260 × 170) = 0.64 Torr. Since we have equilibrium between arterialized blood and alveolar air it follows that P_{aCO} = P_{ACO} = 0.64 Torr. F_{ACO} = P_{ACO}/(P_B − 47) = 0.64/(760 − 47) = ***0.0009 or 0.09%.***
3. *Oxygen supply. The oxyhaemoglobin saturation degree in the venous blood ($S_{\bar{v}O_2}$) is a reliable indicator of the oxygen supply to the most hypoxia-threatened cells, namely those located at the venous end of the capillaries. The arteriovenous oxygen content difference is assumed to be 50 ml STPD l^{-1} for both persons. A [haemoglobin] of 6.6 mM equals: (148 × 6.6/9.18) =* ***106.4 g l^{-1}.***
   The $S_{\bar{v}O_2}$ of the CO-poisoned person is: (170 − 50)/198.3 = 0.605.
   The $S_{\bar{v}O_2}$ of the anaemic person is: (106.4 × 1.34 − 50)/(106.4 × 1.34) = ***0.65****. Such an anaemic person has a severely reduced working capacity, but since his oxygen dissociation curve is switched strongly to the right, his $P_{\bar{v}O_2}$ is much higher than that of the CO-poisoned with a mixed dissociation curve located to the left. This CO-poisoned person is disabled, whereas the anaemic person can function although tired and with* ***reduced working capacity.***

36. Multiple Choice Questions

Answers A, B, and E are true statements, whereas C and D are false.

37. Multiple Choice Questions

37.1: Answers A, C, D, and E are false statements, whereas B is true.
37.2: Answers B, D, and E are false statements, whereas A, and C are true.

38. Case History

1. *The alveolar air equation states:*
   $R = \dot{V}_{CO_2}/\dot{V}_{O_2} = 0.8 = [\dot{V}_A \cdot P_{ACO_2}/(P_B - 47)]/[\dot{V}_A (P_{IO_2} - P_{AO_2})/(P_B - 47)];$
   $R = P_{ACO_2}/(P_{IO_2} - P_{AO_2});\ P_{ACO_2} = 0.8(120 - 100) =$ ***16 Torr.***
2. $\dot{V}_A + \dot{V}_D = \dot{V}_E$
   $\dot{V}_E = 5.6 + (0.12\ 14) =$ ***7.28 l STPD min^{-1}.***

39. Case History

1. *Mass balance:*
   $\dot{V}_{O_2} = \dot{Q} \cdot (C_{aO_2} - C_{\bar{v}O_2});\ \dot{V}_{O_2} = 25 \times 170 =$ ***4250 ml STPD min^{-1}.***
2. *Assuming she is working at or above her* $\dot{V}_{O_2}max$ *we can calculate her* ***fitness number*** *to 4250/62 =* ***69 ml O_2 kg^{-1} min^{-1}****. This is an excellent result for a well-trained athlete.*

40. Multiple Choice Questions

40.1. Answers A, B, C, and D are true statements, whereas E is false.
40.2. Answers A, C, D, and E are true statements, whereas B is false. The patient has a **grave iron deficiency anaemia.**

41. Case History

1. $P_{AO_2} = [760 \times 3 - (47 + 40)];\ P_{AO_2} = 2193$ *Torr.*
   The P_{O_2} *of the blood leaving the lung capillaries is assumed to be equal to* P_{AO_2}. *The physically dissolved oxygen: (0.022 × 2193 × 1000/760) =* ***63.5 ml STPD l^{-1}****. This is more than the arterio-venous oxygen difference of a resting person (50 ml STPD l^{-1}).*
2. *The chemically bound oxygen is (170 × 1.39) =* ***236.3 ml STPD l^{-1}.***
3. *The CO_2 transport from tissues to lungs suffers from lack of* ***oxygen free haemoglobin*** *and thus of* ***carbamino-binding sites.***

42. Case History

1. *The difference between pK and pH is (9.3 − 7.3) = 2 decades or 10^2. Thus, we have* ***100 times as much NH_4^+ as ammonia in the blood.***
2. *12600/0.46 = 27391 mmol oxygen day^{-1}.* $RQ = \dot{V}_{CO_2}/\dot{V}_{O_2};\ \dot{V}_{CO_2} = 0.83 \cdot 27.391 =$ ***22.7 mol CO_2 daily.***

3. *The synthesis of urea in the liver is the major route of removal of NH_4^+: $2\ NH_3 + CO_2 \rightarrow$ urea + water. Thus, NH_4^+ is eliminated by the use of half a mol of carbon dioxide daily.*

43. Case History

1. $F_{AO_2} = 18.45\ kPa/(101.3 - 6.2)\ kPa = 0.194.$
   $F_{AN_2} = 74\ kPa/(101.3 - 6.2)\ kPa = 0.778.$
   $\mathbf{F_{ACO_2} = 1 - (F_{AO_2} + F_{AN_2}) = 0.028.}$
   $F_{ACO_2} = \mathring{V}_{CO_2}/\mathring{V}_A;\ \mathring{V}_A = \mathring{V}_{CO_2}/F_{ACO_2}$
   $\mathring{V}_A = 0.600/0.028 = \mathbf{21.429}$ *l STPD* min^{-1}.
2. *The* **P_{ACO_2} is $(0.028 \times 95.1) = 2.66$ kPa (20 mmHg)**. *The* P_{aCO_2} *is assumed to be similar, when tension equilibrium is established; the normal venous mixing can only increase the tension to a small extent. Such a low* P_{aCO_2} *is a clear indication of* **hyperventilation.**
3. $\mathring{V}_{O_2} = \mathring{V}_A\ [F_{IO_2} \times F_{AN_2}/F_{IN_2} - F_{AO_2}]$
   $\mathring{V}_{O_2} = 21.429\ (0.2093 \times 0.778/079 - 0.194)$
   $\mathring{V}_{O_2} = \mathbf{0.26}$ **l STPD min^{-1}.**
4. $R = 0.600/0.26 = 2.31$
5. *This R value is higher than the highest possible RQ (1 by* ***pure carbohydrate metabolism****).*

44. Multiple Choice Questions

A, B, C, and E are true statements, whereas D is false.

44. Case History A

1. *The extra heat energy stored in his body at the end of the 30-min period is:* $[3°C \cdot 70\ kg \times 3.47\ kJ/(°C\ kg)] = \mathbf{729\ kJ.}$
2. *A minimal metabolic rate is assumed to be just about BMR: 300 kJ* h^{-1} *or 5 kJ* min^{-1} *or 83 W. If the heat loss is assumed to be zero, because all the heat produced in 30 min is stored, it explains an accumulation of no more than:* $(5 \cdot 30\ min) = 150$ *kJ. This does not explain the rise in heat energy stored (729 kJ).*
3. *Only* ***dramatic shivering*** *during the abrupt rise in temperature can explain the large heat energy store.*
4. *The heat capacity of the body is 3.47 kJ* $kg^{-1}\ °C^{-1}$*, and that of water is 1 kcal or 4.2 kJ* $kg^{-1}\ °C^{-1}$*. Intake of 1 l of ice water corresponds to a maximum of* $(40°C \times 4.2)\ kJ = 168$ *kJ. The body weight multiplied with the heat capacity of the body show that* $(70\ kg \times 3.47) = 243$ *kJ must be eliminated from the body in order to reduce the body temperature of the patient 1°C. Accordingly, the ice water can reduce the temperature of the patient by* $168/243 = \mathbf{0.7°C}$.
5. *Evaporation of sweat water* (V_{sweat}) *implies a loss of energy* (Q_E *J* min^{-1}) *according to the equation:*
   $\mathbf{Q_E\ (J\ min^{-1}) = 2436\ (J\ g^{-1}) \cdot \mathring{V}_{sweat}\ (g\ min^{-1})} = (2436 \times 8\ ml\ min^{-1})10^{-3} = \mathbf{19.5\ kJ\ min^{-1}.}$

With a water evaporation of (32 $\times$ 0.25 =) 8 ml min^{-1} and 19.5 kJ min^{-1}, it will take (729/19.5) = ***37 min*** *in order to eliminate the accumulated heat energy. The calculation illustrates the fact that evaporation of only 25% of a sweat volume of about 1 l (37 min, 32 ml min^{-1} = 1187 ml) eliminates much more heat energy than ingestion of 1 l of ice water.*

44. Case History B

1. *The metabolic rate is [400 17.5 + 100 39 + 100 17] =* ***12600 kJ daily*** *=* ***12.6 MJ day^{-1}.***
2. *The metabolic water formed by oxidation of the food is: 32 g MJ^{-1} multiplied by 12600 kJ. This water volume is* ***403 g or ml****. This amount explains the discrepancy between 2100 and 1700 ml of water.*

45. Multiple Choice Questions

Answers A, C, and E are true statements, whereas B and D are false.

45. Case History

1. *The distribution volume of the driver is his total water phase, probably 67–68% of his body weight: (83 0.68) =* ***56.44 kg water.***
2. *The alcohol elimination rate is 0.0025 per mille min^{-1} or 0.15 per mille in 1 h. At 2 p.m. the driver had at least a plasma [alcohol] of (1.3 + 0.15) =* ***1.45 per mille or g kg^{-1}.***
3. *The driver oxidized alcohol in an estimated time period of (8 p.m. to 3 a.m.) 7 h, before the blood sample was taken. The alcohol intake must be: [(7 0.15) + 1.3] 54.44 =* ***127.9 g of alcohol****. With an average content of 12 g alcohol per drink this implies an intake of* ***at least 11 drinks****. The excretion of alcohol is assumed to be minimal at rest, and it is not included in this estimate.*

46. Multiple Choice Questions

I: Answers A, B, D, and E are true statements, whereas C is false.
II: Answers A and C are true, whereas B, D and E are false. Dehydration is frequent in elderly persons.

46. Essay

Write a short essay about gastrointestinal peptides (see chapters 46, 64, 72 and 73).

47. Multiple Choice Questions

Answers A and C are false statements, whereas B, D and E are true.

47. Case History

1. *The heat capacity of the human body is 3.47 kJ kg^{-1}°C^{-1} (units in Chapter 75). The extra heat energy stored in her body is: [3.47 · 60 (41.4 – 36.9)] =* ***937 kJ.***
2. *The diabetic female is in* ***coma caused by hyperthermia and metabolic acidosis****.*
3. *The primary compensation is* ***hyperventilation*** *causing an increased CO_2 elimination through the lungs, shown by the fall in P_{aCO_2}. Later an increased renal excretion of acid takes place.*
4. *The evaporation energy for water is found in Chapter 75: 2436 kJ kg^{-1}. The 937 kJ are eliminated by evaporation of [937/2436] = 0.385 kg of water, corresponding to (5 × 0.385) =* ***1.925 l of water in sweat.*** *The total sweat volume also includes dry substances.*
5. *Upon* ***stimulation of the pancreas****, the [bicarbonate] of the pancreatic juice increases importantly and that of [Cl^-] falls, whereas the [Na^+] and [K^+] remain relatively constant. Pancreatic juice buffers the extremely acid gastric juice and chyme entering the duodenum.*

48. Multiple Choice Questions

Answers A, B, C and D are true statements, whereas E is false.

49. Multiple Choice Questions

Answers A, B, and C are true statements, whereas D and E are false.

49. Case History

1. *The* ***equilibrium potential (e.p.)*** *for a given ion over a membrane with a* ***concentration gradient for the ion****, is defined as the* ***potential difference resulting in a net transport of zero****. Thus, e.p. is the inside minus the outside cell potential. Here, the* ***total e.p.*** *must be* ***plasma minus gastric juice potential****, and* ***luminal e.p.*** *must be* ***cell inside minus gastric juice potential****.*
2. *According to the Nernst equation: Total e.p. = 60 log [H^+]^{out}/[H^+]^{in} mV. Total e.p. = 60[–7.40 – (–1)] =* ***–384 mV****.*
3. *Based on the same calculation the* ***luminal e.p.*** *is 60[–7 – (–1)] =* ***–360 mV****.*
4. *One mol of glucose delivers (180 × 15.5) = 2790 kJ. In order to transport 1 mol of H^+: 85/2790 =* ***0.03 mol of glucose must be oxidized.***
5. *The necessary energy for this active process is (35 × 85)/1000 = 2.975 kJ h^{-1}. With an enthalpy equivalent of 20 for oxygen, this corresponds to: (2.97 × 1000/20) =* ***118 ml STPD for 1 h.***
6. *This result is a major part of the* ***total oxygen uptake*** *(e.g. 300 ml min^{-1}).*

50. Multiple Choice Questions

Answers B, C, D, and E are true statements, whereas A is false.

51. Multiple Choice Questions

Answer A is a true statement, whereas B, C, D, and E are false.

51. *Case History*

1. *The daily water absorption is [1500 − (150 × 0.75)] =* ***1388 ml****.*
2. *The daily net absorption of Na^+ is [1.5 × 120 − (0.150 × 0.75 × 20)] =* ***177.75 mmol****.*
   The daily net absorption of K^+ is [1.5 × 4 − (0.150 × 0.75 × 5)] = ***5.44 mmol****.*
3. *The daily Na^+ loss in the faeces is (0.150 × 0.75 × 20) =* ***2.25 mmol****.*
   The daily K^+ loss in the faeces is (0.150 × 0.75 × 5) = ***0.563 mmol****.*
4. *Following ileostomy:*
   The daily Na^+ loss in the faeces is (1.5 × 120) = ***180 mmol****.*
   The daily K^+ loss in the faeces is (1.5 × 4) = ***6 mmol****.*
5. *Dietary measures are* ***not essential****, because the salt loss is covered by a normal salt intake.*

52. Multiple Choice Questions

Answers A, B, and C are true statements, whereas D and E are false.

52. *Case History*

1. *The elimination rate constant (k) is 0.001 or 1/1000 per 24 h. Half-time = 0.693/k.*
   Half-time = ***693 days****.*
2. *$M_t = M_o\, e^{-tk}$. By insertion: $0.5 = 5\, e^{-t/1{,}000}$. Thus,* ***t = 2302 days or 2302/365 = 6.3 years. Pernicious anaemia is developed over years****.*

53. Multiple Choice Questions

Answers A, B, and C are true statements, whereas D and E are false.

53. *Case History*

1. ***Coma*** *is such a deep state of unconsciousness that the* ***subject does not react, even to the most painful stimuli****. In this case there is an increase in blood [NH_4^+], and it is believed that hepatic coma is caused by intoxication with*

ammonia or closely related substances. However, lack of glucose in the brain is probably involved.

2. *The alcoholic patient has* ***liver insufficiency****, with a minimal glucose production. Within a short time this condition develops into* ***severe hypoglycaemia****, because formation of glucose outside the liver is minimal. Such a profound hypoglycaemia leads to unconsciousness, since the brain only oxidizes glucose. After some time the unconscious patient gradually slides into* ***hepatic coma****.*

54. Case History A

1. *Two physical half-lives equals the 24 h following the injection of radioactive K^+. Thus the remaining activity of the injected $^{42}K^+$ is: 555 000 Bq $\times$ $(\frac{1}{2})^2$ =* ***138 750 Bq.***

   The specific activity, SA, is 55.5 Bq $mmol^{-1}$. SA is equal to the ratio between the total contents of $^{42}K^+/^{39}K^+$, that is in the urine 2220/40 = 55.5 Bq $mmol^{-1}$ after 24 h.

   The exchangeable K^+ pool of her body:

   $$\text{(Injected} - \text{eliminated)/SA} = (138\,750 - 4144 - 2220)/55.5\ \text{Bq/(Bq mmol}^{-1}) = \mathbf{2385\ mmol.}$$

   The normal value is (71kg $\times$ 41) = ***2911 mmol ^{39}K.***

2. *The elimination flux is 40 mmol 12 h^{-1} =* ***3.333 mmol h^{-1}.***
   The elimination flux divided by the K^+ pool is the rate constant, k:

   $$3.333/2385\ (\text{mmol h}^{-1})/\text{mmol} = \mathbf{0.0014\ h^{-1}.}$$

   The biological half-life is 0.693/k or ***495 h.***
   The ratio between physical and biological half-life is 12/495 = ***0.02424****, and the biological half-life is the main determinant of the total half-life.*

3. *The diagnosis for the patient is* ***hypokalaemia with K^+ depletion of the body.*** *Intensive use of* ***diuretics without compensatory K^+ intake*** *is the cause.*

4. *The amount of natural ^{40}K is measured in a* ***whole body counter.*** *This amount divided by the specific activity (0.012%) is the* ***total body potassium.***

54. Case History B

1. *We must calculate the* ***elimination rate constant (k)*** *for the excretion of this substance through the kidneys. The definition of k is the* ***fraction of the total amount of substance in the body eliminated per time unit.***

   The available information suggest that the substances are like inulin which is eliminated by ***glomerular ultrafiltration****, without being secreted or reabsorbed. For such substances the calculation is:*

   $$\mathbf{k_1} = \text{GFR/ECV} = 120/14\,000 = \mathbf{0.0086\ min^{-1}.}$$

2. $\mathbf{k_2}$ = *RPF/ECV = 700/14 000 =* ***0.05 min^{-1}.***

55. Case History

1. ***Clearance for inulin*** *= (2 × 20)/0.225 =* ***178 ml of plasma/min.*** *This is a normal value for a tall male of 100 kg. The clearance of 178 ml min^{-1} is equal to 125 ml min^{-1} in a 70 kg male.*

   Clearance for drug X *= (2 × 400)/1 =* ***800 ml of plasma/min.***

2. *Ultrafiltration of the drug X = Clear$_{insulin}$ × 1 × 0.9 (μg ml^{-1}) =* ***160 μg min^{-1}.***
3. ***Excretion of drug X*** *= 400 × 2 × 120 = 96 000* ***μg or 96 mg.***
4. *X is subject to a secretion process in the tubules. The filtration flux of X is only* ***0.16 mg min^{-1}*** *and 96/120 =* ***0.8 mg min^{-1}*** *is excreted.*

56. Case History A

1. *Mean plasma concentration of inulin = 0.45 mg ml^{-1}.*
   Inulin clearance = [(70/30) · 12]/0.45 = ***62.22 ml plasma/min.*** *This value is approximately* ***50% of normal*** *(125 ml min^{-1}) – actually 62.22/125 or 49.8%.*
2. *T_{max} = (62.22 × 3.8) – (45 × 70/30) =* ***131.4 mg min^{-1}.*** *This value is approximately 50% of normal (270 – 320 mg min^{-1}) – precisely 131.44/270 or 48.7%.*
3. *We only obtained 50% of the normal values for both ultrafiltration and glucose reabsorption. Several solutions are possible including (a) and (b).*
   *(a) A normal relationship may exist between the number of functioning tubuli and functioning glomeruli, but with half the normal number of nephrons preserved (****one kidney, with 1 million nephrons, is blocked****).*
   (b) ***On a statistical basis, only half the glomeruli are connected to functioning tubuli****, if all the normal 2 million glomeruli (nephrons) are preserved.*

56. Case History B

1. *Protein contains 16% nitrogen (N). Urea is $CO(NH_2)_2$ which implies a content of 28 g N out of 60 g urea. The daily urea production is (100 × 0.16)/28 =* ***0.57 mol urea****. Urea production must equal urea excretion, because the male is in urea balance. Urea has a molecular weight of 60, so converted into weight unit the renal urea excretion is (0.57 × 60) =* ***34.3 g of urea 24 h^{-1}.***
2. *Since 60 ml min^{-1} is cleared in the kidneys out of 42 000 g = 42 000 ml, the elimination rate constant (k) is: 60/42 000 =* ***0.0014285 min^{-1}*** *or $T_{\frac{1}{2}}$ =* ***484 min.***

57. Case History

1. *The* ***inulin clearance*** *equals the excretion flux of inulin in the urine divided by its plasma concentration. This is 480 mg l^{-1} of plasma.*
   Inulin clearance = (2030/150)/480 = ***0.028 l of plasma/min.*** *This value is extremely low compared to normal values for her age group (above 0.075 l min^{-1}). Her* ***glomerular function*** *must be seriously damaged.*

2. *The secretion flux of PAH equals its excretion flux minus its filtration flux. The PAH concentration is so high that the secretion flux of this patient must equal her maximal secretion capacity (T_{mPAH}).*
   The excretion flux is equal to 18780/150 = ***125.2 mg min^{-1}.***
   GFR is equal to (0.94 × inulin clearance) = ***26.3 ml min^{-1}.***
   The plasma concentration of PAH is 4 g l^{-1} or 4 mg ml^{-1}.
   The filtration flux of PAH equals

$$[GFR \times F_{free} \times 4] = [26.3 \times 0.75 \times 4] = \mathbf{78.9\ mg\ min^{-1}}.$$

   From these premises the maximal tubular secretion capacity can be calculated:

$$T_{mPAH} = (125.2 - 78.9) = \mathbf{46.3\ mg\ min^{-1}}.$$

   A normal T_{mPAH} is 80 mg min^{-1}. Her ***tubular secretion capacity*** *suffers as well.*

58. Case History

1. *$\mathring{Q}$ = 250/50 = 5 l blood min^{-1}.* *RBF = 0.3 × 5 =* ***1.5 l blood min.***
2. *RPF = RBF × 0.5;* *RPF = 1.5 × 0.5 = 0.75 l plasma/min.*
   GFR = 0.2 × 0.75 = **0.15 l ultrafiltrate/min.**
3. *Inulin characteristics:* ***(A) Inulin clearance*** *is independent of the inulin concentrations in plasma and in urine.* ***(B) Plasma proteins*** *do not bind inulin.* ***(C) Inulin-like substances*** *(such as mannitol, sorbitol and thiosulphate) have the same clearance value as inulin. This is in spite of different molecular weights, and thus different diffusion coefficients.* ***(D) Clearance for other substances*** *will approach the value for inulin with increasing plasma concentrations.*
4. *The excretion flux of inulin in the urine equals the filtration flux through the glomerular barrier:*

   C_u × urine flow rate = C_p GFR. The C_u/C_p ratio is 0.15/0.0075 = ***200/1.***

59. Case History

1. *The clearance for PAH = 100/0.2 =* ***500 ml of plasma/min.*** *This is the definition of clearance: The excretion flux of PAH divided by its plasma concentration. At least 500 ml of plasma must pass the kidneys each minute to accomplish this clearance.*
2. *The tubular secretion flux of PAH equals the excretion flux in the urine minus the filtration flux.*
   The secretion flux: (100 − (125 × 0.2 × 0.8)) = ***80 mg min^{-1} of PAH.***
3. *RPF equals the excretion flux for PAH divided by the renal, arteriovenous concentration difference for PAH (0.2 − 0.02 mg ml^{-1}).*
   RPF = 100/(0.2 − 0.02) = ***556 ml of plasma/min.*** *This is the true RPF, which is 10% higher than the clearance for PAH determined in 1.*
   RBF = 556/0.57 = ***975 ml of blood/min.***

4. The excretion flux for PAH: (80 + (125 × 1 × 0.8)) = **180 mg of PAH/min.**
5. The new clearance for PAH: 180/1 = **180 ml of plasma/min.**

60. Case History

1. pH = 7.40. The arterial H^+ concentration = $10^{-7.40}$ mol l^{-1} = **40 nmol l^{-1}.**
   pH = 7.00. The arterial H^+ concentration = 10^{-7} mol l^{-1} = **100 nmol l^{-1}.**
   pH = 7.80. The arterial H^+ concentration = $10^{-7.80}$ mol l^{-1} = **16 nmol l^{-1}.**
2. **Log scales** differ from **proportional scales**. Logarithms often hide detailed information.
3. Metabolic acidosis is characterized by H^+ excess in the blood. This reduces the standard bicarbonate to values below 20 mmol l^{-1} – an important base deficit. This condition occurs in **diabetics** and in persons working **supramaximally** (see Fig. 34-2).

   Metabolic alkalosis is caused by a **primary fall in the proton/ bicarbonate** ratio in the ECV. Loss of gastric fluid (vomiting with loss of HCl), intake of base, and use of diuretics can cause this form of alkalosis. The low $[H^+]$ reduces ventilation and increases [bicarbonate]. The positive base excess is characteristic (high standard bicarbonate). This is condition 6 in Fig. 34-2). The respiratory compensation raises P_{aCO_2} and pH is reduced, but the compensation is never total, since the rise in P_{aCO_2} and the fall in P_{aO_2} in itself limit the fall in ventilation.

61. Case History A

I:

1. Before going to surgery the patient had (28 × 300) = **8400 mOsmol in his ICV** and (14 × 300) = **4200 mOsmol in ECV.**

   Since NaCl must be lost from ECV only, the residue here must be 4000 mOsmol in 12.25 l after the sweat loss.

   Thus, the osmolality is increased to (4000/12.25) = **327 mOsmol l.**

   This is a hyperosmolal dehydration, which is universal. This is because the cells are hyperosmolal as well.

II:

2. This is an isosmolal loss only from ECV. Its name is **isosmolal or isotonic dehydration. The osmotic gradient does not change at all.**

   Only the ECV is reduced by 3 l, not its osmolality (Fig. 61-2: middle).

   There is no change of ICV or osmolality.

III:

3. The total loss of Na^+ is 150 mmol in each of 5 l, corresponding to 1500 mOsmol, and no water deficit. The new osmolality for both compartments must be: [(42 × 300) – 1500]/42 = **264 mOsmol l^{-1} or kg^{-1}.**

   Almost all NaCl is in the ECV, so its normal content of 4200 mOsmol must be reduced to 2700 mOsmol. Since the common osmolality (Fig. 61-2: below) is 264, the new ECV is 2700/264 = **10.2 l, which is seriously low.**

61. Case History B

1. *We have injected* ***2812000 Bq of $^{24}Na^+$*** *with a high SA. The injected dose is negligible when compared to the total amount of Na^+ in the body. After 30 h we must assume an even distribution between the radioactive Na^+ and the exchangeable Na^+.*

   Accordingly, the relation between the radioactive and the exchangeable Na^+ must be the same all over the body including the urinary tract. After 30 h exactly two physical half-lives have elapsed, so the remaining radioactivity must be:

   $$2812000 \text{ Bq} \times (\tfrac{1}{2})^2 = 703000 \text{ Bq}.$$

   Now urinary SA is: 18890/90 = ***210 Bq mmol^{-1}*** *exactly as found in plasma. The exchangeable Na^+:*

   $$(\text{Injected} - \text{eliminated})/\text{SA};$$

   $$(703000 - (2.5 \times 20350) - (0.06 \times 18870))/210 = \mathbf{3100\ mmol.}$$

   This result is higher than the normal maximum for this patient.
2. *The Na^+ eliminated in the 30-h period:*

   $$(2.5 \times 70) + (0.06 \times 90) = 180.4 \text{ mmol}.$$

   This is equal to ***144.3 mmol per 24 h.***

   Compared to the total pool of exchangeable Na^+, this is ***4.65% per 24 h*** *(144.3/3100), which is less than the normal* ***5.5% per 24 h.***
3. *The high plasma $[Na^+]$ must* ***reduce the secretion flux of aldosterone****. This will* ***reduce the Na^+ reabsorption*** *in the distal renal system.*
4. *The volume of distribution for the exchangeable Na^+ is the total Na^+ pool divided by the $[Na^+]$ in plasma water. Since 0.94 parts of plasma are water, the Na^+ concentration of 146 mmol l^{-1} equals (146/0.94) = 155 mmol l^{-1} water.*

   Thus the volume of distribution is: 3100/155 = ***20 l of water****. Healthy persons have an ECV of up to 0.24 parts of the body weight, which is the distribution volume for Na^+. Here the predicted volume is (0.24 × 73) = 17.52 l, so the distribution volume of this patient is not an acceptable normal value.*
5. *During neutron radiation of the patient a few natural ^{23}Na are activated by uptake of an extra neutron forming ^{24}Na.*

   This amount is measured in a whole body counter and divided by the specific activity. The result is the ***total body sodium.***

62. Case History

1. *The nurse turned her intermediary metabolism into ketosis by fasting, and developed a classical* ***hunger diabetes****. She had a* ***totally compensated (normal pH) metabolic acidaemia.***
2. *The nurse was advised to* ***improve her eating habits*** *(healthy and varied food). Her* ***water without food diet*** *proved to be an insufficient eating pattern.*

63. Case History

1. *Note that the rate constant k equals the amount secreted per unit time divided by the total amount of TSH (mol) present in the TBV. The amount of TSH secreted per unit time is a flux (**J**) or mol min^{-1}. The total amount of TSH in the blood is TBV multiplied by the concentration of TSH in the blood (C). The rate constant (k) is equal to ln2/$T_{\frac{1}{2}}$ or 0.693/$T_{\frac{1}{2}}$.* ***Hence, J = (C × TBV × k) = (C × TBV × 0.693)/$T_{\frac{1}{2}}$.***
2. *From the plasma concentration of TSH, the whole-blood concentration can be calculated:*

$$(100 \times 0.5) = 50 \text{ pmol per l of blood.}$$

$$\mathbf{J = (50 \times 4 \times 0.693)/55 = 2.52 \text{ pmol min}^{-1}.}$$

3. *24 h equals 1440 min. The 24-h secretion flux is: (2.52 × 1440) = 3628.8 pmol day^{-1}.*

$$1 \text{ pmol TSH} = (10^{-12} \times 31\,000) \text{ g.}$$

$$J = (3628.8 \times 31\,000 \times 10^{-12}) \text{ g} = \mathbf{112.5\ \mu g/24\ h.}$$

The daily fractional secretion of TSH is (112.5/300 μg) = ***0.375*** *or* ***one-third of the hormone storage****.*

64. Case History

1. *1 fmol of vasopressin is equal to (10^{-15} × 1084) g. 10^{-13} mol ADH = 100 fmol.*
   A normal 60-kg person secretes the following flux of vasopressin: (100 × 60 × 60)/(10^{-15} × 1084) g h^{-1} or ***0.33 μg h^{-1}.*** *The vasopressin secretion of this patient is only 5% or* ***0.0166 μg h^{-1}.***
2. *The ECV must be 12 kg or l and the distribution volume for vasopressin is* ***14.4 l.***
3. *1084 μg vasopressin equals 1 μmol. Thus, 3 μg equals (3/1084) μmol or 2.768 nmol. The rise in [vasopressin] must be: (2.768/14.4) =* ***0.171 nmol l^{-1}*** *(171 pmol l^{-1}).*
4. *The new concentration is 171.01 pmol l^{-1}, and the ratio is 171.01/2 =* ***85.5.***
5. *The plasma [**vasopressin**] necessary to normalize the diuresis of this patient is* ***much higher than normal.***
   Insensitive or destroyed receptor proteins for vasopressin *on the target cells in the kidney tubules can explain this type of diabetes insipidus. However,* ***abnormal receptor proteins*** *can be secondary to* ***abnormal ATPase mechanisms.***
6. *The urine loss is high, namely (10/60) or* ***one-sixth*** *of the patient's body weight. Such a high water loss is only possible, if she continues to drink. Any uncompensated water loss comprising more than* ***one-fifth*** *of the total body water can be* ***fatal****.*

65. Case History

1. *The specific activity ($SA = c^*/C$) of radio-iodide (^{131}I) in the urine is:*

$$1.6 \times 10^7/(0.2 \times 65) = 12\,308 \text{ Bq}/\mu g.$$

Since SA is the same in urine and in plasma, it follows that the concentration of iodide in plasma is: Activity per ml/SA.

Start-concentration = 38 300/12 308 (Bq l^{-1})/(Bq/μg) = ***3.11 μg l^{-1}.***

Compared to the physical half-life of 8 days (192 h) the test period is short. Assuming first order disintegration of iodine the SA must be adjusted for disintegration during the experiment which is 2 h of the normal half-life (192 h).

$$k = 0.693/192 = 0.0036 \text{ h}^{-1} \times SA_2 = SA_0 \text{ e}^{-2 \cdot 0.0036}.$$

$$SA_2 = 12\,308e^{-0.0072} = 12\,220 \text{ Bq } \mu g^{-1}.$$

At the end of the 2-h test, the plasma [iodide] is (26 100/12 220) = ***2.136 μg l^{-1}.***

2. *Secretion of organic-bound iodide from the thyroid gland is negligible during the test. The net uptake of iodide in the thyroid gland is equal to the rise in activity divided by SA: 77 500/12 220 Bq/(Bq μg^{-1}).*
The result is ***6.34 μg*** *or* ***3.17 μg h^{-1}****, which is above the normal average of* ***2.4 μg h^{-1} and suggests hyperthyroidism.***
3. *The thyroid plasma clearance for iodide equals its absorption flux divided by its plasma concentration: (6.34/120)/(3.11 $\times$ 10^{-3}) =* ***17 ml plasma/min****. This result is higher than the predicted value of* ***10 ml min^{-1}.***
4. *Plasma iodide is cleared only through the kidneys and through the thyroid gland. The* ***total iodide clearance*** *must be the* ***renal*** *plus the* ***thyroid*** *plasma clearance: (36 + 17) = 53 ml min^{-1}. The amount of iodide eliminated per minute is a constant fraction (k) of the amount remaining in the body. This is* ***exponential elimination*** *with a rate constant* ***k****. The distribution volume for iodide is 12 l.*

$$\mathbf{k = 53/12\,000 = 0.00441667 \text{ min}^{-1}.}$$

5. *The* ***biological half-life*** *($T_{\frac{1}{2}}$) is equal to (ln 2)/k or 0.693/0.0037857 =* ***153 min.***
The ***physical half-life*** *for ^{131}I is* ***192 h*** *or 8 days.*

66. Case History

1. *777 ng or 0.777 μg = 1 nmol; serum-thyroxine: (180 $\times$ 777) = 139 860 ng =* ***140 μg l^{-1}.***
2. *The total thyroxine content of the patient's body: (140 $\times$ 8) = 1120 μg.*
The elimination rate:

$$(1120 \times 0.14) = \mathbf{156.8\ \mu g\ day^{-1}\ or}$$

$$\mathbf{156.8/0.777 = 201.8\ nmol\ day^{-1}.}$$

3. *The thyroid uptake of iodide equals the elimination of thyroxine, which is:*

$$(201.8 \times 4) \text{ nmol day}^{-1} \text{ or } (807.2 \times 127) = 102\,514 \text{ ng/24 h or}$$

$$\mathbf{(102.5/1440) = 0.071\ \mu g\ min^{-1}.}$$

The thyroid plasma clearance for iodide is:

$$0.071\ (\mu g\ min^{-1})/0.004\ (\mu g\ ml^{-1}) = \mathbf{17.75\ ml\ plasma/min.}$$

4. *The symptoms and signs suggest* ***hyperthyroidism****. The serum thyroxine concentration is almost twice the normal level. The heart attack is probably atrial fibrillation, which frequently occurs in hyperthyroid patients.*

67. Case History

1. *7.21 = 6.8 + log* $(HPO_4^{2-}/H_2PO_4^-)$; $(HPO_4^{2-}/H_2PO_4^-) = 2.57$. *Since* $[HPO_4^{2-}] + [H_2PO_4^-] = 0.84$ *it can be deduced that* $[HPO_4^{2-}]$ *is (0.84 · 2.57)/3.57 =* ***0.60*** *and* $[H_2PO_4^-]$ *is 0.84/3.57 =* ***0.24****, which is different from the normal.*
2. ***[Total calcium]*** *= 2/0.62 =* ***3.23 mmol*** $\mathbf{l^{-1}}$***.*** *The [total calcium] of this patient is much higher than normal (just as her [inorganic phosphate]).*
3. *The patient has an* ***acidosis (pH = 7.21), probably of metabolic character****, due to loss of base (bicarbonate and secondary phosphate).*

$$\text{Ca-Proteins} + 2\,H^+ = Ca^{2+} + \text{Proteins.}$$

The plasma proteins will bind a smaller and smaller fraction of Ca^{2+} *at increasing* H^+ *concentration in blood plasma, so the free fraction of* Ca^{2+} *(*$[Ca^{2+}]$*) must increase.*

4. ***Hyperparathyroidism*** *(primary).*
5. *She has a substantial renal* Ca^{2+} *excretion, and small* Ca^{2+} *crystals and* ***stones in her urinary tract****. She may have polyuria due to reduced sensitivity to vasopressin.*

68. Case History

1. *The renal excretion flux of* Ca^{2+} *is reduced due to increased reabsorption of* Ca^{2+} *in the distal tubules.* ***The plasma concentration*** $\mathbf{[Ca^{2+}]}$ ***is increased****, due to bone mobilization, increased intestinal absorption and reduced renal excretion flux of* Ca^{2+}*. The phosphate excretion flux is increased by PTH. Hereby* ***the plasma [phosphate] is reduced.***
2. *The diagnosis is* ***hypoparathyroidism****. Patients develop hypoparathyroidism following struma surgery.*
3. ***Hypoparathyroidism leads to a low plasma*** $\mathbf{[Ca^{2+}]}$***, and a large plasma [phosphate].*** *The symptoms and signs of hypoparathyroidism are related to the hypocalcaemia, such as low* $[Ca^{2+}]$*; increased neuromuscular irritability with latent or manifest* ***tetany****. Universal tetany can end in respiratory and cardiac arrest.*

69. Case History A

1. *The menstrual cycle:*
   FSH stimulates the ovary – Ovary secretes oestrogen – Oestrogen in low concentration inhibits release of FSH (and some LH) by negative feedback – Rising oestrogen above a certain concentration threshold stimulates release of LH (and some FSH) by positive feedback – ***Ovulation*** *– Corpus luteum develops as a mini-endocrine organ – It secretes progesterone (and some oestrogen).*
2. *Fertilization:*
   Spermatozoa arrives in time – One is selected – The zygote is formed in the uterine tube – Implantation in uterus – Placenta assumes its endocrine role as a gigantic hormone plant – Corpus luteum persists for some time – Pregnancy continues – Placenta releases hCG with a peak around the 10th week – Placenta releases large quantities of progesterone toward the end of pregnancy – At the very end of pregnancy the progesterone secretion falls – The uterus is sensitized by oestrogen – Oxytocin is released from the neurohypophysis store – Oxytocin is possibly the final trigger for parturition – The fall in progesterone allows the adenohypophysis to release prolactin (LTH) – This turns on the milk production – The milk discharge is due to oxytocin – Oxytocin is reflexly released by the baby's suckling.

69. Case History B

1. *Foetal haemoglobin (F) has an oxyhaemoglobin dissociations curve to the left of the A (Adult) haemoglobin. Accordingly, the $P_{0.5}$ is low or the standard affinity between oxygen (O_2) and haemoglobin is high.*
   The high affinity *increases the loading of foetal blood with O_2 during its passage of the placenta.*
   At a given O_2 tension in the intervillous spaces, the foetal blood leaving these villi will have a higher O_2 saturation than the blood of the mother.
2. *FSH stimulates* ***spermatogenesis****, and LH stimulates the Leydig interstitial cells thereby stimulating their* ***testosterone production****.*
3. *Suckling liberates* ***oxytocin****, which stimulates the myometrium.*

70. Multiple Choice Questions

Answers A, C, and D are true statements, whereas B and E are false.

71. Multiple Choice Questions

Answers A, B, and E are true statements, whereas C and D are false.

72. Case History

1. *The young man was in a hypertermic state due to an acute tonsillitis. This had increased his* ***insulin demand*** *and manifested his* ***juvenile diabetes mellitus****. In his case the first sign of the latent disease is diabetic coma. Juvenile diabetes is a life-threatening metabolic condition. Probably all β-cells are destroyed,*

so insulin treatment is imperative in order to keep him alive. The patient has a low glucose tolerance and burns fat, which explain his ketoacidosis.

The patient developed a metabolic acidemia partly compensated by hyperventilation. The hypocapnia must reduce cerebral blood flow, which contributed to the cerebral condition.

Glucosuria leads to osmotic diuresis. This had caused a serious dehydration. Both the ECV and the TBV were insufficient. The patient was in an imminent shock (circulatory insufficiency). This condition would have killed the patient, if his friend had not found him.

2. ***Insulin treatment*** *is imperative.* ***Antibiotics*** *must be given for the infection. The fever does not demand drastic procedures. Intravenous infusion of isotonic sodium bicarbonate solution (167 mmol l^{-1}) is given for the metabolic acidosis and physiological saline for dehydration.*

   The ECV of this patient is assumed to be 20% of the body weight, i.e. 80 kg · 0.20 = 16 l. According to the principles outlined in Chapter 34, the primary compensation of the base deficit in the ECV necessitates the administration of 15 mmol · 16 l = 240 mmol of bicarbonate.

   The treatment is controlled by frequent blood analyses of glucose, gas tensions and electrolytes. Circulatory collapse may occur, so preparations must be taken for a ***transfusion.***
3. *The rise in P_{aCO_2}* ***increases the P_{CO_2} of the CSF****. Since CSF has a low buffer capacity – and now a low [bicarbonate] – the pH is reduced. The high [H^+] in the CSF is a strong stimulus to the central chemoreceptors, and explains the hyperventilation.*
4. *Treatment with bicarbonate, and the patient's own renal bicarbonate reabsorption for 24 h, leads to a* ***rise in bicarbonate*** *all over the body including the CSF. Its low pH is normalized, and thus the stimulus for hyperventilation is eliminated.*

73. Case History.

1. ***The dilution principle.***
2. *The* ***distribution volume*** *is the relevant mass of indicator divided by its representative concentration.*

   The mass of indicator at the time of distribution was [55 000 000 – 3/2(1 598 400 × 0.095)] = 54 722 228 Bq.

   The distribution volume: 54 722 228/1 520 700 = ***36 l*** *or (36/74) =* ***49%.***
3. *Compared to a sedentary person containing* ***55%*** *water the result is low.*
4. *The normal weight for a female of this height is (152 cm – 110) =* ***42 kg.***

   She is not a dwarf. Female dwarfs are below 1.4 m in Anglo-Saxon populations. Her overweight is 32 kg or in percentage (74 · 100)/42 = ***176%.***

   According to WHO any overweight ***above 120%*** *has serious, clinical consequences (increased morbidity and mortality).*

   When less than 50% of her body weight is water, she must have a very ***low body density*** *and a* ***high body fat fraction.***
5. *Her fasting blood [glucose] is higher than normal. She does not pass her renal plasma* ***appearance threshold*** *(10 mM of glucose) in the fasting state. Glucose is most likely only excreted in the urine after meals.*

   Patients often develop ***multiple boils*** *and mycoses before diabetes is diagnosed.*

CHAPTER 75. SYMBOLS AND UNITS

1. Force is measured in Newtons (N). **One Newton** (kgm s^{-2}) is the force required to accelerate a mass of 1 kg with an acceleration of 1 m s^{-2}. The acceleration due to gravity is generally accepted as g or G = 9.8067 or 9.807 m s^{-2}.
2. Joule established in 1848 that mechanical work and heat energy were interchangeable. The commonly used unit of energy is the calorie, which is the energy required to raise the temperature of 1 g of water from 14.5 to 15.5°C. Work is force times distance, and it is measured as Newton-metre or **Joule (J)**. The **Joules equivalent** has been determined to be 4.187 J cal^{-1}.
3. Finally, **work-rate** or power is calculated as work/s. The power unit 1 W, equals one J s^{-1}.
4. **Pressure** is measured as force per area unit that is in **N m^{-2} or Pascal**. In the gravity field of the earth G or g equals 9.807 m s^{-2}, and blood has a relative density of 1033 kg m^{-3}. A 10-m high water column resting on 1 m^2, corresponds to the following pressure: (10 m · 1033 kg m^{-3} · 9.807 m s^{-2}) = **101306.3 (kg m s^{-2}) m^{-2}. This is 101 306.3 N m^{-2} or 101.3 kPa (=1 atmosphere)**. The classical concept is that 1 atmosphere equals 760 mmHg or 760 Torr. Accordingly, 1 Torr or 1 mmHg equals (101 306.3 Pa/760=) **133.3 Pa**. In this book pressures are given in Pa or kPa (with Torr or mmHg in brackets).
5. **Concentration** is mass per volume unit. Squared brackets around a substance or **C** denote concentration. The international unit is mM = mmol l^{-1} = mol m^{-3}.
6. A prefix scale for different units is used as follows: milli = m = 10^{-3}; micro = μ = 10^{-6}; nano = n = 10^{-9}; pico = p = 10^{-12}; femto = f = 10^{-15}.

International symbols
(Fed.Proc. 9: 602–605, 1950).

This is a precise short-cut for intellectual transfer used by all physiologists.

A **dash** above any symbol ($^-$) indicates a mean value. A **dot** above any symbol ($\dot{}$) denotes a time derivative. **Small letters in a suffix** denote gas dissolved in blood, whereas **large letters** denote gas in air. The symbol is often the first letter in the English word.

α	the Bunsen solubility coefficient (ml STPD per ml fluid per 760 Torr)
A	Alveolar gas
AA	arachidonic acid
Ach	acetylcholine
ACTH	adenocorticotropic hormone
Ad	adrenaline
ADH	antidiuretic hormone
ADP	adenine diphosphate
AIDS	acquired immunodeficiency syndrome
AMP	adenine monophosphate
AMPA	special glutamate receptors

ANF	atrial natriuretic factor
ANH	atrial natriuretic hormone
ANP	atrial natriuretic peptide
AP	action potential
AR	absolute refractory period
ASA	acetylsalicyclic acid
ATP	adenine triphosphate
ATPS	ambient temperature, pressure, saturated with water vapour
AV node	atrioventricular node
BB	buffer base
BD	base deficit
BE	base excess
BMR	basal metabolic rate
BSA	body surface area
C	concentration of gas in blood. Squared brackets around a substance also denote concentration
$C_{\bar{v}CO_2}$	concentration of CO_2 in mixed venous blood
CA	carbonic anhydrase
cAMP	cyclic adenine monophosphate
CBF	cerebral bloodflow
CBG	corticosteroid-binding globulin
CCh	carbacholine
CCK	cholecystokinin
cGMP	cyclic guanosine-monophosphate
CNS	central nervous system
CSF	cerebrospinal fluid
COLD	chronic obstructive lung disease
COMT	catechol-*O*-methyl transferase
C peptide	connecting peptide
CRH	corticotropin releasing hormone
CVP	central venous pressure
D	diffusion capacity
DAG	diacylglycerol
1,25-D_3	1,25-dihydroxy-cholecalciferol
DIT	di-iodinethyronin
DM	diabetes mellitus
DMNV	dorsal motor nucleus of the vagus
DMPP	dimethylphenylpiperazine
DOPA	dihydroxy-phenylalanine
2,3-DPG	diphosphoglycerate
DPPC	dipalmitoyl phosphatidylcholine
E	expiration
E_{net}	mechanical net-efficiency of external work
EAA	excitatory amino acids
ECG	electrocardiogram
ECF	extracellular fluid
ECV	extracellular fluid volume
EDIP	end-diastolic intraventricular pressure
EDRF	endothelium-derived relaxing factor

EDTA	ethylene-diamine-tetra-acetate
EEG	electroencephalogram
EF	excretion fraction
EGF	epidermal growth factor
e.p.	equilibrium potential
EPSP	excitatory postsynaptic potential
ER	endoplasmic reticulum
ERBF	effective renal blood flow
ERPF	effective renal plasma flow
ERV	expiratory reserve volume
ESV	end systolic volume
F	foetal haemoglobin
F	fraction of gas in dry air (or force)
f	respiratory frequency (breath min^{-1})
FABP	fatty acid binding protein
FAD	flavin adenine dinocleotide
$FADH_2$	flavin adenine dinucleotide (reduced)
FEV_1	forced expiratory volume in 1 s
FFA	free fatty acids
FGF	fibroblast growth factor
FRC	functional residual capacity (RV+ERV)
FSH	follicle stimulating hormone
FU	flow units (ml blood/100 g tissue/min)
G	Gibbs energy (free, chemical energy)
GABA	gamma-aminobutyric acid
GFF	glomerular filtration fraction
GFR	glomerular filtration rate (normal 118–120 ml min^{-1})
GH	growth hormone
GHIH	growth hormone inhibiting hormone
GHRH	growth hormone releasing hormone
GIP	gastric inhibitory peptide or/glucose-dependent insulin-releasing peptide
GLP	glucagon-like peptide
GnRH	gonadotropin releasing hormone
GLUT	glucose transporter
GRP	gastrin-releasing peptide
GTP	guanosine triphosphate
H	heat content (enthalpy; all energy when the pressure–volume work is zero)
Hb	haemoglobin
HBF	hepatic bloodflow
hCG	human chorionic gonadotropin
HDL	high density lipoprotein
HGF	hepatocyte growth factor
HGH	human growth hormone
HIP	hydrostatic indifference point
HIV	human immunodeficiency virus
hPL	human placental lactogen
HPLC	high pressure liquid chromatography
HSS	hepatocyte stimulating substance

I	inspired gas
ICSH	interstitial cell stimulating hormone
ICV	intracellular fluid volume
IDDM	insulin-dependent diabetes mellitus
IDL	intermediate density lipoprotein
IGF	insulin-like growth factor
IGF-BP	IGF-binding protein
IP_3	inositol-triphosphate
IRV	inspiratory reserve volume
ISF	interstitial fluid
Iso	isoprenaline
ISS	interpreted signal strength
i.v.	intravenous
J	flux of a substance (mol min^{-1}) through an area
JG	juxtaglomerular
LAT	lactic acid threshold
LBNP	lower-body-negative-pressure
LES	lower oesophageal sphincter
LH	luteinizing hormone
LHRH	luteinizing hormone releasing hormone
LPL	lipoprotein-lipase
LDL	low density lipoprotein
LTH	prolactin
LVET	left ventricular ejection time
MAO	monoamine oxidase
MAP	mean arterial pressure/mean aortic pressure
MeCH	methacholine
MEOS	microsomal ethanol oxidation system
MG	monoglycerides
2MG	2-monoglyceride
MIH	Müller inhibiting hormone
MIT	mono-iodinethyronin
mM	mmol l^{-1}
MR	metabolic rate
MSH	melanocytic stimulating hormone
MW	molecular weight
NA	noradrenaline
NAD	nicotinamide adenine dinucleotide
$NADH_2$	nicotinamide adenine dinucleotide (reduced)
NANC	non-adrenergic, non-cholinergic
NBB	normal buffer base/neutral brush border
NGF	nerve growth factor
NIDDM	non-insulin-dependent diabetes mellitus
NMDA	*N*-methyl-D aspartate
NOS	nitric oxide synthase
NSAID	non-steroid anti-inflammatory drug
25-OH-D	25,hydroxy-cholecalciferol
P	partial pressure of gas in air or blood
PAH	para-amino hippuric acid

PCV	packed cell volume
PDE	phosphodiesterase
PDGF	platelet derived growth factor
PEF	peak-expiratory flow
PG	prostaglandins
PG_2	prostacyclin
PGE_2	prostaglandins
PIF	prolactin inhibiting factor
PIP_2	phosphatidylinositol diphosphate
P_B	barometric pressure
$P_{c'CO_2}$	partial pressure of CO_2 in end-capillary blood
P_{IO_2}	partial pressure of O_2 in inspired air in trachea
P_{aO_2}	partial pressure of O_2 in arterial blood
POMC	pro-opiomelanocortin
PP	pancreatic polypeptide, pulse pressure
PRL	prolactin
PRU	peripheral resistance unit
PTH	parathyroid hormone
PVR	pulmonary vascular resistance
$\mathring{Q}$	cardiac output ($l\ min^{-1}$)
QRS	the ventricle complex of the ECG
R	ventilatory exchange ratio (pulmonic)
R	Gas constant
RAS	reticular activating system
RBF	renal bloodflow
RC	respiratory controller, respiratory centres
REM	rapid eye movements
RES	reticulo-endothelial system
RIA	radio-immunoassay
RMP	resting membrane potential
RPF	renal plasma flow
RPM	revolutions per minute
RQ	respiratory quotient
RR	relative refractory period
RV	residual volume
S	saturation degree
S	entropy (the tendency to spread in a maximum space)
SA	specific activity
SA node	sinoatrial node
SB	standard bicarbonate
SDA	specific dynamic activity
SR	sacroplasmic reticulum
SS	stimulus strength
STPD	standard temperature and pressure, dry (0°C, 760 mmHg)
STN	solitary tract nucleus
sv	stroke volume
T	temperature
T	tension (force)
T_3	tri-iodothyronine

T_4	tetra-iodothyronine
TBA	thyroxine-binding albumin
TBG	thyroxine-binding globulin
TBPA	thyroxine-binding prealbumin
TBV	total blood volume
TCA	tricarbocylic acid
TEV	total erythrocyte volume
TFGF	transforming growth factor
TG	triglycerides
TGF	tubuloglomerular feedback
TH	total haemoglobin content
TLC	total lung capacity (RV + VC)
TP	threshold potential
TPVR	total peripheral vascular resistance
TRH	thyrotropin-releasing hormone
tRNA	transfer RNA
TSH	thyroid-stimulating hormone
TV	tidal volume
$\overline{v}$	linear mean velocity
$\mathring{V}$	velocity of gas
$\mathring{V}_A$	expired alveolar ventilation (l min^{-1})
TxA2	thromboxane A2
VC	vital capacity (IRV+TV+ERV)
VD	dead volume
VIP	vasoactive intestinal peptide
VLDL	very low density lipoprotein
VMA	vanillyl mandelic acid
W	external work (with pressure–volume work zero)

Nutritive equivalents and enthalpy

Nutritive equivalents for oxygen are: carbohydrate 37 mmol oxygen g^{-1}, fat 91 mmol oxygen g^{-1}, and protein 43 mmol oxygen g^{-1}. On a **mixed diet 20 kJ of energy** is transferred per litre STPD of oxygen used; the RQ is 0.8.

Nutritive equivalents for carbon dioxide are: carbohydrate 37 mmol g^{-1}, fats 64 mmol g^{-1}, and protein 34 mmol g^{-1}.

Metabolic enthalpies (heat energy liberated in the body per gram combusted nutrient) in kJ g^{-1} substance: protein 17, fat 39 and carbohydrate 17.5.

Essential atomic and molecular weights

These are given in g mol^{-1} (or Da) throughout the text. Calcium 40; carbon 12; glucose 180; helium 4; hydrogen 1; nitrogen 14; oxygen 16; PAH 194.2; phosphorus 31; potassium 39; sodium 23; xenon 131.

Physical constants and conversion factors

Acceleration due to gravity (standard): 9.81 m s^{-2}.
Avogadro's constant: $6.02 \cdot 10^{23}$ molecules mol^{-1}.
Diffusion coefficients for most molecules: 10^{-10} m^2 s^{-1} per molecule.

Energy (J = N m = Volts Coulomb): 1 cal = 4.187 J.
Farad = Coulomb/Volts.
Faraday's constant: 96 487 (10^4) Coulomb/mol monovalent ion.
Molar gas constant (R): 8.31 J mol^{-1} per degree Kelvin (K).
Specific heat capacity of the human body: 3.47 kJ kg^{-1} $°C^{-1}$.
Energy transfer by evaporation of 1 kg of water at the usual skin temperature: 2436 kJ.
Pressure (Pascal = Pa = N m^{-2}): 1 mmHg = 1 Torr = 133.3 Pa.
Surface tension of body warm water: 0.07 N m^{-1}.
Temperature conversion between degrees of Fahrenheit (°F) and Celcius (°C): **(°F) = 9/5 (°C) + 32.**

Calculated partial pressures

The partial pressures of respiratory gases are calculated in the alveoli and in the surrounding air of a healthy person, resting at sea level (101.3 kPa = 760 Torr = 1 atmosphere).

The water vapour tension in a fluid (air or liquid) of the temperature 310K (37°C) is **6.27 kPa or 47 Torr (mmHg).** At 293K (20°C) the tension is 2.4 kPa or 18 Torr. The alveolar gas fractions are: $F_{AO_2} = 0.15$, and $F_{ACO_2} = 0.056$. The composition of atmospheric air is: **$F_{IO_2} = 0.2093$ and $F_{ICO_2} = 0.0003$.**

$P_{O_2} = F_{O_2}(101.3 - 6.27)$ kPa.

$P_{AO_2} = 13.3$ kPa (100 Torr); $P_{aO_2} = 12.7$ kPa (95 Torr); $P_{\bar{v}O_2} = 6$ kPa (45 Torr).

$P_{ACO_2} = 5.3$ kPa (40 Torr); $P_{aCO_2} = 5.3$ kPa (40 Torr); $P_{\bar{v}CO_2} = 6.1$ kPa (46 Torr).

Index